Springer Series in Optical Sciences   Volume 66

Editor: Theodor Tamir

Springer-Verlag Berlin Heidelberg GmbH

Springer Series in Optical Sciences

Editorial Board: A. L. Schawlow A. E. Siegman T. Tamir

Managing Editor: H. K. V. Lotsch

Volumes 1-41 are listed at the end of the book

H. I. Bjelkhagen

Silver-Halide Recording Materials

for Holography and Their Processing

Second Edition
With 64 Figures

 Springer

Professor Dr. HANS I. BJELKHAGEN
American Propylaea Corp., 555 S. Woodward, Suite 1109
Birmingham, MI 48009, USA

ISBN 978-3-540-58619-7   ISBN 978-3-540-70756-1 (eBook)
DOI 10.1007/978-3-540-70756-1
CIP data applied for

This text was prepared using the PS™ Technical Word Processor
SPIN 10481664   54/3144 - 5 4 3 2 1 0 - Printed on acid-free paper

The time has come
New Lippmann said
To talk of many things:
Of plates and film
And HOEs
And holographic fringe
And that the image may just fade
And whether light has wings.

Teresa Bjelkhagen

Foreword

Holography is done primarily on silver-halide photographic materials. This was true from the very beginning in 1947 and remains true today (except, of course, for the mass-production embossing processes, but even these start with a photographic master). Since 1962, the common emulsions used for conventional photography have been abandoned in favor of specialized, high-resolution emulsions that allow holography to realize its rather astonishing capabilities. Even more unconventional are the various chemical processes for transforming the exposed emulsions into the finished hologram. Sometimes the usual photographic developers and fixers are used, but more often the chemical processes are highly complicated and esoteric. The aim is to produce holograms with low noise (i.e., low scatter) and with high diffraction efficiency, so that the image will be as clear and bright as possible. As a consequence of this research, the quality of holograms has improved spectacularly over the past three decades.

This book deals with the photographic process as it applies to holography. It is the only book dedicated to this topic and is the most complete compilation of this vast and important technology. Dr. Bjelkhagen has done the holography community a worthy service by producing this work. And his credentials are impeccable. He has worked for over two decades on making holograms on silver-halide emulsions and has contributed significantly to the research results that he writes about. This book, made possible by his considerable expertise, will be of great value to anyone seriously interested in the process of making high-quality holograms.

November, 1994   *Emmett N. Leith*

Preface

This book is devoted first and foremost to silver-halide recording materials for holography and their processing methods. Silver halides are commonly used in the recording of holograms and are particularly popular in display holography and scientific imaging where the demands placed on the recording material are extremely high. The fact is that lasers and the holographic equipment of today are capable of producing holograms of the highest possible quality; the limitations on the quality of holograms lie in the recording material itself. For this very reason the author is convinced that a systematic and exhaustive presentation of silver-halide materials - their features and qualities, the various processing methods and techniques that can be chosen from and applied in the holographic process, as well as the relevant chemical formulas and recipes to be used - will be beneficial to any reader interested in holography. It is the author's hope that the book will be able to serve as a source to which the interested reader may refer for advice on any point connected with the above, as well as that it will provide the reader with a general understanding and an overall view of how the materials work and of ways to achieve the best results from them.

This book is based on the author's long experience in the field, combined with careful studies of more than 800 scientific papers and publications dealing with silver-halide materials and their processing techniques. Among the references, the reader will find many publications from the former Soviet Union, where the extensive research in this field has resulted in many remarkable contributions to holography. As most of those contributions have been published only in the Russian language, the author would like to pass on some of that experience in the field to the Western world.

Although this book is intended primarily for readers already somewhat familiar with holography, it may also be useful to students interested in the theoretical and practical aspects of holographic silver-halide recording materials and their processing methods. For scientists working in the field of holography the book is meant to serve as both a general reference book in the area and a laboratory handbook. Finally and hopefully, for the experienced holographer it will provide the inspiration for further research and aid him/her in creating holograms of unsurpassed quality and vividness yet to be seen: the real world captured in a three-dimensional nutshell.

Evanston, IL
January 1993

H.I. Bjelkhagen

Acknowledgements

In the course of writing this book I have received assistance and help from various people who have directly or indirectly contributed to its final shape. I would like to thank each and all of them. Particularly, thanks are due to my two Swedish colleagues, Dr. Nils Abramson and Per Skande, both of whom were involved in the early development of holographic imaging and processing methods for silver-halide materials. In addition, I would like to express gratitude to both Dr. Robert Sekulin, Rutherford Laboratory, and Dr. Homaira Akbari, CERN, for their collaboration in some of the experiments concerning the influence of processing methods on the holographic image resolution. I also wish to gratefully acknowledge many fruitful discussions with Dr. Nicholas Phillips of Loughborough University as well as his comments on an early version of the manuscript. Edward Wesly of the Lake Forest College is gratefully acknowledged for helping me with some experiments on pulsed-holography processing and for many valuable discussions we had together. Likewise, I am grateful to Dr. Vladimir Markov from Kiev for his contribution to my understanding of ultrafine-grained holographic materials, and many useful and stimulating discussions. Dr. Ventseslav Sainov contributed with a detailed account of Bulgarian emulsion making and processing methods and shared with me his experience concerning various practical problems connected with holographic recording materials. At Northwestern University, Gale Hagan helped me with preparing the tables and Nabeel Rasheed assisted me in producing the computer-generated illustrations. I am also indebted to Dr. Helmut Lotsch of Springer-Verlag for his encouragement and support in writing this book. Last but not least, I would like to express my deep gratitude to Teresa Bjelkhagen who helped me enormously to improve both the quality of English as well as that of the scientific meaning of the text. She made it possible for me to take part of the work published in other languages, by providing me with English translations of various publications from Russian and Spanish.

Contents

1. Introduction

When we say "holography" we tend to think of it in terms of a new and fresh discipline that sprang up only a decade or two ago. Many have never even heard the word, and a few more have never yet seen a hologram. Those who have, bestow holography with all the veneration that seems to be its natural due paid by many to this and other technological novelties of this our atomic age. And yet, holography is almost as old as this century - at least in thought, if not in practice.

Long before the holographic theory was formulated as a coherent and comprehensive whole, a number of scientists (W. L. Bragg, H. Broersch and F. Zernike) were considering the possibilities of X-ray microscopy for re-creating the image from the diffraction pattern of a crystal lattice [1.1]. In 1920, the Polish physicist Mieczyslaw Wolfke wrote that if an X-ray diffraction pattern of a crystal is illuminated with monochromatic light, a new diffraction pattern is created. This diffraction pattern is identical with the image of the object [1.2]. An even earlier publication is a paper from 1901 by *Cotton* [1.3], describing Holographic Optical Elements (HOEs). *Cotton* explained the possibility of reconstructing light waves reflected from spherical and multi-angled mirrors recorded in photographic Lippmann layers.

The holographic theory was presented to the world for the first time in 1948 by the Hungarian-born physicist and Nobel Prize winner, Dennis Gabor [1.4-6]. At that time *Gabor* was trying to improve the resolution of the electron microscope by overcoming the spherical aberration of the lenses. Earlier, in 1936, O. Scherzer was able to prove that the spherical aberration cannot be eliminated in any axially symmetrical electron-optical lens system. *Gabor* was trying to find a way around this problem. He asked himself: "Would it not be possible to take first a bad picture, but one which contains the whole information, and correct it afterwards by a light-optical process?" [1.7]. He found that, by adding a coherent background as a phase reference, the original object wave was contained in this "interferogram" which he later called a *hologram*. The lack of coherent X-ray sources, which is a problem even now, made it difficult to demonstrate the new imaging technique in electron microscopy. Instead, *Gabor* concentrated on experimentally verifying his theory, using visible light of the best possible coherence that he could achieve.

Holography had to wait for its first practical applications until the early sixties when the laser was invented and laser light could be used for the production of the first holograms, which happened almost simultaneously in both the USA [1.8-11] and the former USSR [1.12-14]. Since that time, the interest in holography has been growing at an increasing rate and has given rise to numerous applications of holographic techniques in science, industry, medicine and art. Today holography is a part of everyday life. To give just one example of a mass-application of holography today is the use of holograms on credit cards, where they have proved to be a successful security device. Here, a special technique for mass-production is used and the hologram itself is embossed in a plastic material.

Even though this type of hologram represents an important application, the recording materials used most frequently for high-quality, large size holograms, are the silver-halide materials. These are the most widely used ones not only because of their high light sensitivity, but also because of their commercial availability. A more detailed study of these materials and the various processing methods that have been developed throughout the years appears to be an interesting undertaking and one that should bring about a better understanding of the factors that contribute to the quality of the holographic image.

A large number of books dealing with holography, holographic interferometry, holographic recording materials, and processing methods have been published [1.15-33]. Some books, like those by *Saxby* [1.28,29], focus on display holography, whereas the main concern of other publications is the use of various holographic recording materials [1.30,31]. However, up to the present moment, no single book devoted solely to the discussion of silver-halide recording materials for holography has appeared in English. Therefore, a systematic presentation of the above-mentioned materials and their processing methods will be of interest to many scientists and artists working in the field of holography. The study also includes several technical papers and publications from the former USSR, e.g., the books by *Kirillov* [1.31], *Komar* and *Serov* [1.32], and *Sobolev* [1.33].

The importance of the recording material to the quality of the holographic image is recognized by all who know what holography is about. This is what Stephen Benton of MIT said at the OSA meeting in Houston, Texas (1974), when presenting a new holographic processing technique: "Finally, most of us would agree that holography is not a straightforward extension of photography, yet our materials are still evaluated in terms that are closely related to conventional photographic analysis. It is very satisfying, therefore, to find a developer that produces dreadful photographs and quite extraordinary holograms!" [1.34].

A quotation from Yuri Denisyuk is also worth mentioning: "With regard to the future prospects of holography development, it should be noted that the problems of holography as a science will apparently be strongly oriented toward a detailed investigation of the interaction processes of light with light-sensitive materials and nonlinear media." After describing different applications of holography, he continues: "The basis for these applications will be the development of techniques for making and chemically developing various photographic materials." [1.35].

The above quotations are meant to draw the reader's attention to the importance attached by the two pioneers of holography to the recording material itself and its processing techniques. This still holds true - the future of holography is to be found in the improvement of the existing and the development of new materials and processing techniques, where the demands are very different from those of conventional photography. Therefore, it is necessary to compile the knowledge concerning holographic silver-halide materials and their processing techniques that has been accumulated over the years in a variety of publications and put it in one volume. It is the hope of the author that the present publication, with its comprehensive list of references, will act not only as a source for a better understanding of the technology of hologram-making but also as an inspiration and a guide for researchers in the field to further improve the quality of holograms.

This monograph is mainly devoted to the silver-halide recording materials used in holography and their processing methods. However, at the end of this chapter, a short overview of different recording materials for holography is presented.

In Chap.2, silver-halide recording materials are discussed from a theoretical point of view, and the fundamental principles of the recording process are set out. The manufacturing of photographic and holographic emulsions is described. Chapter 3 gives a detailed survey of the existing commercial silver-halide materials for holography, including some Russian and East-European materials.

Chapter 4 deals with the development process of the latent image, describing various chemical and physical development techniques and developers, both from a theoretical and a practical point of view. The technique of producing phase holograms by means of bleaching procedures is a very important field in holography. Chapter 5 is devoted particularly to this subject and includes both theoretical and practical considerations for obtaining high-quality phase holograms. In Chap.6, various special techniques, such as hypersensitization and latensification, problems due to a short exposure time when using Q-switched lasers, reciprocity failure, latent-image fading, etc., are treated. Chapter 7 describes well-working processing schemes for a variety of hologram types, based on the details described in previous chapters.

Chapter 8 describes techniques applied after the main processing, such as washing, drying, emulsion protection, and storage. In the same chapter, several practical considerations concerning various aspects, such as dark room layout, plate or film processing, machine processing, and quality control are discussed. It also contains a section on hologram copying. Chapter 9 describes the various existing methods for recording holograms in true colors as well as ways for creating pseudo-color holograms, which has become popular among artists. It also takes up holography on silver halides and other materials in the IR and UV parts of the spectrum.

Chapter 10 lists a large collection of recipes for developers, fix and bleach baths and other special solutions used in holography. It also contains a chemistry section, where most of the products employed in holographic processing are listed together with the corresponding chemical formulas.

An extensive list of references is cited for each chapter including almost every important paper that has ever been published in the field. Finally, there is also a subject index at the end of the book.

1.1 History

1.1.1 Before Holography

Long before holography came into being, there existed various recording materials which could have been used with good results for the recording of holograms. The origin of processing fine-grained emulsions can actually be traced back to the beginning of this century when Gabriel Lippmann was experimenting with color photography [1.36-38]. His work concerned wavefront reconstruction through the recording of standing waves in a volume medium. To some extent the work was based on earlier contributions from J.T. Seebeck, E. Beckerel and W. Zenker. *Lippmann's* photographic recording procedure (*Lippmann photography*) shows similarities to holography, but it was not very effective in color photography as the exposure times were too long for practical use (Fig.1.1).

Because of the demand for high resolving power, the material had very low sensitivity. The emulsion coated on Lippmann plates was in contact with a highly reflecting surface (mercury) reflecting the light back into the emulsion, then interfering with the light from the other side of the emulsion. The standing waves of the interfering light produced a very fine fringe pattern throughout the emulsion with a periodic spacing of $\lambda/(2n)$, that had to be recorded (λ is the wavelength of light in air, and n the refractive index of the emulsion). The color information was stored locally in this way. The larger the separation between the fringes, the longer the

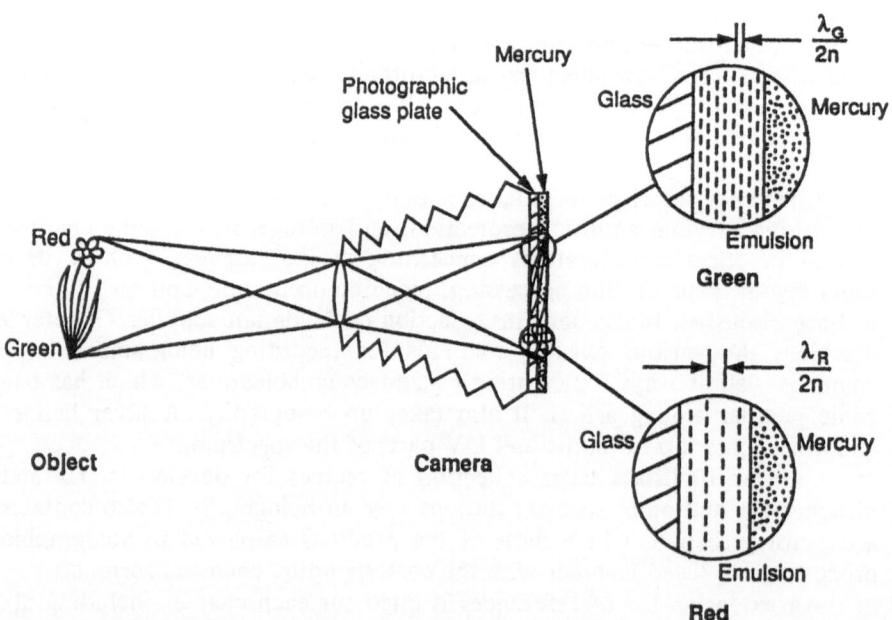

Fig.1.1. The principle of Lippmann photography

wavelength of the recorded part of the image information. When the developed photograph was viewed in white light, different parts of the recorded image produced colors, which was due to the fringe separation in the emulsion. The light was reflected from these fringes and locally created different colors corresponding to the colors that produced them during the recording. Only a narrow part of the spectrum around the original color was reconstructed because the waves scattered from the layers in the emulsion added up in phase only for the original color. It is obvious that there is a high demand on the resolving power to record the fringes separated in the order of half the wavelength of the light. It is also clear that the processing of these plates was critical, as one was not allowed to change the separation between the fringes, which would have created wrong colors. One also had to find ways for obtaining a high efficiency.

Even though *Lippmann*'s technique never became popular, he was awarded the Nobel prize for his invention in 1908. It was his idea of recording fringes throughout the depth of the emulsion that *Denisyuk* used when he introduced the technique of recording single-beam reflection holograms in the early 1960's [1.12-14].

Let us look at how Lippmann plates were made and processed. In 1908 *Ives* published a paper providing detailed information on how to make Lippmann photographs [1.39]. It stated that a relatively thick emulsion was needed, as the information was recorded in depth. Plates were made with "chemically pure" silver nitrate and potassium bromide. The gelatin was, e.g., "a department store gelatine recommended as the best for puddings". It had to be of very hard consistence and free from grease. The recipe is given in Table 1.1.

Table 1.1. Ives' Lippmann emulsion

Solution A	Solution B	Solution C
Gelatin 1 g	Gelatin 2 g	$AgNO_3$ 0.3 g
Water 25 mℓ	KBr 0.25 g	Water 5 mℓ
	Water 50 mℓ	

The solutions A and B are heated until the gelatin melts and are then allowed to cool to 40° C. Solution C is then added to A and after that the mixture is slowly added to solution B while stirring. Thereafter, sensitizers are added. Finally, the emulsion is filtered and the plates coated. The thickness of the emulsion varied between 7 and 70 μm. Isocol, erythosine, and pinacyanol were used as sensitizers for covering the whole visible electromagnetic spectrum.

The plates were developed in developers based on pyrogallol and ammonia. However, developers containing ferrous oxalate, glycin and hydroquinone yielded also good and uniform results. The paper mentions that bleaching with mercury chloride was employed to improve efficiency. Pyrogallol-developed plates did not perform well with this bleach, but, in conjunction with the other developers mentioned above, the results improved.

Exposure times using bright sun light to illuminate the objects were between 1.5 to 5 minutes at f/3.6. The paper also discussed the application of dichromated gelatin - a material with high resolving power which is utilized nowadays in the production of certain types of holograms.

Recent years have seen a revival of interest in Lippmann photography among scientists and holographers [1.40-46]. For example, *Phillips* et al. [1.43-46] found that it is possible to obtain a higher reflectivity in recorded interference layers from an incoherent source (e.g., sodium discharge lamp) than from fringes recorded with laser light. The possibilities of contact-copying holograms using incoherent light is the main topic of the above-mentioned work. Earlier, *Oliva* et al. [1.47,48] utilized a mercury vapor lamp at 405 nm to contact-copy Silver-Halide Sensitized Gelatin (SHSG) master holograms into dichromated gelatin. *Miller* [1.49] reported that a contact-copied hologram was made on an Agfa plate using a sodium-vapor lamp. The quality of the reflection copy was rather good.

1.1.2 Early Holography

The very first holograms recorded by Dennis Gabor in the late 1940's were made on fast, black-and-white, silver-halide material and developed in a high-contrast developer. The demand on resolution is less critical in the type of hologram he was recording at that time. The very low intensity of coherent light achieved by Dennis Gabor and his assistant, Ivor Williams, using a single line from a mercury lamp focussed through a 3 μm pinhole required a fast recording material. It should be mentioned that much later fast conventional photographic materials, e.g., the Kodak Tri-X Pan film, were recommended for the recording of in-line holograms with laser light [1.50]. Exposures of only about 1 nJ/cm^2 were then required to obtain holograms of acceptable quality. Among other well-known conventional photographic materials that have been used for certain special types of holograms and holographic techniques, Ilford FP4 plates [1.51], Kodachrome [1.52], Polaroid's P/N film [1.53-55], and Polaplan CT [1.56] can be mentioned. In a low-resolution holographic camera, mainly intended to be employed for Holographic Non-Destructive Testing (HNDT) applications, a conventional panchromatic microfilm (Agfa Agepan ff) was used [1.57,58].

The demands on the resolving power were much more severe when Emmett Leith and Juris Upatnieks made their first off-axis holograms in the early 60's, utilizing laser light. The same was also certainly true for Yuri Denisyuk, recording reflection holograms.

At the same time in the USA, Kodak was producing a spectroscopic, silver-halide plate with high resolving power (2000 line pairs per mm) - 649-F. This material was then used by *Leith* and *Upatnieks* for their first laser-produced holograms whose quality was remarkably good. The developer used for processing these first holograms was the Kodak D-19 developer. This developer is still used by many scientists and artists for a variety of holograms. Although the above mentioned spectroscopic plate can still be obtained on the market, Kodak produces nowadays silver-halide materials for the exclusive application in holography.

For his first Lippmann reflection holograms, *Denisyuk* employed a modified version of Ives' formula for preparing plates [1.14]. In order to increase the sensitivity of Ives' emulsion to light, potassium iodide was introduced into solution B (Table 1.1) in the amount of 3% of the silver content. *Denisyuk* mentioned two methods for chemical processing of Lippmann plates. The one makes use of a special pyrogallic developer, whereas the other employs an ordinary metol-hydroquinone developer. The paper also describes the way of making phase holograms by bleaching the silver image, and suggests different methods for hypersensitizing, such as the use of TriEthanolAmine (TEA).

Soon after *Leith-Upatniek*'s publications, *Cathey* mentioned in a letter to the Journal of the Optical Society of America that phase holograms could be made on Kodak 649-F plates treated in the bleach used in Kodak's chromium intensifier process. (Kodak In-4, potassium dichromate - hydrochloric acid) [1.59]. The idea of making phase holograms was, however, not entirely new, it had been mentioned in one of *Rogers'* papers as early as in 1951 [1.60]. *Rogers* used the so-called *Carbro process* to create a phase hologram, which "acts like a normal one, but with greater light-gathering power". The fact that this technique was not new at the time when the first laser-produced phase holograms started to appear, was mentioned by *Rogers* in a letter to the editor of the Journal of the Optical Society of America in 1965 [1.61]. Also, as mentioned above, *Denisyuk* and *Protas* had described phase holograms in their paper [1.14].

This short historical survey of holography must also mention the early work of *Stroke* whose contribution to early holography was considerable and who published the first book on holography in 1966 [1.15]. *Stroke* was not only the first to introduce the technique of making lensless Fourier-transform holograms [1.62] but he was also among the first to make reflection holograms which could be reconstructed in white light [1.63]. Earlier, *Hoffman* et al. [1.64] described a reflection hologram which they called an "inverted reference-beam" hologram intended for laser-light reconstruction. The very first reflection holograms were made by C. Schwartz and N. Hartman at the Battelle Memorial Institute in the USA according to *Leith* [1.65]. The first laser-reconstructible color hologram was made by *Pennington* and *Lin* [1.66] and the first white-light viewable color reflection hologram was made by them together with *Stroke* and *Labeyrie* [1.67].

Shortly after the first publications concerning laser-recorded holograms, a tremendous interest in the new imaging technique called *hologra-*

phy sprang up (the word "holography" was coined by G. L. Rogers; D. Gabor used the words "hologram" and "wavefront reconstruction" in his early papers) [1.68]. This interest is reflected in the increasing number of papers published on the subject: Between 1948 and 1964, i.e. until the first publications on laser-produced holograms, the number of papers was quite small, whereas during the following two years over 1000 papers appeared. Since then papers on holography have continued to appear in a variety of journals, with a peak in the 70's. Most of these papers deal with holographic techniques or applications, whereas only a limited number of them is directly concerned with the recording material and the problems associated with its processing.

Among the papers published towards the end of the 60's, one particular publication is especially interesting from the historical point of view [1.69]. The paper describes the white-light transmission holographic technique, introduced for the first time by *Benton*. The "rainbow" or "Benton" hologram is probably the most commonly used hologram type so far, being mass-produced with cheap embossing methods for different commercial and security applications.

1.2 Recording Materials

1.2.1 Silver-Halide Materials

Silver-halide recording materials for holography are interesting for many reasons. Silver halide was the first material used for recording holograms; it is also the most important material for holography in respect of its numerous scientific and artistic applications. In addition, it has high sensitivity in comparison with many other alternative materials, it can be coated on both film and glass, it can cover even very large formats, it can record both amplitude and phase holograms, it has high resolving power, and is easily available. Nevertheless, it does have some drawbacks; it is absorptive, it has inherent noise and a limited linear response, it is irreversible, it needs wet processing, it creates printout problems in phase holograms, etc. Since the silver-halide materials constitute the main topic of this book, these will be discussed in a greater detail later on, while the remaining part of this chapter is devoted to other, non-silver holographic materials.

1.2.2 Non-Silver Materials

A detailed presentation of a wide range of recording materials for holography can be found in the book edited by *Smith* [1.30]. There are also several general survey papers published on this subject [1.70-79]. The following non-silver materials are frequently used in holography for special applications.

a) Dichromated Gelatin

DiChromated Gelatin (DCG) is the material mentioned by *Ives* [1.39] as a potential recording material for Lippmann photographs. Scientists have known for a long time (since 1830) that UV and blue light can cause gelatin molecules to cross-link if the gelatin contains small amounts of dichromate. Dichromated gelatin has a high resolving power and remarkable brightness due to a refractive-index modulation of 0.08, which is the largest among holographic materials known today. During the exposure of a dichromated gelatin emulsion to UV or blue light, the hexavalent chromium ion (Cr^{6+}) is photoinduced to trivalent chromium ion (Cr^{3+}) which causes cross-linking between neighboring gelatin molecules. The areas exposed to light are hardened and become less soluble than the unexposed areas. Developing consists of a water wash which removes the residual or unreacted chemical compounds. Dehydration of the swollen gelatin follows after the material has been immersed in isopropanol which causes rapid shrinkage resulting in voids and cracks in the emulsion, thus creating a large refractive-index modulation. The underlying mechanism is not completely clear because high modulation can also be caused by the binding of isopropanol molecules to chromium atoms at the cross-linked sites. The DCG material has a rather low sensitivity of about 100 mJ/cm^2. The material is employed in the production of Holographic Optical Elements (HOEs) mainly. In display holography it is used for pendants and other jewelry items.

b) Photopolymer Materials

Photopolymer materials can be used for recording phase holograms, where applications in mass-production of display holograms and optical elements are of main interest. A lively interest in these materials was shown at a very early stage of holography. Companies like AT&T Bell Laboratories, du Pont and Hughes have produced photopolymer materials for recording holograms. Normally, the material has a short shelf-life and a rather limited refractive-index change. Its sensitivity is not too bad though, and the advantages are a low noise level and its suitability for applying dry processing techniques. Recently, Polaroid developed a new polymer material for holography named DMP-128. E.I. du Pont de Nemours & Co. has also released an improved version for holography marketed under the name OmniDex.

A photopolymer recording material consists of three parts: a photopolymerizable monomer, an initiator system (initiates polymerization upon exposure to light) and a polymer (the binder). First, an exposure is made to the information-carrying interference pattern. This exposure polymerizes a part of the monomer. Monomer concentration gradients, formed by variation in the amount of polymerization due to the variation in exposures, give rise to diffusion of monomer molecules from the regions of high concentration to the regions of lower concentration. The material is then exposed to regular light of uniform intensity until the remaining monomer is polymerized. A difference in the refractive index within the material is obtained.

The Polaroid polymer DMP-128 requires a developing/fixing bath to produce stable holograms. Wet processing is also needed to control the color and the bandwidth of the material. The sensitivity of the Polaroid material is about 10 mJ/cm^2 and the shelf life is improved compared to the earlier photopolymer materials. The new du Pont material requires only a dry processing technique (exposure to UV light and a heat treatment) to obtain a hologram.

An interesting material has been developed in the former USSR called *Reoxan*. The recording is based on the reaction of sensitized photooxidation of acenes. Reoxan is a glossy polymer with a photooxidated compound and a dye sensitizer for photooxidation, both dissolved in it. The dye absorbs the visible light and anthracene compound absorbs in the near-infrared region. Before the recording, the material is saturated by oxygen under pressure in an autoclave. It is then light-sensitive and has to be exposed before oxygen diffuses out of the material. The phase recording is based on photoinduced transformation of the anthracene compound causing a change of the polarizability of the molecules. The hologram is auto-fixed by self-diffusion of oxygen from the polymer. It is a high-resolution (10000 line pairs per mm) material but its sensitivity is low (~ 1 J/cm^2).

c) Photoresist Materials

Photoresists are well known from the electronic industry, where they are used in the production of circuit boards. In holography, they are employed mainly for the production of master plates for embossed holograms and for manufacturing holographic gratings. A photoresist is a photosensitive material which produces a relief pattern in the material after its exposure and processing. The exposure to actinic radiation produces changes in the photoresist layer that result in a solvency differentiation as a function of exposure. Processing is done with a suitable solvent dissolving either the unexposed or exposed regions, depending on whether the resist is of the negative or the positive type. For optical recording, positive photoresist (exposed resist removed during development) is preferred to the negative type because of their higher resolving power and low scatter. On the resulting surface relief pattern particles of nickel are deposited by electrolysis to make a mold. This mold can than be used as an embossing tool. The photoresist process can be used for making transmission holograms only. If an embossed hologram is mirror-backed by using, e.g., an aluminum coating process, it can be utilized in the reflection-reconstruction mode as well. A typical photoresist for holography (e.g., Shipley AZ-1350) has a sensitivity of about 10 mJ/cm^2. It is sensitive for UV and for visible light up to 500 nm. The Waycoat HPR series of photoresists represent another popular brand for optical imaging purpose. Towne Laboratories, Inc. manufactures coated photoresist plates for holography.

d) Thermoplastic Materials

Thermoplastic recording materials for holography are found mainly in the nondestructive testing field. Several camera types that can be used for real-time holographic interferometry investigations are presently on the market. The material's sensitivity is similar to that of the silver-halide materials', its resolving power is quite high, and it requires only dry processing, all of which makes it rather interesting. Briefly, the version that is used for holography is a multilayer structure coated on glass or film. The substrate is first coated with a conducting layer (e.g., evaporated gold), then a photoconductor (e.g., Poly-n-Vinyl Carbazole, PVC) that has been sensitized with e.g., 2,4,7 TriNitro-9-Fluorenone (TNF) and on top of this layer a thermoplastic coating is deposited (usually a styrene-methacrylate material). The recording of a hologram starts with a uniform charging of the surface of the thermoplastic material using a corona charger. The charge is divided between the photoconductor and the thermoplastic layer. Exposure and the consequent photogeneration in the photoconductor cause charges of opposite signs to migrate to the interface with the thermoplastic layer and the substrate. This will not change the charge, but only decrease the surface potential. Before the image can be developed, the material has to be re-charged with the uniform corona charger. This step adds additional charges to the exposed areas, in proportion to the reduced potential which is also proportional to the exposure. The material is then heated to a softening temperature of the thermoplastic layer which will be deformed due to the electrostatic forces acting on it. The material is then cooled and the image is fixed as a relief surface pattern in the thermoplastic layer. The processing time after the exposure necessary to obtain the image is between ten seconds and half a minute. If the material is heated to a somewhat higher temperature, the pattern disappears and the image is erased. The thermoplastic material can now be used for the recording of another hologram, repeating the recording procedure. Such a cycling technique can be repeated hundreds of times, without any serious affect on the quality of the image. That, and the fast dry processing make the material popular for the use in holographic cameras for nondestructive testing. The sensitivity is between 10 and 100 $\mu J/cm^2$ over the whole visible electromagnetic spectrum.

e) Ferroelectric Crystals

These normally rather small crystals, e.g., of lithium niobate ($LiNbO_3$) and barium titanate ($BaTiO_3$), are used mainly for scientific applications, e.g., phase-conjugation and nonlinear-optical experiments. Holograms recorded in these materials consist of bulk space-charge patterns. An interference pattern acting upon a crystal will generate a pattern of electronic charge carriers that are free to move. Carriers are moving to areas of low optical illumination where they get trapped. This effect will form patterns of net space charge, creating an electrical field pattern. As these crystals have strong electro-optic properties, the electrical field pattern will create a corresponding refractive-index variation pattern, which means a phase hologram. These holograms can immediately be reconstructed with a laser beam different from the beams creating the hologram. The hologram can also be erased and a new hologram can be recorded in the crystal. The advantages of ferroelectric materials are their high resolution, reversibility, instant readout, and rather high sensitivity.

f) Additional Materials

The following materials can also be used for holographic purposes, but, very few practical applications have been reported so far: Chalcogenide glasses, ferroelectric-photoconductors, liquid crystals, magneto-optic films, metal and organic-dye ablative films, photochromic and photodichroic materials, transparent electrophotographic films, and light-harvesting protein (bacteria rhodopsin).

2. Silver-Halide Materials

The present chapter gives a general description of the ways in which holographic image recording is made, describing in particular the application of silver-halide materials when used to store the interference patterns generated during the recording process. What follows here is therefore a short description of the holographic theory, together with some definitions of common photographic properties. Both microscopic and macroscopic characteristics of the holographic recording process are treated. For example, the forming of a latent image in silver-halide grains is explained. The theory of noise limitations, the signal-to-noise ratio, and the diffraction efficiency in the holographic recording process is discussed, with practical examples provided. The manufacturing of both conventional silver-halide emulsions and holographic emulsions is described, too.

2.1 Theory of Silver-Halide Materials

2.1.1 Photographic Silver-Halide Materials

For information on the general principles governing the holographic process the reader is referred to other, more basic books on holography mentioned in Chap.1, which give a more comprehensive presentation of the holographic theory, including detailed definitions of different types of holograms.

In order to record the entire light field scattered from an object, *both the light's amplitude and phase* have to be stored in some way. There is no material which can detect directly both amplitude and phase of a light wave. The solution is therefore to use the interferometric, two-step process introduced by *Gabor*, where the phase information is converted to amplitude variations and thus the recording medium is exposed to intensity variations. In most general terms, the holographic process consists in the fact that the micro-pattern which has been created as a result of the interference between the object beam and the reference beam must be recorded in a light-sensitive material. The time of exposure for a given material will depend on the sensitivity of the material used, as well as on the intensity of the interference pattern. As regards the silver-halide material, this must be processed after the exposure in a specific way for the recorded latent image of the interference pattern to appear. The recorded intensity variations are converted during this processing step to local variations in optical density or refractive index and/or thickness of the recording layer.

A *silver-halide recording photographic material* is based on one type, or a combination of silver-halide crystals embedded in a gelatin layer, commonly known as the *photographic emulsion*. Actually, the photosensitive emulsion is not really an "emulsion" but rather a thin film of silver-halide microcrystals dispersed in a colloid (gelatin). However, the term "emulsion" is commonly used in photography for this perpetual suspension. The emulsion is coated on a flexible or stable substrate material. There are three types of silver halides: *silver chloride* (AgCl), *silver bromide* (AgBr), and *silver iodide* (AgI). Silver chloride is used for low-sensitivity emulsions; chloride/bromide emulsions have high light sensitivity, but the bromide/ iodide emulsions have even higher sensitivity. Silver iodide is never used alone and when employed in a mixture with silver bromide it normally constitutes 5% or less. Adding some silver iodide to fine-grained emulsions at low concentrations gives a higher sensitivity and contrast than pure silver bromide emulsions of the same grain size. Silver-halide crystals are cubical in shape and in each crystal a silver ion (Ag^+) is surrounded by six halide ions. The crystal normally possesses an excess of halide ions originating from the emulsion-manufacturing process. Silver-halide grain sizes vary from about 10 nanometers for the ultra-fine-grained Lippmann emulsions to a few micrometers for high-sensitive photographic emulsions (Table 2.1).

Table 2.1. Emulsion grain sizes

Type of emulsion	Average grain diameter [nm]
Ultra-fine-grained holographic emulsion	10 – 30
Fine-grained holographic emulsion	30 – 50
Fast holographic emulsion	50 – 100
Chlorobromide paper emulsion	200
Lithographic emulsion	200 – 350
Fine-grained photographic emulsion	350 – 700
Fast photographic emulsion	1000 – 2000
Fast medical X-ray emulsion	2500

Silver compounds are sensitive to light at various degrees. Silver chloride is only sensitive to *violet* and *UV light*. Silver bromide absorbs light up to about 490 nm and if silver iodide is added to silver bromide the sensitivity extends up to about 520 nm. Special *sensitizers* (dyes) must be added to the emulsion to make it sensitive to other parts of the spectrum. We say that a photographic material is *orthochromatic* if it is also sensitive to green light. If the material has been sensitized to the whole visible part of the spectrum, including red light, it is said to be *panchromatic*. It is also possible to make the material sensitive to InfraRed (IR) light.

Methods for preparing and coating conventional silver-halide emulsions as well as holographic emulsions will be described in Sect. 2.2. Before we proceed any further with the discussion of the silver-halide materials, a few definitions concerning various general photographic properties may be worth studying by the reader. For more detailed definitions and the mathematical theory of conventional photographic materials, the reader is referred to scientific books on photography [2.1-4].

2.1.2 Photographic Sensitivity

a) Exposure

Exposure H is defined as the incident *intensity* E times the *time* t of exposure of the recording material. If the intensity is constant during the whole exposure time, which is usually the case, then

$$H = Et .\qquad(2.1)$$

In its general use, the term intensity indicates the flux per unit solid angle emitted by a light source. In photography, however, it is the amount of light radiation per unit area falling on the surface. Exposure can be expressed in both *photometric* and *radiometric* units. Photometric units apply to light only, where light means radiant energy within the visible part of the electromagnetic spectrum (400÷700nm). When exposure is expressed in photometric units, intensity means *illuminance* and exposure is then defined as

Exposure = illuminance × time .

Illuminance means luminous flux [lumen] incident on a surface. To be able to calculate the value of illuminance, one must know the candlepower of the light source and the source-to-the-recording-material distance. Since the unit for illuminance is lux [lumen/m^2], the above formula yields the exposure in lux-seconds [lx·s].

Radiation measured in radiometric units applies to radiation over the whole electromagnetic spectrum and is independent of the human eye. The radiometric equivalent of illuminance is *irradiance*, and exposure is then defined

Exposure = irradiance × time .

The unit of irradiance is watt per square meter and the exposure will then be expressed in joule/m^2. Earlier, erg/cm^2 was often used. (1mJ/m^2 is equal to 1erg/cm^2). Holographic materials are usually characterized by radiometric units. The sensitivity of a holographic emulsion is most often expressed in μJ/cm^2 or, at times, in erg/cm^2 (1μJ/cm^2 = 10erg/cm^2). Knowing the sensitivity of the material used and having measured the irradiance at the position of the holographic plate, the exposure time can be calculated using the above formula, i.e.,

Exposure time = sensitivity/irradiance .

b) The Characteristic Curve

The exposure of a silver-halide material will, after processing - i.e. development and fixing - result in a certain optical density of the material, as silver is created during the development. The higher the exposure of the material, the higher the density will be (up to a certain limit). Optical density provides a useful characteristic of the light-absorptive property of a particular region of a photographic negative. The greater the density, the less light is transmitted. A definition of density D or transmission density can be given in terms of transmittance T and opacity O

$$D = \log_{10}(1/T) = \log_{10}O , \quad \text{and} \quad T = 10^{-D} . \tag{2.2,3}$$

The reciprocal of transmittance T is opacity O. Transmittance is measured as the ratio of light transmitted through the material and light incident on the material. If a certain area of the material transmits one-half of the incident light, the transmittance is 0.5 and the opacity 2.0; the density of that area will then be $\log_{10}2$ (≈ 0.30) (Table 2.2). The reason to use density or transmission density rather than transmittance or opacity values to plot the characteristic curves is that the eye judges luminance differences approximately on a logarithmic scale.

One way of presenting the characteristics of photographic materials is the use of the *characteristic curve* also known as the *Hurter and Driffield curve* (H&D curve). A typical H & D curve is shown in Fig.2.1.

The curve is based on the results from a carefully controlled sensitometric test of the photographic material, where the test sample is gradually exposed (in discrete steps) using a sensitometer. It is then uniformly processed and the resulting densities are determined by a calibrated instrument (densitometer). The densities are then plotted versus the logarithm of the exposures that produced them. An ideal recording material should exhibit a linear relationship between exposure H and density D. In real recording materials, though, only a part of the curve shows this linearity. The shape of the curve at its left end indicates that the exposure must exceed a certain threshold value before a noticeable density occurs above the fog level. The initial fog level consists of the base density (density of the substrate on which the emulsion is coated) and the density of the fog generated in the emulsion during processing when a certain amount of reduced silver is formed in the areas which have not been exposed. The nonlinear portion at the beginning of the curve is called the *toe*. The central region of the curve is most interesting because the straight line indicates that the density increases here in proportion to the logarithm of the exposure. The slope of the linear portion of the curve indicates how fast the density increases with the increased exposure, which is an indication of the *contrast* of the materi-

16

Table 2.2. Relation between density, opacity, and transmittance [%]

Density D	Opacity O		Transmittance T
0.0	1.00	$= 10^{0.0}$	100
0.1	1.26	$= 10^{0.1}$	79.5
0.2	1.59	$= 10^{0.2}$	63
0.3	2.0	$= 10^{0.3}$	50
0.4	2.51	$= 10^{0.4}$	40
0.5	3.16	$= 10^{0.5}$	32
0.6	4.0	$= 10^{0.6}$	25
0.7	5.0	$= 10^{0.7}$	20
0.8	6.3	$= 10^{0.8}$	15.8
0.9	7.9	$= 10^{0.9}$	12.5
1.0	10.0	$= 10^{1.0}$	10
1.1	12.6	$= 10^{1.1}$	7.9
1.2	15.9	$= 10^{1.2}$	6.3
1.3	20.0	$= 10^{1.3}$	5.0
1.4	25.1	$= 10^{1.4}$	4.0
1.5	31.6	$= 10^{1.5}$	3.2
1.6	40.0	$= 10^{1.6}$	2.5
1.7	50.0	$= 10^{1.7}$	2.0
1.8	63.0	$= 10^{1.8}$	1.6
1.9	79.5	$= 10^{1.9}$	1.25
2.0	100.0	$= 10^{2.0}$	1.00
3.0	1000	$= 10^{3.0}$	0.10
4.0	10000	$= 10^{4.0}$	0.01

al. The number characterizing the slope of the curve is denoted by the *gamma value*: $\gamma = \Delta D/\Delta \log H$. The angle α between the log H axis and the straight-line portion of the H&D-curve is related to the gamma value as $\gamma = \tan\alpha$. The steeper the slope, the higher the contrast and the gamma value. For photographic purposes the gamma value is normally close to one, since then density variations in the film will directly follow the variations in exposure.

Holographic emulsions are of the high-contrast (high-gamma) type. Small variations in exposure result in relatively large variations in densities. Development procedures can affect the shape of the characteristic curve for certain materials. In addition to that, it is possible to enhance the contrast between light and dark regions by selecting a suitable developer.

If the exposure is increased above a certain value the density-exposure relationship becomes again nonlinear. This portion of the curve is often referred to as the *knee* or the *shoulder*. The slope of the curve decreases un-

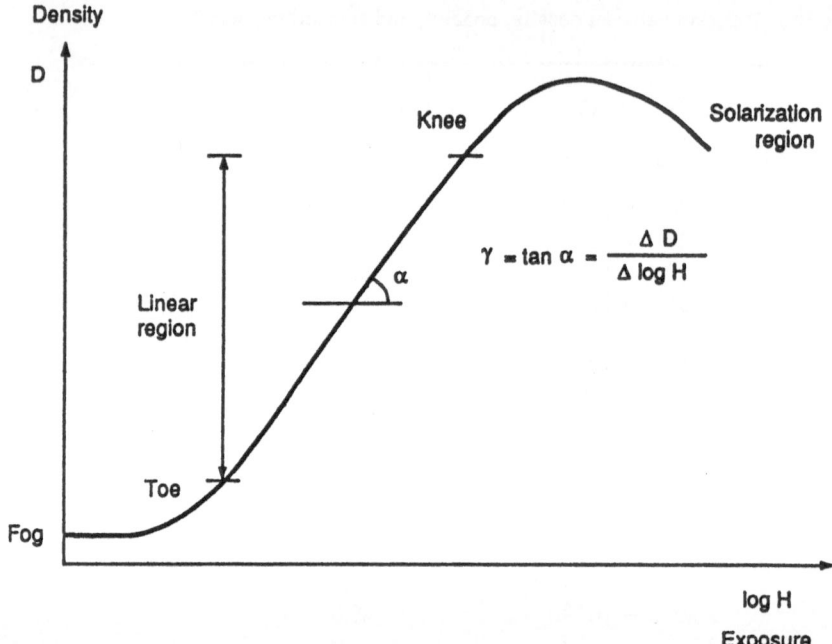

Fig.2.1. The characteristic curve (H & D curve) for a photographic material shows the optical density D versus the logarithm of the exposure H

til the curve parallels the horizontal axis again. At this point, all the silver-halide crystals will produce development centers during exposure so that the density cannot be increased any further. If the exposure is increased above this point, the density will actually be reduced (additional exposure triggers off a number of complex reactions that cause the curve to turn downward). This effect is called *solarization* and is made use of in making reversal film that has been preexposed to the solarization limit, so that additional exposure will result in a positive image after processing.

The linear part of the curve embraces the range of exposures called the *latitude* of the material. In both photography and holography the restricted linear part of the characteristic curve is associated with the problems that are apt to appear when recording objects with a large dynamic range.

c) Photographic Sensitivity

To be able to compare different products and estimate the required exposures, one should properly understand what is meant by *photographic sensitivity* or the *speed* of the recording material. When exposure is expressed in radiometric units, the value obtained is called *radiometric speed*; when it is expressed in photometric units, the value is referred to as *photometric speed*, which is commonly referred to as *photographic sensitivity*. It should be mentioned that both radiometric and photometric speeds depend on the spectral characteristics of the exposing radiation. An indication is important as regards the type of light source (or any other type of radiation) that a certain speed is associated with.

The H&D curve is used to determine the sensitivity of the photographic materials. The *DIN* and *ASA speeds* of conventional films are evaluated from such curves according to special definitions. High-resolution holographic emulsions have low ASA values and their sensitivity is given in radiometric units instead (e.g., $\mu J/cm^2$), as described earlier. *Frecska* [2.5] actually calculated the ASA-speed for the high-sensitive holographic Agfa 10E70 film and found it to be 0.57 ASA. The slow Kodak 649-F emulsion is 0.025 ASA and the fast SO-243 is 1.6 ASA. The sensitivity of a holographic emulsion is normally given for an amplitude hologram exposed to light of a certain wavelength and developed to a density of 0.6 using a special processing method. Depending on the manufacturer, the sensitivity values supplied for a given emulsion may exhibit extensive variations. This is due to the fact that there does not exist one standard, uniform method for measuring the sensitivity of a holographic silver-halide emulsion. This is why sensitivity values supplied by the manufacturer for a particular emulsion should be regarded as approximate. Adequate comparisons between holographic materials produced by different manufacturers are difficult to make.

Spectral sensitivity curves of a given photographic material are often supplied by the manufacturer. These curves show the logarithm of sensitivity versus wavelength and are obtained with the help of spectrosensitometers. The curves show clearly that sensitivity of photographic materials is wavelength dependent. Holographic silver-halide emulsions are sensitized in such a way that they are optimized for laser wavelengths commonly used in holography.

2.1.3 Emulsion Structure

Sensitometric data alone are not sufficient to get a clear picture of the characteristics of a given photographic material. Several other features should also be described for that purpose.

a) Granularity

Due to the fact that the developed silver-halide emulsion consists of millions of discrete silver particles, it is obvious that when a normal photographic image is enlarged enough, the areas which appeared homogeneous to an unaided eye will now show a granular pattern. The subjective sensation of how the human eye perceives this pattern is called *graininess*. The measurable variations in the density of the granular pattern are called *granularity*. Grain sizes in the emulsion will affect granularity very much: the smaller the grains the lower the granularity, i.e. the higher the contrast. Silver-halide crystals of the conventional photographic film can be larger than one micrometer, which results in high granularity. Holographic emulsions have grain sizes between 10 and 100 nm and will therefore show the lowest granularity possible. Granularity is measured with the help of a microdensitometer with a very small aperture scanning the area under investigation. The random density fluctuations will result in variations in the light transmitted through the material which hits a photoelectric cell and is then recorded. The standard deviation of the random density fluctuations is called the *root mean square* (rms) *granularity* σ. R. Selwyn found that the product of the rms granularity and the square root of the aperture used to make the measurements is constant for a given black-and-white material. The *Selwyn granularity* G is defined as

$$G = \sigma\sqrt{2a} \tag{2.4}$$

where a is the area of the scanning aperture, and σ is the rms granularity.

Another characteristic of granularity is the *Wiener spectrum* that expresses granularity as a function of spatial frequency.

Photographic turbidity is caused by light scattering and light absorption in the emulsion, which results in widening of the recorded bright points or lines and will increase with increased exposure.

b) Resolving Power

The resolving power of a photographic material is a measure of the ability to record fine detail. It is defined as: "The ability of a photographic material to maintain in its developed image the separate identity of parallel bars when their relative displacement is small". Normally, the resolving power of photographic materials is tested by using a resolution test chart. The highest number of lines per millimeter that can be resolved in the emulsion corresponds to the resolving power of the tested material. A line in this definition is a line with its adjoining space, and it corresponds to line pairs in electronic images. The resolving power of the holographic material is a very important feature that must be taken into account when defining its characteristics.

c) Modulation Transfer Function

The resolution capability of an image reproduction process is normally described by an *Optical Transfer Function* (OTF). For a given test input, the OTF is defined as the *complex response* (amplitude and phase) of the reproduced image for each spatial frequency ν. In practice, usually only the modulus of the OTF is quoted, which is known as the *Modulation Transfer Function* (MTF). The MTF is a good aid in demonstrating the quality of a particular photographic emulsion as well as a means of comparing different emulsions. Briefly, a test pattern containing a sinusoidal variation in illuminance combined with a continuous variation in spatial frequency along one direction is recorded. The modulation M of the pattern in the test target is

$$M = \frac{H_{MAX} - H_{MIN}}{H_{MAX} + H_{MIN}} \tag{2.5}$$

where H is the exposure incident on the photographic material. When this pattern is recorded in the material, light scattering will take place in the emulsion, which will reduce the original contrast of the pattern. Therefore, the modulation of the pattern will be decreased, in particular at high spatial frequencies. The effective exposure modulation M′ will then be

$$M' = \frac{H'_{MAX} - H'_{MIN}}{H'_{MAX} + H'_{MIN}} \tag{2.6}$$

where H′ is exposure in the emulsion.

The original modulation M is usually constant and accurately known; it is also independent of the spatial frequency. After the tested emulsion has been processed, the corresponding "exposed" modulation is obtained from the density variations. The ratio between modulation M′ in the emulsion and modulation M of the incident exposure is called *modulation transfer factor*, also called *response*

$$R = M/M' . \tag{2.7}$$

If the response is plotted as the function of spatial frequency, this curve will then be the *modulation transfer function* of the material (Fig.2.2). The Fourier transform of the MTF is the light spread function of the emulsion which indicates the width of a line image recorded in the emulsion.

Sometimes another function is used, the *Contrast Transfer Function* (CTF), which is defined in a manner analogous to the MTF. In this case the

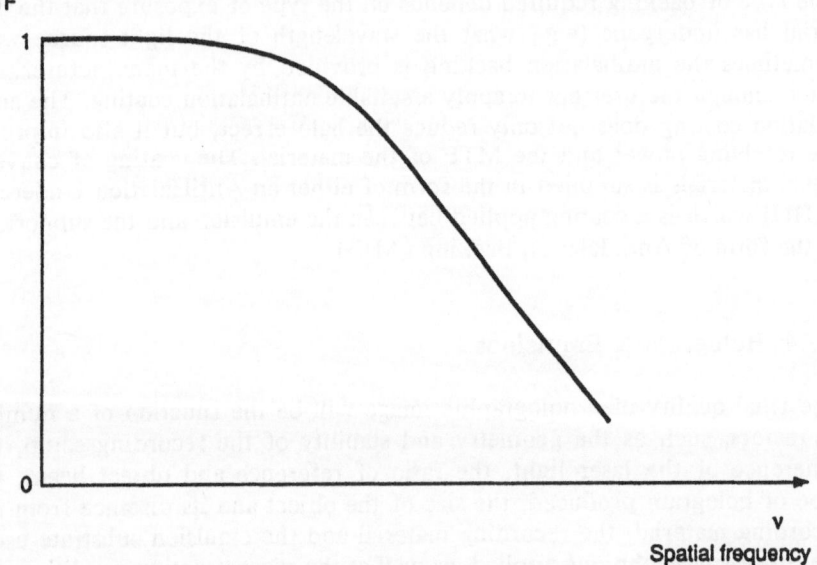

Fig.2.2. Graphical representation of the Modulation Transfer Function (MTF) curve

test pattern represents a binary intensity variation, a black-and-white stripe pattern. The CTF is easier to measure experimentally, since well-defined binary test patterns are readily fabricated. There is, however, a relation between MTF and CTF which in a linear system is

$$MTF(\nu) = \frac{\pi}{4}\left[CTF(\nu) + \frac{CTF(3\nu)}{3} - \frac{CTF(5\nu)}{5} + ... \right].$$ (2.8)

d) Acutance

Acutance is used to estimate the sharpness or quality of a photographic image. It is the spread of a knife-edge exposure in the emulsion and it is usually defined as the mean-square gradient divided by the density scale. For a linear response, acutance is the density difference divided by the square of the distance over which the knife-edge image is spread in the emulsion. Acutance is normally a better measure of the sharpness of a photographic image than is the resolving power.

e) Halation

Halation occurs when during the exposure light passes through the emulsion and the support material, and is reflected back into the emulsion. The reflected light causes some unwanted exposure and produces a halo in the image. The halo around a bright point source will cause the image of a point to look like a circle or an extended disk. The thicker the substrate, the more pronounced the halation effect will be. Under identical exposure conditions, a glass plate will show more halation than a thin film. Halation can be reduced by backing the material with an AntiHalation (AH) coating. The type of backing required depends on the type of exposure that the material has undergone (e.g., what the wavelength of the light source was). Sometimes the antihalation backing is provided by the manufacturer, but often enough the user has to apply a suitable antihalation coating. The antihalation coating does not only reduce the halo effect, but it also improves the resolving power and the MTF of the material. The coating of conventional materials is supplied in the form of either an AntiHalation Undercoat (AHU) which is a coating applied between the emulsion and the support, or in the form of AntiHalation Backing (AHB).

2.1.4 Holographic Emulsions

The final quality of a holographic image will be the function of a number of factors, such as the geometry and stability of the recording setup, the coherence of the laser light, the ratio of reference and object beam, the type of hologram produced, the size of the object and its distance from the recording material, the recording material and the emulsion substrate used, the processing technique applied, as well as the reconstruction conditions. In this and the following chapters we will focus on the way in which the choice of the recording material and the processing method will influence the final quality of the holographic image. In most other scientific books on holography, the other factors mentioned above have been treated in considerable detail, e.g., as regards the magnifications and aberrations that occur when a hologram is reconstructed with a differently shaped reference beam and/or a different wavelength, assuming that the recording material is perfect. We know that if during the reconstruction of the hologram, the reference beam is identical with the recording reference beam, no image aberrations will occur. This applies also to circumstances when the reconstruction reference beam is a conjugate of the original reference beam (time reversed). Theoretically, the holographic technique is the most perfect imaging technique in existence, since both the amplitude and the phase of the light wave scattered from the object are recorded. In practice, the holographic image is subject to certain limitations imposed by the recording material.

Three main factors will determine the resolution of a holographic image: The recording wavelength, the numerical aperture and the properties of the recording material itself. However, the following three points must be considered when discussing the attainable resolution of a holographic image:

- Ideally, the ultimate resolution should be independent of the properties of the recording material, and should depend only on the wavelength that was used for the recording as well as on the size of the recorded area of the material (the aperture) and the object distance.

- In practice, the limit on resolution may be set by the recording material, e.g., if it cannot record spatial frequencies above a certain limit.

- As regards aberrations introduced during the reconstruction of the hologram the following applies: If we assume that no aberrations are introduced by altering the reference beam, the only aberrations that will occur are then caused by the recording material itself.

a) Demands on the Recording Emulsion

A silver-halide emulsion must comply with certain requirements to be suitable for the recording of holograms. The most important one concerns the resolving power of the material. The recording material must be able to resolve the highest spatial frequencies of the interference pattern due to the maximal angle θ between the reference and the object beams in the recording setup (Fig.2.3).

If λ is the wavelength of the laser light used for the recording of a hologram, then the closest separation d_a between the fringes in the interference pattern (in air) is

$$d_a = \frac{\lambda}{2\sin\frac{1}{2}\theta} .$$

(2.9)

In the recording layer the fringe spacing d_e depends on the refractive index n of the emulsion and is

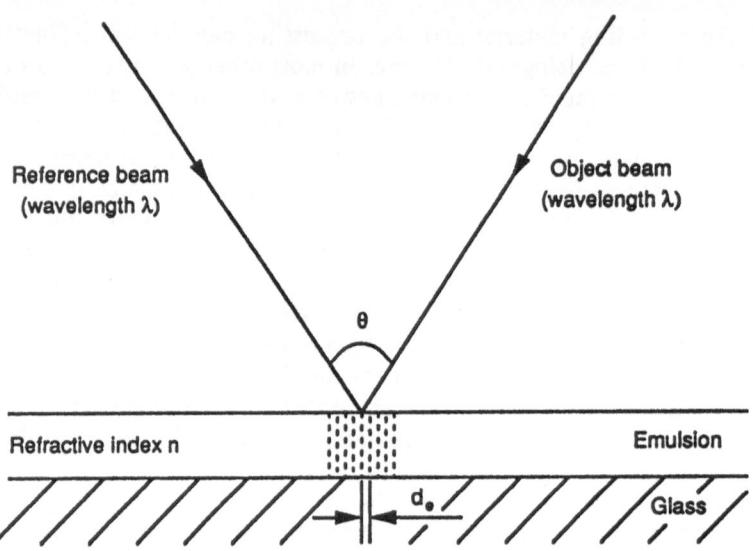

Fig.2.3. Demand on resolution for recording a hologram. The recording material must resolve the highest spatial frequencies of the interference pattern created by the maximal angle θ between the reference and the object beams in the recording setup

$$d_e = \frac{\lambda}{2n\sin\frac{1}{2}\theta} \cdot \qquad (2.10)$$

One example of the resolving power needed in a practical situation employing an emulsion with a refractive index of $n = 1.62$ is the following: A ruby laser with the wavelength $\lambda = 694$ nm and a recording geometry with the maximum angle $\theta = 25°$ between the beams are used. This gives $d_e \simeq 1$ μm, which corresponds to $\lambda = 1/d = 1000$ lines/mm; this is the minimum resolving power required. Close to its resolution limit the material exhibits a low MTF and will thus make a low-quality hologram with poor fringe contrast and a low signal-to-noise ratio. For high-quality holograms the resolution limit of the material must be much higher than the minimum value obtained according to the above formula. For a reflection hologram recorded in blue light ($\lambda = 400$nm) with an angle of $180°$ between the beams a minimum resolving power of 7600 lines/mm is needed..

Sometimes it is more convenient to use the following equation

$$\nu = (\sin\theta_r - \sin\theta_o)n/\lambda \qquad (2.11)$$

where θ_r is the angle between the reference beam and the normal to the emulsion surface, and θ_o is the corresponding angle for the object beam, λ is the wavelength in air, and n the refractive index of the emulsion.

Figure 2.4 illustrates differences in quality (mainly as regards diffraction efficiency and signal-to-noise ratio) between two holographic emulsions (Agfa 10E75, 3000 ℓp/mm and Agfa 8E75 HD, 5000 ℓp/mm) in the

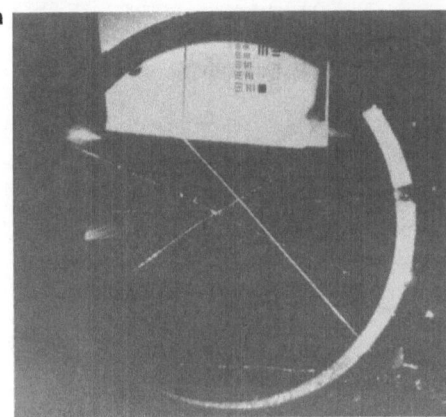

Fig.2.4. Difference in quality between two holographic emulsions used for the recording of a transmission hologram. Agfa 8E75 HD with a resolving power of about 5000 ℓp/mm (a) and the faster Agfa 10E75 with a resolving power of about 3000 ℓp/mm (b)

recording of an in-line hologram, where both emulsions have sufficient resolving power to record an image.

b) Resolution of the Holographic Image

In holography the resolution of the holographic image and the resolving power of the recording material are not directly related in the way they are in photography. Resolution of a holographic image has been discussed in depth in several papers [2.6-15]. According to the way in which the phase is normally stored in a hologram, only the phase of the object wave which is *in phase* with the reference beam is actually recorded. This leads to ambiguity in the reconstructed wavefront and results in two different wavefronts. This fact was pointed out by *Gabor* and *Goss* [2.14] who also suggested a method to record the entire phase information in a hologram. The influence on the image resolution caused by recording only one half of phase information in a hologram has been discussed from the theoretical point of view by *Russell* [2.15].

Equation (2.10) gives the minimum resolving power needed *to actually record* an off-axis hologram. This figure is not directly related to the *resolution of the image* recorded in or reconstructed from a holographic plate. If no lenses are involved in the holographic image-formation process, the resolution of the image will be dependent on the area of the recording material, the recording-laser wavelength and the distance between the recording material and the object (Fig.2.5).

Image resolution dependent on the recording material

Theoretically, the resolution of the holographic image should be the true diffraction-limited resolution that can be obtained when the information is collected over an aperture equal to the size of the recording holographic

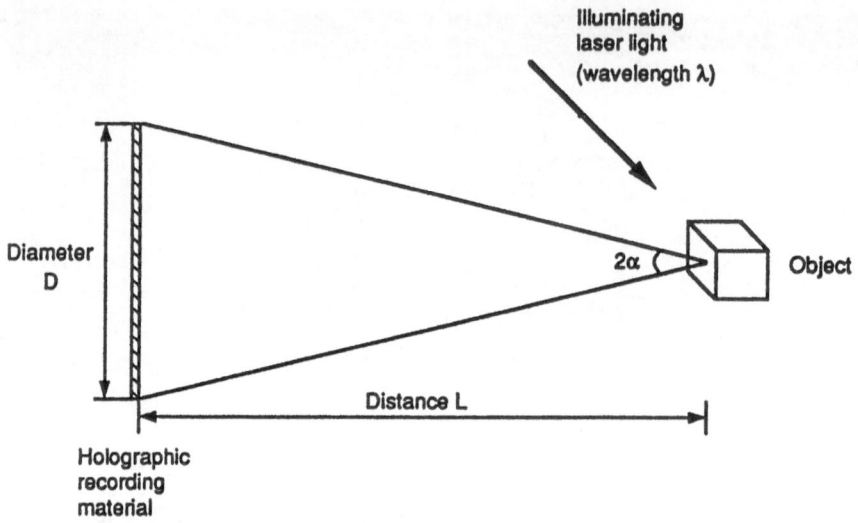

Fig.2.5. Resolution of a holographic image depends on the area (diameter D) of the recording material, the recording laser wavelength λ, and the distance L between the recording material and the object. The resolution that can be obtained in a holographic image will, theoretically, be the diffraction-limited resolution

plate. In principle, the larger the holographic plate the better the resolution will be. The limit is of the order of the wavelength λ when the dimension of the plate is infinite ($D \rightarrow \infty$). If the holographic plate is very large, however, the resolving power of the recording material will eventually limit the resolution of the image. If the resolving power of the recording material is sufficient, the diffraction-limited resolution can be obtained under the assumption that the high-resolution recording material is also perfect in that the position of the recorded interference fringes will not be changed during the processing of the material. In practice, a stable support for the emulsion (like a glass plate) will be needed and the processing methods applied must be such as not to affect the recorded fringe position in the emulsion (e.g., no fixing). The factor most limiting the resolution of a holographic image is, however, in the form of distortions appearing in the emulsion. These aberrations are introduced by

- variations in the thickness of the recording medium before processing;
- variations in the thickness of the recording medium produced during processing;
- variation in the refractive index of the recording medium produced during processing; and
- deformations of the recording medium that will occur between recording and reconstruction.

An ideal holographic-recording plate should consist of a uniformly thick emulsion coated on a perfectly flat plate of homogeneous glass and

26

uniform thickness. In practice, the recording plate must be treated as one of the optical elements in the holographic system that will eventually affect the image quality. Thus, if the variation in the thickness of the emulsion is approximated by a quadratic curve, the recording medium will act as a lens in the system. *Kubota* [2.16] investigated both theoretically and experimentally the influence on diffraction efficiency and angular sensitivity of inbuilt stresses in the emulsion. By pre-stressing the emulsion, *Kubota* found a bending of the interference fringes which was zero at the emulsion surface. Its shape was shown to be approximately represented by a quadratic curve in the direction of the emulsion thickness. This causes a decrease in diffraction efficiency as well as changes the direction of reference illumination during the reconstruction compared to the recording. To introduce artificial stress into the emulsion *Kubota* soaked the material in tepid water for 10 minutes and then dried it under various magnitudes of the centrifugal force.

The variation in the thickness of the emulsion has been studied by *Gara* and *Yu* [2.17], and *Majkowski* and *Gara* [2.18]. The emulsion stresses caused by the coating process can be released before the recording is performed, which will reduce the variations between exposure and reconstruction. The plate is washed for two minutes in distilled water with a wetting agent, after which it is slowly air-dried before the actual exposure. The present author also recommends this method. *Butters* et al. [2.19] soaked the plate in water for two minutes, and to speed up drying, used a bath of industrial methylated spirits. *Friesem* and *Walker* [2.20] applied the pre-soaking method combined with an optimization of the recording geometry to reduce the influence of emulsion changes. *Majkowski* and *Gara* [2.18] annealed the material in 100% humidity at 30° C for 8 hours before recording on it. Reducing the influence of the emulsion variations is of greatest importance in applications concerning high-resolution imaging, real-time holographic interferometry, and holographic optical elements and filters. *Butters* et al. [2.19] demonstrated the improvements of real-time fringes by using pre-exposure processing to relieve emulsion stresses. *Duffy* [2.21] combined this method by eliminating the fixing step as well, which further improved the real-time interferometric investigation. Omitting the fixing step is certainly an important improvement as it will affect the emulsion thickness variation much more than the inbuilt stresses will. Not only the image resolution is affected by thickness variations but also the diffraction efficiency, as discussed by *Dzyubenko* et al. [2.22]. The influence of different processing methods on the resolution of the holographic image will be further discussed in Chap.4.

The use of a *liquid gate* will solve some of the problems caused by the lensing effect of the recording plate. Liquid gates can be applied for both the recording and the reconstruction of holograms or at the reconstruction stage only. High-resolution holography requires, as a rule, the use of glass plates combined with liquid gates for optimal results. Phase disturbances caused by commercial film substrates are severe (much more than the emulsion variations are), as shown by *Ingalls* [2.23]. Therefore, holograms recorded on film require most often index-matching techniques. The use of liquid gates and suitable index-matching liquids are further discussed in Sect.6.5. *Matsumura* analyzed the wave-front aberrations caused by deformation of the recording media [2.24] and evaluated the deformation tolerance for small holograms [2.25]. The main conclusions are that the inclination and curvature of the medium produces astigmatism in the image. Using symmetrical beam arrangements during recording will reduce the effects of emulsion deformation. According to *Matsumura*, Fourier-transform hologram will be the type of hologram that is least affected by emulsion variations. *Soares* and *Leite* [2.26, 27] have studied the degradation of the holographic image due to wavefront defects, in particular, the influence of the *emulsion supporting substrate*. The problem was analyzed for the wedge phase error and the sinusoidal perturbation. They also presented interferometric methods for testing the optical quality of the plate substrate.

The aberrations caused by *emulsion shrinkage* have further been considered by *Verbovetskii* and *Fedorov* [2.28] and methods of measuring emulsion variations have been published [2.29, 30]. The earlier discussed real-time interferometry tests using stress relieve methods have confirmed the reduced effect of emulsion distortions. *Gupta* and *Aggarwal* [2.29] used quantative speckle methods to study emulsion swelling and shrinkage rates during processing and drying. *Jaroszewicz* [2.30] used a phase-difference amplification method to evaluate minute emulsion movements. The method is much more sensitive than the real-time hologram interferometric evaluating technique. The smallest measurable emulsion movement is equal to 0.05λ, which means about 30 nm if a HeNe laser wavelength is used. *Jaroszewicz* found that roughly a triple reduction in the range of emulsion movement can be achieved using the pre-exposure water soaking strain relieve techniques.

The influence of a drastic nonuniform shrinkage on the resolution of a photographic emulsion is illustrated in Fig.2.6.

The aberrations caused by the *reduced refractive index* of the emulsion must be considered, too. Even if the thickness of the emulsion is the same at the reconstruction (unchanged or corrected by post-process swelling) as at the recording, the change of the refractive index from n_1 to n_2 ($n_2 < n_1$) will affect the reconstruction angle. The Bragg condition for reconstruction; $\sin\phi = \pm\lambda/(2n_2d)$, indicates that $\phi > \alpha$ (recording angle), when the emulsion thickness is constant. To make $\phi = \alpha$ the thickness of the emulsion can be increased by swelling to obtain a fringe spacing of dn_1/n_2 in a direction perpendicular to the emulsion. Only corrections for the perpendicular case can be performed by swelling. To avoid aberrations caused by the index change an increase of the refractive index to its original value is needed, which may be difficult in practice. The *desensitization* of holographic emulsions instead of fixing them is not only helpful in avoiding emulsion shrinkage but also in reducing the refractive-index change of the gelatin layer. Concerning further details about the refractive-index variations of silver-halide gelatin emulsions refer to the index-matching techniques treated in Chap.6.

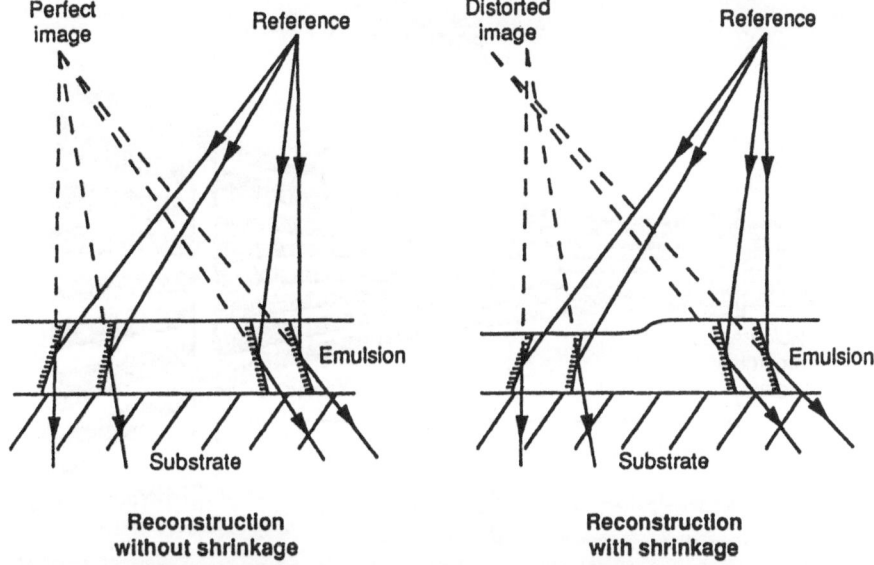

Fig.2.6. Resolution degradation due to nonuniform emulsion shrinkage. If the position of the recorded interference fringes is changed during the processing of the material, image aberrations are introduced. To obtain highest possible resolution a stable support for the emulsion (like a glass plate) is needed and the processing methods applied must be such so as not to affect the recorded fringe position in the emulsion (e.g. no fixing)

Image resolution dependent on other factors than material

The resolution that can be achieved in a holographic image independently of the restrictions that otherwise apply to the recording material are discussed in the following section. Resolution can be regarded as a coherent observation (Young's fringes of observation) and therefore it is almost identical to the separation between Young's fringes projected onto the surface of the object at the distance L from the recording plate, when the fringes are created from two diagonal points situated at the periphery of a circular holographic plate (the recording aperture) of diameter D (Fig.2.7).

The coherent sources are emitting light of wavelength λ. The separation δ is then

$$\delta \approx \frac{\lambda L}{D} \quad \text{for} \quad L \gg D , \quad \text{or} \quad \delta = \frac{0.5\lambda}{\sin\alpha} \qquad (2.12)$$

where α is the angle indicated in Fig.2.7. However, if a limiting aperture is part of the system, diffraction occurs, which means that the resolution is slightly smaller than it would be if only the two points were used. For a circular aperture in *incoherent light the diffraction-limited resolution* is [Ref.2.31, p.419]

$$\delta \approx 1.22 \frac{\lambda L}{D} \quad \text{for} \quad L \gg D , \quad \text{or} \quad \delta = \frac{0.61\lambda}{n\sin\alpha} . \qquad (2.13)$$

29

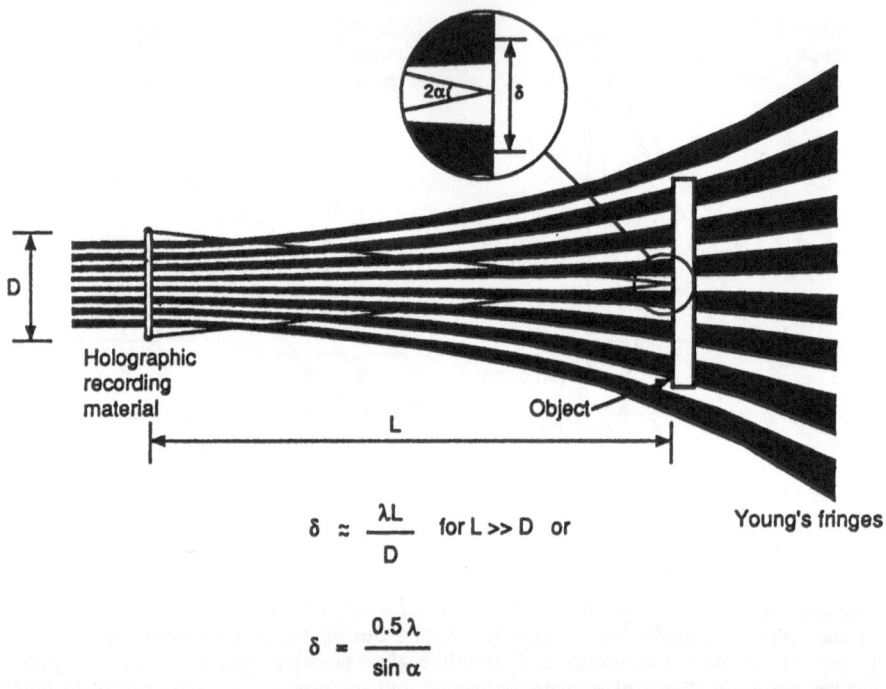

$$\delta \approx \frac{\lambda L}{D} \quad \text{for } L \gg D \quad \text{or}$$

$$\delta = \frac{0.5 \lambda}{\sin \alpha}$$

Fig.2.7. Obtainable image resolution in a hologram. Resolution can be regarded as coherent observation (Young's fringes of observation) and therefore it is almost identical with the separation between Young's fringes projected onto the surface of the object at distance L from the recording plate when the fringes are created from two points opposite to each other at the periphery of a circular holographic plate (the recording aperture) of diameter D. The wavelength of the laser light is λ. Image resolution $\delta = 0.5\,\lambda/\sin\alpha$, where α is the angle indicated in the figure

Due to aperture diffraction, a point source is viewed in any imaging system as the *Airy disk pattern* (a central area of maximum intensity surrounded by concentric diffraction rings of zero intensity). Normally, two points are said to be resolved when the maximum of one Airy disk falls on the first minimum ring of the second point's Airy disk. According to the *Rayleigh resolution criterion* two points can be considered resolved if the illumination difference between them is 20% of the maximum value. Using coherent light the phase difference between the waves coming from two points midway between them will be twice as big as the phase difference obtained in incoherent light. This leads to a resolution degradation by a factor of $\sqrt{2}$.

In *coherent light the diffraction-limited resolution* is therefore [Ref. 2.31, p.424]

$$\delta \approx 1.54 \frac{\lambda L}{D} \quad \text{for} \quad L \gg D , \quad \text{or} \quad \delta = \frac{0.77\lambda}{n\sin\alpha} . \tag{2.14}$$

The quantity nsinα is called the *Numerical Aperture* (N.A.) used as a measure of the "light-gathering power" for microscope lenses in particular. The refractive index of the space between the lens and the object is n and λ is the wavelength of the light in vacuum. In most cases n = 1. Although modified expressions are used for different types of optical instruments, the resolving power of any such instrument (as well as that of a hologram) can always be found by applying the above formula and the corresponding numerical aperture. At the recording of a hologram the limiting aperture and the recording material are in the same plane. Thus, no diffraction occurs.

The following example illustrates the maximum image resolution that can be obtained from a circular holographic plate of 50 mm diameter, using laser light of 633 nm and a distance of 2 meters between the object and the plate:

$$\delta = \frac{633 \cdot 10^{-9} \cdot 2}{0.05} \simeq 25 \ \mu m \ . \tag{2.15}$$

If the hologram is reconstructed with light of the same wavelength as was used for the recording and with the same reference beam geometry, and if the whole recorded area of the plate is used (reconstruction aperture), then the calculated resolution of 25 μm in (2.15) will also be the resolution obtainable in the image at reconstruction, provided that the recording material is perfect. If, for imaging purposes, a limiting aperture of the same size like the recording area on the holographic plate is employed, diffraction occurs and the resolution will be reduced by the factor 1.54 ($\approx$40μm), which will also be the size of the speckles in this reconstruction process of the recorded image. It is obvious that the image resolution depends a great deal on how the hologram is reconstructed. The above example is only meant to indicate the best possible resolution that can eventually be obtained. If a smaller aperture is used at reconstruction (only a part of the recorded area of the plate is utilized) the resolution will be lower. For example, if only a 5 mm aperture is used at reconstruction, the image resolution will then be about 400 μm only. If a different reference beam or a different wavelength is used at reconstruction, the image will normally be also affected by aberrations that can drastically limit the resolution.

In all imaging systems using coherent light *speckles* are present [2.32]. Speckles usually indicate the fundamental resolution of a particular optical system. We can therefore say that speckles and resolution are one and the same thing. The speckle size will be equal to the image resolution calculated above. As an illustration of the relation between speckles and resolution, the common spatial-filtering technique applied to "clean" the illuminating laser beams employed in holography will be described. At the focal point of the microscope lens in the spatial filter equipment a pinhole is placed. The diameter of the aperture is chosen in such a way that only one "bright speckle" is covering the entire illuminating field. This field is used to illuminate, e.g., the surface of the object to be holographed (Fig.2.8). This is equivalent to the highest resolution obtainable through the pinhole aperture.

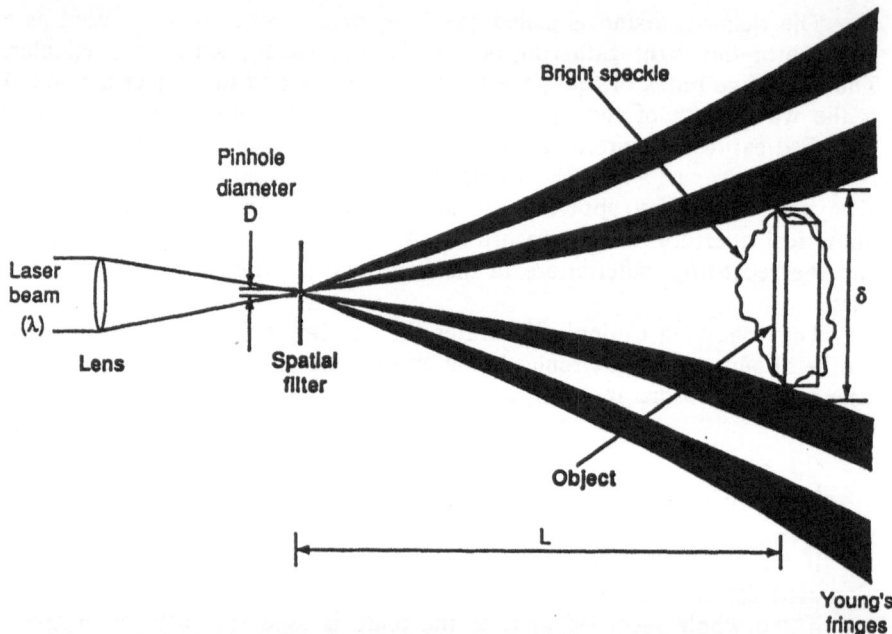

Fig.2.8. Spatial filtering and resolution. Speckles indicate the fundamental resolution of a particular optical system. The diameter of the aperture in a spatial filter is chosen in such a way that only one "bright speckle" covers the entire illuminating field which is used to illuminate the object. If the object surface is observed through the aperture, no details can be resolved over the illuminated area, which means that the resolution is equal to the diameter of the total illuminated area at the position of the object

If the object surface is observed through the aperture, no details can be resolved over the illuminated area, which means that the resolution (separation of the finest details) is equal to the diameter of the total illuminated area at the position of the object. Young's fringes generated from two diagonally situated points at the pinhole aperture will be separated at a distance corresponding to the diameter of the illuminating field at the position of the object. This fact can be used to calculate the diameter of the pinhole in the spatial filter in a given case.

Several different *speckle-reduction methods* [2.33] can be used, in particular, at the reconstruction stage. Reduction of speckle noise makes it possible for the observer to see the image details close to the resolution limit – the details that would otherwise be embedded or hidden in the speckle background. If the speckle noise background is present in a coherent image, an effectively poorer resolution, lower than its theoretical limit, is normally experienced. In other words, image details must often be 3 to 4 times larger than the existing speckle size (representing the diffraction-limited resolution) in order to be distinguishable.

It is also important to realize the influence of speckles in the object illumination during the recording of a high-resolution image hologram. The best illumination is generated from a spatial filter (only one bright objective speckle across the object). If a ground glass is introduced in the illuminating beam (to make a more uniform illumination), fine objective speckles are generated all across the entire object. The size of these speckles have to be finer than the desired image resolution. If these speckles are too large, they will actually limit the final resolution of the holographic image. These speckles are *recorded* in the hologram and by, in this case, increasing the size of the hologram area will not improve the image resolution. The larger the illuminated area of the ground glass is the finer the objective speckles will be. Therefore, either the illuminating light has to be generated from an extremly small area (the spatial filter) or from a rather large ground-glass area, to generate very fine speckles in the object illuminating field, in order to obtain holograms with high image resolution.

c) The Space-Bandwidth Theorem

The space-bandwidth product of a one-dimensional object or a recording medium can be useful in judging the combined effect of the object resolution and the resolving power of the recording emulsion. *The product of the object width and the object bandwidth cannot exceed the product of the hologram width and the hologram recording-material bandwidth.* Assume that the highest spatial frequency that the recording material can resolve is ν_h and the width of the hologram is L_h. Assume that the finest details that can be resolved in an object positioned at distance D from the holographic plate are separated at a distance δ_o corresponding to the spatial frequency $1/\delta_o = \nu_o \simeq L_h/(\lambda D)$. For the maximum width of the object L_o at the distance D which corresponds to the highest spatial frequency ν_h of the recording material, we have the following relation: $L_o \simeq \lambda D \nu_h$. Combining this relation with the previous one by eliminating the distance D, we obtain the following equation for the two space-bandwidth products:

$$L_o \nu_o \simeq L_h \nu_h \ . \tag{2.16}$$

2.2 Emulsion Manufacturing

2.2.1 Conventional Photographic Emulsions

The preparation of photographic emulsions is described in various books dealing with photographic processes [2.1-4]. A rather old book written by *Baker* is devoted exclusively to the process of emulsion preparation and contains numerous practical considerations and hints [2.34]. A good publication is an English translation of a Russian book by *Zelikman* and *Levi* [2.35]. Another book, by *Duffin*, on the photographic emulsion chemistry is also recommended [2.36]. In a paper by *Hill* a good bibliography on this topic can be found [2.37].

Silver-halide emulsions can be manufactured and coated in a holographic laboratory, just like dichromated gelatin emulsions for holography are. However, the preparation of silver-halide emulsions is a tedious and difficult process and is therefore not recommended for normal applications of holography. Moreover, a number of manufacturers do produce silver-halide holographic emulsions (Chap.3). However, as mentioned by *Phillips* [2.38], the grain size of a holographic Lippmann emulsion should be of the order of 10 nm and not more, in particular if the emulsion is to be used for recording in blue light. At a grain size of 10 nm the Rayleigh scatter off the grain does not impede the transit of the light through the emulsion. On the other hand, such an emulsion would be very slow and a compromise between speed and grain size must be made. Currently, it is difficult to manufacture the ultra-fine-grained emulsions (like some of the Russian materials) on an industrial scale. Such emulsions are best prepared and coated just before they are used. Likewise, emulsions exhibiting characteristics different from those typical of the usual commercial products as regards grain structure, thickness, or spectral sensitivity, must be made in the laboratory. A brief description of how a holographic silver-halide emulsion is made is therefore presented below.

a) Silver-Halide Emulsion Manufacturing

The technique of making silver-halide photographic emulsions has been well known for over a hundred years now. As mentioned earlier, Lippmann experimented with high-resolution emulsions of this type to record color photographs. Later on, *Denisyuk* and *Protas* prepared an emulsion [2.39] which was a modified version of emulsions made according to *Ives'* formula [2.40] for reflection holograms (Chap.1). Despite this, the number of papers dealing with the preparation of holographic silver-halide emulsions is quite small, especially when compared to the amount of literature published on dichromated gelatin plates for holography. The reason for this may be that although commercial companies have extensive knowledge of the field, much of it is regarded as industrial secrets and does not appear in the scientific literature.

Briefly, the technique of making a conventional photographic silver-halide emulsion is as follows:

I) Emulsification. A solution of silver nitrate is mixed with another solution containing gelatin and an alkali halide, e.g., potassium bromide according to the following *double decomposition reaction*:

$$Ag^+ + NO_3^- + K^+ + Br^- \rightarrow AgBr + K^+ + NO_3^- .$$

As regards the molecular weight, 170 parts of silver nitrate combined with 119 parts potassium bromide will form 188 parts of silver bromide and 101 parts of potassium nitrate. The formation of solid crystals from a solution in this way is called *precipitation*, and the chemical compound itself the *precipitate*. Depending on the mixing technique and the gelatin concentration, different grain sizes can be obtained in the final emulsion. At this stage the sensitivity is low.

II) Physical ripening. At the next step the emulsion is ripened or digested (*first ripening* or *Ostwald ripening*) to make it more sensitive to light. This is done by heating the emulsion for about an hour at the temperature of 55° C. During this process the smallest grains are dissolved and redeposited on some of the larger grains. In this first ripening process grain growth takes place, which means that the sensitivity of the emulsion increases. Another reason for the increase of sensitivity is the fact that the ripening takes place in the presence of gelatin. Often, more gelatin may be added at this step.

III) Washing. After the ripening the emulsion is cooled to a stiff jelly which is shredded into "noodles". The noodles are then washed in order to remove excess alkali halide and alkali nitrate formed in the diffusion process. The presence of excess potassium bromide in an emulsion will have a depressive effect on its sensitivity, whereas potassium nitrate, if not washed out, will craze the emulsion by crystallizing out when coated. The noodles can be stored until they are ready to be used for the final product. Often, instead of noodle-washing, salting-out by coagulation or flocculation of the liquid emulsion is performed.

IV) After-ripening. Before using the emulsion for coating, it is remelted and kept at a steady temperature while being continuously agitated (*second ripening* or *digestion*). At this stage, sensitizers, hardening agents, stabilizers, antifoggants, anti-bacterial preservatives and other products are added. In this process no grain growth will take place but the surface environment around the crystals will change, which will give increased light sensitivity. Sulfur from the gelatin adsorbed to the surface of the grains decomposes during this ripening process to form silver sulfide, which causes a greatly increased sensitivity. Excessive after-ripening results in an excess of silver sulfide which gives rise to fog. Therefore, the process must be carefully controlled.

V) Coating. The final step consists in coating the emulsion on a suitable substrate. The emulsion is melted, after which certain solutions, known as *doctors*, are often added to facilitate coating and to modify the physical and sensitometric properties of the film. Alcohol, for example, may be added to reduce froth formed during coating; glycerin, to make the dry film more pliable; surface active agents, such as saponin, to make the film more easily wetted by the processing solutions, etc. The doctored emulsion in liquid form is then piped to a coating machine where it is applied to the substrate material.

Photographic films are often coated in rolls up to 1.2 meter in width and thousand meters in length by passing the substrate under a roller which dips into the liquid emulsion contained in a shallow trough. This is, however, not the only method used today to apply a uniform emulsion layer to one side of the base. The final step consists in cooling the film in a chill box and cutting it to different formats.

Photographic plates are normally coated in a machine where the glass plates, butted end to end on a rotatory band, pass through a narrow slot across the track, just above the band, through which the emulsion is fed on to them. After the coating, the plates pass a section where they are cooled and then removed from the band, after which they are transferred onto racks in drying cupboards.

Depending on the type of emulsion desired, some variations in the procedure described above can occur. Also, depending on the production volume, emulsions can be manufactured in a batch or a continuous process.

In general, industrial manufacturing of high-quality photographic materials is a difficult process, mainly because of the following factors. For most operations total darkness is required and at the same time, to guarantee consistence from batch to batch, a meticulous control of chemical and physical conditions is imperative, especially as regards gelatin. The presence of impurities (water- or air-borne) in speck form requires that the water and air are very accurately filtered. Dust can cause severe problems, which is why the personnel have to be dressed from head to foot in lintless clothing. Actually, the standard of cleanliness in a photographic factory is often much higher than in the food industry, so that a photographic plant resembles a biochemical laboratory or an operating theatre rather than a factory in the traditional sense of the word.

b) Gelatin

One of the components used for the preparation of silver-halide emulsions is gelatin, an organic material which is a complex compound of carbon, hydrogen, oxygen, nitrogen, sulfur, and other elements. Gelatin is prepared from collagen which is a fibrous protein found in certain animal tissue. In collagen, three polymer chains are twisted around one another to form a three-stranded helix.

Photographic gelatin is made from carefully selected hides and ears of calves, which are then processed according to a special technique where the strands are separated and converted to water-soluble protein. The medium by which gelatin is extracted from collagen may be aqueous acid or aqueous lime, $Ca(OH)_2$. Compounds, such as gelatin, which form viscous solutions and jellies are known under the name of *colloids*. Chemically, gelatin is a polymer consisting of monomeric α-amino acids. A single gelatin molecule may contain more than 2000 amino acid units and there can be up to 19 different units in each molecule. The gelatin structure can be represented as

$$\ldots -CO-NH-CH-CO-NH-CH-CO-NH- \ldots$$
$$\qquad\qquad | \qquad\qquad |$$
$$\qquad\qquad R \qquad\qquad R'$$

where R, and R' represent a variety of groups.

Gelatin has proved to be a most satisfactory material for the production of photographic emulsions as it possesses certain special features, as well as a unique combination of chemical and physical properties, which makes it difficult to replace it by any other material. And thus, gelatin can keep silver-halide crystals uniformly dispersed; it is permeable so that it allows penetration of the processing solutions; it also contains natural sensi-

36

tizers, such as allyl thiourea and allyl isothiocyanate. In addition, dry gelatin is rather stable which means that unexposed and finished products can be stored over long periods of time. It is also believed that gelatin can act as bromine acceptor, which means that it will prevent the recombination of silver and bromine, and thereby fully guarantee that the produced latent image is as stable as possible.

To measure the stiffness or rigidity of a gelatin emulsion the shear modulus G can be used. Gelatin of high quality produces moduli of about $1 \cdot 10^5$ dynes/cm^2. However, in practice, more common is the use of *Bloom gel strength*. In a commercial apparatus the force in grams required to produce a specific distortion of the gelatin surface by a plunger is determined. This value is the *Bloom strength*. Bloom values for photographic gelatin are generally 200 to 300 g under specific conditions (6.66 wt.%, 10°C, and 17 hours aging).

The quality of gelatin is very important for the preparation of holographic emulsions. Normally, photographic quality gelatin can be employed, such as Sigma G 9282 [2.41] of Bloom strength 225. Another company producing commercial photographic gelatin is Kind & Knox Gelatin [2.42].

The use of certain additives to the gelatin such as high molecular weight synthetic and natural polymers has been discussed by *Croome* [2.43]. He has also reviewed some attempts at the total replacement of gelatin as the binding agent for silver-halide photographic emulsions. Saponified cellulose esters, such as cellulose acetate butyrate binders is one example of an alternative material to gelatin.

2.2.2 Preparation of the Emulsion

a) Lippmann Emulsions

In the following section various publications describing the production of fine-grained silver-halide emulsions for photographic or holographic applications are discussed. Holographic silver-halide emulsions are generally emulsions of the fine-grained type. They are also of the black-and-white version of photographic emulsions. First, the "old" Lippmann emulsions from the beginning of this century are treated, and later the modern publications on fine-grained emulsion making.

The following formulae are taken from *Valenta's* book [2.44] and the Encyclopedia of Photography from 1911 [2.45] where the Valenta, Senior and Lippmann formulae for making Lippmann emulsions are found, comparing them with the previously mentioned *Ives'* formula (Table 2.3) [2.40].

Neuhauss is probably the most experienced scientist as regards Lippmann photography and he is also believed to have made hundreds of Lippmann photographs. His book [2.46] and laboratory notes [2.47] contain important information on emulsion making. *Neuhauss* stressed the importance of gelatin quality for making successful emulsions. The emulsion recommended by him consists of the following:

Solution A: Gelatin   2.5 g
                    Distilled water   70 mℓ

Solution B: Silver nitrate   1.5 g
                    Distilled water   5 mℓ

Solution C: Gelatin   5 g
                    Distilled water   75 mℓ
                    Potassium bromide   1.5 g .

First, the gelatin is dissolved in cold, distilled water which takes about 10 minutes. Then the gelatin-water solutions A and C are heated until the gelatin is completely melted, after which solution C is cooled to 35° C and solution A to 37° C. The book stresses the fact that the temperature of the solutions *must not exceed* 40° C. Solution B is then mixed with solution A

Table 2.3 Silver-halide emulsions used for Lippmann photography

Emulsion	Lippmann	Vatenta	Senior	Ives
Part A				
Gelatin		10 g	5 g	1 g
Silver nitrate	0.75 g (dry)[a]	6 g	3 g	0.3 g
Water		300 mℓ	225 mℓ	30 mℓ
Part B				
Gelatin	4 g	20 g	5 g	2 g
Potassium bromide	0.53 g	5 g	2.1 g	0.25 g
Water	100 mℓ	300 mℓ	225 mℓ	30 mℓ
Sensitisers used	Cyanine 1:500 6 mℓ	Cyanine 1:500 4 mℓ		Isocol
	Chinoline red 1:500 6 mℓ	Erythrosine 1:500 2 mℓ		Erythrosine Pinacyanol
	Mix the two & add 1+2 mℓ/ 100 mℓ emulsion			
Hypersensitized in		AgNO$_3$ 5 g Silvereoside Acetic acid 5 mℓ 0.2 g/100 mℓ Alcohol 1 ℓ emulsion		

[a] Part A is added to Part B
The mixing of the emulsions is done in a similar way for all the above mentioned emulsions: Mix at 35° C, add Part A to Part B slowly, filter, wash, and coat.

under vigorous stirring. This mixture is poured *drop by drop* into solution C under stirring. When finished, sensitisers are added to the mixture:

Erythrosine - alcohol solution (1:500) 1 mℓ
Cyanine - alcohol solution (1:500)   2 mℓ .

After that the emulsion (which should appear completely transparent) is filtered in a coating bottle, which constitutes a part of the coating process. The preheated plates (previously cleaned in a 50% HNO_3 solution for 24 hours) are coated as soon as possible after the solution is ready. The emulsion is coated by letting it float over the surface of the leveled plates until the entire plate is completely covered. The amount of the emulsion prepared this way will be sufficient for 8 to 12 plates 9×12 cm^2. The plates are then quickly cooled by placing them on a leveled marble table. It is important that the plates do not dry out completely after cooling as there is a danger of potassium nitrate crystallizing within the emulsion before they can been washed to remove the unwanted salt. The plates are then rinsed and placed in a tray filled with water for 15 minutes, during which time the water bath must be changed once. After the bath, the plates are dried. *Neuhauss* recommended to use a centrifuge for this purpose in order to avoid leaving drop marks on the finished plates. The emulsion must appear completely clear, otherwise the plates will not perform well. He also mentioned another important point, concerning the mixing of the emulsion, which is the mixing and the coating of the plates without delay. *Neuhauss* claims that the plates prepared this way can be stored for a long time. However, his test is not very convincing perhaps, since, as he said, the bad plates prepared 1894 gave equally bad (but not worse) results in 1897 as when they were first used three years earlier.

The following description for the preparation of a Lippmann emulsion has been taken from the book by *Baker* [2.34]:

Solution A: Gelatin   20 g
          Distilled water   390 mℓ .

Gelatin is dissolved at 35° C and then filtered. Take 80 mℓ of solution A and add solution B to it:

Solution B: Silver nitrate   4 g
          Distilled water   10 mℓ .

To the balance of solution A add and dissolve 3.2 g potassium bromide. When potassium bromide has been completely dissolved, the solution is slowly poured into the gelatin silver-nitrate solution. Make sure that all the solutions are at 35° C. After precipitation is completed, stir for three and a half minutes. After that, add slowly the following sensitizing dye solutions (warmed to 30° C):

Pinacyanol   (1:1000 alcoholic solution) 4 mℓ
Orthochrom T (1:1000   "   ") 4 mℓ
Acridine orange (1:500   "   ") 4 mℓ .

This should be done in about 45 seconds. No further heating of the emulsion is allowed. Well cleaned glass plates are then coated and cooled. A ten-minute wash is needed to remove potassium nitrate. Finally, the plates are dried in a horizontal position.

The particular sensitizing dyes mentioned above and their combinations were discovered and used by H. Lehmann to produce the best correct-color sensitivity ever achieved in Lippmann photography. However, he kept secret his good formula for Lippmann plates and it was not revealed until after his death [2.34]. If one wants to try this old photographic technique today, these dyes are definitely the first choice.

A paper on the preparation of Lippmann emulsions published by *Crawford* [2.48] in 1954 as well as a small booklet by the same scientist published in 1960 [2.49] contain many practical aspects on making fine-grained emulsions. *Crawford* described the difficulties that one encounters in the industrial production of such emulsions. The usual technique of first preparing the emulsion and later remelting it for the coating, will result in grain growth so that the fine-grained structure of the original mixture will be lost. Therefore, *Crawford* gave a rather detailed description of the necessary steps for preparing emulsions with a grain size of about 30 nm. The components needed here are: gelatin, potassium bromide, silver nitrate and water, all of high purity. First, gelatin in granulated form is dissolved in water at 37° C; this process can take one hour or even more. It is then filtered and poured into a clean beaker. Potassium bromide and silver nitrate are both separately dissolved in water. The potassium bromide and the silver nitrate solutions are then simultaneously poured very slowly (drop by drop) at the rate of approximately two drops per second into the beaker with the gelatin solution (*double-jet precipitation technique*) [2.50]. The temperature of the gelatin solution must be kept at 37° C and the solution must be stirred vigorously during the entire process which takes five to ten minutes. *Crawford* recommended the use of two all-glass hypodermic syringes coupled together so that the pistons can be moved simultaneously. The solutions are pressed out from the two syringes into the gelatin solution through narrow quill glass tubes drawn out at the tips to fine jets. The important thing to remember is that the two jets should be at *two different levels in the gelatin solution*, so that mixing of the reactant solutions does not take place until each solution has been well mixed with the gelatin solution. Mixing of the emulsion must be done in a darkroom with red safelight. *Demers* [2.51] designed a double-syringe machine to produce very fine-grained nuclear emulsions. He also recommended that the emulsion washing temperature should not exceed 2° C. *Berriman* [2.52] has investigated the factors affecting crystal growth during the formation of silver bromide using the double-jet method. The present author recommends a peristaltic pump for mixing the emulsion.

After the precipitation of silver bromide it is also necessary to provide the emulsion with a proper color sensitivity, and harden it. For that reason, sensitizing dyes and chrome alum are added immediately after the precipitation, while the emulsion is still being stirred. *Crawford* used pinaflavol for the sensitizing dye. The emulsion with the highest resolving power suitable for casting Lippmann reflection filters was composed of

Gelatin	12 g in 80 mℓ water
KBr	0.281 g in 10mℓ water
AgNO$_3$	0.375 g in 10mℓ water
Chrome alum	3 mℓ of 2 water solution (0.5% of the dry weight of gelatin)
Pinaflavol	3 mℓ of 0.1 alcoholic solution .

Immediately after the sensitizer and the hardener have been added and mixed with the emulsion, the emulsion is coated, using a casting technique worked out by *Dew* and *Sayce* [2.53]. The glass plates used for coating are now cleaned and the emulsion is casted between the top (the final plate) and the base plate. The top plate is equipped with three spacers whose thickness must be equal to the emulsion thickness desired. The base plate is precoated with chlormethyl-silane and cleaned. The two plates must be warmed up before a puddle of emulsion is put on the horizontal base plate and the top plate is gently pressed into contact with the three spacers. The emulsion sandwich is then cooled in a horizontal position. After a period of time between half an hour and one hour, the plates can be separated and the emulsion-coated plate is washed in running water until the soluble salts (mainly potassium nitrate) have been removed. The plates are then air-dried. This method, slightly modified, was used by *Denisyuk* et al. [2.54] to coat holographic plates with an emulsion of high surface quality usable for holographic optical elements. The molding plate was made of silicate glass, the surface of which was polished with high optical accuracy. For the hydrophobization process they utilized dimethyl dicloro silane.

b) Holographic-Emulsion Preparation

In a more recent paper, *Thiry* [2.55] described how he has used the *Crawford* technique for *making a holographic emulsion* with a grain size of about 20 nm. In principle, *Thiry* made use of the same emulsification process but he applied a different coating technique. The casting technique described above requires a high proportion of gelatin (12%), which means a high concentration of silver nitrate and which, according to *Thiry*, may conflict with obtaining a very fine-grained structure. *Valenta* and *Lumière* used only 5% gelatin solution and a gelatin/silver nitrate ratio of 5 to make their emulsions. *Thiry* recommended the following method: A 2% gelatin solution is cooled by keeping the beaker with the solution in running water for 20 minutes. After that the solution is transferred into a 2-liter beaker covered with a fine cloth. The beaker is then placed under a cold-water tap for half an hour. The next step is to heat the solution to 37° C. Water is added to decrease the gelatin concentration to 1.5% The coating that follows next is a flow-coating technique. The horizontal plate to be coated is heated to 26° C. After the plate has been coated, it is cooled to 20° C and left to dry at room temperature for 18 hours. The plate is then washed for a few minutes and dried.

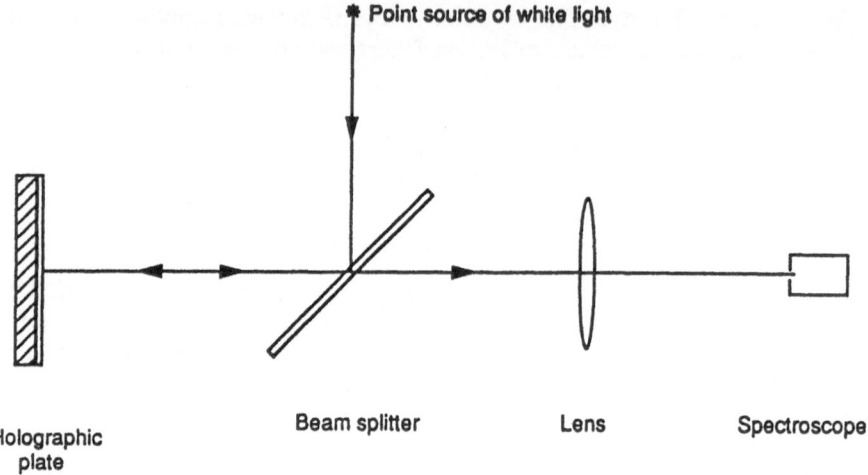

Point source of white light

Holographic plate   Beam splitter   Lens   Spectroscope

Fig.2.9. Optical setup to measure the emulsion thickness. A white spotlight illuminate the emulsion to be measured perpendicularly through a beam splitter. The reflected light passes through the beam splitter and an image of the layer is created by a lens. The image is focused on the entrance of a spectroscope. Black lines appear in the spectrum which are used to calculate the emulsion thickness [2.55]

The tested emulsion formula giving 20 nm grains was the following:

Gelatin   2 g in 80 mℓ water
KBr   3.5 mℓ of a 0.76 g per 10 mℓ solution
AgNO$_3$ 3.5 mℓ of a 1 g per 10 mℓ solution .

The advance of 1mℓ of the silver nitrate solution over the potassium bromide solution can result in finer grains according to *Thiry*. Sensitizing dies used here were pinaflavol (0.5% of the silver-nitrate amount) and "dye I" of a concentration between 0.5÷0.7% of the amount of silver nitrate.[1] Dye I is a red sensitizer used by Agfa and is not described in detail in *Thiry*'s paper. This dye is suitable for holographic work with a HeNe laser.

In order to increase the sensitivity of these types of "primitive" emulsions the material should be treated in a halogen acceptor or an anti-oxidant for about one minute. 0.5% solutions of hydroquinone or resorcinol are suggested for this purpose.

Thiry's paper also describes an optical technique of measuring the thickness of the emulsion. A white spotlight (12V, 20W) illuminates the emulsion to be measured perpendicularly through a beam splitter (Fig.2.9).

The reflected light passes through the beam splitter and an image of the layer is created by a lens. The image is focused on the entrance of a spectroscope. Black lines appear in the spectrum corresponding to the extinction condition $2nd = k\lambda$, n being the refractive index, d the thickness of the emulsion, k an integer, and λ the wavelength. By counting the number

---

[1] The symbol ÷ is used throughout the text as a shorthand for "from - to" or "between".

of black lines (m) between two known wavelengths (λ_1 and λ_2) the thickness d can be obtained from

$$d = \frac{m\lambda_1\lambda_2}{2n(\lambda_2 - \lambda_1)} \ . \tag{2.17}$$

More experimental work on making holographic emulsions has been done by *Liu* [2.56], and *Shave* and *Liu* [2.57] in the present author's laboratory at Northwestern University. Here, the technique introduced by *Denisyuk* and *Protas* [2.39] was studied and improved upon. The following emulsion was made:

Solution A: Gelatin (Bovine - 225 Bloom)   1 g
            Distilled water   25 mℓ

Solution B: Gelatin   2 g
            Potassium bromide   0.25 g
            Sodium iodide   0.0126 g
            Distilled water   50 mℓ

Solution C: Silver nitrate   0.3 g
            Distilled water   5 mℓ .

The preparation and mixing of the solutions was done at 40° C. First, solutions A and C were mixed and the mixture was added in safelight to solution B. This final mixture was stirred for 2.5 minutes.

Both chrome alum and chrome acetate were used to harden the emulsion. In particular, chrome acetate was very effective as solutions of 3÷5% chrome acetate by weight of gelatin gave best results. Plates used for coating were cleaned in an ultrasonic bath. The undercoating of the clean glass plates is very important since if the plates are not treated properly the emulsion can come off the plates during the processing. Here, silane (Dow Corning Z-6020) was used for that purpose giving good results. A solution of 40% of silane, 55% of isopropyl alcohol, and 5% of water was prepared. This solution acted as a stock solution and was further diluted with isopropyl alcohol to a 2÷5% working solution after 24 hours. The plates were then treated in this bath. It is important to filter the emulsion after the precipitation and to coat it soon thereafter. (The filtering can be done in a heated Buchner funnel). The next step was to coat the emulsion on glass plates using a Meyer bar. (Meyer bars can be obtained in the USA from, e.g., R.D. Specialties [2.58]). An emulsion thickness of about 5 μm was obtained. When the emulsion was dry, the plates were washed in five trays of distilled water, three minutes in each tray, for a total of fifteen minutes. The plates were then dried. Chemical sensitization was achieved by dipping the plates in the solutions of the following dyes prepared at an earlier stage: 1,1'-diethyl-2,2'-cyanine, and 3,3'-diethyl-thiocarbocyanine. The supersensitized spectral response was achieved by the following combination: 400 mg of the first dye and 40 mg of the second per mole of silver. This gave a uniform spectral response from 400 to 600 nm.

A paper by *Bonmati* et al. [2.59] also describes the preparation of a holographic emulsion and applies mainly to transmission holograms since here a slightly larger grain size ($\simeq$50nm) is obtained than in the emulsion discussed previously. The following procedure is given in the paper.

Potassium bromide and silver nitrate (10mℓ, 0.7mol of each) are mixed with gelatin (200g at 1% concentration). A halide excess of 0.2 to 0.5 mℓ KBr is recommended. The mixture is prepared under continuous agitation during two minutes at a speed of 5 mℓ/min. To avoid grain growth, congelation and then decongelation is performed. A congelator is introduced into the mixture at 26°C during 24 hours, which helps to avoid physical ripening when the water is eliminated from the gelatin. The emulsion is then chipped and washed to remove the byproducts from the precipitation process. The washing takes place at a temperature of 0°C during 30 minutes. Before the emulsion is coated a sensitizer is added (pinacyanol, 10mg/ℓ), which takes place at 40°C. The glass plates are treated in sodium hydroxide. To coat a plate of 6.5×9 cm^2, 2 mℓ of the emulsion is needed. The thickness of the dry emulsion will then be 10 μm. In order to increase the sensitivity of the emulsion, additives such as sodium thiosulfate and/or acid tetra gold chloride reduced with potassium thiocyanate (10mg/ℓ) can be added before coating. The usual hypersensitization using TriEthanolAmine (TEA) (2% solution for 5 minutes) is also suggested. The influence of excessive KBr concentration during precipitation is mentioned in the paper. An excess of 0.2 mℓ KBr gives a faster emulsion than when the emulsion contains 0.5 mℓ. The use of a phenidone-hydroquinone developer instead of conventional developers such as Kodak D-19 increases the speed of the emulsion.

c) Holographic Emulsions from Russia

Now a few words on the work of Russian scientists. A good deal of information on that can be found in *Denisyuk's* work in which he described various techniques used in the former USSR for manufacturing fine-grained emulsions [2.60-62]. In that country, the main interest has been in the production of reflection holograms of predominantly single-beam type, also called *Denisyuk holograms*. In this type of hologram, the grain size is of primary importance, which is why Russian holographers concentrate on emulsions containing very fine grains. In the emulsion which *Denisyuk* used for his first holograms, grain growth was slowed down by high concentration of KBr [2.39]. This emulsion had grain sizes of about 30 nm. TEA was utilized to sensitize the emulsion prior to use. The next step was to improve the so-called "Valenta layers" [2.44]. In this case it was possible to reduce grain sizes by instantaneous emulsification at a low temperature. A paper by *Zagorskaya* [2.63] describes this process. Here, the emulsification temperature was 32°C and chilled alcohol was used. Rapid melting of the coagulated emulsion in a steam bath provided reproducible results. A diffraction efficiency between 20÷24% was obtained and the grain size was about 30 nm. The basic parameters of *Zagorskaya's* emulsion are

Silver bromide	0.06 mol/ℓ
Gelatin	5 Wt.%
Excess bromide (Br$^-$)	0.01 mol/ℓ .

The optimal exposure for the TEA-sensitized (0.2%) emulsion was 20 mJ/cm^2.

The experience gained from Valenta emulsions enabled *Protas* to make the high-quality plate LOI-2 (now called *PFG-02*). *Protas* was able to slow down grain growth during emulsification by increasing the number of growth centers and introducing special growth inhibitors. The best Russian emulsion ever made is probably the one achieved by *Kirillov* et al. [2.64, 65]. In their case, grain growth was hampered by the fact that in the emulsification process, a highly diluted solution was used and the emulsion concentration was increased by applying the method of gradual freezing and thawing. They used a rather diluted solution of emulsion containing 0.5÷1% gelatin and the following method: Just after the emulsion has been mixed, it is poured into a beaker and frozen at a temperature between -10° C and -20° C or even lower. The emulsion is kept in the beaker for 10 to 15 hours. A rapid freezing method is also suggested. Here, a thin layer of emulsion is poured into an already chilled tray. The frozen emulsion is then chopped into small noodles and put on a grid for thawing. To speed up this process, the frozen emulsion noodles are showered using cold water (3÷5° C). During this process, the emulsion is at the same time washed in order to remove unwanted salts. When the temperature of the emulsion increases (about 25÷30° C) and becomes liquid, it undergoes a gentle ripening. The process of freezing and thawing can be repeated several times in order to increase the concentration of the emulsion. It has been verified that the silver content in the emulsion typically increases about ten times in this repeated process. The silver content of the emulsion is about 2÷2.5 g/ℓ at emulsification and 20÷30 g/ℓ after concentration. Sodium thiosulfate is added at a temperature of 30° to 32° C for 5 to 10 minutes. Then, gold sensitizing takes place for 5 minutes as well as optical sensitizing. A 20% gelatin solution is added to the emulsion before it is coated on glass plates. This type of emulsion is used for the PE-2 plates (now called *PFG-03*) which has a grain size of about 10 nm. *Kirillov* et al. [2.64, 65] recommended a special developer for this material, which contains a high level of phenidone:

Phenidone	7 g
Hydroquinone	20 g
Sodium sulfite	40 g
Sodium hydroxide	20 g
Potassium bromide	2 g
1-phenyl-5-mercaptotetrazole	0.01 g
Distilled water	1 ℓ .

Development time 0.5 to 1 minute.

At the Kurchatov Institute of Atomic Energy in Moscow, *Ryabova* et al. [2.66] published a report on making emulsions for pulsed-laser holography with a sensitivity about 60 to 70 times higher than the PE-2 sensitivity. They discussed the following factors important in emulsion making:

- The concentration of gelatin in the emulsification process.
- The concentration of silver halide.
- Excess of halides and silver nitrate during emulsification.
- The presence and quantity of potassium iodide.
- The presence of silver-halide Microcrystals Growth Inhibitors (MGI).

In particular, the influence of MGIs was experimentally investigated. A MGI in the emulsion acts both as a grain growth inhibitor and a stabilizer. Normal stabilizers (e.g., a sta-salt) are not needed in an emulsion with a MGI. The MGI reduces both the grain size and the distribution as well as increased the gamma-value of the emulsion. The MGI emulsions are also more stable and almost no fog is developing over several months of storage. A 12% drop in sensitivity was experienced after the first month but then it was stabilized. The content of metallic silver in the dry emulsion layer amounts to about 1.5 g/m^2. The sensitivity of the new IAE material is typically about 1 $\mu J/cm^2$, the resolution between 2000 to 3000 lines/mm and different versions of the emulsion have spectral sensitivity both for green and deep red (including infrared) pulsed lasers.

An interesting investigation concerning emulsions of the Russian type has been described by *Crespo* et al. [2.67]. They have studied the influence of the silver-ion concentration in the emulsion on the diffraction efficiency of colloidal silver-halide holograms. The most common procedure used for these types of holograms is based on rather diluted emulsions processed in semi-physical developers in such a way that silver particles of the order of 20 nm are obtained. A semi-physical developer contains not only a reducing agent but also a silver-halide dissolver (Chap.4). In principle, during the development process Ag^+ ions are reduced to metallic silver in the latent image centers. The number of these centers grows with increased exposure, whereas the number of silver ions in a given emulsion is constant, i.e. it is dependent solely on the concentration of silver in the emulsion and is independent of the exposure time. If the exposure time is short, few centers will be formed and silver ions will produce silver particles in the same way as when a conventional developer is used. The density of the emulsion will increase with exposure. However, above a certain exposure value (the number of latent image centers is high), due to the action of the semi-physical developer, the number of silver ions transferred to every center diminishes considerably, which results in silver particles of a smaller size than for conventional processing, and which, in turn, causes a lower density of the emulsion, i.e. low absorption (colloidal silver).

Reflection holograms in the work by *Crespo* et al. were made using different emulsions with a silver-ion concentration between 0.8 and 4.7 g/m^2. The sensitizer employed was pinacyanol and exposures were made with a 633 nm laser wavelength. The plates were hypersensitized being heated for 24 hours and then treated in a 0.5% TEA solution for 2 minutes before they were used. The developer applied for the processing was the Russian GP 2. Two development times were studied, namely 12 or 24 minutes at 20° C without agitation. The influence of an additional heat treatment before exposure was also investigated (30 minutes at 95÷100° C).

The following conclusions were made in the paper:

- The optimum silver concentration in the emulsion was found to be 1.1 g/m² with an emulsion thickness of about 10 μm.
- The development time in the GP 2 developer was 12 minutes without agitation at 20° C.
- The highest optical density was obtained at the exposure of about 0.2 mJ/cm², while the optimum diffraction efficiency was found at the exposure of about 0.6 mJ/cm² (3 times higher) when the optical density is reduced and colloidal silver is formed.
- The additional heat treatment also improved the diffraction efficiency, hardening the emulsion and thus making it less prone to shrinkage during processing.
- In general, the lower the concentration of silver ions in the emulsion, the longer the development time must be in order to obtain a high diffraction efficiency.

From the discussion above it is clear why little success can be expected from the application of the semi-physical developing technique to the Western types of silver-halide materials which have a high silver content and rather large silver-halide grains. Normally, if this technique is applied to, e.g. Agfa materials, problems with silver precipitation on the emulsion surface combined with dispersion due to dichroic effects may occur [2.38]. It is also clear that due to the combination of very small grains in the emulsion and the demand for higher exposure in order to create colloidal silver during processing, Russian types of emulsions are much less sensitive compared to commercial silver-halide emulsions produced by Western photographic companies.

d) Sensitizing Dyes

The most difficult part of emulsion-making is to find suitable sensitizers for, in particular, fine-grained materials. Commercial companies are usually very secretful about sensitizers they apply to their emulsions, which is why it is so hard to find any publication that gives comprehensive description of dyes or sensitizers. In particular, sensitizers for the deep-red part of the spectrum are difficult to find. There are two types of sensitization of photographic emulsions.

1. **Chemical sensitization** which refers to methods to obtain the highest possible sensitivity of the silver-halide crystals. Here, sulfur, gold and reduction sensitization are common.

2. **Spectral sensitization** which refers to methods to sensitize the silver-halide grains to light in a region of the spectrum in which they would normally not absorb. Here, special dyes are used.

In this section the second type of sensitization is of main interest. Chemical sensitization of holographic silver-halide emulsions has, in general, been discussed by *Pantcheva* et al. [2.68]. Emulsions intended for short exposure times using pulsed lasers have been treated by *Pangelova* et al. [2.69].

As already mentioned, pure silver-halide crystals are only sensitive to light in the UV and violet parts of the spectrum. According to the Grotthus-Draper law, only light which is absorbed can cause a chemical change. The formation of a latent image is a chemical change. Up to a certain photon energy, the silver-halide crystal alone can absorb the radiation energy according to Planck's law ($E = h\nu$). The energy is high enough to raise the silver-halide molecule from the ground state to a higher state to free an electron which can combine with a silver ion and thus form a latent-image speck. For light of longer wavelengths the energy of individual photons are not high enough and no photon absorption takes place. Therefore, in order to record at other wavelengths within the electromagnetic spectrum the emulsion has to absorb the radiation in some way and transfer the energy to the silver-halide grain to produce a latent image. There are special dyes that can absorb light of different wavelengths. At certain wavelengths the molecules of the dye can absorb the radiation and the molecule is then raised to a higher state. This means that an electron is raised from the valence band to the conduction band. Now two things could happen:

Either the electron is directly transferred to the silver-halide crystal to form metallic silver by combining with a silver ion (*electron transfer*); or the electron formed by the dye will bring about an excitation of the silver-halide crystal, causing a bromide ion to part with an electron (*energy transfer*). The duration of the exited state of the dye molecule which is responsible for the sensitization is very short, about 10^{-11} s. Both electron and energy transfer can take place depending on type of dye. By such mechanisms it is possible to affect the silver-halide crystals to form a latent image at radiation of various wavelengths.

Not only must such a dye absorb the light of a certain wavelength but also must it adsorb to the silver-halide crystal surface. Only if the dye is in intimate contact with the grain can it transfer the effect of the light. The force acting between the dye and halide molecules is of the Van der Waal type. If the dye is not in good contact with the grain or if it is located in the gelatin matrix the effect of it is a light-filtering function which actually reduces the intensity of the illumination in this region.

The dyes used for the sensitization are mainly the *cyanine dyes*. They can be divided into two classes: the *true cyanines* and the *merocyanines*. Often the chalkocarbocyanines are used. More about these dyes can be found in *Duffin's* book [2.36]. Some of the dyes employed in Lippmann emulsions and holographic emulsions have already been mentioned and are listed in Table 2.4.

A special sensitizer mentioned in one of the Russian papers that is claimed to be very good for holograms at the ruby-laser wavelength (694nm) is the: 3-allyl-3'- ethyl-4',5'-diphenyl-4-keto-5(1"-ethyl-dihydroquinolilidene-4"-ethylidene) thiazolinothiazolocyanine bromide [2.70].

Table 2.4. Sensitizing dyes for silver-halide emulsions

Acridine orange	3,6"-(dimethyl amino)-acridine hydrocloride hydrate
Auramine O (Basic yellow 2)	4,4'-(imidocarbonyl)-"-(N,N,-dimethylaniline)hydrochloride
Dithiozanin iodide	3,3'-diethyl-thiacarbocyanine iodide
Eosin Y (Eosin yellowish)	2',4',5',7'-tetrabromofluorescein disodium salt
Erythrosin B (Iodesine)	tetraiodfluoresine disodium salt
Isocyanine iodide	1,1'-diethyl-2,4'-cyanine iodide
Kryptocyanine (Rubrocyanine)	1,1'-diethyl-4,4'-carbocyanine iodide
Orthochrom T	1,1'-diethyl-6,6'-dimethylisocyanine
Pinachrome	1,1'-diethyl-6-ethoxy-6'-methoxyisocyanine bromide
Pinacyanol bromide	1,1'-diethyl-2,2'-carbocyanine bromide
Pinacyanol chloride (Quinaldine blue)	1,1'-diethyl-2,2'-carbocyanine chloride
Pinacyanol iodide (Sensitol red)	1,1'-diethyl-2,2'-carbocyanine iodide
Pinaflavol	1-methyl-2-p-dimethylaminostyryl pyridine
Pinaverdol (Sensitol green)	1,1'-6-trimethylisocyanine iodide
Pseudocyanine iodide	1,1'-diethyl-2,2'-cyanine iodide
Rhodamine B	N,N,N',N',tetra ethylrhodamine hydrochloride

2.2.3 Substrates for Holographic Emulsion

The material on which the emulsion is coated has a strong bearing on the final quality of the hologram. The best choice is often a *glass plate* as it is mechanically stable and optically inactive. Also, the light-scattering noise level in clear glass is very low. In many applications of holography glass is actually the only possible support material. High-resolution imaging, hologram interferometry, Holographic Optical Elements (HOEs) and spatial filters are a few examples where a very stable emulsion support is important. In display holography it is also often convenient to use glass plates, mainly because of the need for stability when using CW lasers. Producing master plates for hologram-copying is another example when most holographers choose glass plates. Yet another example is the use of glass in the recording of expensive art holograms where it is important to protect the emulsion well (if sealed with another glass plate after processing) against detrimental environmental effects (humidity, air pollution, etc.).

The use of *film substrates* has been growing steadily in recent years, especially in display and industrial holography. In many cases, the use of film has many advantages as compared to that of glass (breakage, weight, cost, size, etc.). For example, for industrial applications, such as Holographic NonDestructive Testing (HNDT), film substrates are often sufficient and more economical than glass. In display holography, hologram copying in larger quantities is done mainly on film (sometimes the copies are laminated to a stable substrate after processing). The increased use of pulsed lasers has made hologram recording simpler when utilizing film substrates. Finally, film substrates are exclusively employed in the production of large-format holograms.

In Fig.2.10 the difference between holographic emulsions coated on glass or film substrates is illustrated.

a) Glass Plates

Holographic glass plates are commonly made of soda-lime glass of high quality (free from graininess and molecular orientation) with the help of the flat-drawn sheet or the float process method. The refractive index (refractive dispersion) of glass varies depending on the light wavelength and is about 1.516 for $\lambda \simeq 600$ nm. Good optical quality and high mechanical and thermal stability of glass are the main advantages for glass as a substrate for holograms. Young's modulus for glass is $70 \cdot 10^9$ N/m^2 and the thermal coefficient of expansion is only $8.1 \cdot 10^{-6}$ cm/cm/° C.

The glass thickness varies depending on the format and is between 1.0 mm and 6 mm. Some manufacturers offer plates of different flatness. The standard plate format is between 50 mm×50 mm and 500 mm×600 mm.

The emulsion coated onto untreated glass plates tends to peel off when dry, or frill off when it is wet. Therefore, a well cleaned glass plate is often precoated with an extremely thin substratum of gelatin hardened with chrome alum, or sometimes with a layer of chrome-alum solution alone, for

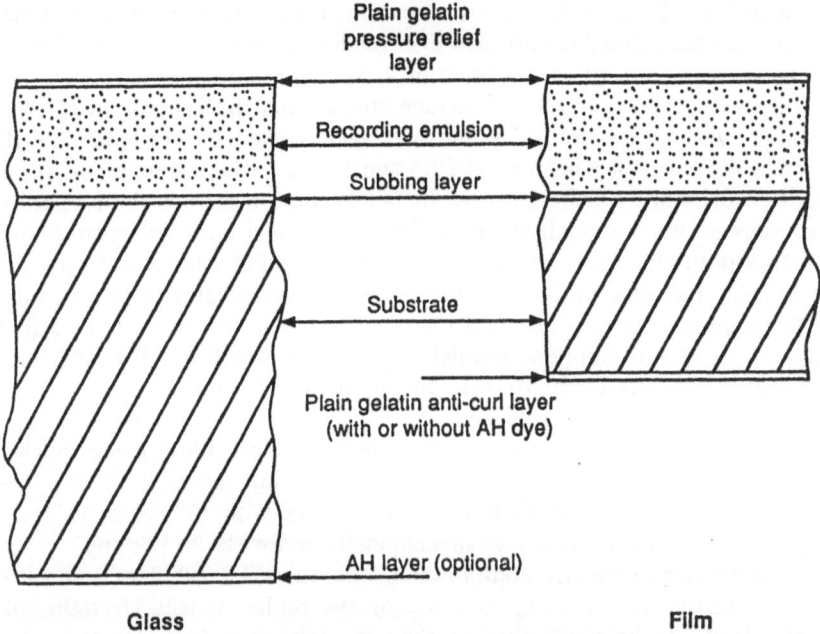

Plain gelatin
pressure relief
layer

Recording emulsion

Subbing layer

Substrate

Plain gelatin anti-curl layer
(with or without AH dye)

AH layer (optional)

Glass   Film

Fig.2.10. Holographic silver-halide material. Holographic emulsion coated on glass and film substrates, including additional layers used in holographic materials

the sake of making the underlayer of the subsequently applied emulsion very hard. This process is referred to as *subbing*.

If one wants to remove the gelatin emulsion from existing plates in order to use the glass substrate for coating a new emulsion the old has to be dissolved away. This can be done using, e.g., a solution of sodium hypo-chlorite (Clorox, household bleach). A biologic detergent, such as Biotex, can also be taken advantage of. Soak the plate for about 30 minutes in a solution at a temperature of between 30° and 40° C. A faster way is to use a 1÷2% solution of ammonium or sodium bifluoride in which the plates should be soaked for only 10 to 30 seconds otherwise the glass surface can be affected by the etching bath. An acid solution of sodium fluoride can al-so be used.

b) Film Substrates

Film substrates are of two types mainly: a *polyester* (polyethylene tere-phthalate) or a cellulose ester, commonly *triacetate* (cellulose triacetate) or acetate-butyrate. In addition to the light-sensitive emulsion coating (and the necessary subcoating), curl control and antihalation coating are often added here. Film may curl badly due to the variations in gelatin concentration caused by humidity. Therefore, a coating of pure gelatin is often applied to the back of the substrate to counteract the curl of the emulsion. If an absorbing dye is added to such gelatin coating, it can serve as an antihalation layer at the same time. A film can also receive a coating for static protection. The coating can be in the form of a layer containing matting particles, which prevents close surface-to-surface contact and the generation of static electricity upon separation. The matting layer (coated on both sides of the substrate) also prevents individual film sheets from sticking together due to humidity variations during storage. In holographic materials these coatings are sources of noise explaining why the quality of a film hologram (especially on a polyester base) not always can be as good as a hologram recorded on glass. As a matter of fact, a thin layer of pure gelatin is often coated over the light sensitive emulsion (super coating). The reason for this is that, if this layer is not applied, emulsion grains affected by, e.g., pressure marks, could produce image defects during development.

Various important aspects must be considered when choosing a film-base material for a given holographic application. *Sprackling* has published an investigation on the mechanical response of film materials [2.71]. The mechanical behavior of the base material, with or without the emulsion, is strongly viscoelastic. Polyester is mechanically more stable (Young's modulus $4.5 \cdot 10^9 \, N/m^2$) than triacetate (Young's modulus $3.8 \cdot 10^9 \, N/m^2$) and it is also less sensitive to humidity. Because of the higher tensile strength, the polyester film can be made thinner than the triacetate film. On the other hand, it is birefringent and can cause many problems when recording reflection holograms (where the reference beam has to pass the substrate). It also has a higher inherent scattering level. During the manufacture of polyester, the polymer is biaxially oriented when it is drawn and tentered so that it has different refractive indices for each of the three orthogonal directions (α, β, γ). Polyester shows larger wavelength-dependent variations in refractive dispersion than triacetate does. For polarized light entering the polyester material at normal incidence the refractive index for light at $\lambda \simeq 600$ nm is $n_\gamma = 1.66$ and $n_\beta = 1.65$, respectively, depending on whether the electric field vector is oriented parallel or perpendicularly to the major axis. For polarized light propagating in the plane of the support the effective refractive index is typically $n_\alpha = 1.50$. The refractive index for triacetate is about 1.48 at $\lambda \simeq 600$ nm. This material is optically inactive and has a low inherent scattering level.

In general, polyester is recommended for transmission holograms in cases when a mechanically stable base is important, whereas triacetate is more suitable for reflection holograms, where birefringence causes severe problems if polyester is used. The inherent scattering levels in these materials are discussed in Sect. 2.3.3a.

The thickness of film substrates for holography varies between 64 and 200 μm. Thinner film substrates apply to polyester only. Film materials are normally manufactured in sizes from 4"×5" and up to 1.25×10 m^2 in rolls.

2.2.4 Antihalation Reduction for Holography

Transmission holograms need sometimes to be protected by AntiHalation (AH) coating. Without that, the highly coherent laser light reflected (Fresnel reflection) from all interfaces within the material and, in particular, from the back of the support material, will interfere with the recorded interference pattern and modulate it. This will result in a coarse pattern resembling a piece of grainy wood [2.72]. Since the highest refractive-index gradient occurs at the glass-air interface at the back of the plate, the strongest reflection is also produced there. The spurious interference pattern is not only aesthetically disturbing but it also gives a variation in the diffraction efficiency of the holographic image. The modulation can be quite strong so that a reflection of only 0.25% can create interference fringes with a modulation of 10%.

a) Brewster-Angle Method

Under certain conditions the reflected part is minimized, which happens when polarized light is incident at the so-called *Brewster angle*. This condition is frequently used in holography to minimize the internal reflections in the plates, when the plane-polarized reference beam is directed so that it hits the plate at the Brewster angle (Fig.2.11).

When the incident light is plane-polarized with the electric vector parallel to the plane of incidence, it is referred to as *p-polarization*. When the electric vector is perpendicular to the plane of incidence, it is called *s-pol-*

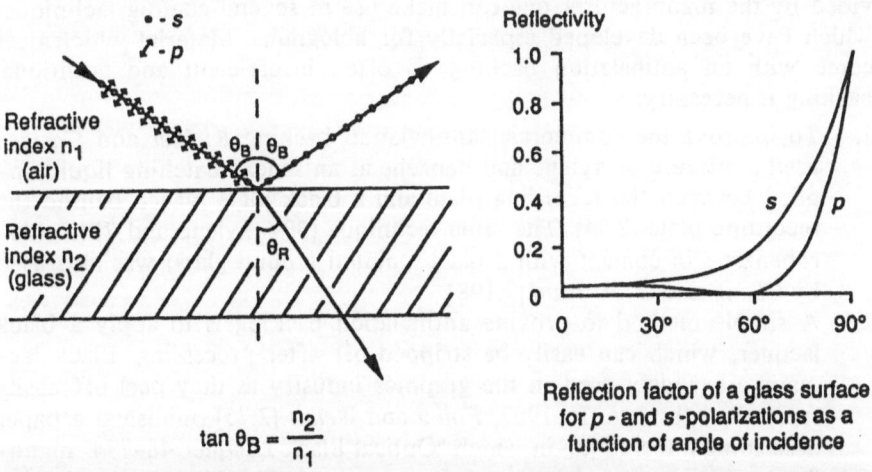

$$\tan \theta_B = \frac{n_2}{n_1}$$

Reflection factor of a glass surface for *p*- and *s*-polarizations as a function of angle of incidence

Fig.2.11. Polarized laser light incident on a plate at the Brewster angle. The incident light is plane-polarized with the electric vector parallel to the plane of incidence (*p*-polarization). At a certain angle of incidence the reflected *p*-polarized light will be zero. This angle θ_B is called the Brewster angle and is given by $\tan \theta_B = n_2/n_1$, where n_1 is the refractive index of the medium surrounding the material and n_2 the refractive index of the material

arization. The letters p and s stand for parallel and senkrecht (German). At a certain angle of incidence the reflected p-polarized light will be zero. This angle is called the *Brewster angle* and is given by

$$\tan\theta_B = n_2/n_1 \ . \tag{2.18}$$

For a glass plate ($n_2 = 1.52$) in air ($n_1 = 1.0$) the Brewster angle θ_B is about $56.7°$.

b) Additional Methods for Antihalation Reduction

In addition to the Brewster-angle method, there are two other methods to reduce or eliminate the antihalation effect:

- Matching the refractive-index difference at the interface and absorbing the transmitted radiation.
- Using antireflection coating at the interface.

The fact that internal reflections (caused by light reflections from the boundaries: glass/air and emulsion/air) in holograms affects the quality of the recording interference patterns was pointed out by *Phillips* et al. [2.101]. Actually, boundary reflections are also a problem when recording reflection holograms which have been discussed by *Owen* et al. [2.73]. The recording of volume-reflection gratings surrounded by an index-matching liquid was performed in that investigation. Concerning reflection holograms, the index-matching technique or the Brewster-angle incidence can reduce the negative influence of the unwanted reflections.

Concerning transmission holograms, if a suitable backing is not provided by the manufacturer one can make use of several coating techniques which have been developed especially for holograms. Material which does come with an antihalation backing, is often insufficient and additional backing is necessary:

1. To improve the commercial antihalation backing *Richter* and *Carlson* used a mixture of xylene and benzene as an index-matching liquid applied between the recording plate and a black plate placed behind the recording plate [2.74]. The same technique (30% xylene and 70% chlorobenzene in contact with a black painted ground glass) was also used by *Chang* and *Bjorkstam* [2.198].
2. A simple method to provide antihalation backing is to apply a black lacquer, which can easily be stripped off after processing. Black lacquers are widely used in the graphics industry as they peel off easily from the substrate. In 1967, *Foley* and *Wendt* [2.75] published a paper describing a technique in which "Optical Black Lacquer 48-774" manufactured by Pratt and Lambert, Inc. was used. This lacquer remains intact during processing and is easily peeled off after drying. In a more recent paper *Wesly* [2.76] mentioned that the "33 Metal Blocking Spray" from Universal Photonics, Inc. [2.77] is a good antihalation coating spray for holograms. The Universal X-59, Black Stripable Coating, which is applied with a brush, can be used, too.

3. An efficient but rather messy technique employed mainly for film is the use of water-soluble black ink for block printing, (e.g., Speedball Screen Print Ink) [2.76] mixed with water and applied to the back of the film. The ink also acts as an adhesive to attach the film to a suitable support, such as a glass plate, for example, during exposure, which provides the necessary stability for CW laser holograms. After the material has been exposed, it is removed from the supporting glass plate and the ink is washed off by showering the back of the material. The ink is easily removed this way, but it must not come in contact with the gelatin, which makes it tricky to handle in a darkroom.

4. *Biedermann* [2.78] recommended a coating consisting of a mixture of 100 g polyvinylalcohol dissolved in one liter of boiling water. After the solution has cooled down, a suitable non-scattering dye is added. For the argon-laser wavelengths metanil yellow can be utilized, whereas methylene blue is used for red laser wavelengths. The dry coating can easily be peeled off after exposure before processing. *Phillips* [2.79] suggested also the use of non-scattering dyes, such as, e.g., a blue glass lacquer to be employed with red lasers for antihalation purposes.

5. A simple dry method that causes sufficient reduction in the halation is the use of a black self-adhesive PVC masking tape method described by *Soares* [2.80].

6. Another dry method is to utilize the new black laminating materials for holography. They have an index matching adhesive and can be removed from glass plates before processing. The MACTAC MACal 9800 series or MACbond B2978 can be applied for this purpose.[2.81].

2.3 The Holographic Recording Process

2.3.1 Microscopic Characteristics. Latent-Image Forming Process

For a comprehensive description of the formation of the latent image in a silver-halide emulsion the reader is referred to the *Gurney-Mott* concentration theory [2.82, 83] or the *Mitchell* concentration theory [2.84-86]. Recently, *Tani* published a paper on the physics of the latent image [2.87]. Briefly, the theory of forming a latent image in a silver-halide emulsion which can be later developed to a silver image, can be summarized in the following way.

As the reader already knows, gelatin is a necessary component of photographic emulsions. The reason for this is gelatin's unique combination of features mentioned earlier. One of the features not mentioned before is that gelatin contains labile sulfur compounds which easily decompose when heated, producing silver sulfide (Ag_2S). The sensitivity specks that silver-halide grains exhibit on their surface and which are formed during the emulsion manufacturing process are made up of sulfur sulfide. According to *Mitchell*, sensitivity specks play an important roll in forming the latent image.

The silver-halide crystal is an n-type photoconductor with a valence band of electrons and with a conduction band in which injected electrons are free to migrate throughout the crystal until trapped by a lattice defect. During the exposure of an emulsion, photons are absorbed by the crystals. When a photon of sufficient energy is absorbed, an electron from the crystal is promoted to the conduction band, leaving behind a positive hole which is a free halogen atom:

$$Ag^+X^- \text{ (silver-halide crystal)} + h\nu \rightarrow Ag^+X^0 + e^- \ . \tag{2.19}$$

In *Mitchell's* theory the photogenerated hole is believed to be trapped at a surface sensitivity site by partial S^{2-} charges from the adsorbed silver sulfide specks. This results in a positively charged Ag_2S^+ particle, which dissociates into AgS and Ag^+:

$$Ag_2S^+ \rightarrow AgS + Ag^+ \ . \tag{2.20}$$

The silver ion will then attract the photogenerated electron to form a silver atom, the so-called *prespeck*:

$$Ag^+ + e^- \leftrightarrow Ag^0 \ . \tag{2.21}$$

The *Gurney-Mott* theory describes this process in a slightly different way. Here, the free electron is first trapped by a positively charged surface lattice defect. Once trapped, the electron will attract an interstitial silver ion (Ag^+) to the sensitivity site to form the silver atom prespeck (Ag^0).

One isolated silver atom has an average lifetime of about one second, statistically. The lifetime can be calculated knowing the binding energy for the electron to the subspeck which has been experimentally measured to be about 0.70 eV. The Boltzmann statistical lifetime is then

$$t = \tau e^{E_0/kT} \tag{2.22}$$

where $\tau = 10^{-12}$ is the estimated electron collision period, E_0 is the binding energy, k Boltzmann's constant ($1.3805 \cdot 10^{-23}$ J/K), and T the temperature [K]. For T = 300 K and $E_0 = 0.70$ eV we have $t \simeq 1$ s.

In order to create a *sublatent image speck* on the silver-halide crystal where a diatomic silver molecule is formed by the process of *nucleation*, a second silver atom is needed at the site of the first silver atom during its lifetime:

$$Ag + Ag^+ + e^- \rightarrow Ag_2 \ . \tag{2.23}$$

A subspeck of two atoms is stable at room temperature ($E_0 = 1.74$ eV, t $= 1.7 \cdot 10^{17}$ s). The sublatent image speck grows larger with further photon absorption, resulting in photogenerated electrons. The *latent image* is usually regarded as a collection of a few silver atoms at one site produced by the reduction of silver ions in the process of *photolysis*. Silver formed in this way is known as *photolytic silver*. A latent image of *at least three to four silver atoms* is needed for *developability*. Developability means the formation of a latent image which has the *catalytic* property of increasing the development rate of silver-halide grains reduced to metallic silver by the reducing agent called *developer*. For all the exposed grains, chemical development will then reduce the entire silver-halide grains to metallic silver.

The chemical sensitization of an emulsion is similar to the doping process of a semiconductor. The introduction of such impurities as sulfur, gold, or silver - alone or in combinations - into the emulsion increases the grain's sensitivity (*finishing*). The chemical reduction is induced by raising the temperature. Depending on the impurities which have been introduced into the emulsion the emulsion is called *sulfur sensitized*, *gold sensitized*, *sulfur plus gold sensitized*, or at times, *reduction sensitized*. The sensitivity of a grain is defined as the reciprocal of the number of absorbed photons necessary to produce developability of the grain. A highly sensitized grain requires fewer photons than a less sensitized grain to be developable. Grain size is, however, even more important for sensitivity: the larger the grain, the higher the sensitivity of the material. A typical large silver-halide grain with the volume of 1 μm^3 ($=10^{-12} cm^3$) contains about $2 \cdot 10^{10}$ silver ions. In such a grain just a few photons are expected to produce a stable latent image, which can later be used to trigger off the process of converting the entire grain to silver atoms. The overall amplification, from the quanta absorbed to the silver atoms produced, can be greater than 10^9 in this process. For a typical holographic emulsion with the grain size of about 50 nm, the amount of silver ions in the grain is about $2.6 \cdot 10^6$, which means an amplification of about one million. However, this constitutes only about 1/1000 of the sensitivity of a conventional high-speed photographic film.

The latent-image-formation theory and its finer points will be further discussed in Sect. 2.6.3. The mechanism behind reciprocity failure of photographic materials and latent-image fading will be treated in Chap. 6.

2.3.2 Macroscopic Characteristics

Hologram recording on photographic materials has been given a more or less full account in many publications [2.88-105]. In brief, when recording the interference structures in an emulsion the following macroscopic aspects of the holographic recording process must be observed:

The *complex amplitude transmittance* $T_a(x)$ of a holographic recording material can be written in a general manner as

$$T_a = |t_a(x)| e^{-\alpha(x)d} e^{-i\phi_t(x)} = |t_a(x)| e^{-\alpha(x)d} e^{i2\pi nd/\lambda(x)} \tag{2.24}$$

where α is the absorption constant of the material, d the thickness, n the refractive index of the material, and λ the wavelength of the laser light. The amplitude transmittance T_a is the square root of the transmittance T

$$T_a = \sqrt{T} = 10^{-D/2} .$$ (2.25)

There are two main types of holograms: *amplitude* and *phase* holograms. In a *pure amplitude hologram* ($\phi_t(x)$ = const.) *only the absorption* α varies with the exposure (after processing), whereas in a *pure phase hologram* ($\alpha = 0$, $|t_a(x)| = 1$) *either n or d changes* with the exposure. For the phase hologram the phase factor is

$$\phi_t = (2\pi/\lambda)\,nd ,$$ (2.26)

$$\Delta\phi_t = (2\pi/\lambda)[d\,\Delta n + (n-1)\,\Delta d] .$$ (2.27)

If the hologram is thin, $d \simeq 0$: Phase variations are then caused by surface relief variations only

$$\Delta\phi_t = (2\pi/\lambda)(n-1)\,\Delta d .$$ (2.28)

If the hologram is thick and has a negligible surface relief ($\Delta d = 0$), phase variations are caused by index variations only

$$\Delta\phi_t = (2\pi/\lambda)\,d\,\Delta n .$$ (2.29)

In many cases the phase modulation in a hologram is a combination of the two different extreme types (a *complex* hologram).

For amplitude holograms the amplitude transmission T_a against exposure [2.199] (or log exposure [2.206]) is used instead of the H&D curve described in Sect.2.1.2b (Fig.2.12). For a phase hologram the corresponding curve is the phase shift against a log exposure relation (Fig.2.13).

One should also get familiar with the following definitions in order to understand the recording process. The *beam ratio* (K-value) between the intensity of the reference beam E_r and the object beam E_o is expressed as $K = E_r/E_o$. This ratio may be considered the best if $K = 1$, but that is only true when making gratings and other optical elements. Using extended objects and operating at a K-value close to 1, intermodulation will take place resulting in increased noise and running into nonlinearities of the recording process. Hence, for extended objects the K-value has to be greater than 1 - in holography it is normally somewhere between 4 to 20. Even very large K-values can produce holograms of good quality but the diffraction efficiency is then low.

The K-value determines the modulation M_i of the interference fringes created between the reference and the object waves:

$$M_i = \frac{2\sqrt{E_r E_o}}{E_r + E_o} , \quad K = \frac{E_r}{E_o} = \frac{A_r{}^2}{A_o{}^2} .$$ (2.30,31)

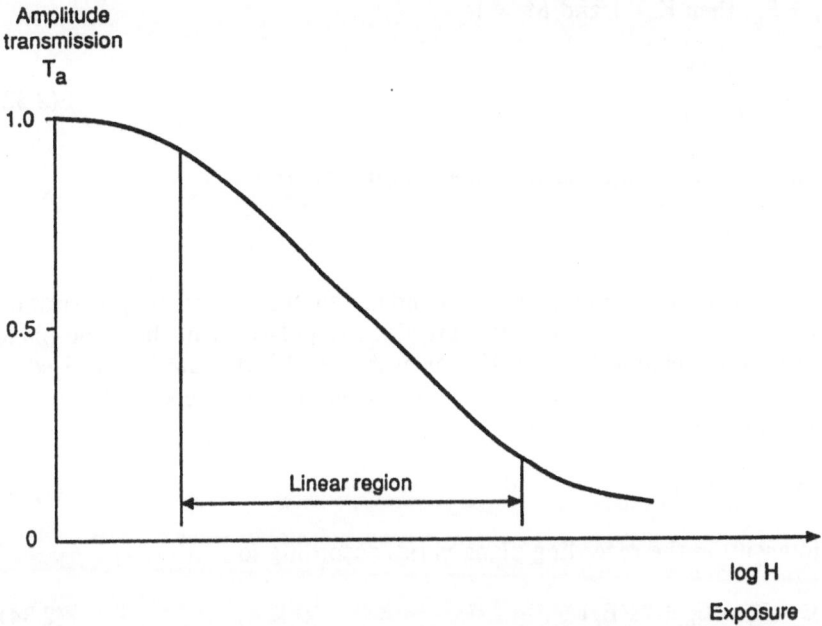

Fig.2.12. Amplitude transmission T_a versus the logarithm of exposure H used for the characterization of amplitude transmission holograms

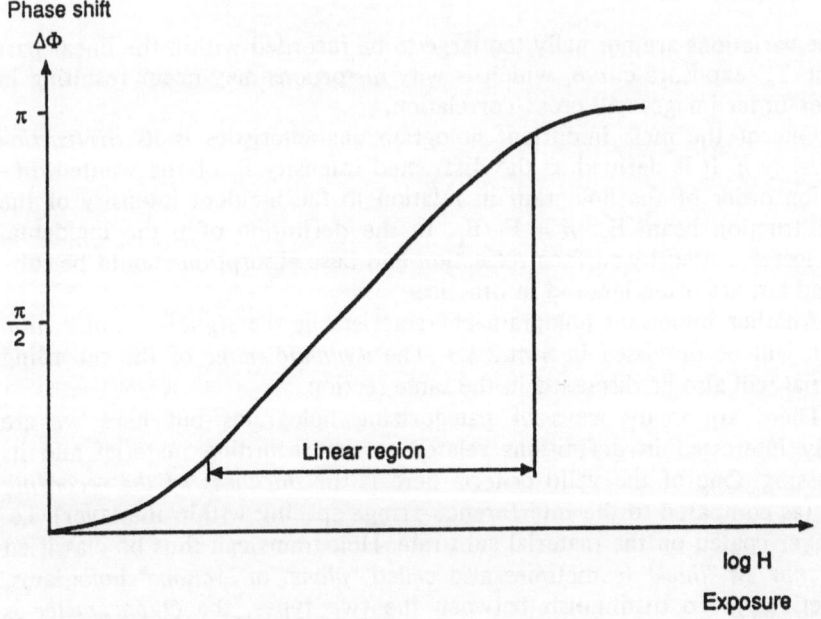

Fig.2.13. Phase shift $\Delta\phi$ versus the logarithm of exposure H used for the characterization of phase transmission holograms

If $E_r = E_o$, then $K = 1$ and $M_i = 1$:

$$M_i = \frac{2\sqrt{K}}{K + 1} .$$ (2.32)

For small values of M_i, and K-values larger than 1:

$$M_i \simeq 2/\sqrt{K} .$$

Note that the modulation is dependant on the degree of polarization between the beams. If both beams are linearly polarized in the same plane then the above relation is valid. If, however, the object beam is polarized in a plane at an angle ψ to the polarization plane of the reference beam, the modulation will be

$$M_p = M_i \cos\psi .$$ (2.33)

The intensity in the recording plane varies according to

$$E = E_r + E_o \pm 2\sqrt{E_r E_o} .$$ (2.34)

At the highest possible modulation ($M_i = 1$, and $E_r = E_0$):

$$E_{M_i = 1} = 0 \text{ and } +4E_o .$$

These variations are normally too large to be recorded within the linear part of the T_a-exposure curve, which is why distortions may occur resulting in higher-order images and cross-correlation.

One of the most important hologram characteristics is its *diffraction efficiency* η. It is defined as the diffracted intensity E_i of the wanted diffraction order of the hologram in relation to the incident intensity of the reconstruction beam E_r: $\eta = E_i/E_r$. In the definition of η the incidental light losses caused by surface reflection and base absorption should be subtracted but are often ignored in practice.

Another important hologram characteristic is the *signal-to-noise ratio* which will be discussed in Sect.2.3.3. The *dynamic range* of the recording material will also be discussed in the same section.

There are many ways of categorizing holograms but here we are mainly interested in definitions related to the recording material and its processing. One of the valid criteria here is the *thickness of the recording layer* (as compared to the interference-fringe spacing within the layer), i.e. the layer coated on the material substrate. Holograms can thus be classified into "*thin*" or "*thick*" (sometimes also called "*plane*" or "*volume*" holograms, respectively). To distinguish between the two types, the *Q-parameter* is normally used; it is defined in the following way:

$$Q = \frac{2\pi\lambda d}{n\Lambda^2} \qquad\qquad\qquad (2.35)$$

where λ is the wavelength of the illuminating light, d the thickness of the layer, n the refractive index of the emulsion, and Λ the spacing between the recorded fringes. A hologram is considered thick if $Q \geq 10$, and thin when $Q \leq 1$. Holograms with Q-values between 1 and 10 are sometimes treated as thin and at other times as thick.

Holograms can also be classified into *amplitude holograms* (holograms in which intensity variations of the interference pattern are converted to density variations in the finished hologram) and *phase holograms* (holograms in which intensity variations of the interference pattern are converted to variations in thickness and/or the refractive index in the finished hologram). If a hologram affects both amplitude and phase simultaneously it is termed a *complex hologram*. In the following sections the above mentioned hologram types are described in brief. For a complete description of the above-mentioned hologram types the reader is referred to the more general books on holography mentioned in Chap. 1.

a) Hologram Type: Thin Holograms

The following short exposition of the theory of the *thin holograms* starts with the description of the recording process of thin amplitude holograms. This is because with this type of hologram it is easy to understand just how the information is recorded and what properties the reconstructed image will possess.

For a given input modulation $M_i(\nu)$ of the interference fringes created between the object and the reference fields at a certain spatial frequency ν, reduced modulation $M_e(\nu)$ will be recorded in the silver-halide emulsion due to scattering that will occur within the emulsion. This effect increases with increased spatial frequencies. The Modulation Transfer Function of the material is normally used to describe this relation.

Figure 2.14 shows how the input modulation signal is recorded in the material. Both the theoretical D-logH and the T_a-logH curves are presented. The derivative of the T_a-logH function is indicated as $|\alpha|$. An important characteristic of diffraction efficiency is the function $\alpha(H)$ which is the derivative of T_a with respect to logH, i.e.,

$$\alpha(H) = \left\{\frac{dT_a}{d\log H}\right\}_H \doteq -\frac{\ln 10}{2}T_a(H)\gamma(H) . \qquad\qquad (2.36)$$

The second relation indicates that the diffracted flux is proportional to the square of the local gradient $\gamma(H)$ of the D-logH curve. The diffraction efficiency of the hologram is dependent on both $\alpha(H)$ and $\gamma(H)$. The two will actually move in two opposite directions with increasing exposure. It is obvious that the maximum of the diffraction efficiency will be obtained in the toe region of the D-logH curve. This means that a steep bend in the toe

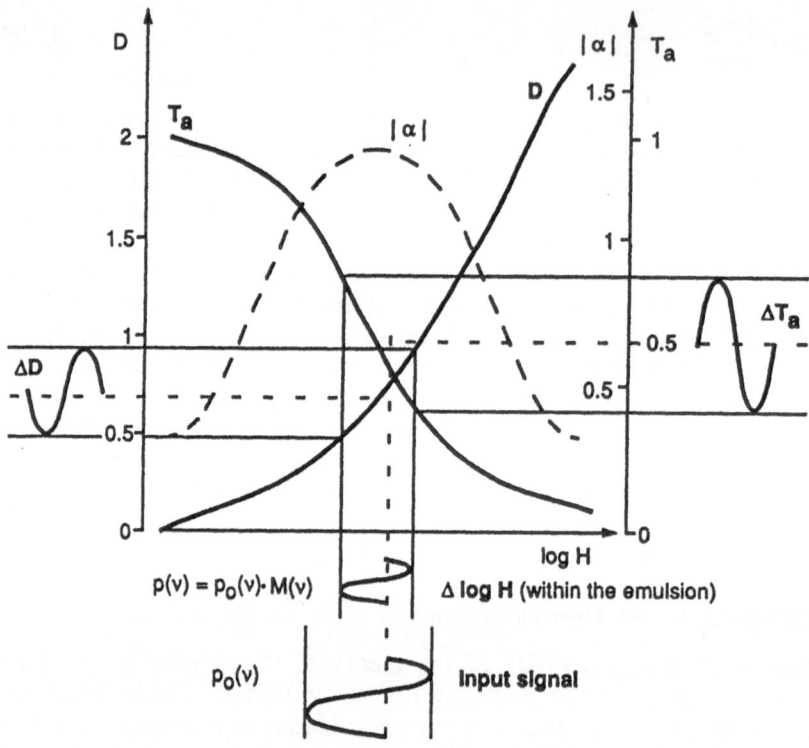

Fig.2.14. Holographic recording process. Recording of the input modulation signal in the material. The relation between log exposure H and density D or amplitude transmittance T_a in the developed hologram are shown. The derivative of the T_a-logH function is the $|\alpha|$-curve. This is an important characteristic of diffraction efficiency. The diffracted flux is proportional to the square of the local gradient $\gamma(H)$ of the D-logH curve. The diffraction efficiency of the hologram is dependent on both $\alpha(H)$ and $\gamma(H)$. The maximum diffraction efficiency will be obtained in the toe region of the D-logH curve [2.93]

region is more important than a high γ-value in the straight-line part of the D-logH curve. Operating close to the toe requires that the material has a low fog (and base fog) level. *Biedermann* showed that the α^2 values varied between 1 and 3 for various holographic materials present at the time of his investigation [2.100]. At the maximum α^2 values the T_a values are about 0.4 to 0.5, which corresponds to density values of about 0.6 to 0.8.

Intensity variations of an interference pattern are recorded as variations in exposure and they increase proportionally to the exposure time. There is a choice between two different operating points. A *minimum of harmonic distortion* is obtained at the inflection point of the T_a versus H curve where the gradient $\beta(H) = dT_a/dH$ has its maximum. The *maximum diffraction efficiency* will occur at the point where the gradient $\alpha(H)$ of the T_a versus logH is largest:

$$\beta(H)H = H(dT_a/dH) , \qquad\qquad (2.37)$$

$$\beta(H)H = \log_{10}e(dT_a/d\log H) = 0.434(dT_a/d\log H) . \qquad (2.38)$$

The density curve normally indicates that the density increases with exposure, which takes place when the negative developing technique is applied. However, *Gabor* showed that a *gamma of -2* is needed for a perfect linear recording using a positive interference pattern [2.104]. The reason for this is the following. Generally speaking, in photography a linear recording is such that the difference in density is proportional to the difference in the corresponding log exposure, with the constant of proportionality being the gamma value (γ) when operating along the straight-line portion of the H&D curve

$$D_2 - D_1 = \gamma(\log H_2 - \log H_1) . \qquad\qquad (2.39)$$

In holography, as indicated above, it is the amplitude transmittance versus exposure which is of importance instead. Therefore, if transmittance is substituted for density in the above formula, we obtain

$$\log(1/T_2) - \log(1/T_1) = \gamma(\log H_2 - \log H_1) , \qquad (2.40)$$

$$\log T_1 - \log T_2 = \gamma(\log H_2 - \log H_1) \qquad\qquad (2.41)$$

and with the amplitude transmittance

$$\log(T_{a1})^2 - \log(T_{a2})^2 = \gamma(\log H_2 - \log H_1) , \qquad (2.42)$$

$$\log(T_{a1}/T_{a2}) = \log(H_2/H_1)^{\gamma/2} . \qquad\qquad (2.43)$$

The above indicates that a linear recording can be obtained only when $\gamma = |2|$. $\gamma = +2$ describes the negative recording process and $\gamma = -2$ the positive one.

In a pure in-line hologram, a positive image is obtained if a positive interference pattern is used in the hologram and the hologram is processed with a gamma value of -2. In this case the image intensity will be added to the bright background. If a negative pattern is used, the object contrast is reversed (the image intensity will then be subtracted from the bright background) and a negative image is obtained, which will produce a correct negative image if a gamma value of +2 is used.

Using the negative developing technique for an off-axis amplitude hologram, a negative image of the intensity distribution of the interference pattern is obtained in the final amplitude hologram. It is, however, possible to reconstruct the original wavefront from this negative recording. *Both the positive and the negative recording of the interference pattern reconstruct the same original wavefront according to Babinet's principle.* The invariance of the reconstructed holographic wave front with the sign of the gamma value can be considered to be an exact form of this principle. The image reconstructed from the hologram will therefore always be a positive image of the object. Nevertheless, there are ways of obtaining a *negative, reconstructed image* from such holograms. *Nishida* [2.105] has shown that using K < 1, i.e., when the object beam is of a higher intensity than the reference beam, one can produce a negative image or at least a partial negative image in direct transmission or image-plane holograms. Normally, recording with K < 1 is not advisable, which is why this technique has limited applications. *Nishida* also claimed that a negative image can be produced using a high bias exposure level, which will resemble photographic solarization.

The modulation transfer function. Holographic silver-halide materials have high resolving power, which normally means that the *Modulation Transfer Function* (MTF), denoted by M(ν), is close to one even at quite high spatial frequencies ν. The MTF describes the results of light scattering in the emulsion during exposure. Figure 2.15 shows a typical holographic MTF compared with the MTF of conventional photographic materials.

Several papers have appeared dealing with holographic MTF's as well as methods for obtaining an MTF of holographic materials [2.106-114]. Emulsions for conventional photography show an approximately exponential line spread function which decreases to 1/10 at a width k of about 20 μm [2.106]. The thinner the layer is the better the MTF will be for a given type of emulsion. The *mean diffusion length* is important, but it would be unrealistic to try to make very thin emulsions in order to reduce scattering. The way to improve the mean diffusion length would rather be to make

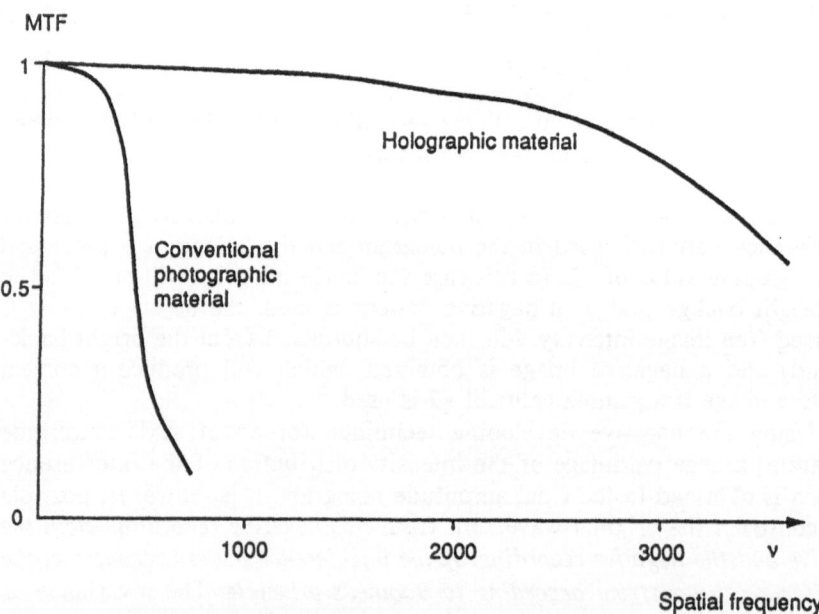

Fig.2.15. Modulation Transfer Function (MTF) curves for a conventional photographic material and a holographic material

very transparent emulsions with extremely fine grains. The photons entering the emulsion should be either absorbed directly upon their first contact with a silver-halide grain or else leave the emulsion. They should not be allowed to get scattered by the grains and be absorbed later on. The directly absorbed photons will be represented by a δ-function of the spread function, which means that their Fourier transform is a constant extending the MTF to high spatial frequencies forming a plateau. The mean diffusion length in a typical holographic emulsion is increased by at least one order of magnitude compared to conventional photographic materials. The MTF can be described in the following way according to *Frieser* [2.106]

$$M(\nu) = \rho + \frac{1 - \rho}{1 + (\pi k\nu/2.3)^2} \tag{2.44}$$

where ρ is the fraction of photons absorbed at the first contact with a grain and $(1-\rho)$ describes the photons absorbed after scattering. The parameter for the exponential line-spread function is k, the width at 1/10 of the maximum, or k/2.3 at 1/e.

The MTF is wavelength dependent, a fact to be well aware of. And so, a material that performs very well in red laser light, may not perform so well at a blue laser line. The scattered part $(1-\rho)$ that has not been used for information recording in the hologram is still coherent and can interfere with the reference field, producing a coherent noise pattern referred to as "diffusion mottle" by *Biedermann* [2.100]. When the exposing light fields and the scattered light are reflected (Fresnel or total) at the boundaries of the medium, more spurious patterns are exposed in the volume of the emulsion, which *Biedermann* refered to as "reflection mottle" [2.100].

Biedermann has also introduced the so-called "as-if-MTF" or "apparent MTF" [2.93, 100]. Due to the *adjacency* or *neighborhood developing effect* during the processing of holograms it is possible to enhance the contrast of the fringes which can actually improve the MTF of an emulsion. The adjacent effects will occur only if the developer has some silver-solvent compound added to it. The edge effect works in such a way that a fringe with high exposure will be developed to an increased density compared to a uniform area exposed of the same level. The reason is that neighboring (less exposed) silver-halide grains will be slightly dissolved during the processing and thus contributing liberated silver ions which will be reduced to silver atoms at highly exposed grains in the fringe. Therefore these grains will contribute to a higher density of the fringe than motivated by the received exposure. These phenomena will be discussed in a greater detail in Chap.4.

Diffraction efficiency. The recording material is exposed to the intensity of the interference pattern during a certain time t. If the intensity distribution in the pattern is constant during the exposure, the total exposure of the material is

$$H = Et \tag{2.45}$$

where E is the average intensity plus the intensity modulation (depending on the beam ratio, the phase, the original object intensity variations, and the MTF of the material). If linear recording is assumed, these intensity variations will be recorded as density variations in the processed emulsion for an amplitude hologram. The density variations are causing an amplitude modulation of the reference beam, when the processed hologram is reconstructed. The amplitude modulation of the primary, first-order diffraction light is mainly of interest, since it contains the original scattered wavefront from the recorded object. This is the holographic image when recording diffuse wavefronts. The intensity of this light over the intensity of the light illuminating the holographic material is known as the *diffraction efficiency* η of the hologram. Normally, light reflected from the interfaces are first subtracted, to obtain a more accurate value of the real diffraction efficiency.

Theoretical evaluation of the diffraction efficiency for different types of holograms (gratings) is straightforward and has been published in many papers and books. Therefore, in the following only a brief explanation will be given and the results concerning the diffraction efficiency under optimal conditions (e.g., Bragg condition satisfied) will be presented for various hologram (grating) types.

Diffraction efficiency for amplitude holograms. The diffraction efficiency η of a thin amplitude hologram is

$$\eta = [\tfrac{1}{4} \log_{10} e \, \alpha(H) M_i M(\nu)]^2 \ . \tag{2.46}$$

It is proportional to the square of the gradient of the T_a vs $\log H$ curve, $[\alpha(H)]$, as well as to the squares of the input modulation M_i and the Modulation transfer function $M(\nu)$. The input modulation M_i is K-dependent which is why the diffraction efficiency can be expressed in the following way as well

$$\eta = 0.188 \alpha(H)^2 M^2(\nu)/K \ . \tag{2.47}$$

In the expression above, the input modulation is assumed to be not too high so that the recording will occur within the linear part of the T_a-$\log H$ curve.

To estimate what the maximum diffraction efficiency can be (for a diffraction grating with an input modulation $M_i = 1$) the following has to be considered. The amplitude transmittance of the exposed and processed grating is

$$T_a(x) = T_a(H_0) + T_v \cos(\omega x - \phi_0) \ . \tag{2.48}$$

The first term represents the average transmittance dependent on the average exposure, whereas T_v is the amplitude of the spatially varying portion of the exposure $H = tE$

$$H = t[\langle O^2(x)\rangle + R^2]M_i M(\nu)\cos(\omega x - \phi_0) . \qquad (2.49)$$

In order to achieve the maximum diffraction efficiency for the grating, M_i and $M(\nu)$ must both be equal to 1, and $T_a(x)$ should vary between 0 and 1. Thus

$$T_a(x) = \tfrac{1}{2} + \tfrac{1}{2}\cos(\omega x - \phi_0) , \qquad (2.50)$$

$$T_a(x) = \tfrac{1}{2} + \tfrac{1}{4}e^{i(\omega x - \phi_0)} + \tfrac{1}{4}e^{-i(\omega x - \phi_0)} . \qquad (2.51)$$

The above equation shows that only one-fourth of the illuminating amplitude is diffracted into either the primary or the conjugate wave. The intensity in the primary diffracted wave is then only 1/16 of the intensity illuminating the grating, which means that the *maximum diffraction efficiency for a thin amplitude hologram is 6.25%*. The diffraction efficiency obtainable in an amplitude hologram of an extended object is always lower than the maximum value, normally 1 to 2% only.

Diffraction efficiency for phase holograms. For a thin, lossless phase grating $|T_a(x)| = 1$, which means that the complex amplitude transmittance is only phase dependent

$$T_a(x) = e^{-i\phi(x)} . \qquad (2.52)$$

If the phase shift produced by the recording medium is linearly proportional to the intensity in the interference pattern

$$\phi(x) = \phi_0 + \phi_v\cos\Phi \qquad (2.53)$$

where $\Phi = \varphi_0 - \beta x$.

The complex amplitude transmittance of the grating is

$$T_a(x) = e^{-\phi_0} e^{-i\phi_v\cos\Phi} . \qquad (2.54)$$

If the constant phase factor on the right-hand side of the equation above is neglected, this side can be expanded into the Fourier series

$$T_a(x) = \sum_{n=-\infty}^{+\infty} i^n J_n(\phi_v) e^{in\Phi} \qquad (2.55)$$

where J_n is the Bessel function of the first kind of order n. When a thin phase grating is illuminated, the light is diffracted into several orders. The diffracted amplitude in the nth order is proportional to the value of the Bessel function $J_n(\phi_v)$. Since only the first order is of interest in holography, the amplitude of the light diffracted into order 1 is proportional to

$J_1(\phi_v)$ which has a maximum value of 0.58. Thus, the *maximum diffraction efficiency for a thin phase hologram* is $(0.58)^2 = 33.9\%$. This value is about five times higher than the diffraction efficiency obtainable for a thin amplitude grating, which makes the production of phase holograms a desirable and attractive venture. An even more dramatic difference is experienced in volume holograms which will be studied in the next subsection.

b) Hologram Type: Thick Holograms

The *thick* or *volume hologram* represents a very important hologram type since it is here where, at least theoretically, the highest possible diffraction efficiency can be obtained. A simple theory of the first-order properties of thick holograms was presented by *Gabor* and *Stroke* [2.115]. A special theory known as the *coupled-wave theory* has been developed by *Kogelnik* [2.116]. In the analysis of a hologram formed with two plane waves, a sinusoidal recording is used to find the different diffraction efficiencies. Only some results concerning the maximum diffraction efficiency predicted by this theory will be presented here. In general, the coupled-wave theory can be used to study, e.g., the angular and wavelength selectivity of a hologram. There are many papers specifically on the subject of volume gratings and volume holograms [2.117-160]. Recently, an alternative theory of diffraction by modulated media was presented by *Harthong* [2.161]. A modulated medium is a medium in which the refractive index varies in such a manner that it is approximately periodic in small regions. Typical modulated media are holograms, because holograms record interference fringes, which are always contour surfaces. The theory is devoted to the analysis of the wave propagation in modulated media with a period of the same order of magnitude as the wavelength. The purpose of the paper was to give a method for treating the complete, three-dimensional problem in which an arbitrary shaped wave is diffracted by a medium of arbitrarily shaped modulation. A book on practical volume holography written by *Syms* has also been published [2.162].

Bragg's law and the *Bragg condition* are of importance for thick holograms (Fig.2.16):

$$2d\sin\theta = \lambda_a/n \tag{2.56}$$

where θ is *half the angle* between the reference and the object beams at the recording stage (as well as the angle between the illuminating and the diffracted beams and the scattering planes in the emulsion at reconstruction), d the spacing between the interference planes in the emulsion, λ_a the wavelength in air, and n the average refractive index of the hologram medium.

The following four types of thick holograms will be discussed, assuming that the recorded fringes are the sinusoidal variations of the refractive index or the absorbtion constant: *Thick amplitude transmission, thick phase transmission, thick amplitude reflection, and thick phase reflection holograms.*

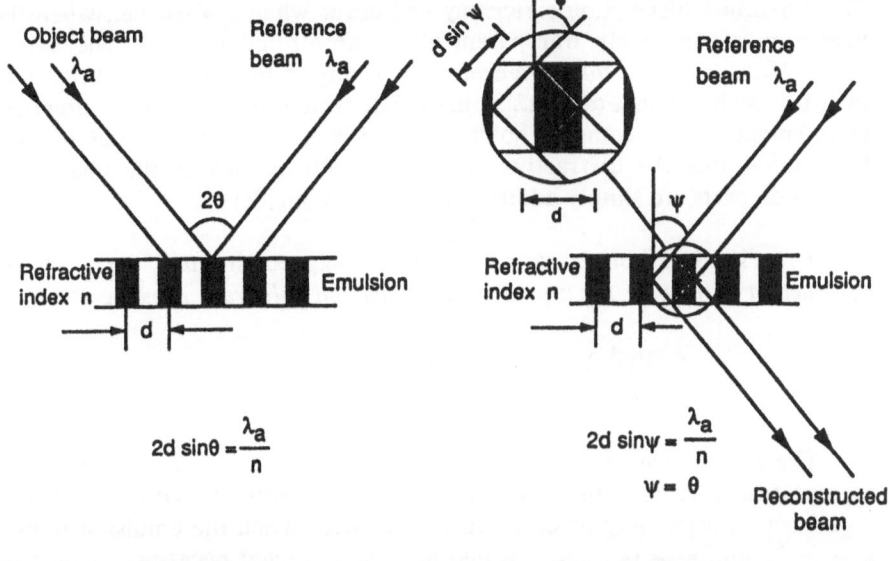

a. Recording **b. Reconstruction**

Fig.2.16. Bragg condition. θ is half the angle between the reference and the object beams at the recording stage, ψ is the angle between the normal and the reference beam at reconstruction, d is the spacing between the interference planes in the emulsion, λ_a is the wavelength in air, and n the average refractive index of the hologram medium. The Bragg angle ($\psi = \theta$) is then given by $2d\sin\theta = \lambda_a/n$

For pure phase holograms $n_0 = n + n_1\cos(2\pi x/\Lambda)$ and for pure amplitude holograms $\alpha_0 = \alpha + \alpha_1\cos(2\pi x/\Lambda)$, where Λ is the period of the grating, and α and n are the average values of the refractive index and the absorption constant, respectively. In the following equations d is the thickness of the recording layer, λ the wavelength of the light, and θ_0 half the angle between the recording reference and object beams.

Thick transmission holograms. Providing that the Bragg condition is satisfied the *diffraction efficiency for a pure phase transmission hologram* is expressed by

$$\eta_{tp} = \sin^2\left(\frac{\pi n_1 d}{\lambda \cos\theta_0}\right). \tag{2.57}$$

The *diffraction efficiency of a thick, phase hologram will be theoretically 100%* when $n_1 d = \frac{1}{2}(\lambda\cos\theta_0)$.

When the Bragg condition is satisfied the *diffraction efficiency for a pure amplitude transmission hologram* is given by

$$\eta_{ta} = e^{-2\alpha d/\cos\theta_0}\sinh^2\left(\frac{\alpha_1 d}{2\cos\theta_0}\right). \tag{2.58}$$

69

The maximum diffraction efficiency will occur when $\alpha = \alpha_1$, i.e., when the hologram is completely transparent at the absorption minima. The *maximum efficiency for a thick transmission grating* at the Bragg condition will then be reached when $\alpha d/\cos\theta_0 = \ln3 = 1.1$, which means 3.7%. In practice, most holograms are complex holograms and it may therefore be of interest to mention that the diffractive-efficiency contributions of the phase and amplitude parts are simply additive at the Bragg incidence.

Thick reflection holograms. When the Bragg condition is satisfied the *diffraction efficiency for a pure phase reflection hologram* is expressed by

$$\eta_{rp} = \tanh^2\left(\frac{\pi n_1 d}{\lambda \cos\theta_0}\right). \tag{2.59}$$

The *diffraction efficiency for a thick reflection hologram* approaches asymptotically 100% with increasing $n_1 d$. For example, if $n_1 d/(\lambda\cos\theta_0) = 1$, a diffraction efficiency of 60% can be obtained. When the emulsion thickness is 20 μm, $\theta_0 = 0°$, and $\lambda = 600$ nm, the expected necessary change in the refractive index amounts to $n_1 = 0.01$. Such a change in the refractive index can be achieved for a bleached silver-halide material.

The *diffraction efficiency for a pure amplitude reflection hologram* when the Bragg condition is satisfied, can be written as

$$\eta_{ra} = \left(\frac{-\alpha_1}{2\alpha + \sqrt{4\alpha^2 - \alpha_1^2}\coth\left[d\sqrt{\alpha^2 - \tfrac{1}{4}\alpha_1^2}/\cos\theta_0\right]}\right)^2. \tag{2.60}$$

The maximum diffraction efficiency will occur when $\alpha = \alpha_1$, i.e. when the hologram is completely transparent at the absorption minima. *The maximum efficiency for a thick reflection grating* at the Bragg condition is obtained when $\alpha d/\cos\theta_0 \geq 2$, which corresponds to an optical density ≥ 1.7, resulting in the maximum value of 7.2%.

c) Summary of the Diffraction Efficiency

The highest possible diffraction efficiencies obtainable for gratings are shown in Table 2.5.

According to the efficiencies obtainable, predicted by the coupled-wave theory, it is obvious that for some applications only thick holograms of the phase type are of interest. Holographic optical elements possessing thick emulsion layers are usually extremely efficient and are therefore of great importance to practical holography. Diffraction efficiencies that can be obtained from extended, diffuse objects recorded in thick phase holograms have been studied by *Upatnieks* and *Leonard* [2.122]. They found that the theoretical maximum efficiency for a diffuse wavefront is 64%. The analysis of diffuse objects and a plane reference wave based on the coupled-wave theory can be found in a publication by *Korzinin* and *Sukha-*

Table 2.5. Theoretical maximum diffraction efficiency for a grating

Hologram type	Thin transmission		Thick transmission		Thick reflection	
Modulation	Amplitude	Phase	Amplitude	Phase	Amplitude	Phase
Efficiency	6.25%	33.9%	3.7%	100%	7.2%	100%

nov [2.147]. *Lokshin* et al. [2.148] gave an account of a quantitative experimental investigation on diffuse objects recorded on some Russian materials.

The question about the influence of the thickness of the emulsion on the diffraction efficiency from a theoretical as well as a practical point of view is important. *Smith* [2.149] has studied the diffraction efficiency of amplitude transmission holograms as a function of emulsion thickness. He found that (under the assumption that the K-value is moderately high, at least >5) the diffraction efficiency is independent of the emulsion thickness. Experimentally, he found no change in diffraction efficiency above about 7-μm emulsion thickness. Recently, *Fimia* et al. [2.150] carried out experiments of the Agfa 8E75 material with thicknesses of 7 and 14 μm and found no difference in the diffraction efficiency between the two materials. This was observed for diffuse-object holograms. However, an increased signal-to-noise ratio was noted for the thicker emulsion for prebleached densities between one and four. For densities above five no difference in the signal-to-noise ratio was found. *Fimia* et al. concluded that the intermodulation noise of diffuse-object holograms is dependant on the nonlinearity of the developer and the photochemical processes at the high density in multiplexed holograms. *Kiemle* [2.151] has adopted the electrical ladder-network model to find an expression for the diffraction efficiency of the reflection phase hologram, which is actually dependent on the emulsion thickness. The theory predicts for a certain emulsion having a diffraction efficiency of 15% for a 7-μm thick emulsion it will increase to 50% at 20 μm thickness and 76.4% at 50 μm. Another contribution to the treatment of thick phase reflection holograms has been made by *Hariharan* [2.152]. The *Kogelnik* theory [2.116] indicates that the diffraction efficiency is directly proportional to the emulsion thickness and the refractive-index modulation. *Hariharan* has shown the existence of an optimal value for the emulsion thickness achieved as a result of a compromise between the ratio of peak diffraction efficiency and bandwidth at hologram reconstruction in white light. A thickness between 5 and 7 μm seems to be a good compromise according to the paper. Another fact mentioned is that an increase in the emulsion thickness will also cause an increase in the scattering within the emulsion which is due to the finite size of the grains. This is the reason why fine-grained emulsions are in high demand as well as improved processing methods giving higher refractive-index modulation for holograms of higher luminance. It is interesting to state that for thin or thick, amplitude or phase, transmission or reflection holograms the optimum thickness of a silver-halide emulsion is about 7 μm.

Dammann [2.153] has demonstrated that the diffraction efficiency is limited to 18.4% for a thin phase hologram of a diffuse object compared to 33.9% for the sinusoidal thin diffraction phase grating. *Upatnieks* and *Leonard* [2.123] found the diffraction efficiency to be 22% for a diffuse object recorded in a thin phase hologram. More information concerning diffuse objects including experimental work done in this area can be found in a paper by *Clausen* and *Dammann* [2.154]. *Sidorovich* [2.134] has analyzed the diffraction efficiency of three-dimensional phase holograms for the case of an arbitrary ratio of the beam intensities (K-values). Diffraction efficiencies obtainable from holograms made on the recording material treated as a discrete-carrier material (here, the interference fringes are recorded by isolated particles dispersed in a binder; this procedure applies to silver-halide materials) have been presented by *Kovachev* et al. [2.155]. The paper contains information concerning the influence of the size, the optical properties and the concentration of the light-sensitive particles on the diffraction efficiency. The results are presented as topograms. *Zeldovich* et al. [2.141] have shown that energy characteristics of both transmission and reflection speckle-field holograms differ considerably from the characteristics of plane-wave holograms. A classification of speckle-field holograms was proposed and volume holograms were investigated in detail. *Korzinin* and *Sukhanov* [2.147] have presented interesting aspects of recording diffuse objects in volume phase holograms. They found that the optimum K-value from an energy point of view depends on the geometry of the recording scheme. When recording a hologram of a diffuse object one should use an obliquely incident reference beam. If the angle of incidence of the object beam is reduced, the efficiency of the hologram is increased. Diffusely scattering objects recorded in thick layer transmission holograms have been treated by *Staselko* and *Churaev* [2.156]. In Chap.5, some volume effects directly related to phase holograms will be described in some detail.

A special phenomenon that occurs in volume holograms with strong coupling, is a secondary scattering effect that can cause scattering rings or general noise. *Ragnarsson* [2.157] has explained the factors causing these effects using a modification of the Ewald-sphere concept, previously explained by *Forshaw* [2.158]. *Ragnarsson* used extremely thick silver-halide emulsions (about $400\,\mu$m) in order to experimentally verify how internally scattered light is affected by the Bragg condition producing image disturbing scattering rings and Kossel lines. The secondary scattering effect has also been treated by *Yakimovich* [2.159].

Recently, *Dubois* et al. [2.160] developed a new integral model of the diffraction process for thick holograms that makes possible the derivation of analytical solutions. The new theory predicts the same holographic behavior as does the coupled-wave theory when the hologram size is large with respect to the thickness of the material and for diffraction efficiencies up to 25%. When the size of the hologram is smaller and when the recording and reconstructing wavelengths are different the new theory is more accurate. *Dubois* et al. claimed that the new theory makes it possible to handle holographic problems more easily than using the coupled-wave theory. The earlier-mentioned modulation theory introduced by *Harthong* [2.161] can also be compared with this new theory as well as *Kogelnik's* theory for volume holograms [2.116].

d) Holographic Exposure Index

The fact that a certain optical density in an amplitude holographic plate has very little meaning as regards the reconstructed image quality, led to the introduction of the *Holographic Exposure Index* (HEI) by *Biedermann* [2.93]. HEI is a figure which should be a better means of describing the holographic recording process. The HEI value S_η is defined as the sensitometric speed S_i divided by the holographic expenditure factor X

$$S_\eta = S_i/X \,. \tag{2.61}$$

In a holographic system a certain amount of radiant flux has to be used to illuminate the object in order to achieve a desired diffraction efficiency η. Hence, exposure time t_h necessary for a hologram to undergo the average required exposure H_i becomes X times as long as the exposure time (t_{min}) which corresponds to the hypotethical case of sending all the laser flux through the reference beam alone. For a given material the t_{min} exposure time is correlated with the sensitometric speed value S_i for that material. In order to estimate the value of the expenditure factor X, the reflectivity r of the object must be known. The flux scattered from the object is normally considerably reduced, compared to the object illuminating flux, which is why $r \ll 1$. If q is the ratio by which the beam splitter divides the laser beam into the reference and the object beams, the reflectivity r is

$$r = q/K \tag{2.62}$$

where $K = E_r/E_o$.

The polarization of the light scattered from the object will also affect the recording. If ψ is a fraction of the object-field flux linearly polarized parallel to the reference field, ψ will normally be ≤ 1. The expenditure factor X is a function of diffraction efficiency η, and therefore

$$X = \frac{t_h}{t_{min}} = 1 + \frac{1 - r}{r\psi[\log^2(e)\,\alpha^2_{max}\,M^2(\nu)/\eta - 1/\psi]} \,. \tag{2.63}$$

In the above formula the α-value (amplitude holograms) can be substituted by the corresponding Γ-value at the operating point for phase holograms. A decrease of the expenditure factor gives an increase of S_η, which indicates a shorter exposure time. Possibilities to increase the HEI value of holographic materials, which involve hypersensitization techniques, as well as other methods for increasing the sensitivity of holographic materials will be discussed in Chap.6.

Leonard and *Smirl* [2.163] presented two other figures of merit: Q and A. Q is defined as the maximum diffraction efficiency divided by the signal beam exposure, valid only for high values of K. (K > 100). A is defined similarly for the maximum contrast instead. *Leonard* and *Smirl* also used the following model: $T_a = [b/(b+H^d)]^c$, from which the A and Q values could be calculated for a given recording material. In the formula, b, c, and d are constants depending on the material and its processing, and H is the total exposure. Q and A values presented in the above mentioned work apply to different commercial materials existing at the time of the investigation.

Finally, it should be mentioned that *Lin* [2.94] described a rather general and interesting method of characterizing recording materials for holography. He used the following relation: $\sqrt{\eta} = S\langle H \rangle M_i$, where η is the diffraction efficiency, $\langle H \rangle$ the average exposure, M_i the modulation (or visibility) of the interference fringes and S a constant ("holographic sensitivity") for a given material. For a *perfect recording material*, the curves of $\sqrt{\eta}$ versus $\langle H \rangle$ with M_i being constant as a parameter, as well as curves of $\sqrt{\eta}$ versus M_i with $\langle H \rangle$ being constant as a parameter are straight lines (Fig.2.17).

Such curves can easily be plotted for different real recording materials, since η, $\langle H \rangle$, and M_i are all measurable quantities. In this case, the curves are straight lines only within limited ranges of η, $\langle H \rangle$, and M_i. Within these ranges the material has a linear response, which is why it can be regarded as ideal. Outside these regions, the curves indicate nonlinear recording and suggest other unwanted behavior. The following information concerning the particular recording material can be obtained from such curves:

- The range of modulation M_i (or K-value) within which the recording is linear, which is indicated by the straight-line part of the $\sqrt{\eta}$ vs M_i curve at an appropriate average exposure $\langle H \rangle$.
- The maximum achievable diffraction efficiency η.
- The exposure sensitivity, i.e., the exposure value needed to obtain a certain diffraction efficiency η at a fixed modulation value M_i or for a fixed K-value.
- The optimum average exposure $\langle H \rangle$ for the best compromise between a large range of linearity and high diffraction efficiency.

The holographic sensitivity expressed as $S = \sqrt{\eta}/(\langle H \rangle M_i)$ is constant only for a perfect material having any combination of η, $\langle H \rangle$ and M_i. For a real material, the holographic sensitivity can be determined by taking the slope of the linear region of a $\sqrt{\eta}$ vs M_i curve and dividing it by the average exposure $\langle H \rangle$ at the center of the linear region. These S-values can be expressed in m^2/J for real materials, where these values represent the largest area (m^2) of a given material that can be exposed with the energy of one Joule.

Some S-values for the following materials were presented in the paper:

Kodak 649-F, amplitude hologram $0.26 \ m^2/J$
Kodak 649-F, phase hologram $0.04 \ m^2/J$

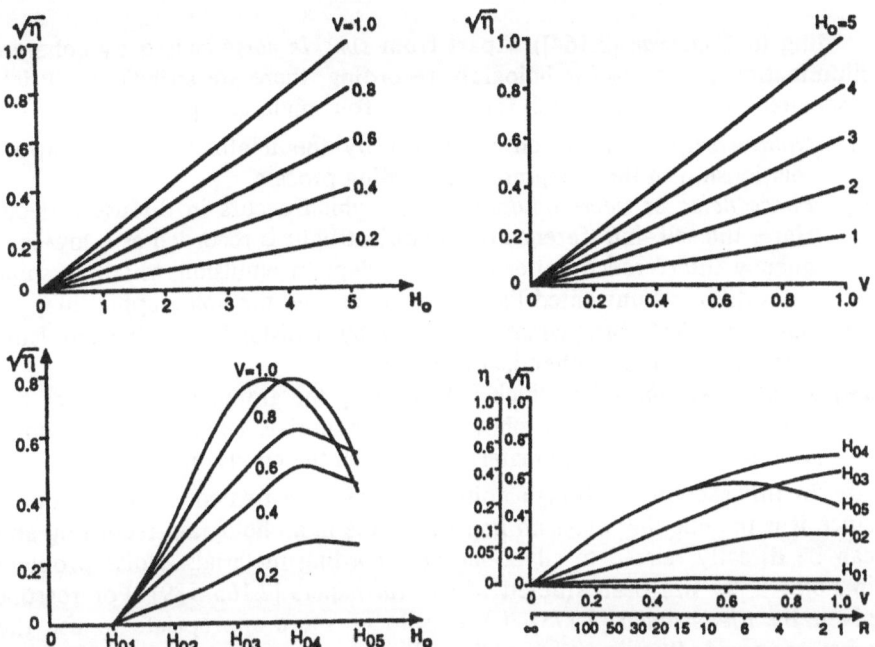

Fig.2.17a-d. Lin's holographic sensitivity curves for both ideal and real recording materials [2.94]. (a) $\sqrt{\eta}$-H_0 characteristics of the ideal recording material, (b) $\sqrt{\eta}$-V characteristics of the ideal recording material, (c) composite $\sqrt{\eta}$-H_0 characteristics of real recording materials, and (d) composite $\sqrt{\eta}$-V characteristics of real recording materials (η: diffraction efficiency, H_0: average exposure value, R: irradiance ratio of interfering plane waves, V: fringe visibility)

| Agfa 10E70 amplitude hologram | 7.3 m^2/J |
| Dichromated gelatin hologram | 0.0065 m^2/J . |

The most practical side to this way of characterizing holographic materials (using the $\sqrt{\eta}$-M_i curve) is that different materials or hologram types can be compared directly to each other and that they can also be related to an ideal material. However, there are two important properties of the recording material not reflected in these curves, namely the spatial frequency response (the MTF of the material) and the influence of noise on the recording due to scattering within the emulsion.

2.3.3 Noise

a) Noise Factors

Noise is often the most important image-degradation factor, particularly when the object wave is weak as compared to the reference field (this according to *Goodman* [2.164]). Apart from *speckle noise* caused by coherent illumination necessary for hologram recording, there are specific material-associated noise factors which fall into the four main groups:

I) *Nonlinearity noise* which is caused by the nonlinear input-output relationship in the holographic recording process.

II) *Phase noise* or *intermodulation noise* which occurs in diffuse objects. Here, the self-interference of the object light is recorded as a low frequency interference pattern in the hologram emulsion which in some cases diffracts unwanted light into and around the holographic image.

III) *Surface-relief noise* which is caused by a distortion of the emulsion surface due to, e.g., chemical processing.

IV) *Emulsion-grain noise* which is caused by the granular structure of the photosensitive layer in photographic materials (including also noise caused by the substrate material on which the emulsion is coated).

In this section the fourth noise factor will be treated in greater detail since it is the one that plays an important role in all hologram recording and can be directly related to silver-halide recording materials. Noise problem in holography has been studied in several papers [2.165-189]. For reasons given by *Biedermann* [2.173] it seems to be more appropriate to use the term *scattered flux spectrum* rather than the common term grain noise. Arguments have also been raised against the usage of the term *intermodulation noise*, since, as it is casual rather than truly random in nature, it might more accurately be called *intermodulation degradation* [2.171]. The upper limit of the dynamic range of a holographic recording is set by noise from Group I and II, and the lower limit is set by noise from Group IV.

The noise background limits the weakest signal that can be recorded in a holographic system. *Goodman* [2.164] pointed out that the spatial fluctuations in the background coherent light should be regarded as noise in a holographic emulsion. *Biedermann* [2.173] has shown that the noise level in a coherent system is different from the noise level in an incoherent one due to diffusion mottle in the emulsion produced by the coherent light in the first case. The diffusion mottle increases the noise level by a factor of 1.5 in a coherent optical system as compared to the incoherent system.

Helstrom [2.165] has computed the average value $\langle E_n \rangle$ of the background noise light in emulsions. *Goodman* [2.164] defined the relevant signal-to-noise ratio as the ratio between the deterministic reconstruction image intensity E_i at a given point and the standard deviation σ of the total image intensity in the vicinity of that point, i.e.,

$$\frac{E_i}{\sigma} = \frac{E_i}{\langle E_n \rangle} \left(1 + 2 \frac{E_i}{\langle E_n \rangle} \right)^{-1/2} . \tag{2.64}$$

If $E_i / \langle E_n \rangle \gg 1$, which means that the image point is easily detectable in the background-noise field, then

$$\frac{E_i}{\sigma} = \sqrt{E_i / 2\langle E_n \rangle} . \tag{2.65}$$

In general, the signal-to-noise ratio is a monotonically increasing function of the signal-to-background ratio $E_i/\langle E_n \rangle$. In order to calculate the grain noise in the emulsion simple models can be used, e.g., the checkerboard or the overlapping circular grain model [2.190]. Both models predict that the maximum signal-to-noise ratio for amplitude transmission values will be in the range of 0.5 to 0.6. These models will be discussed again later, when a comparison of experimental results is made with what can be predicted from these models.

Burckhardt [2.166] presented some experimental data on Wiener spectra for holographic emulsions. The Wiener spectrum also measured by *Biedermann* [2.173], *Vilkomerson* [2.176], and *Smith* [2.181] for holographic materials, describes noise distribution in space from all points. The noise level at a particular point in space is measured over a particular area (bandwidth). The early papers treated the only holographic plate of the time – the Kodak 649-F spectroscopic plate – which is why most of the information presented concerns this particular material. For 649-F, *Burckhardt* showed that the storage capacity of this plate was two orders of magnitude lower than that of a hypothetical noiseless emulsion. *Kozma* [2.168] as well as *Urbach* and *Meier* [2.170, 171] presented interesting calculations based on noise characteristics for Kodak 649-F. For a point object (an object that emits light in all directions) we have

$$\frac{E_i}{\langle E_n \rangle} = \frac{K_2 \chi^2 H_r H_0 \alpha_h}{\phi(\nu)} \tag{2.66}$$

where K is the MTF $M(\nu)$ for a Fourier transform hologram, χ the slope of the T_a-H curve, H_r the reference exposure ($\propto |r|^2$), H_o the object exposure ($\propto |o|^2$), α_h the recording area of the holographic plate, and $\phi(\nu)$ the value of the Wiener spectrum at an appropriately chosen spatial frequency ν.

In this case we consider the distribution of noise in one dimension only. The function $\phi(\nu_x; \nu_y)$ is in reality two-dimensional but ν_y can be regarded to be zero for a simpler analysis.

Sometimes it is more convenient to use

$$\frac{E_i}{\langle E_n \rangle} = \frac{K^2 \xi^2 m^2 \alpha_h}{\phi(\nu)} \tag{2.67}$$

where $\xi = \chi H_r$ is the slope of the normalized T_a-H curve, and m^2 is H_o/H_r in this case. For Kodak 649-F at an amplitude transmission of 0.529, $\phi(\nu)$ is according to *Kozma* [2.168]

$$\phi(\nu) = e^{-0.00248\nu} \cdot 10^{-8} \text{ mm}^2 . \tag{2.68}$$

It is now possible to calculate $(H_o/H_r)_{min}$, i.e. the weakest point-source object that can be recorded by the Kodak 649-F material. Say that we wish E_i/σ to equal 5 in order to easily identify the object image in the hologram. If $E_i/\sigma = 5$ then

$$\frac{E_i}{\langle E_n \rangle} = 50 \ . \tag{2.69}$$

If we choose the spatial frequency ν to be 300 lines/mm, for example, which corresponds to the angle of $\simeq 10°$, the MTF for the Kodak 649-F plate will be 0.95 at this frequency. $M(300) = 0.95 \ (=K)$. The slope ξ of the normalized T_a-H-curve at $T_a = 0.5$ will then be 0.56 for 649-F developed for 5 minutes at a temperature of $20°\,C$ in the Kodak D-19 developer. By combining (2.67 and 68) we obtain

$$\left(\frac{H_o}{H_r} \right)_{min} = m^2 = \frac{(E_i/\langle E_n \rangle)\,e^{-0.00248\nu} \cdot 10^{-8}}{K^2 \xi^2 \alpha_h} \ , \tag{2.70}$$

$$\left(\frac{H_o}{H_r} \right)_{min} = \frac{50 e^{-0.00248 \cdot 300} \cdot 10^{-8}}{(0.95)^2 (0.56)^2 \alpha_h} \tag{2.71}$$

which gives

$$\left(\frac{H_o}{H_r} \right)_{min} = (8.4 \cdot 10^{-7})/\alpha_h \quad (\alpha_h \text{ in mm}^2) \ .$$

For a 10×10 mm^2 hologram

$$\left(\frac{H_o}{H_r} \right)_{min} = 8.4 \cdot 10^{-9} \ ,$$

and for a 100×100 mm^2 hologram

$$\left(\frac{H_o}{H_r} \right)_{min} = 8.4 \cdot 10^{-11} \ .$$

If, however, the angular distribution of the object light is limited, we will not gain anything by increasing the holographic plate area α_h. If the illumination from the object is limited to the area α_o in the plane of the holographic plate then

$$\frac{E_i}{\langle E_n \rangle} = \frac{K^2 \xi^2 m^2 \alpha_h}{\phi(\nu)} \left(\frac{\alpha_o}{\alpha_h} \right)^2 \ . \tag{2.72}$$

If we now consider an object with a diffuse surface containing N resolved object points H_o, then $H_s = NH_o$

$$\frac{E_i}{\langle E_n \rangle} = \frac{K^2 \xi^2 H_s / H_r}{\phi(\nu)} \left(\frac{\alpha_h}{N} \right). \tag{2.73}$$

If the object is a square with side L_o positioned at the distance d_i from a square holographic plate with side L, then the object consists of

$$N = \frac{L_o^2}{(\lambda d_i / L)^2} \tag{2.74}$$

resolved points. The area is $\alpha_h = L^2$. Then

$$\frac{E_i}{\langle E_n \rangle} = \frac{K^2 \xi^2 \lambda^2 (H_s / H_r)}{\phi(\nu)(L_o / d_i)^2}. \tag{2.75}$$

$E_i / \langle E_n \rangle$ is then *independent of the size of the holographic plate but it is inversely proportional to the square of the angular size of the object.*

Knowing the sensitivity of the Kodak 649-F plate (at a certain wavelength) to be $\simeq 110 \ \mu J/cm^2$ for $T_a = 0.5$, $(H_o)_{min}$ will be $9.2 \cdot 10^{-9} \ \mu J/cm^2$ for a point-source object recorded in a 100 cm^2 hologram.

b) Dynamic Range

The upper limit of the object intensity guarantees that the holographic recording is confined to the linear region of the T_a-H-curve. According to one of the early investigations on the Kodak 649-F material, the K-value should be larger or equal to 13.5 for linear recordings [2.90].

The maximum value of the object/reference exposure ratio is

$$\left(\frac{H_o}{H_r} \right)_{max} = 7.4 \cdot 10^{-2}.$$

Knowing the upper and the lower limits for the object light it is possible to calculate the *dynamic range* of the recording in a hologram. *King* has discussed the dynamic range of holographic materials in some detail and described methods for measuring it [2.91,92]. He has also introduced a figure of merit, which is found by dividing the dynamic range by the energy required to bias the recording material at the proper exposure point. This ratio, which is called the *dynamic-range efficiency* of the material, can be interpreted as the dynamic range of a hologram per unit optical-energy input. Materials with high dynamic-range efficiency are preferable to the ones whose ratio is low. Compare also the HEI-values presented in Sect. 2.3.2f.

The dynamic range for the Kodak 649-F material in the example above is then

$$\frac{H_{o,max}}{H_{o,min}} = 7.4 \cdot \frac{10^{-2}}{(8.4 \cdot 10^{-9})/\alpha} = 8.8 \cdot 10^6 \alpha.$$

For a hologram recorded on a 125 cm^2 plate the dynamic range is 1.1·10^9 (~90dB). The Kodak 649-F plate used for the above calculations is coated with a silver-halide emulsion that can be considered representative of the contemporary materials for holography as regards scattering. The dynamic range that can be captured in a hologram exceeds the range of any other imaging technology known today. The reason for this is that the high-intensity parts of the holographic object are not focussed on any particular part of the emulsion (as is the case in photography, which means running into nonlinearity of the recording process) but that they are spread over the entire recording area of the holographic material. *Leith* [2.88] has investigated the dynamic range for conventional film materials used in coherent optical systems and found that the dynamic range for Kodak Tri-X film, for example, is 19 dB for a 1-mm aperture, and 35 dB for a 40 mm aperture. The dynamic range is increased by 3 dB each time the aperture is doubled. In photography it is not practical to increase the recording aperture too much because of the limitations on the imaging lens system. In holography, in principle, the problem associated with the use of very large recording apertures (very large plates) does not exist.

Biedermann [2.100, 173] and several others [2.168, 176, 181] investigated the relationship between emulsion and scattering (scattering actually limits the dynamic range of a holographic recording) in both the Kodak 649-F material and other emulsions. Let us compare some of these results for ν = 300 lines/mm (corresponding to the angle of ~10° between the beams).

Table 2.6. Scatter noise in different holographic recording emulsions and substrates

| Material | Scattering values ($\times 10^{-9}$ mm^2) | | | | | |
	Kodak 649-F	Agfa 10E70	Agfa 8E70	DCG Glass	Brewster window	Polyester substrate
Kozma [2.168]	4.7					
Smith [2.181]	1.7			0.12		10
Vilkomerson [2.176]	8.0	25		0.8	0.01	
Biedermann [2.100]	1.5	3.0	2.0	0.5		

Interesting experiments regarding scattering in different film substrate materials as well as various holographic recording materials from Russia were reported by *Stozharova* [2.183]. In particular, it should be noticed that for the high-resolution materials the scattering from the base material (cellulose acetate or polyester bases) is the biggest contributor to noise. The thicker the base material, the more scattering is observed.

c) Models of Holographic Emulsions

The checkerboard model, already mentioned, can be used for computing the noise level in an emulsion for a holographic recording. The emulsion grains are assumed to be regularly spaced squares (side L, area A_g: L^2). O'Neill [2.190] has shown that the predicted noise-power spectrum has the following form

$$S_{theory}(\nu_x;\nu_y) = T(1 - T)L^2[sinc(\tfrac{1}{2}\nu_x L)]^2[sinc(\tfrac{1}{2}\nu_y L)]^2 \qquad (2.76)$$

where T is the transmission of the emulsion, L the side of the square grain in the checkerboard, and ν the radian frequency of the spectrum.

The length L of the grain in the above model is not the physical measurement of the silver-halide grain size but a correlation length, i.e. a measure of the distance which is parallel to the surface of the emulsion over which the transmission does not change appreciably. The model does not include multiple scattering, overlapping, etc., that occur in a real, three-dimensional emulsion.

For the 649-F plate, *Vilkomerson* [2.176] compared the theoretical values predicted by the model with the measured values. Using the checkerboard model he found, for example, that for

$T = 0.75$, $L = 0.2 \cdot 10^{-3}$ mm,

$\nu_x = 300$ lines/mm and $\nu_y = 540$ lines/mm,

$S_{theory} = 7 \cdot 10^{-9}$ mm^2, $S_{measured} \sim 10^{-8}$ mm^2.

A more accurate model for the emulsion is the overlapping circular grain model [2.190] in which one assumes that the grains are of circular shape (diameter L). They are either perfectly opaque or perfectly transparent. The area of a single grain is then $A_g = \pi(L/2)^2$. Grain centers are assumed to be randomly distributed with the uniform probability over the emulsion area. *Goodman* [2.164] has shown that

$$\langle E_n \rangle = \frac{E_p \alpha_h A_g}{\lambda^2 d_i^2} G(T_b) \qquad (2.77)$$

where

$$G(T_b) = c \int_0^1 (T_b^{2-F(\xi)} - T_b^2)\xi d\xi \qquad (2.78)$$

with $F(\xi) = (2/\pi)[cos^{-1}\xi - \xi\sqrt{1-\xi^2}]$, $\qquad (2.79)$

and E_p is the intensity over the area of the holographic plate, α_h the area of the holographic plate, A_g the area of the grain in the model, λ the wavelength of the light, d_i the distance to the image, c a constant depending on the material (for Kodak 649-F: $c = 8$), and T_b the bias transmittance.

Both models disregard the fact that in a real emulsion grains would vary in size and shape and would have different transmittance values, also that the emulsion's thickness is not zero.

Phillips et al. [2.191] and *Phillips* [2.192, 193] discussed scattering in silver-halide emulsions from the concept of the *mean diffusion length*. Photographic layers for commercial holographic materials are normally about 7 μm thick. The emulsion consists of small silver-halide crystals in a gelatin matrix. In 1871, Lord Rayleigh suggested that *light scattering from these crystals would be proportional to the sixth power of their radius for a given wavelength and would increase with the inverse fourth power of the wavelength as the wavelength decreases*. If the radius of the silver-halide crystals is a and the number per unit volume of the layer is N, then

$$\frac{4 d N \pi a^3 \rho}{3} = m_{AgH} \tag{2.80}$$

where ρ denotes the density of silver halide, m_{AgH} is the mass of silver-halide unit area, and d the thickness of the emulsion. The atomic weight of silver is 108 and that of bromine is 80. Then

$m_{AgH} = 188/108 \; m_{Ag}$,
$m_{Ag} \; = 5 \; g/m^2$ (for Agfa materials) ,
$\rho_{AgBr} = 6.47 \; g/cm^3$.

Therefore, an emulsion with the grain size of 2a = 30 nm will give

$$N = \frac{3}{4\pi}\left(\frac{188}{108}\right)^4 \frac{2}{3}\cdot 10^{-4} \; \frac{3}{6.5}\cdot 10^{-4} \simeq 10^{16} \; grains/cm^3 \; ,$$

i.e., the gelatin layer has approximately 10^{16} grains/cm^3. A change in grain size usually means that the same amount of silver bromide would be shared between fewer but larger grains which, in turn, means that Na^3 is constant. For the following Agfa materials we have

8E-materials: $2a = 30 \; nm$ [2] $\rightarrow N_{8E} - 10^{16}$ grains/cm^3 ,

10E-materials: $2a = 90 \; nm$ $\rightarrow N_{10E} - 10^{11}$ grains/cm^3 .

The scattered intensity I_S of light off these small particles is $S \propto Nda^6$. But since Na^3 is constant, I_S is therefore $\propto a^3$, which means that light-scattering varies with the cube of the grain size. *Phillips* et al. [2.191] have introduced the parameter f denoting the ratio of scatter mean free path to the emulsion thickness, which can be used as a figure of merit to describe the holographic recording layer, namely

---

[2] The real grain size of the Agfa 8E emulsion is about 44 nm (precoated about 35nm).

$$f = 1/(N\sigma_{RS}d) \qquad (2.81)$$

where N denotes the number of grains/unit volume, d the emulsion thickness, and σ_{RS} the Rayleigh scatter cross section being

$$\sigma_{RS} = \frac{\pi}{12}\left(\frac{2\pi}{\lambda_a}\right)^4 \frac{n_G^4(n_H^2 - n_G^2)^2}{(n_H^2 + 2n_G^2)^2}\delta^6 \qquad (2.82)$$

with λ_a being the wavelength of the light in air, n_G the refractive index of the gelatin ($n_G = 1.54$), the n_H the refractive index of the halide grain ($n_{AgBr} = 2.236$), and δ the grain diameter.

The numerical value of f for the Agfa 10E-materials is $\simeq 0.6$ and for the 8E-materials $\simeq 5$, which indicates that the 8E-material is just on the border of acceptability for the use in holographic recordings when the emulsion thickness is ~6 μm. The 10E-material will have a higher scattering level which makes it difficult to be used for high-quality holographic imaging, especially, in the reflection regime (Denisyuk holography).

2.3.4 Recording Materials
from the Quantum-Theoretic Point of View

a) Theoretic Considerations

The process of photographic exposure was first considered from the quantum-mechanical point of view by *Silberstein* [2.194, 195]. *Goodman* [2.164] and *Goodman* et al. [2.169] have also treated the holographic recording process in this way.

The physical limitations on the signal-to-noise ratio in the reconstructed image are related to the mechanism by which the object exposure variations are transformed into variations of transmittance in the recording material. The sensitivity of a grain from the quantum point of view is described by two numbers:

- The probability ϵ that the photon will be absorbed by a grain, which is referred to as *quantum efficiency* of the grain.
- The minimum number m of absorbed photons required to make a grain developable, which is called the *quantum threshold* of the grain.

Using the *checkerboard model* and on the assumption that the arrivals of photons are Poisson-distributed, the probability that a grain will become opaque during the development is

$$\text{Prob}(t=0) = \sum_{k=m}^{\infty} \frac{(\epsilon\overline{N})^k}{k!}\exp(-\epsilon\overline{N}) \qquad (2.83)$$

and the probability that the grain remains transparent is

$$\text{Prob}(t=1) = \sum_{k=0}^{m-1} \frac{(\epsilon\overline{N})^k}{k!} \exp(-\epsilon\overline{N}) \qquad (2.84)$$

where t is the transmittance of an individual grain, and $\overline{N}$ the average number of photons incident on the grain during exposure. $\overline{N}$ is then

$$\overline{N} = \frac{H_T A_g}{h\nu} \qquad (2.85)$$

with H_T being the total exposure, A_g the area of the grain, h the Planck constant, and ν the light frequency. The expected transmittance $\overline{t}$ of the film at any point is then given by

$$\overline{t} = 0 \cdot \text{Prob}(t=0) + 1 \cdot \text{Prob}(t=1) , \qquad (2.86)$$

$$\overline{t} = \sum_{k=0}^{m-1} \frac{(\epsilon\overline{N})^k}{k!} \exp(-\epsilon\overline{N}) . \qquad (2.87)$$

The standard deviation of the grain transmittance is

$$\sigma_t = [\overline{t} - (1-\overline{t})]^{1/2} . \qquad (2.88)$$

The bias transmittance T_b is determined by the exposure from the reference H_R mainly, when considering a weak object field, i.e.,

$$n_R = \frac{H_R A_g}{h\nu} , \qquad (2.89)$$

$$T_b = \sum_{k=0}^{m-1} \frac{(\epsilon\overline{N})^k}{k!} \exp(-\epsilon\overline{N}) , \qquad (2.90)$$

$$T_b = \sum_{k=0}^{m-1} \frac{(\epsilon H_R A_g/h\nu)^k}{k!} \exp\left(-\epsilon \frac{H_R A_g}{h\nu}\right) . \qquad (2.91)$$

Direct differentiation of T_b with respect to H_R gives the slope of the T_a-H curve

$$\chi = \frac{\epsilon A_g (\epsilon H_R A_g/h\nu)^{m-1}}{h\nu(m-1)!} \exp\left(-\epsilon \frac{H_R A_g}{h\nu}\right) . \qquad (2.92)$$

84

Substitution of (2.91 and 92) into the expression for $E_i/\langle E_n\rangle$ for the checkerboard model gives after some mathematical steps

$$\frac{E_i}{\langle E_n\rangle} = (Q_m\epsilon)N_o , \tag{2.93}$$

$$Q_m = \frac{(\epsilon n_r)^{2m-1} e^{-\epsilon n_r}}{[(m-1)!]^2 \left\{ \sum_{k=0}^{m-1} [(\epsilon n_r)^k/k!] \right\} \left\{ 1 - \sum_{k=0}^{m-1} [(\epsilon n_r)^k/k!] e^{-\epsilon n_r} \right\}} , \tag{2.94}$$

$$N_o = \frac{H_o \alpha_h}{h\nu} . \tag{2.95}$$

N_o is the average number of photons from the object point striking any part of the entire holographic film α_h. Q_m depends on the quantum threshold, but is independent of ϵ. The product (ϵn_r) can be set to any desired level regardless of ϵ (the reference beam exposure). For a perfect recording material ($\epsilon=1$, $m=1$) $Q_m = 1$. However, it can be shown that $Q_m > 2/\pi$, regardless of how large m may be. Therefore, the signal-to-noise ratio (S:N) is approximately

$$S:N = \frac{(2/\pi)\epsilon N_o}{[1 + (4/\pi)\epsilon N_o]^{1/2}} . \tag{2.96}$$

It should be noted that the effective sensitivity in the holographic recording technique is independent of the quantum threshold m. It is proportional only to ϵ. In holography the reference beam does more than just exposing the film to a certain bias level. The fact that it is coherent with the object beam is of great importance. The total exposure of the film is

$$H_{tot} = H_R + H_o + 2(H_R H_o)^{1/2} \cos[\omega x + \theta(x)] \tag{2.97}$$

where ω is the spatial frequency due to the angle between the object and the reference beams, $\theta(x)$ is the remaining phase difference between the object and the reference beams at each point. The information-bearing fringes have an amplitude that is proportional to $\sqrt{(H_R H_o)}$ rather than to H_o alone. In heterodyne detection this is referred to as the *conversion gain* and it occurs only when the reference field is coherent with the signal field.

For radiation detectors the *Detective Quantum Efficiency* (DQE) is often introduced; it is the squared value of $(S/N)_{real}/(S/N)_{ideal}$ [2.196]. For a photographic film it has been shown that DQE is

$$DQE_{film} = \frac{0.434 e_p}{H(A\sigma^2)} \left(\frac{dD}{d\log_{10} H}\right)^2 \tag{2.98}$$

85

where e_p denotes the energy of the photon in ergs, H the total exposure, $dD/ \, dlog_{10}H$ the gradient (γ) of the H&D-curve, and $A\sigma^2$ the Selwyn granularity. For most of the photographic silver–halide materials the DQE-value is only about 1%.

b) Practical Applications of the Quantum Theory

Let us look at a practical example of the theory just presented, in which holograms are exposed on a 60 mm diameter area of a recording material, for example. The area will then be $\pi \cdot 30^2$ mm². If the material used is an emulsion coated on a polyester substrate, the limiting scattering level will be about 10^{-8} mm$\mp$2. From the scattering point of view, it does not matter which emulsion is used as long as the scattering level within the emulsion itself is lower than the scattering from the base material. The modulation transfer function $M(\nu)$ can vary from emulsion to emulsion and it should be as close to one as possible for the actual spatial frequency ν of interest. Equation (2.67) gives (for $E_i/\sigma = 5$, K = 0.95 and $\xi = 0.56$)

$$\left(\frac{H_o}{H_r}\right)_{min} = 50 \cdot \frac{10^{-8}}{(0.95)^2(0.56)^2\pi(30)^2} = 6.25 \cdot 10^{-10} \ .$$

Let us assume that the Agfa 10E75 is used. The lowest detectable light energy from a point object can be calculated here, knowing the sensitivity of this material which is $\simeq 0.5 \ \mu J/cm^2$. This means that the energy density from the object must be only $\simeq 1.5 \cdot 10^{-18}$ J/cm² (corresponding to $7.5 \cdot 10^{-11}$ W/cm², for a 20 ns pulse from a ruby laser) at the recording plane.

Using 10E75 as the recording material it is possible, in principle, to re-cord a point object down to $\simeq 1.5 \cdot 10^{-18}$ J/cm² if the present background-light level caused by other factors than the scattering within the material is well below this value during the exposure of the film. The Agfa 10E75 emulsion has about 10^{11} grains/cm³ [2.191]. The thickness of the emulsion is 7 μm and the exposed area A_{HOLO} is 28.3 cm² assuming a circular holo-gram with a diameter of 6 cm. The emulsion volume used is then $V_{HOLO} = 0.02$ cm³. The total number of grains (size: 90nm) in this volume is $N_{HOLO} = 7.4 \cdot 10^9$ grains. If all these grains were located in one plane and were placed side by side in the emulsion, the area they would cover an area A_{GRAIN} which is then $\simeq 10$ cm². The ratio between this area and the actual hologram area would then be $A_{GRAIN}/A_{HOLO} = 1/3$. Only a maximum area of one-third of the total recording area of the hologram would then really be able to detect the photons emitted from the object. In practice, the area is much smaller since the grains are distributed in depth within the emul-sion. The scattering calculations have shown us that the minimum energy density from the weakest detectable point object must be

$$H_{o,min} = 1.5 \cdot 10^{-18} \ J/cm^2 \ .$$

For the ruby-laser wavelength $\lambda = 694$ nm, the corresponding frequency ν is $4.28 \cdot 10^{14}$ Hz. The energy of such a photon is

$$E_{ph} = h\nu = 6.6 \cdot 10^{-34} \cdot 4.28 \cdot 10^{14} = 2.8 \cdot 10^{-19} \text{ J} .$$

The minimum object exposure for the hologram area A_{HOLO} which has been used then $4.2 \cdot 10^{-17}$ J. The amount of photons emitted from the weakest object point is then $4.4 \cdot 10^{-17}/(2.8 \cdot 10^{-19}) \simeq 150$. The bias exposure needed to secure the hologram with the appropriate density is provided by the reference beam. Since the sensitivity of the Agfa 10E75 is $\sim 0.5 \cdot 10^{-6}$ J/cm^2, the energy that is needed for the hologram is then $14.2 \cdot 10^{-6}$ J. The amount of the photons emitted in the reference beam is then

$$14.2 \cdot 10^{-6}/(2.8 \cdot 10^{-19}) \approx 5 \cdot 10^{13} .$$

The total amount of grains in the recording area of the hologram was $7.4 \cdot 10^9$ ($\sim 10^{10}$ grains). This means ~ 5000 photons/grain on average. The DQE for the recording material is normally about 1%. Therefore, ~ 50 photons/grain can exert certain influence on the grain. In the case of short exposure (for a Q-switched pulse of 20 ns duration), the material may suffer from high intensity reciprocity failure (HIRF) and the required amount of photons may be higher.

It is remarkable that only about 150 photons from the weakest object point can be detected. But, as mentioned earlier, there is no quantum threshold for the object beam and as long as the reference-beam bias exposure makes it possible to operate in the linear region of the T_a-H curve, the object information will actually be recorded.

The recording materials and the scattering properties described in *Goodman*'s investigation [2.169] show slight differences when compared to the above example, but on the whole the two agree to a large extent, which is not surprising as most of the theories presented here are based on *Goodman*'s analysis. However, the most important thing is that his experimental investigation is in good agreement with the theory, which shows that holographic recording technique is really unique in its ability to store information.

Goodman et al. [2.169] have also presented some experimental results which can be directly compared with the calculations above. They found, e.g., that 120 photons were emitted from each fundamental resolution cell of the object surface in their test. Due to depolarization effects in the object field, only about 90 photons per resolution cell were co-polarized with respect to the reference beam. This was the very limit in the experiment. In the same investigation they also compared the holographic recording technique with the conventional photographic recording technique and found, e.g., that about 900 photons per resolution cell were needed to yield a barely recognizable effect on the original negative (Kodak Plus-X film).

The influence of the polarization of the scattered light from the object during recording is important since it affects the obtainable signal-to-noise ratio. The decrease in fringe visibility due to depolarization is causing this influence on the signal-to-noise ratio, as shown by *Ghandeharian* and *Boerner* [2.189].

For other investigations concerning the information capacity of image recording media when using conventional photographic film the reader is referred to an interesting paper by *Jones* [2.197] as well as the previously mentioned investigation by *Leith* concerning photographic film as one element of an optical system [2.88].

Chang and *Bjorkstam* employed a threshold model for the photographic detection process to predict suitable bias exposure values for thin amplitude and phase holograms [2.198,199]. They found that for holograms recorded with a small modulation depth the optimum linearity for a thin amplitude hologram is obtained at an average T_a value = 0.66 and the maximum efficiency at T_a = 0.49, which is in good agreement with the experiments performed by *Thomas* [2.200]. The corresponding biasing points for optimum linearity and maximum efficiency of a thin phase hologram are found at pre-bleached average densities of 0.99 and 1.59, respectively, assuming a maximum possible density of 2.8 in the threshold model. *Chang* and *Bjorkstam* later investigated the case when the modulation depth was arbitrary, using a more sophisticated but still realistic model of the photographic emulsion [2.201]. The results that were obtained are somewhat in conflict with *Kogelnik*'s coupled-wave theory, mainly because of the fact that a more realistic model of the emulsion was used, taking, e.g. nonlinearities into consideration.

2.3.5 Ordered Grain Structures

Phillips [2.202] has pointed out that the irregular grain structure in holographic emulsions is the main source of noise. His considerations are based on the original ideas of E. Land at Polaroid and experiments performed by *Cowan* [2.203], *Cowan* and *Slafer* [2.204], and *Cowan* [2.205]. *Phillips* was seeking ways of arranging the grains in the emulsion in a regular array as a potential means of reducing noise. *Cowan* has applied a three-beam recording process to create hexagonal isophotic patterns. These high contrast patterns are recorded onto a layer of photoresist. Using nonlinear processing technique the high spots of the isophotic pattern are enhanced. A symmetric relief image is the result, with very little random noise.

2.3.6 Nonlinear Recording

The holographic recording theory assumes sometimes that the recording process is linear, which is not the case when using, e.g., silver-halide materials. Here, due to the existence of a developable threshold in silver-halide emulsion grains and a saturable optical density for the emulsion, *the photographic recording process is nonlinear*. Various effects brought about by the nonlinear recording of holograms have been studied in several papers [2.206-228]. *Kozma* [2.206] was the first to publish the results of an investigation on nonlinear recording of holograms. The effects of a nonlinearity applied to simple objects were discussed by *Friesem* and *Zelenka* [2.207], and for diffuse objects by *Knight* [2.209], and *Goodman* and *Knight* [2.210]. *Bryngdahl* and *Lohmann* [2.212] studied the influence of nonlinearities assuming that amplitude transmittance can be represented by a polynomial.

Not always is the nonlinear effect decreasing the performance of the recording process, as shown by *Bendall* et al. [2.217]. They improved the diffraction efficiency of thick amplitude gratings to exceed the efficiency predicted from the coupled-wave theory (3.7%) [2.116]. They also modified the coupled-wave theory to include nonlinearities. They treated the grating produced by nonlinear recording as a superposition of sinusoidal gratings. The modified theory predicts a diffraction efficiency of 6.4% in the nonlinear case. Experimentally, they obtained an efficiency of 5.8% using the Kodak 649-F material.

Goldmann [2.220] presented a theory not only for the nonlinearity of amplitude holograms but also of phase holograms. Converting amplitude holograms to phase holograms by bleaching can reduce nonlinear effects introduced during development. *Phillips* and *Heyworth* [2.225] explained that the effect of high contrast in development can be counteracted by a bleach process that modulates the hologram by diffusion transfer. In fixation-free rehalogenating bleaching the diffusion transfer mechanism is an important feature that will be described in Chap.5.

As may be seen from Fig.2.18, providing that a sinusoidal fringe signal has been recorded in the linear part of the T_a-H curve, the intensity distribution of the optical field will be stored as a distortion-free, sinusoidal amplitude-transmission distribution in the emulsion. But if the sinusoidal input is allowed to record in the nonlinear regions of the curve, then the output signal will be distorted. In the terms of the Fourier analysis this implies the introduction of further harmonics into amplitude transmission, which means that *nonimage waves components are created at the reconstruction*. Nonlinearities can be divided into two classes; *intrinsic* and *material* nonlinearities.

In a thin phase diffraction grating, produced by two intersecting laser beams of equal intensity, a first-order image or spectrum will be generated when the grating is illuminated with laser or white light. In addition to the first order, higher orders will appear simultaneously due to wavefronts leaving the grating with two or more times the tilt of the first order relative to the zero order. Often the higher orders are well separated from the first order and will thus not affect the first-order image. The diffraction grating is a simple example of a HOE with intrinsic nonlinearity affecting the holographic recording process. Material related nonlinearity to the holographic process is also important, in particular for volume holograms.

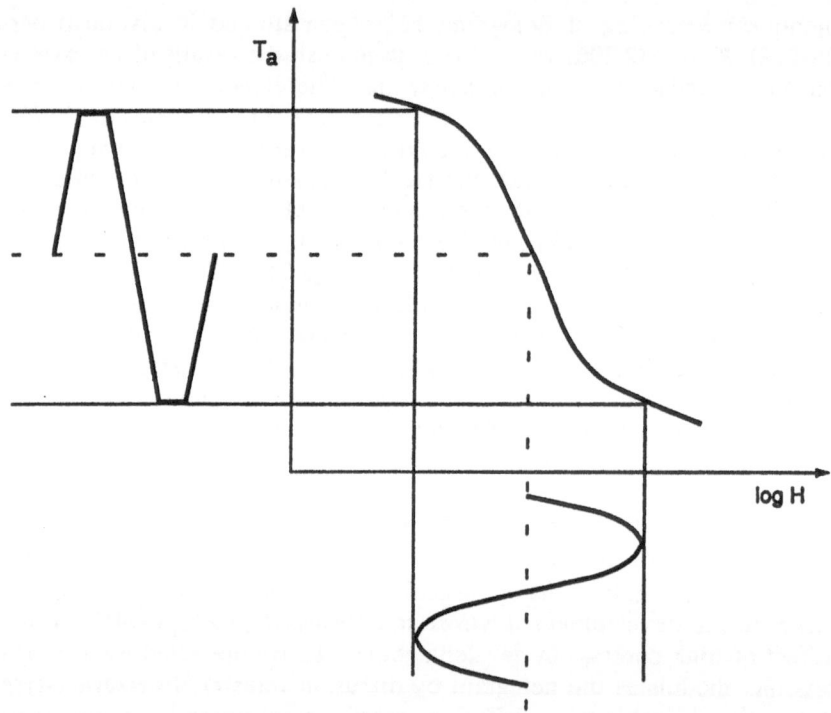

Fig.2.18. Nonlinear recording. If a sinusoidal input signal is allowed to record in the nonlinear regins of the T_a-logH curve, then the output signal will be distorted, which implies the introduction of further harmonics into amplitude transmission, which means that nonimage waves are created at the reconstruction

Intrinsic nonlinearities

For a linear recording the amplitude transmittance T_a of an amplitude hologram is

$$T_a = \langle t \rangle + \beta T[rr^* + oo^* + r^*o + ro^*] \tag{2.99}$$

where $\langle t \rangle$ is the average transmittance for the average exposure $\langle H \rangle$, $\beta = dt/dH$ at $\langle H \rangle$, T is the exposure time, r the reference-wave amplitude, and o the object-wave amplitude.

In practice, if the fluctuation terms on the right-hand side of the transmittance equation (2.99) are comparable with the bias term, the recording will no longer be linear when using photographic silver-halide materials for the recording.

The complex amplitude transmittance $T_a(x, y)$ for a phase hologram is

$$T_a(x, y) = \exp[-i\phi^t(x, y)] . \tag{2.100}$$

To show the effect of nonlinearities we can make a Taylor series expansion of $T_a(x,y)$

$$T_a(x,y) = 1 - i\phi - (1/2)i\phi^2 + (1/6)i\phi^3 \dots . \tag{2.101}$$

The expansion is taken around the phase point zero, for simplicity. The series is infinite but can be terminated after a number of terms if the exponent is small compared to 2π. If the phase modulation is increased during holographic recording, the higher-order terms of $T_a(x,y)$ cannot be neglected. At reconstruction the reference wave is multiplied with the hologram transmittance, $T_a(x,y)$ and the reconstructed wave will contain all the corresponding terms. All terms containing absolute squares of o and r constitute the zero-order wave. Terms containing or^* and ro^* constitute the first-order images, terms containing $(or^*)^2$ and $(ro^*)^2$ constitute the second-order images, etc. The first-order terms contain not only the primary and conjugate waves but also spurious terms [e.g., $(oo^*)^2$] proportional to $|o|^2$ propagating in the same direction as the image wave, which will affect the image quality. The effect is primarily the generation of a halo surrounding the directly transmitted first-order beam. Such a halo is also generated at reconstruction due to the oo^*-term in the linear recording case. However, the spatial frequency spectrum of the quadratic term in the nonlinear case has twice the width of the spectrum of oo^*, which means doubling the angular width of the halo. There are also first-order intermodulation terms [e.g., $2o^2o^*$] in the reconstruction field that will give rise to false images and contribute to the noise halo around the true image. The degradation of the image is caused by both the nonlinearity of the recording process as well as the object field itself. The image of a single point source will not be disturbed by nonlinear effects, as only higher-order images will be created. Diffuse objects will suffer from the degradation, which is why the ratio between the reference and object beams (K-value) has to be increased when recording these objects holographically.

Due to the intermodulation effects it is difficult to produce high-quality images of diffuse objects using thick phase holograms. However, as described by *Güther* and *Kusch* [2.219] due to the angular selectivity of such holograms the intermodulation noise can be significantly suppressed, which means that the Bragg condition is not satisfied as regards the (rather low) intermodulation frequencies.

De Belder [2.213] has shown that, because of emulsion shrinkage, second-order images can be generated even in the absence of second harmonics.

Kasprzak et al. [2.226] published recently an analytical approximation of the recording-material characteristic curve based on the Fermi formula for energy distribution. The formula is easy to use for all kinds of holographic materials and it also provides the possibility to calculate the influence of nonlinearity on Fourier transform hologram recording parameters.

The approximation formula for the D-logH is

$$D = \frac{D_\infty}{1 + \exp(a - b\log H)}$$ (2.102)

with

$a = 2(1 + 2\gamma \log H_i / D_\infty)$,

$b = 4\gamma / D_\infty$,

D_∞: maximum optical density.

A somewhat better representation of the D-logH curve in the toe-region can be obtained by adding a correction term to the Fermi formula which is presented in the paper by *Kasprzak* et al. [2.226]. From the D-logH formula the T-H characteristics can be easily obtained. These researchers claimed that the approximation formula introduced by them is more convenient to use than polynomial approximation methods which are often employed for studying nonlinearities.

Sultanova and *Staneva* [2.228] have compared the linearity of Agfa and Bulgarian holographic emulsions, and presented optimal biasing points (T_a values) for these materials.

Material	Optimum biasing point
Agfa 10E56	0.72
Agfa 8E75	0.66
Agfa 10E75	0.69
HP490	0.75
HP650	0.66 .

The main observation here seems to be the fact that green-sensitive materials have a slightly higher T_a value than red-sensitive recording materials.

3. Commercial Silver-Halide Materials

In this chapter the existing commercial silver-halide materials for holography are treated. Plate and film materials manufactured by well-known Western photographic companies are described. After the manuscript was completed, Ilford ceased the production of holographic materials. Instead of removing the Ilford section, it was kept for comparison and historical reasons. Included also are many Russian and some East-European materials.

3.1 Manufacturing Companies

The market offers various types of holographic silver-halide recording materials. The holographic materials produced by Agfa-Gevaert, Ilford and Kodak are easily obtainable in most countries of the world.

In addition to the above-mentioned, holographic materials are also produced in Russia and in some Eastern European countries. These, however, are still difficult to purchase outside the former iron curtain. The main reason to present them in this chapter is for comparative purposes, as they are often quite different from the Western materials. Holographic materials similar to the Russian products are also manufactured in China, e.g., the Tianjing-I and HP-633P plates.

Kodak materials were used, as mentioned earlier, for the very first laser-produced holograms in the USA. Nowadays the company has a selection of holographic materials suitable for different recording regimes.

Nevertheless, since Agfa-Gevaert introduced their materials in 1968, these have become the leading products for recording purposes in holography all over the world, including the USA. This is especially true for reflection holograms. On the other hand, for transmission holograms, some holographers prefer Kodak materials. In the mid 80's, Ilford introduced a new line of materials for holography. In the early days of holography Ilford manufactured a red-sensitive material (a plate named He-Ne 1) which, however, had never become very popular and was therefore discontinued. Today, Ilford holographic products are of high quality, in particular their green-sensitive emulsion. Ilford's main interest is to produce fine-grained recording materials coated on film intended for mass production of holograms.

At the present moment, Japan has no holographic silver-halide materials to offer on the international market - the reason being probably that the market for holography is not big enough for the Japanese to be seriously interested in competing with the "old" photographic companies. (Locally, Sakura manufactures a holographic HRP plate, e.g.) A rather limited market may also explain why Kodak has not really made any serious effort to produce a material that could eventually become the leading, number-one·material for holography.

On the following pages materials from different manufacturers are presented. The information indicated here is based mainly on the data provided by the respective companies. For comparing different holographic materials and their parameters, no uniform set of rules has been worked out yet, (like in the ASA and DIN systems in photography) which is why one must be very careful when making comparisons between the parameters of various materials, such as sensitivity and the resolving power, for example. The sensitivity indicated in the printed tables is the sensitivity corresponding to the exposure of the material resulting in an optical density of 0.6 (amplitude transmission holograms). A special, controlled processing method is often also used which, however, differs from manufacturer to manufacturer. The sensitivity of an emulsion depends on many factors, e.g., the laser wavelength, the exposure time (reciprocity failure), the development, (developer type, processing time, temperature, agitation, etc.) and the storage conditions. Holographic sensitivity can vary to a certain degree from batch to batch, which is rare in conventional photographic materials. It is therefore recommended to make exposure and processing tests each time an important holographic recording is to be performed.

3.2 The Agfa-Gevaert Materials

The Agfa-Gevaert products are manufactured in Belgium, where holographic materials are a part of the NDT-product line (NonDestructive Testing) of the company [3.1-5]. The address is

Agfa Gevaert N.V.
Septestraat 27
B-2510 Mortsel, Antwerp, Belgium.

The first Agfa-Gevaert's holographic recording materials were found among materials for scientific applications, called *Scientia*. Following the growth of holography, holographic materials became a separate product group and received the name *Holotest*. The name carries a hint as to the main application of Agfa-Gevaert's holographic materials, which is in the field of holographic nondestructive testing. The current materials have been improved in comparison with the early products, as regards both sensitivity and the resolving power as well as in regard to the signal-to-noise ratio. The Agfa materials for holography are the ones to have been most frequently used so far, they have been described in many papers which are referred to in other chapters of the book.

Agfa's coding system used for the Holotest materials is somewhat complicated and should therefore be explained here. The product can be labelled something like this:

Holotest 8E75-T3-HD

8   indicates the relative sensitivity. It can be either 8 or 10. Eight stands for a slower emulsion and 10 for a faster one.

E   indicates contrast. A scale from A to E is used, where E stands for the material of highest contrast ($\gamma \simeq 5$).

75   refers to the spectral sensitivity. The number 75 indicates that the material is sensitive up to 750 nm. The green sensitivity material is marked 56, which means a sensitivity of up to 560 nm.

T   indicates the base material, namely T for triacetate and P for polyester.

3   is the thickness of the base. 3: 190 μm for triacetate. A thinner polyester base, marked 1, is used for roll films. 1: 100 μm.

HD   High Definition. A material with the smallest grain size (about 35 nm).

If an Anti-Halation backing has been provided, only for films never for plates, this is indicated as AH, while NAH means No Anti-Halation. If there is *no indication*, there is no anti-halation backing. For roll films, the indication P means perforated and NP, not perforated.

3.2.1 Emulsion Characteristics

Table 3.1 gives a review of Agfa materials. Currently, there are three different emulsions to choose from: Two red-sensitive materials (10E75 and 8E75 HD) and one green-sensitive (8E56 HD). Only the 10E75 film offers

Table 3.1. AGFA products

Material	Emulsion thickness [μm] plate/film	Spectral sensitivity [nm]	Sensitivity [μJ/cm^2] at 514 633 694			Resolving power [ℓp/mm]	Grain size [nm]
8E75 HD	6/5	<750	–	10	20	<5000	35/44[a]
10E75	6/5	<750	–	1	2	<2800	90
8E56 HD	6/5	<560	25	–	–	<5000	35/44[a]

[a] The mean grain size before coating the emulsion is about 35 nm. Investigatins on the actual grain size in a coated emulsion indicate the mean grain size for the HD material of about 44 nm.

a choice of AH backing or NAH backing. Here, anti-halation is frequently required as the 10E75-film is used for the recording of transmission holograms. The 35 and 70 mm film rolls are the ones that have the anti-halation backing as a rule. The 8E materials are normally not anti-halation protected, since they are mostly employed for the recording of reflection holograms.

3.2.2 Base Substrate and Formats

The customer has a choice of glass or film for the material. The emulsion is usually coated on normal photographic glass.

Glass plates are produced in the following formats:[1]

2.5" × 2.5"	thickness	1.5 mm
4" × 5"	"	1.5 mm
8" × 10"	"	3.3 mm
30 cm × 40 cm	"	3.3 mm
32 cm × 43 cm	"	3.3 mm .

The film material used is either triacetate (T) or polyester (P). The refractive index for triacetate is 1.485 and about 1.6 for polyester. The material can be ordered as sheets or rolls of film.

Sheets of film are produced in the following formats:

4" × 5"	thickness	190 μm
8" × 10"	"	190 μm
30 cm × 40 cm	"	190 μm
50 cm × 60 cm	"	190 μm .

Rolls of film are produced in the following formats:

35 mm × 100'	thickness	100 μm
70 mm × 100'	"	100 μm
9½" × 200'	"	190 μm
114 cm × 10 m	"	190 μm .

In Fig.3.1, the characteristic curves for Holotest materials are shown. Red exposures were made at 627 nm and green ones at 514 nm. The processing was carried out in G 282 (diluted 1 + 2) for 4 minutes at 20° C. After an intermediate rinse in water (1 minute at 20° C) fixing was performed in G 321 for 4 minutes, followed by washing for 15 minutes. Figure 3.2 exhibits T_a versus exposure curves for Agfa materials using the same processing technique, and in Fig.3.3 absolute color sensitivity curves are presented.

---

[1] Nonmetric units are used, for example, when materials are actually produced in exact inch formates. Note that 1″ = 2.54 cm, and 1′ = 30.5 cm.

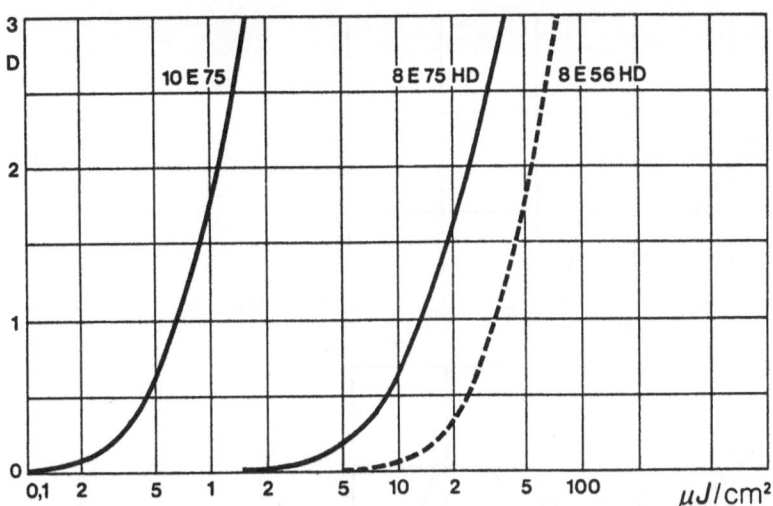

Fig.3.1. The characteristic curves for Agfa Holotest materials. Red exposures were made at 627 nm and green at 514 nm. The density of the developed emulsion layers was measured by parallel light. (Reprinted courtesy Agfa Gevaert N.V.)

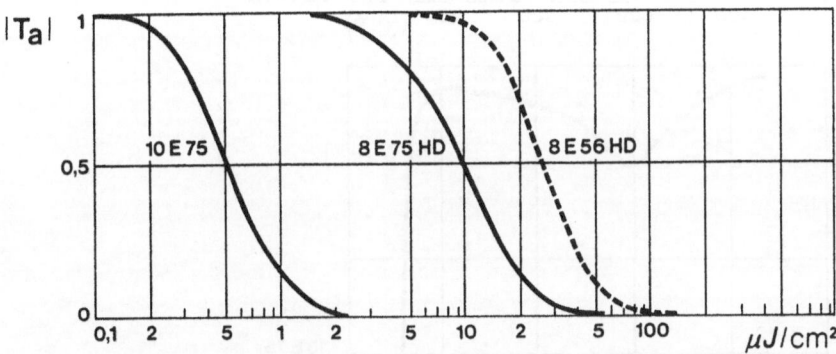

Fig.3.2. Amplitude transmission T_a versus log exposure curves for Agfa Holotest materials. Red exposures were made at 627 nm and green at 514 nm. (Reprinted courtesy Agfa Gevaert N.V.)

The emulsion side of the sheet film is marked by a notch along one side, as indicated in Fig.3.4. This way of marking the film is not only used by Agfa, but also other manufacturers apply the same convention.

In Table 3.1 the most characteristic parameters of the Holotest emulsions are listed. At the end of the chapter, Table 3.7 compares different commercial holographic recording materials.

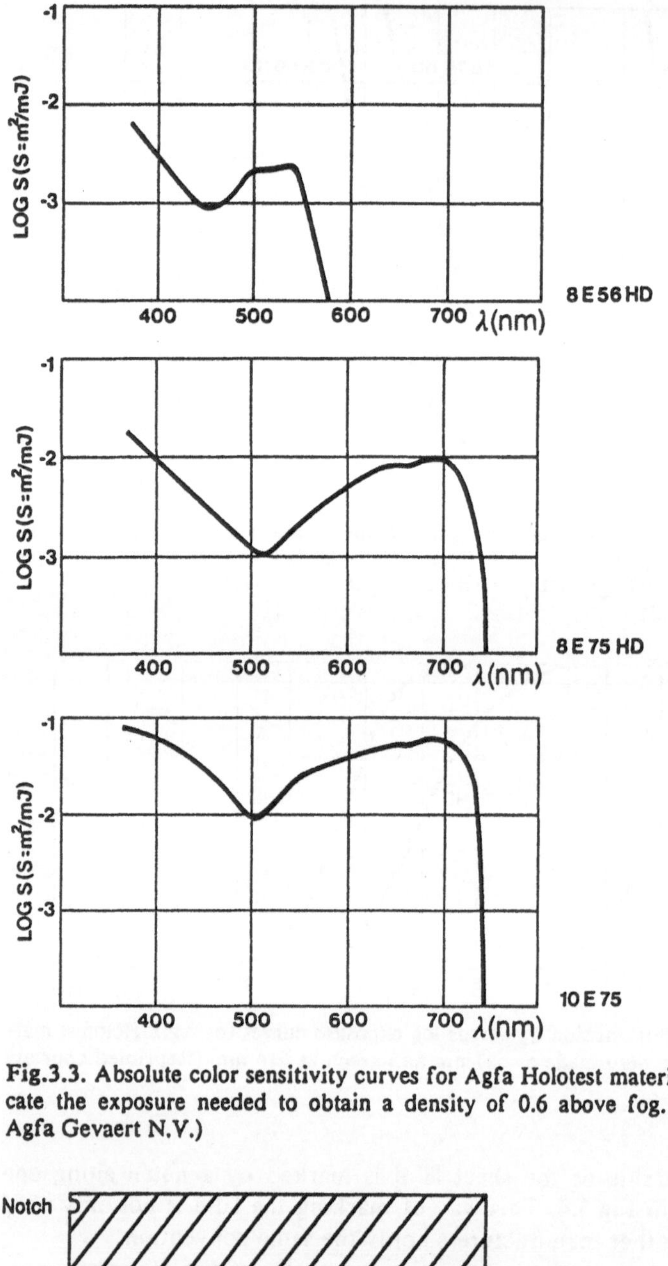

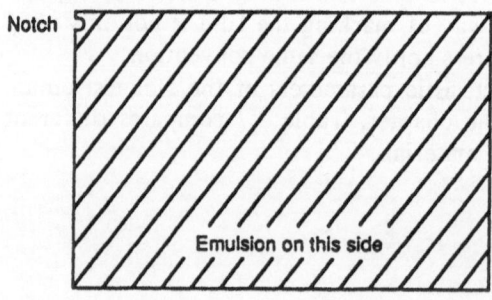

Fig.3.3. Absolute color sensitivity curves for Agfa Holotest materials. The curves indicate the exposure needed to obtain a density of 0.6 above fog. (Reprinted courtesy Agfa Gevaert N.V.)

Fig.3.4. Emulsion identification marking. The emulsion side of sheet film is marked by a notch along one side as indicated in the figure

3.2.3 Safelight Recommendations

For the red-sensitive material 8E75 HD a dark green filter such as Agfa G V 505 or Kodak Safelight No.3 (panchromatic) can be used. It must be noted that the 10E75 material is too sensitive even to this illumination, for which reason it should be treated in complete darkness. More information on how to make a suitable filter for the red-sensitive materials using a Kodatrace and the Ilford gelatin spectrum blue-green (603) filter can be found in a paper by *Rogers* [3.6]. Instead of the Ilford filter a Kodak Wratten 75 can be employed. Safelight filters from EncapSulite International [3.7] have been recommended by *Feroe* [3.8] to be utilized with fluorescent light fixtures.

The green-sensitive material 8E56 HD can be used with dark red safelight illumination without any problems. An Agfa R 4 filter or Kodak Safelight No.2 (orthochromatic) are recommended. Safelight filters are normally used with a 15 W bulb. For low-sensitive holographic materials $25 \div 40$ W bulbs can be used, depending on the distance between the lamp and the material. The Deep Ortho Red, No. R-20 filter from EncapSulite was mentioned by *Feroe* [3.8] to be suitable for the green material.

3.2.4 Processing

Agfa has developed special chemistry for the processing of their materials. In the following chapters processing regimes will be described in great detail, including the chemistry recommended by Agfa. The recipes from Agfa, in general, give good results, but in many cases various other procedures and processing regimes produce better results on Agfa materials than the processing methods recommended by Agfa. More details regarding this are given in Chaps.4 and 5, where Agfa developers and bleach solutions are discussed. Agfa recipes are also found in Chap.10.

3.3 Materials from Ilford

Ilford is a company based in England, with a new line of holographic silver-halide products, as well as special processing chemicals for manual or machine operation [3.9-10]. The address of the company is

Ilford Limited
Mobberley
Knutsford, Cheshire WA16 7HA, England.

The development of the Ilford holographic material is based on their UFG (Ultra Fine Grained) emulsions. Two different emulsions are available - a red- and a blue-green-sensitive product. The red material is the Ilford holographic glass plates FT340T/SP696T and the film Hotec R, while the blue-green version is named SP 695T (plates)/SP 672T (film). SP stands for Special Products. When first introduced in the early 80's, the red material (SP673) was produced mainly for recording with a Q-switched ruby laser, as Ilford was then part of a joint venture with Applied Holographics, PLC., developing a new mass-replication system for holograms. It was soon noticed that the new material was also in demand among the people working with CW lasers, e.g., HeNe lasers. The company has therefore modified the emulsion's sensitivity so that it could be useful for the HeNe laser recordings as well. The material has a high resolving power and the same sensitivity as Agfa's corresponding red-sensitive material but the Ilford green version is faster than the green Agfa 8E56 HD material. The reader may wish to refer to some of the papers on Ilford holographic materials [3.11-16].

The red-sensitive material has a built-in pre-swell treatment (BIPS factor), which means that the material will shrink about 8.8% after processing. This unique feature is of special interest to people working with reflection holograms. Holograms recorded at 633 nm will then replay at 577 nm, which gives an orange-yellow color, suitable for a variety of holographic images. A reflection master recorded with a ruby laser at 694 nm will reconstruct at 633 nm, which can be useful for film copying, using, e.g., a scanning HeNe-laser beam. The predetermined shrinkage will only occur if non-tanning processing chemicals are used such as products recommended by Ilford. Tanning processing solutions containing, e.g., pyrogallol or PBQ will partly over-ride the BIPS factor, producing holograms which will reconstruct at the wavelength close to the one used for the recording. If non-tanning processing is applied and no shrinkage wanted, Ilford recommends washing the unexposed material in water prior to exposure, which will remove the swelling agent in the emulsion. In addition to eliminating the BIPS factor, this treatment will also increase the sensitivity of the material by a factor of two.

3.3.1 Emulsion Characteristics

The characteristics of the two Ilford emulsions are presented in Table 3.2. Since the Ilford material is intended mainly for the use in reflection holography, it is not treated with anti-halation backing.

Figure 3.5 shows the characteristic curve and Fig. 3.6 the spectral sensitivity.

3.3.2 Base Substrate and Formats

Ilford materials can be ordered on both glass and film. The glass plates are of the following formats:

4" × 5"	thickness	1.6 mm
8" × 10"	"	3.0 mm

Table 3.2. Ilford products

Material	Emulsion thickness [μm]	Spectral sensitivity [nm]	Sensitivity [μJ/cm^2] at 442 514 633 694	Resolving power [ℓp/mm]	Grain size [nm]
FT340T/SP696T	6+0[a]	<700	– – 200	<7000	30[b]
HOTEC R (film)	5+2[a]	<700	– – 20 50	<7000	30[b]
SP695T (plate)	6+0[a]	<560	200 100 – –	<7000	30[b]
SP672T (film)	6+1[a]	<560	200 100 – –	<7000	30[b]

[a] The first figure indicates the thickness of the active emulsion and the second the gelatin supercoat.
[b] Mean grain size in a coated emulsion.

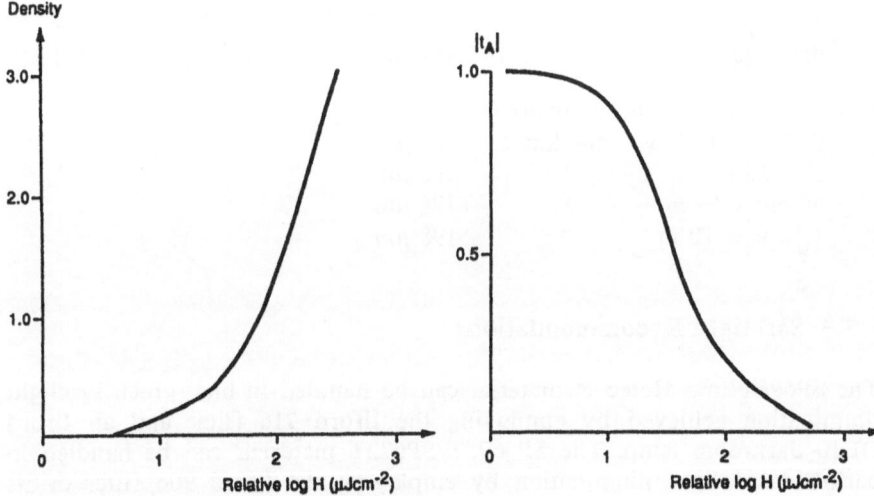

Fig.3.5. The characteristic curve and the amplitude transmission curve for Ilford's red-sensitive emulsion. (Reprinted courtesy Ilford Ltd.)

     30 cm × 40 cm   thickness   3.0 mm
     50 cm × 60 cm   "   5.0 mm .

Ilford coated the holographic emulsion on either polyester or triacetate base. The polyester base is a 63 μm thick (2.5-mil) film and the triacetate base has a thickness of 198 μm (8-mil). The existing sheets of film have the following formats:

     4" × 5"   thickness   198 μm
     8" × 10"   "   198 μm

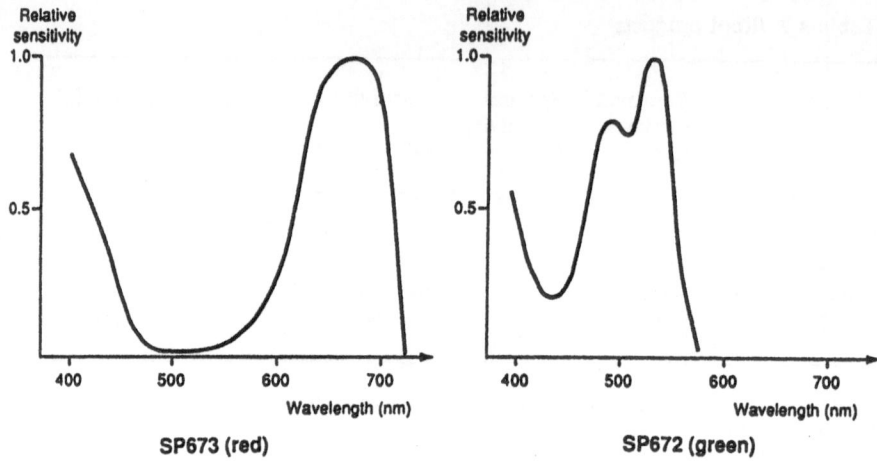

Fig.3.6. Spectral sensitivity curves for Ilford's red- and green-sensitive materials. (Reprinted courtesy Ilford Ltd.)

| 12" × 16" | thickness | 198 μm |
| 20" × 24" | " | 198 μm . |

Rolls of film have the following formats:

9.5" × 200′	thickness	63 μm
5" × 10 m	"	198 μm
50 cm × 10 m	"	198 μm
1.10 m × 10 m	"	198 μm .

3.3.3 Safelight Recommendations

The SP696T and Hotec R material can be handled in blue/green safelight illumination achieved by employing the Ilford 916 filter and an Ilford DL10 darkroom lamp. The SP 695T/SP672T materials can be handled in dark-red safelight illumination by employing the Ilford 906 filter or its equivalent.

3.3.4 Processing

Ilford produces holographic developers and has also developed bleaching chemistry for their products. The chemical products from Ilford intended for the red materials are all marketed under the name *Hotec*. The developer and the bleach can be used for the producion of both transmission and re-flection holograms. The processing chemicals can be employed for both machine processing (at a temperature of 30° C) or in a tray at room temper-ature (20° C). The Hotec developer is used diluted 1+4 and the developing time is 3 minutes with constant agitation. It is based on sodium-L-ascorbate and pentasodium DTPA (DiethyleneTriaminePentaAcetic acid). The capacity of one-liter working strength developer is about 2500 cm^2 emulsion area.

The Hotec bleach is of the rehalogenating type. The bleach is of a completely new type containing: dipotassiumferric DTPA, hydrobromic acid and sulfuric acid. It is supplied as a working strength solution. The normal bleaching time is between 1 and 2 minutes. The capacity of the bleach is about 12500 cm^2 emulsion area, which corresponds to about 100 4"×5" plates.

Ilford claims that Hotec products as well as other processing chemicals manufactured by the company are safe to use, the only exception being the SP715C-light stabilizer. As already mentioned, Hotec products can be used both in machines and trays. However, the SP holographic chemicals are recommended mainly for machine processing of reflection holograms exposed with helium-neon, krypton, or ruby lasers. The following processing baths for machine processing can be obtained from Ilford:

SP678C developer/replenisher part A
SP716C developer/replenisher part B
SP617C developer starter
SP718C stop bath
SP679C bleach
SP715C rinse and light stabilizer

The SP products are all optimized to work at 30° C processing temperature but they can also be used at a lower temperature in a tray. The SP679C is a reversal bleach based on dichromate. The rehalogenating bleach, Hotec bleach, can also be employed in processing machines if less shrinkage is desirable. The SP715C rinse and light stabilizer is not freely available. It is the only chemical product from Ilford that is really toxic and cannot be disposed of down the drain. A customer has to sign an agreement concerning waste solutions to be collected by licenced contractors to be able to order this product.

In addition to the processing solutions manufactured by the company, Ilford has also published several developer and bleach recipes for their materials which are given in the references as well as in the recipe chapter of this book. More details on Ilford processing can be also found in Chaps.4 and 5.

3.4 Materials from Kodak

Kodak's holographic materials are produced in the USA. The plates and film are part of the Scientific Imaging Products and Graphic Imaging Systems Division [3.17-20]. The address is

Eastman Kodak Company
343 State Street
Rochester, NY 14650, USA.

As mentioned earlier, Kodak was producing a spectroscopic plate (649-F), long before the first holograms were recorded, which came in very handy when the very first laser-produced holograms were to be recorded in the USA. The plate is still being manufactured, in spite of the fact that Kodak has a variety of products made especially for holographic purposes nowadays. Kodak materials are described in many papers refered to in other chapters of this book.

Holographic materials from Kodak have grain sizes from 45 nm to 85 nm and a resolving power from 1500 line pairs/mm to >2500 line pairs/ mm. The spectral sensitivity of the material covers the whole visible part of the spectrum.

3.4.1 Emulsion Characteristics

Kodak Spectroscopic Plate and Film, Type 649-F

The 649-F products are characterized by high contrast, high resolving power and panchromatic sensitivity from UV to almost 700 nm. The letter F indicates that the plate is sensitive within the whole visible part of the spectrum. The thickness of the emulsion is 17 μm when coated on glass, and 6 μm when coated on film base (ESTAR) which is a polyester base of 100 μm (4-mil) thickness. The sensitivity is rather low compared to other holographic emulsions. The material can be ordered with or without the AH backing.

Kodak Holographic Plate, Type 120-01 or -02
and Holographic Film, Type SO-173

The Kodak holographic plate/film is covered with an emulsion coated on glass or on the ESTAR base, and is 6 μm thick. The spectral sensitivity is inherent in the blue range and reaches its optimum in the 580 to 750 nm range. The indicators -01 or -02 show the way Kodak marks the presence or the absence of the anti-halation layer in the holographic material. Materials with no anti-halation layer are marked -02 and the ones supplied with it -01. Note that this convention does not apply to spectroscopic plates. The 120 plates are Kodak's best choice for reflection work.

Kodak Special Plate, Type 125-02

This is a high-speed, 6 μm thick emulsion with the spectral sensitivity in the blue and green parts of the spectrum. The product is intended to be used mainly in hologram interferometry or educational transmission work.

Kodak Special Plate, Type 131-01 or -02
and Kodak High-Speed Holographic Film, SO-253

Compared to most of the other holographic materials the above two are very fast materials, especially when exposed with the helium-neon or krypton lasers in the red part of the spectrum. They also have strong sensitivity in the blue and green parts of the spectrum. The 131-01 & 02 plates have a resolving power of up to 1500 line pairs/mm and an emulsion thickness of 6 μm, whereas the SO-253 film is 9 μm thick. The above materials are intended for the use in hologram interferometry mainly, but they are also suitable for other applications when low-power, red CW lasers are employed and for educational purposes.

Kodak High-Resolution Film, SO-343 and Kodak Spectrocsopic Film, Type 649-GH

These films have orthochromatic spectral sensitivity of up to 560 nm and the emulsion thickness of 6 μm. SO-343 is coated on a 175 μm (7-mil) ESTAR base and has a dyed gelatin backing, while Type 649-GH is coated on the regular 100 μm (4-mil) ESTAR base, and has an antihalation backing.

Kodak High-Resolution Plate, Type 1A

This plate has the same emulsion as the SO-343 and 649-GH materials, but is coated on a 1.5 mm (0.060-inch) thick, ultra-flat quality glass in order to achieve high dimensional stability and a high degree of flatness. The plate is used mainly for graphic application.

Kodak High-Resolution Plate, Type TE

This plate is the same as Type 1A, but its emulsion thickness is only 3 μm.

3.4.2 Base Substrate and Formats

Like the other companies, Kodak offers glass or film substrates for their products. Holographic glass plates are made of the best quality soda-lime glass and can be up to 750 mm (30 inches) wide. Only a few popular sizes are factory stored, the other sizes being made on a special-order basis with 30 to 90 days delivery time. The thickness varies from 0.8 mm (0.030") to 6 mm (0.250") The glass plates are grouped in the following categories in regard to their flatness: Selected-flat, Ultra-flat, Precision-flat, and Micro-flat.

Formats of standard holographic plates:

2" × 2"	thickness	1 mm
6.5 cm × 9 cm	"	1 mm
4" × 5"	"	1 mm or 6 mm
9 cm × 12 cm	"	1 mm
8" × 10"	"	2 mm .

The standard film base used by Kodak is a polyester base, named ESTAR, of 100 μm (4-mil) thickness. Products with ESTAR base are marked "E". If, instead of a polyester base, an acetate base is used, the product is marked "A". A thicker base, 175 μm (7-mil) is another alternative for some films. Films may be ordered on different supports at a premium price. A product can be marked E4AH, e.g., which means a 4-mil ESTAR base and a 0.1 neutral density in the base to suppress lightpiping and halation. With the exception of "AH", the letters following the thickness number describe the backing applied to the film base:

B Dyed gelatin pelloid
CB Clear gelatin pelloid
C Clear; i.e. unbacked .

Standard sheet film size:

4" × 5"	thickness	100 μm (4 mil) .

Standard roll film sizes:

35 mm × 125′	thickness	100 μm (4 mil)
70 mm × 125′	"	100 μm (4 mil) .

Figure 3.7 shows Ta-DLogE curves for 649-F, 120-02 and SO-173, and Fig.3.8 depicts spectral-sensitivity curves. The materials are developed in Kodak D-19 developer (8 minutes at 20° C) under continuous agitation.

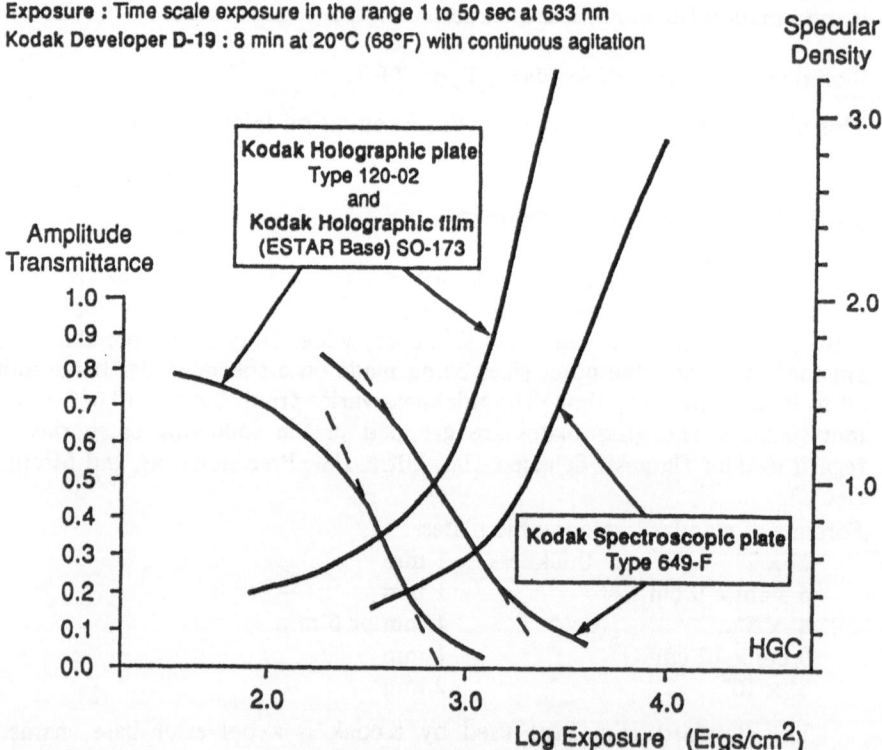

Fig.3.7. Characteristic curves and amplitude transmission for Kodak materials: 649-F, 120-02 and SO-173. (Reprinted courtesy Eastman Kodak Company)

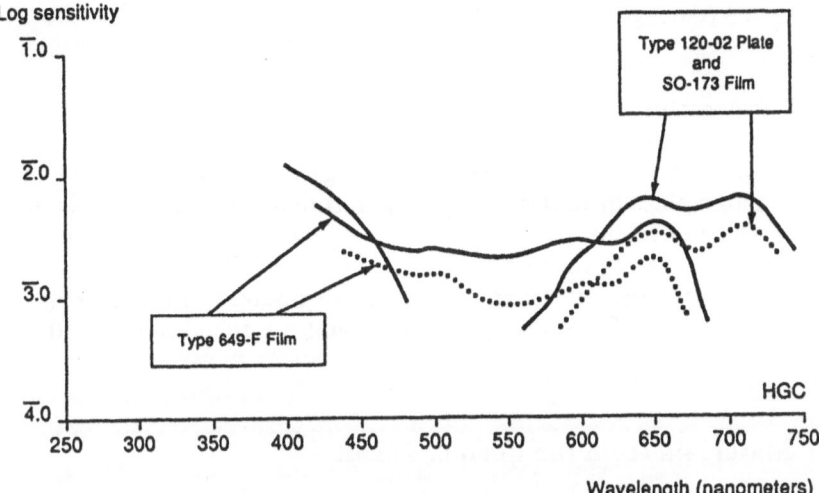

Effective Exposure : 24 sec
Kodak Developer D-19 : 8 min at 20°C (68°F) with continuous agitation

━━━ D = 0.3 above Gross Fog

⋯⋯ D = 1.0 above Gross Fog

Kodak Holographic plate Type 120-02 and
Kodak holographic film (ESTAR Base) SO-173
Kodak Stereoscopic film Type 649-F

Log sensitivity

Type 120-02 Plate
and
SO-173 Film

Type 649-F Film

HGC

Wavelength (nanometers)

Fig.3.8. Spectral-sensitivity curves for Kodak materials: 649-F, 120-02 and SO-173. Sensitivity is defined as the reciprocal of exposure, in ergs/cm^2, required to produce a density D above fog. [Reprinted courtesy Eastman Kodak Company]

Table 3.3. Kodak products

Material	Emulsion thickness [μm]	Spectral sensitivity [nm]	Sensitivity [μJ/cm^2] at 442	514	633	694	Resolving power [ℓp/mm]	Grain size [nm]
649-F (plate)	17	<700	80	80	90	500	>2000	58
649-F (film)	6	<700	80	80	90	500	>2000	58
649-GH (film)	6	<560	100	100	-	-	>2000	58
120 (plate)	6	<700	50	40	40	40	>2500	50
125 (plate)	6	<560	8	5	-	-	>1250	70
131 (plate)	6	<700	5	3	0.5	100	>1250	70
SO-173 (film)	6	<700	50	-	40	40	>2500	50
SO-253 (film)	9	<700	5	3	0.5	100	>1250	70
SO-343 (film)	6	<560	100	100	-	-	>2000	58
HR-1A (plate)	6	<560	100	100	-	-	>2000	58
HR-TE (plate)	3	<560	300	200	-	-	>2000	58

In Table 3.3, the most characteristic parameters of the Kodak holographic emulsions are listed. A comparison between these materials and other commercial holographic products is given in Table 3.7 at the end of the chapter.

107

3.4.3 Safelight Recommendations

Kodak recommends handling all their red-sensitivity materials in total darkness. For the orthochromatic materials Kodak's safelight filter 1A (light red) can be used, whereas for the low-sensitivity red materials, a safelight filter No.3 (dark green) can be selected.

3.4.4 Processing

Kodak supplies very little information as regards the processing methods of their holographic materials. The only recommendation given is to use Kodak D 19 developer for most of their materials. This developer is not a bad choice in many processing situations, although in some cases better alternatives may exist. Processing methods are discussed in greater detail in Chaps.4,5. Recipes for Kodak processing baths are found in Chap.10.

3.5 Materials from the Former USSR

In the former Soviet Union several different types of silver-halide materials are produced for holographic purposes, which were described in Chap.2. The two main differences between these and the Western materials are the grain size and the silver content in the emulsion. The former Soviet recording emulsions can have grain sizes as small as 10 nm and the silver content is usually one-half (about $0.25 g/cm^3$) of the normal silver content in the Western materials. Materials such as PE-2 and LOI-2 posses the above-mentioned characteristics. The LOI-2 materials were developed by R. Protas and are now manufactured under the name of PFG-02. The PE-materials were developed by N. Kirillov and are of the highest quality - their grains do not exceed 10 nm. They are now manufactured under the name of PFG-03. Another type of silver halides manufactured in the former USSR are the VR-M and the VR-P plates which have similar characteristics to those of Agfa 8E and 10E materials. Yet another type of materials with high sensitivity are the FG- emulsions which are coated on film. The new FG-690 materials are specially made for recording holograms with pulsed ruby lasers.

Materials produced in the former USSR cannot be exported as a rule and are therefore intended for domestic use in the country. The company producing these is Soiuzhimfoto. Further information concerning these holographic materials can be obtained from the following address:

Mezhdunarodnaya
KNIGA
Dimitrova 39
Moscow 113095, Russia.

Table 3.4. Russian products

Material	Emulsion thickness [μm]	Spectral sensitivity [nm]	Sensitivity [μJ/cm²] at			Resolving power [ℓp/mm]	Grain size [nm]
			514	633	694		
FG-690 (film)	8+12	694	-	-	100	1000	40
FG-KR (film)	8+12	633/694	-	20	30	1000	40
FG-KC (film)	8+12	633	-	5	-	500	-
FP-GT (film)	8+12	633	-	1000	-	3000	20
PFG-01 (VRM)	10+12	694	-	-	100	5000	30
PFG-02 (LOI-2-500)	10+12	530	1200	-	-	5000	26
PFG-02 (LOI-2-633)	10+12	633	-	1000	-	5000	26
PFG-02 (LOI-2-694)	10+12	694	-	-	1000	5000	26
PFG-03 (PE-2)	7+8	633/694	-	500	500	10000	12
PFG-04	10	530	1200	-	-	5000	30
VR-P		530	100	-	-	1200	100
IAE-530		530	10	-	-	3000	-
IAE-630		630	-	1+5	-	3000	-
IAE-690		690	-	-	1	2000	-

Logak et al. [3.21] have described existing holographic materials on triacetate film base currently produced in Russia. In Table 3.4 the characteristics of different materials are presented.

3.6 Materials from Eastern Europe

3.6.1 Bulgarian Materials

Bulgaria is a producer of one type of commercial holographic recording material which has similar characteristics to those of the Russian materials just described [3.22-23]. It is manufactured by the Bulgarian Academy of Sciences through the Central Laboratory of Optical Storage and Processing for Information located at the following address:

Central Laboratory for Optical Storage and Processing of Information
Bulgarian Academy of Sciences
P.O.Box 95
Sofia 1113, Bulgaria.

The material is produced in two main versions, one red-sensitive, HP-650, and one green-sensitive, HP-490. The red-sensitive material comes in two versions, the second being used in interferometry which is harder than the regular type. It is labelled HP-650I. *Sainov* [3.24] has published a paper on the characteristics of the Bulgarian materials, in which also theoretical aspects of colloidal silver recording technique are discussed. The grain thickness in the emulsion is about 10 nm. The size of the colloidal silver particles after semi-physical development (in the FHP-3 developer) is about 30 nm with a typical concentration of about $2.5 \cdot 10^3$ μm^{-3}. According to the paper, the diffraction efficiency reaches a peak at a developed grain size of 30 nm.

The performance of the green-sensitive material including its processing with the FHP-3 developer has been described by *Katsev* et al. [3.25] Similar to the Russian materials, it is normally not possible to purchase these materials outside the country.

Emulsion Characteristics

The parameters for the two Bulgarian emulsions are presented in Table 3.5. As the material is mainly intended for reflection holography, it is not treated with any anti-halation backing.

Figure 3.9 shows the characteristic curve and Fig.3.10 the spectral sensitivity.

Table 3.5. Bulgarian products.

Material	Emulsion thickness [μm]	Spectral sensitivity [nm]	Sensitivity [μJ/cm^2] at 514 633 694	Resolving power [ℓp/mm]	Grain size [nm]
HP650	7	<680	- 500 -	>6000	10
HP490	7	<520	1500 - -	>6000	10

Base Substrate and Formats

The emulsion is coated on glass plates with the so-called "*selected flat*" only. The standard plate size is 14×14 cm^2. Larger plates can be delivered on special order.

Processing

Bulgarian materials can be processed with help of various Russian developers; the manufacturer recommends, however, a modified version of the GP-3 developer which is supposed to give the best results with the Bulgarian emulsion. A Bulgarian FHP-3 developer can also be employed. A more detailed description regarding the processing methods of the above-mentioned materials can be found in Chap.4.

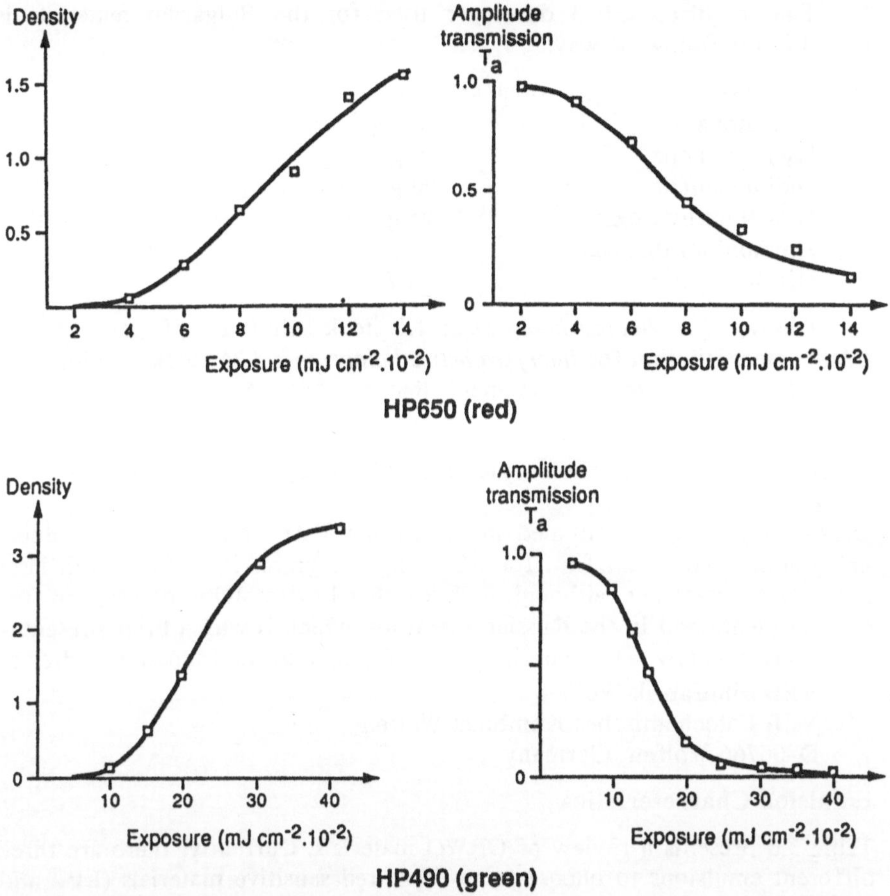

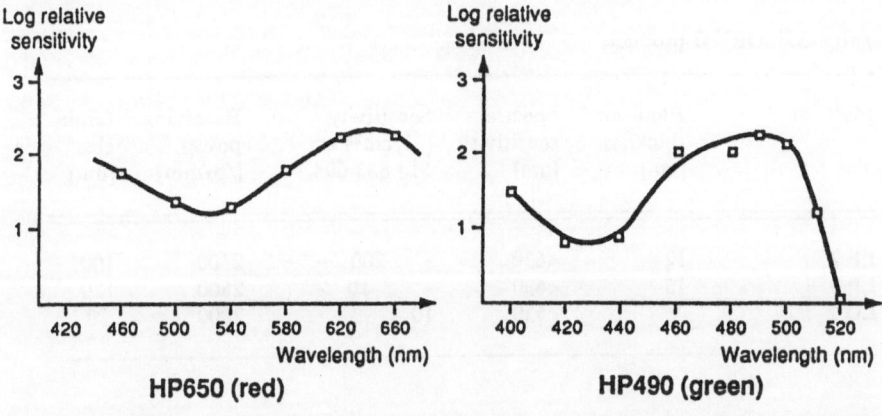

Fig.3.9. Characteristic curves and amplitude transmission curves for Bulgarian red (HP650) and green (HP490) sensitive materials

Fig.3.10. Spectral sensitivity curves for Bulgarian red (HP650) and green (HP490) sensitive materials

111

The modified GP-3 developer used for the Bulgarian materials is mixed in the following way:

Bulgarian **GP-3** developer, stock solution:

Phenidone	0.2 g
Hydroquinone	5 g
Sodium sulfite	100 g
Potassium hydroxide	25 g
Ammonium thiocyanate	45 g
Distilled water	1 ℓ.

Dilution for *display holograms*: 1.5 stock solution + 40 parts of distilled water. Dilution for *interferometric holograms*: 3.75 stock solution + 40 parts of distilled water. Development time at 20° C: >5 minutes.

3.6.2 Materials from the former East Germany

ORWO is a company situated in the former East Germany that produces holographic plates and film for recording holograms in red or green laser light. These materials, although with a rather low resolving power, are frequently mentioned in the Russian literature, which is why a brief presentation is given below. The material is manufactured at the following address:

VEB Filmfabrik Wolfen
VEB Fotochemisches Kombinat Wolfen
D-06766 Wolfen, Germany.

Emulsion Characteristics

Table 3.6 presents a review of ORWO materials. Currently, there are three different emulsions to choose from: Two red sensitive materials (LP2 and LP3) and one green sensitive material (LO2). The company offers an antihalation backing as a rule. The parameters for the ORWO emulsions are presented in Table 3.6.

Table 3.6. ORWO products

Material	Emulsion thickness [μm]	Spectral sensitivity [nm]	Sensitivity [μJ/cm^2] at 514 633 694	Resolving power [ℓp/mm]	Grain size [nm]
LP 2	12	<630	− 200 −	2500	100
LP 3	15	<630	− 10 −	2500	120
LO 2	9	<530	10 − −	2800	90

Base Substrate and Formats

The emulsion is only coated on glass plates. Glass plates are produced in the following formats:

5 cm × 5 cm	thickness	1.3 mm or 1.6 mm
6.5 cm × 9 cm	"	1.3 mm
9 cm × 12 cm	"	1.3 mm
10 cm × 15 cm	"	1.3 mm
13 cm × 18 cm	"	1.6 mm
18 cm × 24 cm	"	1.6 mm .

Safelight Recommendations

The LP2 and LP3 emulsions can be handled in green safelight illumination achieved by employing the ORWO 108 filter in a darkroom lamp with a 15-W bulb. The LO2 emulsion can be treated in dark-red safelight illumination by employing the ORWO 104 filter or its equivalent.

Processing

The ORWO materials are recommended to be developed in the developer ORWO 71:

Metol	5 g
Hydroquinone	6 g
Sodium sulfite	40 g
Potassium carbonate	40 g
Potassium bromide	3 g
Distilled water	1 ℓ .

Development time at 20° C: 4 to 5 minutes.

In Table 3.7 a comparison of all the existing holographic silver-halide materials is given.

Table 3.7. Commercial holographic silver halide products

Material	Emulsion thickness [μm]	Spectral sensitivity [nm]	Sensitivity [μJ/cm²] at 442	515	633	694	Resolving power [ℓp/mm]	Grain size [nm]
Agfa								
8E75 HD	7	<750	–	–	10	20	<5000	35/44
10E75	7	<750	–	–	1	2	<2800	90
8E56 HD	7	<560	25	–	–		<5000	35/44
Ilford								
FT340T (plate)	6	<700	–	–	200	–	<7000	30
Hotec R (film)	7	<700	–	–	20	50	<7000	30
SP695T (plate)	6	<560	200	100	–	–	<7000	30
SP672T (film)	7	<560	200	100	–	–	<7000	30
Kodak								
649-F (plate)	17	<700	80	80	90	500	>2000	58
649-F (film)	6	<700	80	80	90	500	>2000	58
649-GH	6	<560	100	100	–	–	>2000	58
120	6	<700	50	40	40	40	>2500	50
125	6	<560	8	5	–	–	>1250	70
131	6	<700	5	3	0.5	100	>1250	70
SO-173	6	<700	50	–	40	40	>2500	50
SO-253	9	<700	5	3	0.5	100	>1250	70
SO-343	6	<560	100	100	–	–	>2000	58
HR-1A	6	<560	100	100	–	–	>2000	58
HR-TE	3	<560	300	200	–	–	>2000	58
Former USSR								
FG-690 (film)	8+12	694	–	–	–	100	1000	40
FG-KR (film)	8+12	633/694	–	–	20	30	1000	40
FG-KC (film)	8+12	633	–	–	5	–	500	
FP-GT (film)	8+12	633	–	–	1000	–	3000	20
PFG-01 (VRM)	10+12	694	–	–	–	100	5000	30
PFG-02	10+12	530	–	1200	–	–	5000	26
PFG-02	10+12	633	–	–	1000	–	5000	26
PFG-02	10+12	694	–	–	–	1000	5000	26
PFG-03 (PE-2)	7+8	633/694	–	–	500	500	10000	12
PFG-04	10	530	–	1200	–	–	5000	30
VR-P	530		–	100	–	–	1200	100
Bulgaria								
HP650	7	<680	–	–	500	–	>6000	10
HP490	7	<520	–	1500	–	–	>6000	10
ORWO								
LP 2	12	<630	–	–	200	–	2500	100
LP 3	15	<630	–	–	10	–	2500	120
LO 2	15	<530	–	10	–	–	2800	90

4. Development

Development is a processing technique by which the latent image recorded during the exposure of the material is converted into a silver image. In chemical development, this processing technique is called *chemical reduction*. From the chemical point of view, reduction is a process in which oxygen is removed from a chemical complex. The acting compound is called a *reducer*. Reduction is always accompanied by a reciprocal process called *oxidation*.

What happens during development is that a silver-halide crystal containing a latent-image speck (a few silver atoms) is converted into a silver grain by the reduction process triggered off by the *reducing agent* contained in the *developer*. The reducing agent contained in the developer is then oxidized. The development in which the developing agent acts as a reducer is called *chemical development*. This is, indeed, the most common way of developing photographic materials, which also applies to the majority of holographic silver-halide materials.

Another type of development, rarely used in conventional photography but sometimes used in holography, is called *physical development*. It is particularly useful in the processing of ultra-fine-grained emulsions. Both methods will be treated here in great detail, in addition to the developing processes which are a combination of the two.

4.1 Theory of Development

4.1.1 Different Types of Developers

The theory of the photographic development of silver-halide emulsions has been presented in the literature [4.1-9]. In addition, some publications supply information related directly to the development of holographic silver-halide materials [4.10-32, 43-56, 60-62, 64-88].

The development of the exposed silver-halide emulsion can proceed along at least two distinct paths which can, at times, occur even simultaneously. In the first, the silver ions situated at the interface between the silver-halide crystal and a latent-image speck (which is the developing silver center) are reduced to silver atoms by the action of the developer. The development that proceeds along these lines is called *chemical* or *direct development*. In the second type of development, the silver ions from the processing solution (developer) are transferred to the silver centers, such as the initial latent-image nuclei or the silver centers already developed by chemical reduction. The silver ions are then reduced to silver by the action

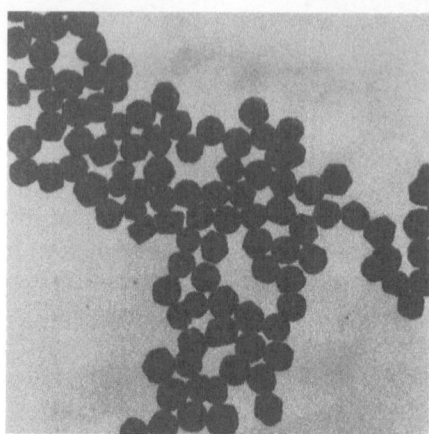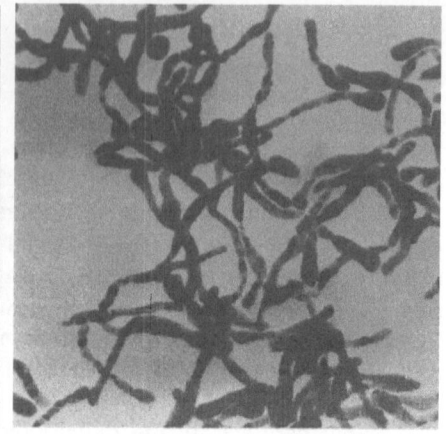

Fig.4.1. Silver structure caused by different development types. Physical development normally yields compact plates of silver (a), whereas chemical development yields mostly filament-like structures of silver (b)

of the developer. This mechanism is called *physical development*. A slightly different mechanism applies if a developer containing no soluble silver salt is used. Here, the silver ions involved in the physical development will be provided from the emulsion's own silver-halide grains. This type of physical development is referred to as *solution-physical development* or *semi-physical development*. Physical development normally yields compact plates of silver, whereas chemical development leads to mostly filament-like structures of silver (Fig.4.1). In holography all the three types of developing mechanisms are utilized in different processing regimes which are described below.

4.1.2 Chemical Development

Chemical development is based on the reduction of silver-halide grains by a soluble developer substance. The deposition of silver atoms occurs in the next step. The following equations describe the process:

$$\text{Oxidation:} \quad \text{Dev}_{red} \rightarrow \text{Dev}_{ox} + e^- , \qquad (4.1)$$

$$\text{Reduction:} \quad Ag^+ + e^- \rightarrow Ag . \qquad (4.2)$$

The reaction will take place only if the equilibrium *redox potential* of the system developer/oxidized developer, see (4.1), is more negative than that of system Ag^+/Ag, see (4.2).

The latent image consists of "specks" residing in silver-halide crystals or grains. Each speck is a collection of a few submicroscopic silver atoms. The specks have a catalytic effect in that they trigger the process of chemi-

116

cal reduction in which each speck acts as a microelectrode which brings the developer molecules into electrical contact with the silver ions. A large number of silver atoms is thus created in all the silver-halide crystals with latent-image specks. There is a tremendous amplification in this process - for conventional photographic materials it is about 10^9.

4.1.3 Physical Development

In pure physical development, silver is supplied externally from the developing solution. The silver from the developer is deposited on the silver specks in the latent image. The developer is thus the source of silver ions. Post-fixation in physical development is a processing technique in which the material is first fixed, which means that the silver-halide grains are totally dissolved and that only the latent-image silver nuclei are left in the emulsion. During the succeeding development process silver is deposited onto these nuclei. This is a rather time-consuming technique since development can take hours. Physical development creates a very fine silver structure (colloidal) in the emulsion, which is why this technique is suitable for the processing of fine-grained materials. The exposure needed for physical development of fine-grained emulsions must be about five times higher than that required for chemical development of the same material.

4.1.4 Solution-Physical Development

The solution- or semi-physical development technique differs slightly from pure physical development. The silver ions involved in solution-physical development will be provided from the emulsion's own silver-halide grains. The developer triggers off the dissolution of the previously exposed grains and even of the nearby unexposed grains. The developer used in this type of processing contains silver-halide solvents, such as, e.g., ammonium thiocyanate or high amounts of sodium sulfite.

4.2 Photographic Developers

A developer for conventional black-and-white photographic materials consists of the following components:
• Developing agent.
• Preservative (or antioxidant).
• Weak silver-solvent agent.
• Accelerator (or activator).
• Restrainer.
• Addenda.
• Solvent.

These seven groups of a developer's chemical building blocks will be discussed in consecutive order.

4.2.1 Developing Agents

The developing agent is the most important constituent in the developer. Chemical reducing agents are only suitable as photographic developing agents if they are capable of differentiating between the exposed and non-exposed silver-halide crystals. A large variety of synthesized developing agents can be obtained on the market. The most commonly used agents are derivatives of organic *benzene* $(C_6 H_6)$. The majority of developing agents may be classified as regards their structure by the Kendall-Pelz rule

$$a - (A = B)_n - a' \tag{4.3}$$

where A denotes carbon, B carbon or nitrogen, a and a' designate OH, NH_2, NHR_1 or $NR_1 R_2$.

Benzene is a ring compound consisting of six carbon atoms linked together with a hydrogen atom attached to each carbon atom in the ring (Kekulé structure). To simplify the notation benzene is often represented as a hexagon, with identification numbers at the six positions on the ring (Fig.4.2).

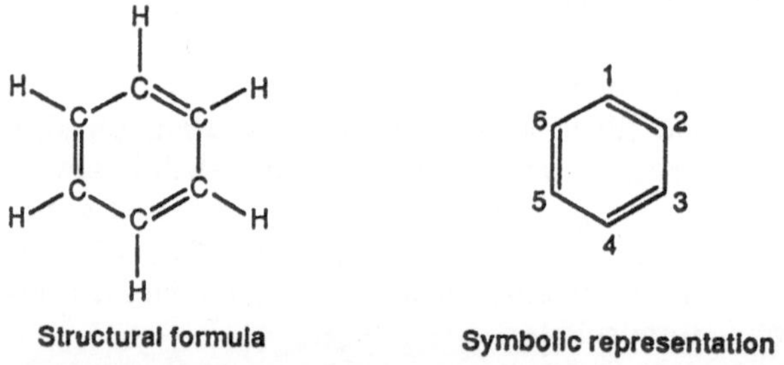

Structural formula   Symbolic representation

Fig.4.2. Structure of the benzene molecule. Often the benzene ring of six carbon atoms is shown simply as a hexagon. Sometimes the hydrogen atoms are not shown, but only other groups attached to the ring

Hydrogen atoms on the ring can be replaced by other atoms. When two substitutions are made in position 1 and 2, the substitutions are said to be *ortho* to each other; in 1 and 3, the substitutions are *meta*, and in 1 and 4, they are in a *para* configuration. Developing agents are substitutes of benzene in *ortho* and *para* positions. The *meta*position does not, as a rule, produce active developing agents. With the exception of the ascorbic acid and phenidone, organic developing agents are derivatives of benzene. Many developing agents are benzene derivatives consisting of three inorganic compounds: hydrogen peroxide (HO-OH), hydroxylamine (HO-NH_2), and hy-

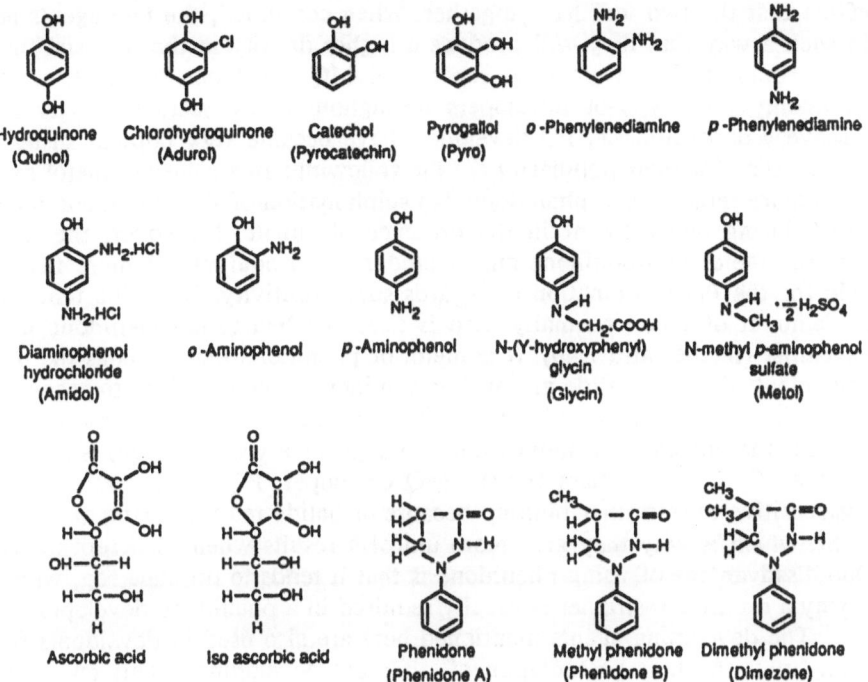

Fig.4.3. Structure formulas of common developing agents

drazine (NH_2-NH_2) in *ortho* or *para* positions. For example, the well-known developing agent *hydroquinone* is a derivative of benzene where the two hydroxyl groups (OH groups) are in the *para* position on the ring. If they are in the *ortho* position, the agent is called *pyrocatechol*. The same developing agent can appear under many different names, both chemical and trade names. Hydroquinone can be named, e.g., *para*-dihydroxybenzene or 1,4-dihydroxibenzene, while pyrocatechol will be named, by the same token, *ortho*-dihydroxybenzene or 1,2-dihydroxybenzene.

Sometimes more than two substitutions on the benzene ring positions are made so that when, e.g., three hydroxyl groups are in the 1, 2, and 3 positions, the compound is called *pyrogallol* or *1,2,3-trihydroxybenzene*. Figure 4.3 illustrates various modern developing agents used in common developers.

4.2.2 Conventional Black-and-White Developers

Today practically all standard Black-and-White (B/W) photographic developers are of two main types. They are compounded of either *metol* and *hydroquinone* (MQ) or else of *phenidone* and *hydroquinone* (PQ). The reason for using a combination of two agents in a developer is the *superadditivity* effect that the two will have together. When combined, the two agents act in such a way that they will produce a higher density in the material than they would if they were used separately. MQ developers were the most commonly used type of developers throughout many years but after the discovery of phenidone, PQ developers have became very popular instead. The reasons for their popularity are the following: In a solution, metol oxidizes more rapidly than phenidone. No sulphonation of the phenidone takes place during development in the presence of sulfite. Moreover, the ratio between metol-hydroquinone and phenidone-hydroquinone is more favorable for the PQ combination as regards superadditivity. Only one tenth of the amount of metol normally used is needed when using phenidone in a developer (in the ratio 1:40). It is metol or phenidone that initiates the development of silver-halide grains; hydroquinone is used mainly to regenerate these agents to their active form. The *induction period* (the time necessary for the developer to induce the first signs of a visible image) is shorter for the PQ developer than for the MQ developer. PQ developers are also less sensitive to restrainer buildup because of halide release during development, which is why they give more uniform results when used repeatedly. One disadvantage of using phenidone is that it tends to produce fog, which is why a bromide restrainer is usually required in a phenidone developer.

The developing agents mentioned here are also used in developers for holography. A short presentation of different developing agents currently utilized in holographic developers is given below.

Di-Aminophenol: $(NH_2)_2 C_6 H_3 OH \cdot 2HCl.$ (Other names: 2,4-diaminophenol dihydrocloride, 4-hydroxy-1,3- phenylenediamine dihydrocloride, Acrol, *Amidol*, Diamol, Dolmi)

This is a developing agent that can work without an alkali at low pH-values and which causes less swelling of gelatin than other developers usually do. It is used for the development of nuclear emulsions where a low level of gelatin swelling is important. When an amidol developer is acidified to the point where no neutral sulfite is left in the solution, it will act as a depth developer, and the development will start at the bottom of the emulsion layer. Amidol in the solution oxidizes fairly rapidly in the air and if alkalis are added the oxidation is even more rapid. If, on the other hand, a weak acid (bisulfite or boric) is added, the solution becomes quite stable.

Ortho-**Aminophenol:** $NH_2 C_6 H_4 OH.$ (Other names: 2-hydroxyaniline, 2-amino-1-hydroxybensen, 2-aminophenol, O.A.P.)

This developing agent is used in developers for ultra-fine-grained materials.

Para-**Aminophenol:** $NH_2 C_6 H_4 OH.$ (Other names: 4-hydroxyaniline, 4-amino-1-hydroxybensen, 4-aminophenol, *para*midophenol, P.A.P., Azol, *Rodinal*, Kodelon)

Para-Aminophenol may be isolated as a free base, but its hydrochloride salt ($NH_2C_6H_4OH \cdot HCl$), is normally used in developers. It is employed in concentrated developers, as it is easily soluble in water and has excellent keeping properties. A developer based on this agent is free from stain and has a low fog level. It is useful for high-temperature processing. The pH-value of the developer must be >8.0 for a developer based on this agent to work well.

Ascorbic Acid: $CH_2OHCHOH_{CHCOH:COHCOO}$.   (Other names: L-ascorbic acid, vitamin C, cevitamic acid)

Ascorbic acid may be thought of as a dihydroxydihydrofuran developing agent which forms almost colorless oxidation products. This is one of the most useful developing agents for holography and it is also non-toxic. It is often combined with metol or phenidone. Conventional sulfite-free metol-ascorbic acid developers are used for latent-image studies since they develop only the surface latent image. They are typical depth developers which, when combined with bromide, provide high contrast and low fog level. Their superadditive effect, when combined with phenidone, is utilized in holographic developers. Iso-ascorbic acid (d-iso-ascorbic acid) has very low antiscorbutic activity and may be substituted for ascorbic acid. Another complex related to ascorbic acid is sodium iso-ascorbate (sodium erythorbate) which has become common in monobath developers with rapid developing activity.

Glycin: $HOC_6H_4NHCH_2COOH$.   (Other names: *para*-oxyphenylglycin, *para*-hydroxyphenylaminoacetic acid, *para*-hydroxyphenylglycin, Athenon, Glyconiol, Iconyl, Kodurol)

Note that glycine (amino acid) is not a developing agent and must not be confused with photographic glycin described here. Glycin is a slow acting but a powerful developing agent, producing fine-grained silver structures. The pH-value of the developer must be >7.5 for a developer based on this agent to work well. It is an agent that oxidizes very slowly in the air, even in alkaline solutions. It is therefore often used in tank developers. Its tendency to fog, even in the absence of bromide, is very low.

Metol: $(HOC_6H_4NHCH_3)_2 \cdot H_2SO_4$.   (Other names: monomethyl-*para*-aminophenol sulfate, *para*-methylaminophenol sulfate, *Elon*, Genol, Pictol, Rhodol)

Metol is a derivative of *para*-aminophenol. It is a popular developing agent, used in many photographic and holographic developers. Used alone in a developer, it gives low contrast and is therefore often combined with hydroquinone. Note that metol has to be dissolved before other components, such as, e.g., sodium sulfite, are added to the developer. The pH-value of the developer must be >6.9 for a developer based on this agent to work. Highly alkaline metol baths will tend to produce fog, unless restrained. This agent (or impurities of it) is prone to cause dermatitis in allergic users.

Hydroquinone: C_6H_4-1,4-$(OH)_2$. (Other names: *para* dihyroxybenzene, 1,4-dihydroxybenzene, 1,4-benzenediol, hydroquinol, *quinol*)

This is perhaps the most popular developing agent used in many different kinds of developers. If employed alone with a strong alkali and low concentrations of sulfite, it will produce a high-contrast developer used for infectious development, eliminating tones intermediate between black and white. It is almost always applied together with metol or phenidone for continuous-tone photographic work. The pH-value of the developer must be >9.7 and the temperature above 15° C for the developer based on this agent to work well.

Chlorohydroquinone: ClC_6H_3-1,4-$(OH)_2$. (Other names: 2-chloro-1,4-dihydroxybenzene, *Adurol*, Chloroquinol, Quinotol)

This agent is actually better than hydroquinone as it is more stable, has better solubility, a smaller tendency to create fog and is easier to adjust (using potassium bromide) than the regular hydroquinone. Its reduction potential is higher than that of hydroquinone and it has about six times the developing energy of hydroquinone because of the presence of a chlorine atom in the quinol structure. Under normal photographic conditions it gives a blacker image. Its solubility in water is higher than that of hydroquinone, which is why it is often used in concentrated developers. It can be used as a preservative in amidol developers. This agent was frequently used in the early days of photography and has later been proven useful for developing nuclear and holographic emulsions.

Phenidone: C_6H_5-$C_3H_5N_2O$ (Other names: 1-phenyl-3-pyrazolidone, 1-phenyl-3-pyrazolidinone, Graphidone, Phenidone A)

Phenidone is one of the most useful developing agents for holography, in particular as regards holograms recorded at very short exposure times (e.g., with Q-switched lasers). Its superadditivity effect, when used together with hydroquinone, is strong, and it also shows a superadditive effect when combined with ascorbic acid, which makes it the fastest holographic developer of today. Phenidone is difficult to dissolve in water but very soluble in alkaline solutions. For best results, phenidone should be added to the developer just before it is used. Unlike the aromatic developing agents, phenidone is not indefinitely stable in alkaline solutions, as it will be affected by hydrolysis. A phenidone developer cannot be stored over long periods of time. Degradation of phenidone developers during storage has been studied by *Alletag* [4.34]. It has a low toxicity.

Methyl Phenidone: $C_{10}H_{12}N_2O$. (Other names: 1-phenyl-4-methyl-3-pyrazolidone, Phenidone B, Phenidone Z)

Methyl phenidone and dimethyl phenidone (other names: 1-phenyl-4,4-dimethyl-3-pyrazolidone, dimezone) are derivatives of phenidone. They are slightly better than phenidone in that they are more alkali-stable in a mixed developer. Dimethyl phenidone does not undergo hydrolysis. The methyl group (CH_3) has an increasing effect on the development speed in a high-pH developer. Methyl phenidone has been strongly recommended for many Russian holographic developers. Its additive effect is believed to be even stronger than that of phenidone. The solubility of phenidone derivatives in water is similar to that of regular phenidone.

Ortho-Phenylenediamine: $C_6H_4(NH_2)_2$. (Other names: *ortho*-diaminobenzene, 2-aminoaniline, O.P.D.)

This developer agent is similar to the *para*-version, but does not have the staining effect on the emulsion. It is also weaker, which is why it is usually combined with other agents in developers for producing fine-grained results. It is a toxic developing agent.

Para-Phenylenediamine: $C_6H_4(NH_2)_2$. (Other names: *para*-diaminobenzene, 4-aminoaniline, P.P.D., Diamine, Paramine)

This is a developing agent used for fine-grained materials. It works in a semi-physical way, has a tendency to stain the emulsion, and is very toxic. A developer with this agent has no induction period. The pH-value of the developer must be >10 for a developer based on this agent to work well. In plain sulfite solutions, *para*-phenylenediamine works very slowly, reducing a part of each grain only, which results in low contrast and granularity. It is often combined with glycin or metol in developers for fine-grained materials.

Pyrocatechol: C_6H_4-1,2-$(OH)_2$. (Other names: *ortho*-dihydroxybenzene, 1,2-dihydroxybenzene, *catechol*, pyrocatechin, Dinol)

This hydroquinone isomer is another agent commonly used in holography. It has the advantage of producing no fog. Employed without sulfite, it hardens gelatin just like pyrogallol does. It is utilized in high-quality developers for reflection holograms. Similarly to pyrogallol, it also stains the emulsion, but to a lesser degree. This agent is outstanding in the ability to retain its activity in the presence of large quantities of sodium thiosulfate, which is why it is often used in monobath developers.

Pyrogallol: C_6H_3-1,2,3-$(OH)_3$. (Other names: 1,2,3-trihydroxybenzene, 1,2,3-benzenetriol, *pyro*, sometimes improperly called pyrogallic acid)

Pyrogallol is a very important developing agent in holography. Used in a developer without sulfite, it has a hardening effect on the gelatin. It also has a staining effect on the emulsion, which is good from the point of noise suppression. The staining of gelatin (to a brownish color) results from the oxidation products in the development of the silver image. Since pyrogallol solutions are stable only when they are acid, they are usually prepared by mixing two stock solutions: one containing the pyrogallol with bisulfite, and the other containing an alkali. If the developer is mixed with sulfite, the staining and hardening effects are reduced. The pH-value of the developer based on this agent should be >8.2. When combined with di-aminophenol the developer works without an alkali. The behavior of a pyrogallol developer depends upon its dilution: high concentration produces density and contrast readily, whereas low concentration is softly working and produces density and contrast slowly. The monomethyl ether of pyrogallol (3-methoxycatechol, Rubinol) may be used instead, as it is less readily auto-oxidized and has a smaller staining effect.

4.2.3 Preservatives

Sulfite is the most common preservative in photographic chemistry. Sulfite's main role is to prevent the developer from getting oxidized by the oxygen in the air. Sulfite usually reacts with the oxidation products to form sulfonates. However, it also serves as a weak silver-solvent compound. Normally, *sodium sulfite* (Na_2SO_3) is used but at times *potassium sulfite* (K_2SO_3) can be preferred because of its greater solubility. It is particularly found in liquid concentrated developers. Although sodium sulfite is invariably used in practice, organic compounds, such as cysteine or thioglycolic acid have also been reported to be effective in inhibiting the oxidation of developing agents.

If a developer does not contain sulfite, it is called a *sulfite-free developer*. This type of developer is important in some holographic processing regimes. In general, a developer that contains no sulfite oxidizes very fast. It must therefore be prepared just before being used. During the development, such a developer can cause oxidation fog in the emulsion.

A gradual increase of the amount of sulfite in the developer will make the developer more and more oxidation resistant, but, on the other hand, high concentrations of sulfite will also cause increased silver-solvent action, which in some holographic processing schemes can become a problem. Too high concentration of sulfite can also cause the appearance of fog in the emulsion. Sulfite is alkaline enough to be able to function as both a preservative and an accelerator in some weak developers.

Sodium sulfite can be dehydrated (anhydrous) (Na_2SO_3) or crystalline ($Na_2SO_3 \cdot 7H_2O$). For developers intended for the processing of fine-grained materials sodium sulfite of the highest quality must be used. If the crystalline version is used instead of the dehydrated one, the quantity of sodium sulfite must be doubled.

In some developers advantage is taken of *sodium metabisulfite* ($Na_2S_2O_5$) (other names: sodium acidsulfite, sodium pyrosulfite) or *potassium metabisulfite* ($K_2S_2O_5$, another name: potassium pyrosulfite). In aqueous solutions sodium metabisulfite undergoes the process of hydrolysis and forms sodium bisulfite ($NaHSO_3$). These compounds are used particularly in developers for fine-grained materials to decrease the developer's alkalinity and thus to reduce their intensity. Metabisulfite is also employed as a preservative in high-energetic developers.

4.2.4 Silver-Solvent Agents

Sulfite is a weak silver solvent and can be used in developers as such, as well as combining the actions of a solvent and a preservative. In a compensating developer for fine-grained materials, the sulfite concentration is higher than what is necessary for bare preservation. Silver solvents play an important role in many holographic developers. Strong silver solvents are *ammonium thiocyanate* (NH_4SCN) (other names: ammonium sulfocyanate, ammonium sulfocyanide), *potassium thiocyanate* (KSCN) (other names: potassium sulfocyanate, potassium sulfocyanide), and *sodium thiocyanate* (NaSCN) (other names: sodium sulfocyanate, sodium sulfocyanide). Identical to these are *ammonium rhodanide* (NH_4CNS), *potassium rhodanide* (KCNS) and *sodium rhodanide* (NaCNS), respectively, which sometimes appear under slightly different formulas. These compounds are used in, e.g., physical developers. They are very tricky to control and many problems can occur, e.g., dichroitic fog may form as a result of a too high concentration of these substances. *James* and *Vanselow* [4.35] investigated the rate of solution of silver-halide grains caused by these compounds in a developer.

4.2.5 Accelerator

With few exceptions a developer becomes active only when it is alkaline. Moreover, its activity increases with the increase in alkalinity. The alkaline compound acts like an accelerator in a developer. Alkalines used in developers can be classified into three main groups:
(i)   Strong alkalines.
(ii) Medium alkalines.
(iii) Weak alkalines.

Strong alkalines include *potassium hydroxide* (KOH) (other names: caustic potash, potassium hydrate), *sodium hydroxide* (NaOH) (other names: caustic soda, sodium hydrate), and *ammonium hydroxide* (NH_4OH) (other names: aqua ammonia, ammonia water). Ammonium hydroxide has silver-solvent properties and is therefore seldom used. Strong alkalines are used in rapid developers. Normally, they produce relatively large silver grains in a developed emulsion. Developers mixed with these alkalines have a pH-value of >12.

Medium alkalines include *sodium carbonate* (Na_2CO_3) (other names: soda, washing soda, soda ash, solvay soda) and *potassium carbonate* (K_2CO_3) (other names: potash, pearl ash). Sodium carbonate can be anhydrous (Na_2CO_3), monohydrate ($Na_2CO_3 \cdot H_2O$), or decahydrate ($Na_2CO_3 \cdot 10H_2O$). Potassium carbonate comes as anhydrous (K_2CO_3) or crystalline ($K_2CO_3 \cdot 2H_2O$). At equal concentrations, potassium carbonate will render a developer more alkaline than sodium carbonate will. *Phosphate* is used sometimes. *Sodium orthophosphate* ($Na_3PO_4 \cdot 12H_2O$) or *sodium pyrophos-*

phate ($Na_4P_2O_7 \cdot 10H_2O$) can be employed, in particular, for fine-grained material development. Developers containing these alkalines have pH-values between 10 and 11.5.

Weak alkalines include *sodium tetraborate* (decahydrated) ($Na_2B_4O_7 \cdot 10H_2O$) (other names: sodium borate, sodium biborate, sodium pyroborate, *Borax*) and sodium metaborate ($NaBO_2 \cdot 4H_2O$) (other names: Kodak's balanced alkali, *Kodalk*). Developers with weak alkalines produce fine grains and will also level out the contrast. Note that borates should not be used in developers containing pyrocatechin or pyrogallol developing agents. Sodium metaborate does not give rise to bubbles of gas when carried into an acid stop bath introduced directly after development. *Sulfite* is a very weak accelerator but it can be used in some slow developers. For weak alkalines the pH-values are between 8 to 10.

4.2.6 Restrainer

Certain chemical processes are responsible for the fact that some of the unexposed grains of the emulsion will also be developed. That phenomenon causes fog in the photographic emulsion. In order to prevent this, restrainers are added to the developer. Some low-alkalinity developers do not produce fog and there is no need for a restrainer. *Bromide* and, in particular, *potassium bromide* (KBr) are the most commonly employed restrainers found in most developers. Occasionally *sodium bromide* (NaBr) is used.

In stock solutions, potassium bromide should be mixed with the alkali part as it has a tendency to oxidize the developing agent. Other restrainers include citrates, tartrates and borartrates, among which alkali iodides should be mentioned particularly. Potassium iodide can also accelerate the development process (the Leiner effect). Antifoggants can also be considered as restrainers from the functional point of view. They are treated next.

4.2.7 Addenda

Various compounds that can influence the development process in one way or another can be found in the recipes for developers. Antifoggants [4.36] are one example of such compounds, e.g., *benzotriazole* ($C_6H_4NHN_2$) (other names: 1-2-3 benzotriazole, 1-H-benzotriazole, azimidobenzene, Kodak Anti-Fog 1) and *6-nitro-benzimidazole* (another name: Kodak Anti-Fog 2), as well as *ethylenediaminetetraacetic acid tetra sodium salt* (other names: EDTA NA4, Squestrene, Trilon B). One gram of the last compound chelates 215 milligrams of calcium carbonate, which is why it is often used in developers as a calcium sequestering agent. Yet another compound, *2,4 dihydroxybenzophenone* [$(HO)_2C_6H_3COC_6H_5$] (other names: DHBH, Kodak Antical No.3) which acts as an antistain and antisludge agent is often added to developers. Powerful organic compounds containing sulfur can also be utilized, such as *1-phenyl-5-mercaptotetrazole*. The chemical structure of some of these compounds as well as common wetting agents discussed in the next subsection are shown in Fig.4.4.

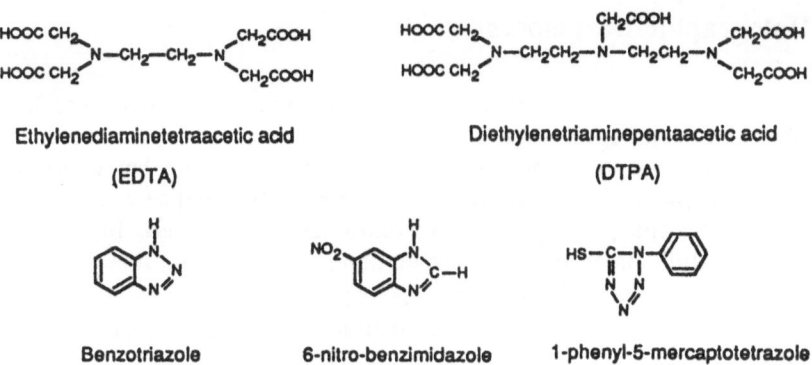

Fig.4.4. Structure formulas of common sequestering and anti-fog agents

4.2.8 Solvent Agents

The main solvent employed in developers as well as in other photographic processing baths is water. Certain things should be kept in mind when mixing developers with water. Regular tap water is not recommended for holographic processing baths. However, if tap water is used, it is advisable to add 1 g/ℓ of *sodium hexametaphosphate* $[(NaPO_3)_6]$ (other names: Graham's salt, Calgon) to the developer to prevent the precipitation of insoluble calcium, magnesium, or other salts usually present in tap water. *EDTA* (Sect.4.2.7) is often used as a calcium sequestering agent in photographic processing solutions and especially in developers. To be on the safe side it is recommended to take advantage of *distilled water* for the preparation of processing baths, or at least *deionized water*. Deionized water is tap water that has been passed through a bed of certain synthetic resins to remove ions from the water. It is also a good idea to boil the water before mixing a developer, in order to reduce the air-content in the water.

A suitable *wetting agent* can be added to the developer to make the wetting action on the material during processing more uniform. Wetting agents are also used in the final wash of holograms. Suitable wetting agents are compounds containing a hydrophilic part and a hydrophobic part. There are three groups of wetting agents:

(i) Anionic compounds, such as sodium alkyl sulphonates (RSO_3Na) or sodium alkyl sulfates (RSO_4Na). R represents a long-chain alkyl group, which may contain cyclic substituents.

(ii) Non-ionic compounds, such as polyethyleneglycol derivatives, saponin, etc.

(iii) Cationic compounds, such as hexadecyltrimethyl-ammonium bromid, for example.

Only small quantities (about 0.1%) are required. Wetting agents are supplied by all the photographic companies. Often Kodak Photo-Flo is used in holographic processing baths.

4.3 Holographic Developers

Silver-halide materials for holography are all of the fine-grained type. Such materials require the application of special processing techniques to achieve the best possible results. Therefore, a great deal of research has gone into the problem of processing methods of holographic silver-halide emulsions.

In general, the processing of conventional fine-grained, black-and-white silver-halide materials is performed along the following lines:

- Overexpose the material and underdevelop it.
- Use slow developers with a weak alkaline and a high concentration of sulfite.
- Use a developer with high silver-solvent action, containing, e.g., ammonium thiocyanate.
- Use a semi-physical developer containing, e.g., *para*-phenylenediamine.
- Use physical development technique.

Since a hologram is something very different from the conventional photographic picture, strict adherence to the processing recommendations just mentioned will not guarantee that a high-quality hologram will be obtained in each and every case. Back in Lippmann's times (Chap.1) one was faced with a similar problem, namely, that the processing parameters had to be adjusted to suit the development of color photographs on silver-halide materials which are quite similar to the holographic materials of today. And thus, the processing techniques used for the development of *transmission* holograms will differ from those used for *reflection* holograms. The same goes for the production of *amplitude* holograms as opposed to *phase* holograms.

Besides considering the type of hologram being produced, one must also consider other parameters, such as, for example, the grain size of the holographic material used. Thus, the processing technique will be different for the ultra-fine-grained emulsions and for a material with a somewhat larger grain structure. Another example of the importance of matching the processing methods with the type of hologram to be obtained comes from display holography. Here, normally a phase hologram is desired in which bleaching is done after development. This calls for a developing technique that will match the subsequent bleaching. Bleaching procedures for phase holograms are discussed in Chap.5.

In the early days of holography the already existing conventional developers were used for the processing of amplitude holograms - the basic hologram type at that time. To develop a high-resolution plate such as Kodak 649-F which was employed for the recording of early holograms in the USA, a high-contrast, fast developer of the X-ray type was required. Kodak D-19 is such a developer and it still gives good results in the processing of different types of holograms. Agfa-Gevaert recommended for developers: G3P, G150 and Metinol-U (similar to D-19) at the time they introduced the first Scientia materials for holography.

The first attempts to produce phase holograms created a problem with obtaining high diffraction efficiency coupled with a low noise level. This triggered off extensive research into the improvement of the performance of holographic materials and led to the presentation of various processing regimes. In this chapter the developing process will be described, in Chap.5 the bleaching process, and in Chap.7 complete well-working processing schemes will be presented.

4.3.1 Conventional Developers Used in Holography

Kodak D-19 black-and-white developer is the most frequently used conventional developer for both transmission and reflection holograms. The D-19 and the D-19b versions consist of

	D-19	D-19b
Metol	2 g	2.2 g
Sodium sulfite (anhydrous)	90 g	72 g
Hydroquinone	8 g	8.8 g
Sodium carbonate (monohydrated)	52.5 g	48 g
Potassium bromide	5 g	4 g
Distilled water	1 ℓ	1 ℓ.

Developing time 4 to 5 minutes at 20° C.

This MQ developer is a high-contrast, low-fog developer that will produce clean amplitude transmission holograms on the majority of the existing materials. Often, the b-version is recommended for holographic processing, mainly because of its lower sulfite and bromide contents.

Biedermann and *Stetson* [4.13] mentioned that the *confluence point* (the point where the extensions of the linear parts of the D-logH curves intersects) may lie below or on the logH axis depending on whether or not the developer has a high concentration of potassium bromide. The confluence point is situated on the logH axis for a surface developer (a developer without KBr). If the confluence point is situated below the axis, the developer is a depth developer (a developer with high KBr concentration). *Phillips* [4.24] has actually recommended to reduce sodium sulfite to 30 g and to completely omit the potassium bromide in the D-19 developer. The development time for holograms is about half of the time recommended by the manufacturer for conventional materials, indicating the earlier-mentioned overexpose/underdevelopment approach to reduce grain size in the final hologram. *Pernick* et al. [4.12] investigated the influence of the developing time on the slope and linearity of the T_a-logH curve for the Kodak 649-F emulsion developed in D-19. *Biedermann* and *Stetson* [4.13] did a more fundamental test and found that higher-quality holograms could actually be made using shorter developing times. The slope of the linear region of the D-logH curve changes with development time. They also found that the noise increased faster with increasing development time than does the diffraction efficiency. This indicates that shorter development times tend to favor signal-to-noise ratios in hologram reconstructions. The effect can be further enhanced by reducing the development time down to one to two minutes with the corresponding increase of exposure, of course. If the time is too short, however, the developer has no time to penetrate to the bottom of the emulsion and the development will take place mainly at the surface of the emulsion. The development will not be very uniform, which will result in poor-quality holograms. The influence of the developing temperature is also affecting the quality of the holograms. For example, *Cox* and *Buckles* [4.14] obtained higher diffraction efficiency of holograms recorded in Kodak SO-243 material and developed in D-19 at 30° C compared to 24° C. They also compared the performance of the D-19 developer with two other Kodak developers: D-76 and Microdol-X.

The MQ-developer can also be used for the processing of reflection phase holograms, provided that it has the high concentration of sodium sulfite and potassium bromide. The b-version is very often the best choice. However, certain modifications that have been made in order to adjust the MQ-developer for the particular use in holography make it easier to use. These will be discussed separately at a later stage.

The **Agfa 80** developer is similar to D-19 but according to *Biedermann* and *Johansson* [4.17] Agfa 80 is slightly better than D-19 for transmission holograms. The developer is mixed in the following way:

Metol	2.5	g
Sodium sulfite (anhydrous)	100	g
Hydroquinone	10	g
Potassium carbonate	60	g
Potassium bromide	4	g
Distilled water	1	ℓ .

Developing time 5 minutes at 20° C.

The main difference between D-19 and Agfa 80 is the use of potassium carbonate in the latter, instead of sodium carbonate, which in some processing regimes is believed to give slightly better results. Nowadays, Agfa recommends the use of the conventional developer *REFINAL* for developing their holographic materials if their own special holographic developers GP 61 or GP 62 are not used.

D-19 is not the only Kodak developer for holographic purposes. Another useful developer is **Kodak D-8**, which is supposed to give slightly better results than D-19 in some processing techniques [4.15, 37]. This developer has been designed to suit very high contrast work. Note the high concentration of KBr in this developer, indicating that it is a depth developer:

Sodium sulfite (anhydrous)	90	g
Hydroquinone	45	g
Sodium hydroxide	37.5	g
Potassium bromide	30	g
Distilled water	1	ℓ .

Normally, this developer is diluted: 2 parts developer + 1 part water.

A developer recommended by *Smith* [4.37] for the development of phase transmission holograms is **Kodak SD-48**. It consists of two parts:

Solution A:

Sodium sulfite (anhydrous)	8 g
Ascorbic acid	1 g
Pyrocatechol	40 g
Sodium sulfate	100 g
Potassium bromide	30 g
Distilled water	1 ℓ

Solution B:

Sodium hydroxide	20 g
Sodium sulfate	100 g
Distilled water	1 ℓ .

Mix solution A and solution B for use. Developing time 5 to 8 minutes.

This standard developer produces reduced-noise phase holograms when combined with the Kodak R-9 bleach. The influence of the relief image on the surface is such that it eliminates the noise produced by the refractive-index variation in the bleached hologram. *Lamberts* and *Kurtz* [4.15] found that by *removing potassium bromide* completely from SD-48 the diffraction efficiency of phase holograms on 649-F plates with a reversal bleaching technique was increased three to six times. Bleaching techniques will be discussed in Chap.5. It often seems that a surface developer is more suitable for processing transmission holograms, which explains the better performance in this case when the potassium bromide was omitted. Remember that *Phillips* [4.24] also recommended to omit KBr from the D-19 developer since he regarded D-19 to be a developer for primarily transmission holograms. A depth developer is obviously not a good choice for the processing of phase holograms of the reversal type. In this case, the developer must not contain KBr. This problem will be discussed in Chap.5, where *Fimia* et al. [4.23] reported that the D-8 developer (with KBr) could only be used for rehalogenated phase holograms. The D-8 applied to the reversal-bleaching method produces very low-efficiency holograms.

It should be mentioned that occasionally the following Kodak developers were also used: HRP, Microdol-X, D-76, D-82 and D-165. They are mentioned in some of the early publications on holography.

Another developer, **Neofin Blue** from Tetenal [4.38] was introduced by *Phillips* and *Porter* [4.19] and is considered to give good results in *transmission holograms*. It works particularly well for pulsed holography. It is a proprietary developer intended for one-shot processing. The developer comes in as a concentrated liquid to be diluted before processing conventional films. For holography, however, it was originally suggested to be used undiluted, which meant an extremely concentrated solution. The reason for this is clear - the concentration of the developing agent is much higher when undiluted and it will also have strong alkaline concentration. These factors have been proved to be important for some holographic developers. A dilution of 1:1 developer/distilled water is what the present author recommends as it will favorably affect the viscosity of the developer and ensure a better penetration into the emulsion. The developer can be applied at temperatures between 20° and 26° C depending on the exposure of the hologram. Processing time is between 2 and 5 minutes.

For transmission holograms best results are obtained if the developer has a temperature of 20° C and a processing time of 3 to 4 minutes (when used without additives and diluted 1:1). A longer time at higher temperatures will produce some fog. It should always be used fresh. The original paper suggests certain additives to the developer, such as 0.3 g/ℓ of benzo-triazole in order to further decrease the fog level which is already low even without it. This will, however, reduce the speed of the developer. Adding 120 g/ℓ of sodium metaborate will slightly increase the pH-value, which is rather modest (pH = 10.5) in the original developer. The developer can be purchased in half-liter bottles, which makes it more economical than to buy it in small ampules normally found in the photographic stores. Neofin Blue is definitely recommendable for transmission holograms.

Another category of developers not so well known among holographers are developers used for processing *nuclear emulsions*. Nuclear emulsion processing is, of course, a rather special technique and cannot be regarded as "conventional". These developers were originally not intended for holography and are therefore interesting to study, especially because there are some similarities between the two fields. The author has been working in the field of high-energy physics where these emulsions are used, and tried some of the techniques applied there for holography. A nuclear emulsion is an extremely thick photographic silver-halide material in which particle tracks are recorded directly as the particles pass through the emulsion [4.39-41]. The emulsion is very thick, $(25 \div 600 \mu m)$ which is why it is difficult to obtain a uniform development in depth through the entire emulsion. One is faced with the same problem when processing Lippmann plates used for re-flection holography even though the thickness of the emulsion in Lippmann plates is much smaller. One solution to this problem as regards nuclear emulsions is to use *amidol* as a developing agent, since it can develop at low pH-values, which reduces the swelling of the emulsion during processing. Another trick is to adjust the *solution temperature* in such a way that the developer can penetrate through the emulsion without having a developing effect on it. (At low temperatures the amidol developer is almost inactive). This is achieved if the developer is kept at a low temperature during the penetration stage and if at the offset of development the temperature is increased. This procedure produces very uniform results. The reader may want to refer to the test on an ice-cold developer for transmission holograms described in a paper by *Kaspar* et al. [4.42]. It should be remembered, however, that this technique is applicable mainly for the processing of reflection holograms.

A typical *developer for nuclear emulsions* is mixed in the following way:

Amidol	4.5	g
Sodium sulfite (anhydrous)	18	g
Boric acid	35	g

| Potassium bromide | 0.8 g |
| Distilled water | 1 ℓ . |

Processing is done by soaking the emulsion in distilled water at 20° C for 30 minutes and then for one hour at a lower temperature (5° C). After that, the cold water is replaced by the developer also at 5° C, and then the emulsion is soaked for one hour. After that the developer is replaced by another developer with a lower amidol concentration ($2.25 g/\ell$) which acts upon the emulsion for 50 minutes at 24° C. The cold developer previously absorbed by the emulsion will become active when the temperature rises and very uniform development will take place throughout the entire emulsion. The processing continues with a stop bath and fixation, using similar cycles and long processing times. Sometimes the warm developer must contain 12 g/ℓ sodium sulfate (Na_2SO_4) to prevent too much swelling of the emulsion.

Holographic processing using this technique with a much shorter processing time (15 minutes instead of one hour) results in reflection holograms of a more narrow spectral bandwidth which is caused by a more uniform processing of the emulsion layers. However, when the plate was illuminated with white light, one could see that the diffraction efficiency was not so high which was due not only to the more narrow bandwidth, but also to the fact that amidol could not provide the necessary high contrast. To obtain better results some modifications of the existing nuclear emulsion developers will be necessary and further investigation of this interesting approach to holographic processing is needed. Compare also the new microheterogenous volume recording Focar material described in Sect.6.9.

4.3.2 Divided Development

In addition to the just described processing techniques used for nuclear emulsions, another possibility to get a very uniform development throughout the entire emulsion is to utilize the *divided development technique*. In this case the developer consists of two parts. The first solution contains the developing agent(s) and preservatives (sodium sulfite and potassium bromide). The second solution contains the alkali. These two baths are used in succession *without* a rinse in between. The developing process starts with soaking the emulsion in the first solution. The time here is not critical; one to two minutes will assure that the entire layer is filled with the developing agents. When the emulsion is immersed in the second solution the development process starts. The temperature of this bath and the processing time here controls the development. This type of developing technique is very interesting for the processing of, in particular, reflection holograms. Often, holographic developers consist of two parts (A and B). To apply the divided development technique; simply soak the holographic emulsion first in part A and then in part B. However, some modifications can be suggested here. The part A of the developer can be made stronger, i.e., higher concentration of developing agent(s). This stronger solution is used as the first bath in the process. The same will also apply to part B; it can contain an alkali of higher concentration. Instead of using part B (regular strength or stronger) as the second solution, a regularly mixed (part A + part B) developer can be used for the second part of the process. This means that some fresh developing agents are introduced during the actual development. This technique works with many developers but some are performing better than others. For example, MQ developers are not recommended. In these developers it is difficult to increase the concentration of developing agents because of the poor solubility of hydroquinone. Instead, adurol (chlorohydroquninone) based developers are recommended, in particular, since such a developer works better with harder emulsions to guarantee a good penetration of the solution into the emulsion.

One advantage of this process is that the part A is not becoming oxidized and therefore it is long lasting. Only part B has to be replaced more frequently, but if it contains only the alkali it is cheap.

The divided development technique is a good solution when very unstable developers are used, such as the sulfite-free pyrogallol developer. For example, in holographic processing this technique has been applied using the popular CW-C2 developer with good results.

4.3.3 Special Holographic Developers

There are many developers created specifically for holographic purposes and satisfying the particular requirements of a given hologram type. The choice of a particular developer depends, firstly, on the type of hologram to be made, and, secondly, on whether bleaching will or will not be performed after the hologram has been developed. The following presentation of the existing developers discusses them according to the *processing method* for a given *hologram type*. A full presentation of a given processing technique, including bleaching procedures, is given in Chap. 7 since the choice of a developer depends also on whether or not a bleaching process will be performed after development. In general, a processing scheme which is suitable for reflection holograms normally works equally well with transmission holograms; but the reverse is not always the case.

a) Theory of Holographic Developers

The theory of holographic developers is slightly different from that of conventional photographic materials. Here, a short presentation will be given, highlighting some of the important factors to consider when formulating developers for holograms. Formulating developers which will work well for processing holograms exposed with *high intensity - ultra- short* pulses will differ from the recipes for developers with *low intensity - long exposure* recordings. Processing of holograms exposed with short pulses will be treated in Chap. 6. The references to the present chapter contain more details on the subject of hologram processing.

The main concern in selecting a suitable processing technique is how to best develop the latent image that was formed during the exposure of the holographic emulsion. Normally in holography, a high contrast combined with fine silver grains is desired. A distinction between faster/coarse-grained holographic materials and ultra-fine-grained materials must be made. The latter can be processed using physical and solution-physical development techniques, while the former as a rule can only be processed using chemical development.

b) Developing Agents Used in Holographic Developers

The most common developing agents used in different holographic developers are: ascorbic acid, catechol, hydroquinone, metol, phenidone, and pyrogallol. For transmission holograms often MQ or PQ developers are used and for reflection holograms ascorbic acid, pyrogallol or catechol developers.

c) Silver-Halide Solvents in Holographic Developers

In early papers on holographic development *Usanov* and *Yermolayev* [4.16], and *Biedermann* and *Johansson* [4.17] found that adjacency or edge effects could improve the quality of holograms. In the first paper a special edge-enhancement developer was used to increase the diffraction efficiency of holographic gratings. In the second paper the MTF function of holographic materials was improved by taking advantage of the adjacency effect during development. In both cases developers having certain amount of silver-halide solvent action were employed. The question concerning the use of silver-halide solvents in holographic developers has been treated in a paper by *Smith* and *Callari* [4.18]. In general, the use of sodium sulfite in holographic developers was discussed by *Austin* [4.43] and for conventional developers *van Veelen* and *Villems* [4.44] described its influence on the super-additivity effect. Since sulfite is a weak silver solvent it may eventually dissolve the exposed fine silver-halide crystals before they have been developed. A slight reduction of the exposed grains size can actually be useful to reduce noise or even to create a reflection hologram using a coarse-grained emulsion demonstrated, as by *Thiry* [4.45]. However, for solution-physical development a controlled silver-solvent action is desired, where sulfite will serve this purpose better than, e.g., ammonium thiocyanate. The effects of the presence of sulfite in a developer intended for processing phase holograms will be further discussed in Chap.5. It is believed, in general, that holograms intended to be bleached should be developed in a sulfite-free developer. However, the opinions concerning this problem are by no way uniform and the subject will be treated again in Chap.5.

Austin [4.43] explained that sulfite acts like a silver solvent which is based on the fact that sulfite ions form complex argento-sulfite ions. A solution containing 66 g/ℓ of sodium sulfite can dissolve 1.9 g/ℓ of silver bromide. Since a 4" × 5" Agfa 8E75 plate contains about 0.07 g of silver bromide, the entire silver content of the plate could be dissolved, given sufficient time. A sulfite solution is actually a very slow fixing bath. Sulfite also reacts with the oxidation products of the developer, which can affect the contrast and image brightness. In particular, a positive effect in this respect has been experienced in developers based on metol.

Comparing the PQ and the phenidone-ascorbic acid developers (PA), the following must be observed. As already mentioned, *van Veelen* and *Willems* [4.44] have studied the influence of sulfite on the superadditivity effect of these developers. *If there is no sulfite in a PQ developer the superadditive effect will be drastically reduced. A PA developer, however, will be almost independent of the sulfite content.* The reason for the sulfite dependance in a PQ developer is that the regeneration of phenidone by hydroquinone can only occur if sulfite is present. In a PA developer, regeneration occurs even without sulfite. Therefore, a sulfite-free PA developer for holography is much faster than a PQ version. The same study showed that a PQ developer containing sulfite is more active at higher phenidone concentrations. Therefore, the phenidone content should be rather high in holographic PQ developers containing sulfite.

The importance of silver solvents in holographic solution-physical developers will be treated in Sect.4.3.6.

d) Holographic Developers for Transmission Holograms

In general, development of transmission holograms of the *amplitude* type is quite simple and straightforward. Using modern holographic materials exposed with CW lasers, MQ developers, such as Kodak D-19, for example, work rather well. Conventional photographic developers applied to holography have already been discussed. Some other developers are listed in Chap.10. A high-contrast, low-fog developer which also shows some adjacency effect, would be, however, ideal for this type of holograms. Most conventional developers contain hydroquinone which is not the optimal developing agent for transmission holograms. *Austin* [4.43] found that metol and phenidone can produce higher MTF values at spatial frequencies common in transmission holograms, based on adjacency effects. Hydroquinone reveals no such effects when used alone in a developer. Since these effects are important in holography, developers containing metol combined with phenidone are worth considering for this type of processing. However, phenidone's superadditivity effect combined with hydroquinone is very strong, which explains why hydroquinone often is found in holographic developers.

For creating *phase* holograms special developers have been formulated to work in different processing regimes and with different materials.

Below, Agfa's recipe for a developer for *transmission holograms*, the GP 61, which is formulated for their Holotest materials:

Metol	6 g
Hydroquinone	7 g
Phenidone	0.8 g
Sodium sulfite (anhydrous)	30 g

Sodium carbonate	60 g
Potassium bromide	2 g
Tetrasodium EDTA	1 g
Distilled water	1 ℓ .

Developing time: 2 min at 20° C.

This developer give rather good results, but often some of the developers discussed in the following give slightly better results on Agfa materials, e.g., the adurol developer introduced by *Phillips* [4.32].

e) Holographic Developers for Reflection Holograms

With the exception of applying some semi-physical and solution-physical processing methods, *reflection holograms* recorded on commercial Western materials are seldom processed in the way of creating amplitude holograms because of the low diffraction efficiency of the latter. Therefore, all developers employed in chemical development are used for producing phase holograms. The ultra-fine-grained emulsions of the Russian type are normally developed by using the physical or the semi-physical development technique producing colloidal silver. A developer, primarily intended for reflection phase holograms is often working very well for transmission phase holograms, too.

A *depth developer* (a long-incubation time developer) is preferred for processing reflection holograms. One extreme of such developer is the hydroquinone - potassium carbonate developer. The opposite to the depth developer is the *surface developer*. One example of such a developer is the metol-potassium carbonate developer. The acid amidol developer is a real depth developer. Development, instead of starting at the top of the emulsion, works from the bottom up. Unfortunately, this developer produces rather poor contrast, another important factor in hologram processing.

In general, it is important that the developer can penetrate the entire emulsion before the process starts. This gives a more uniform development throughout the depth of the emulsion. Compare also the divided development technique in Sect.4.3.2.

4.3.4 Chemical Development of Reflection Holograms Exposed with CW Lasers

a) Developers Based on Pyrogallol

Going back to Lippmann's days, it is interesting to study again the way in which the researchers of those days approached the problems encountered then when dealing with materials of the fine-grained type. The Lippmann plates were mainly of the ultra-fine-grained type since they were supposed to work in the whole visible part of the electro-magnetic spectrum including the blue part. This was, and still is, the most difficult problem in the recording and processing of reflection holograms made on silver-halide materials and exposed to short wavelengths. The scientists processing Lippmann photographs used most often the Lumière **pyrogallol-ammonia developer**. Small variations concerning the concentration used, could be observed, but generally it was mixed in the following way:

Solution A		Solution B	
Pyrogallol	1 g	Potassium bromide	15 g
Distilled water	100 mℓ	Ammonia (saturated)	30 mℓ
		Distilled water	150 mℓ .

Mix: 3 mℓ part A + 6 mℓ part B + 100 mℓ water. Developing time: 1 to 3 minutes.

This developer is based on pyrogallol as the developing agent. Ammonia in the developer is acting as the alkali and a silver solvent at the same time. This means that the developer is of the solution-physical type. This type of developer was first used by *Denisyuk* and *Protas* [4.10] for developing the first single-beam reflection holograms recorded in the former USSR.

Van Renesse [4.22] introduced tanning development as an important part of the bleaching process which will be described in the next chapter. *Spierings* [4.46] employed pyrogallol developers *without solvent action* for processing reflection phase holograms, combined with a subsequent reversal bleaching process known as **"pyrochrome processing"**. The name of the process is a combination of "pyro" from the pyrogallol developer and "chrome" from the oxidizer (potassium dichromate) utilized in the bleach bath. See further details in Chap.5. The developer is mixed in the following way:

Solution A		Solution B	
Pyrogallol	10 g	Sodium carbonate	60 g
Distilled water	1 ℓ	Distilled water	1 ℓ .

Mix: 1 part A + 1 part B (pH about 10.6). Developing time: 2 min at 20° C.

This developer both hardens and stains the gelatin. The tanning effect is most beneficial in processing high-speed, coarse-grained holographic materials such as, e.g., Agfa 10E or Kodak SO-253 products. *Saxby* [4.47] added phenidone to this developer to increase the effective emulsion speed of the Agfa 8E75 materials. The following recipe for **Pyrochrome Plus** gives a threefold increase compared with the previous solution:

Solution A		Solution B	
Pyrogallol	20 g	Sodium carbonate	130 g
Phenidone	1.2 g	Distilled water	1 ℓ
Sodium metabisulfite	30 g		
Distilled water	1 ℓ	.	

Mix: 1 part A + 1 part B. Developing time: 2 to 6 minutes at 20° C.

In addition to phenidone combined with a doubling of the pyrogallol and carbonate contents, the developer also contains sodium metabisulfite which will slightly reduce the tanning effect. This means that a color shift can be obtained when processing reflection holograms with a reversal bleach. Sodium metabisulfite has also a preserving effect which means that the mixed developer will last longer.

138

Blyth [4.28] has recommended a special version of the pyrogallol developer used in his improved reversal bleach process:

Pyrogallol	35 g
Sodium phosphate (dibasic)	80 g
Sodium hydroxide	25 g
Distilled water	1 ℓ .

Developing temperature 30° C.

The main difference between the holographic, pure pyrochrome developer and the old Lippmann pyrogallol developer is that the former is a nonsolvent *chemical developer*. Since it contains no sulfite (or metabisulfite), it will oxidize very quickly when the alkali is added. Therefore, the solutions A and B should be mixed just before being used.

However, the use of a silver solvent in a developer can have some advantages for holography as well. *Thiry* [4.45] has shown that applying an ammonia-pyrogallol developer (Valenta's formula) it was possible to obtain a better-quality reflection hologram on the Agfa 10E75 emulsion than that obtained with the pyrochrome processing technique. Normally it is difficult to record reflection holograms on the 10E75 material because of the marginal resolving power and a high *scatter within the 10E75 emulsion during exposure*. The reason it could be done at all in *Thiry's* experiment was that the developer actually reduced the grain sizes *during the developing process*, which, in turn, produced interference fringes in the emulsion of sufficient quality for the holographic image to be reconstructed. The quality of the hologram obtained by *Thiry* is nonetheless not comparable to that which can be achieved by a material with a higher resolving power.

A developer formulated in the present author's laboratory in Sweden in the early 70's was based on a similar approach in order to produce low-scatter Denisyuk reflection holograms, using the highest-resolution plates of that time from Agfa. Instead of using ammonia, the developer contains a rather high amount of sodium sulfite, which also has a solvent effect. At the same time the hardening effect of pyrogallol was reduced. A good color control was achieved using an EDTA-based rehalogenating bleach. This was quite an unusual approach to take at that time. For more details concerning bleaching refer to Chap.5.

The **Holodev 602** developer composed according to the following recipe was used:

Solution A		**Solution B**	
Pyrogallol	50 g	Sodium carbonate	85 g
Sodium sulfite (anhydrous)	130 g	Distilled water	1 ℓ
Potassium metabisulfite	50 g		
Distilled water	1 ℓ .		

Mix: 1 part A + 1 part B + 1 part water. pH = 9.4. Developing time: 2.5 min at 22° -25° C.

The above-mentioned developer was used to produce reflection holograms of the phase type on the existing Agfa 8E75 materials (grain size of about 50nm) with a high signal-to-noise ratio. Because of the high content of sodium sulfite, this pyrogallol developer is stable even when mixed as a working solution. The developer produces high-quality Denisyuk-type holograms on the modern materials as well. When processing today's finer-grained materials, such as the Agfa 8E75 HD, it is recommended to reduce the sodium-sulfite content from 130 g to somewhere between 30 to 60 g. This developer can produce results similar to the more well-known CW-C2 catechol developer. Typically, an exposure energy of 240 $\mu J/cm^2$ gives optimal results in combination with the reversal Pyrochrome bleach for HeNe laser recordings on Agfa 8E75 HD.

In general, pyrogallol-based holographic developers are very popular in processing reflection holograms recorded with CW lasers.

Agfa has formulated a developer based on the combination of metol and pyrogallol, which produces *reflection holograms* on their materials. The developer is the **GP 62**:

Solution A		Solution B	
Metol	15 g	Sodium carbonate	60 g
Pyrogallol	7 g	Distilled water	1 ℓ
Sodium sulfite	20 g		
Tetrasodium EDTA	2 g		
Potassium bromide	4 g		
Distilled water	1 ℓ .		

Mix 1 part A + 1 part B + 2 parts distilled water. Developing time: 2 min at 20° C.

However, most holographers today prefer developers, such as, e.g., CW-C2 for processing reflection holograms.

b) Developers Based on Catechol

The catechol (or pyrocatechol) developing agent for reflection holograms was introduced by *Lamberts* and *Kurtz* [4.15] in an early processing technique for phase holograms. It became popular after the publication of a paper on that subject by *Cooke* and *Ward* [4.48]. Catechol processing has also been treated by *Aliaga* and *Chuaqui* [4.49]; they have found that the tanning effect produced by catechol developers is not as important for obtaining high efficiency as was often believed before. They also showed that improved efficiency could be obtained with the CW-C2 developer as compared to that obtained by Agfa GP 62 developer. The **CW-C2** developer by *Cooke* and *Ward* consists of

Catechol	10 g
L-ascorbic acid	5 g
Sodium sulfite (anhydrous)	5 g
Urea	50 g
Sodium carbonate (anhydrous)	30 g
Distilled water	1 ℓ .

Developing time: 2 min at 20° C (continuous agitation). pH = 8.8.

The best way of mixing this developer is in two-stock solutions with the sodium carbonate in the second part.

The CW-C2 developer has become one of the most successful developers for the processing of the CW-laser-exposed reflection holograms on, in particular, Agfa materials. The use of urea which softens gelatin serves to increase the developer's penetration by the developer by combating hardening of the emulsion due to tanning. The developer therefore introduces some shrinkage of the emulsion, which does not happen when a sulfite-free pyrogallol developer is employed. CW-C2 can be used in combination with rehalogenating bleaches as well as solvent bleaches. The divided development technique (Sect.4.3.2) works well with this developer. It competes with such developers as Holodev 602 and PAAP for the CW-laser-recorded reflection holograms. Typically, an exposure energy of 240 $\mu J/cm^2$ gives optimal results in combination with the reversal pyrochrome bleach for HeNe-laser recordings on Agfa 8E75 HD.

Although the CW-C2 developer gave the best results when combined with the PBQ-2 bleach, *Cooke* and *Ward* presented yet another version of the catechol developer in the earlier-mentioned paper [4.48], namely the CW-C1 developer:

Catechol	10 g
Sodium sulfite (anhydrous)	10 g
Sodium carbonate (anhydrous)	30 g
Distilled water	1 ℓ .

Development time 2 minutes at 20° C.

For better storing purposes it is advisable to mix this developer in two parts where the sodium carbonate is dissolved in the second solution.

c) Developers Based on Other Developing Agents

Among other developing agents encountered in holographic developers metol and hydroquinone stand out as the ones frequently used in various recipes. The Kodak D-19 developer mentioned earlier, which is just one of several types of MQ developers, is normally used for the processing of amplitude transmission holograms, but from time to time it is used in the processing of reflection phase holograms as well. This developer will produce high-quality holograms, provided the right bleaching procedure is chosen for hologram processing.

Phillips [4.32] introduced for holographic processing the chlorohydroquinone (adurol) developing agent as an alternative to the regular hydroquinone. The **adurol developer** was mixed in the following way:

Ascorbic acid	10 g
Chlorohydroquinone	2 g
Sodium sulfite (anhydrous)	30 g
Potassium bromide	5 g
Sodium metaborate (Kodalk)	10 g
Sodium carbonate (anhydrous)	60 g
Distilled water	1 ℓ .

This developer is used at a temperature between 21° and 23° C.

It was mainly introduced as a low-noise transmission-hologram developer which overcomes some peculiar properties of ascorbic acid employed in holographic developers. It works very well both for Agfa and Ilford materials.

If a nonsolvent developer is desired, the developing agent(s) must create photographically inert oxidation products. The best agent for this purpose is probably *ascorbic acid*. Conventional developers without nonsolvents are known as surface developers. *Benton* [4.50] has introduced such a developer, known as the **PAAP-developer:**

Phenidone	0.5 g
Ascorbic acid	18 g
Sodium hydroxide	12 g
Sodium phosphate (dibasic)	28.4 g
Distilled water	1 ℓ .

Developing time: 4 minutes at 20° C, (pH slightly higher than 7).

Nonsolvent developers are believed to produce the highest-quality phase holograms if used with the correct rehalogenating bleaching technique and without fixing before bleaching [4.32].

Another ascorbic acid developer introduced by *Crespo* et al. [4.29] is applied for fixation-free rehalogenating bleaching. The **AAC developer** is mixed in the following way:

Ascorbic acid	18 g
Sodium carbonate to give a pH of	10.5
Distilled water	1 ℓ .

4.3.5 Monobath Developers for Holography

A *monobath developer* is a processing bath where development and fixation are carried out at the same time. Since these two are competing reactions, such baths used in regular photography are balanced in such a way that the development is rapid and the fixation is slow. The composition of a monobath developer is usually prepared especially for each emulsion, taking into consideration even such factors as agitation and temperature. Highly alkaline solutions of phenidone and hydroquinone are popular. The fixing agent is very often sodium thiocyanate. The monobath development technique for conventional photography has been described in a book by *Haist* [4.51].

In holography monobath developers are required mainly within interferometry and Holographic NonDestructive Testing (HNDT). Here, it is important to reduce the time between the exposure and the reconstruction of the hologram in order to speed up the evaluation in many industrial applications.

Ragnarsson has investigated the standard Agfa Dokufix monobath developer for holograms [4.52]. He found that the useful exposure interval for an amplitude hologram was about five times longer when the above-mentioned developer was used in comparison with a conventional developer. For a phase hologram the useful exposure interval was also considerably increased and it was almost independent of exposure over a large interval. The holograms developed with Dokufix were bleached in an Agfa/Orwo 710/I bleach bath (copper sulfate/sodium chloride).

Hariharan described another monobath developer for holography [4.53, 54] - the result of certain modifications to the regular monobath developers which could be made thanks to the fact that holographic emulsions have a high surface-to-volume ratio of the grains. His developer containing lower than usual concentrations of developing agents, alkalis, and sodium thiosulfate produced very satisfactory results. He also achieved a low fog level without adding antifoggants:

Stock solution A

Metol	3.5	g
Sodium sulfite (anhydrous)	50	g
Hydroquinone	15	g
Potassium alum	5	g
Sodium thiosulfate	50	g
Distilled water to make	800 mℓ .	

Stock solution B

Sodium hydroxide	10	g
Distilled water to make	200 mℓ .	

Mix 4 parts of solution A with 1 part of solution B before use.

For Agfa 10E75 materials an exposure of 0.016 J/m^2 at 633 nm and a processing time of 2 minutes at 20°C are recommended. More details about this technique, including a design of the liquid gate, are found in a publication by *Hariharan* and *Ramprasad* [4.55].

A monobath developer for the Kodak 649-F materials was discussed by *Dietrich* et al. [4.56]. It is intended to be used in real-time holography in a liquid gate, where the material, surrounded by water, is both recorded and reconstructed. According to *Dietrich* et al. the following developer gives high-quality holograms on 649-F materials, without silvering the plate holder, which is often a problem with monobath developing techniques:

Stock solution A

Sodium sulfite (anhydrous)	65	g
Hydroquinone	15	g
Phenidone B	0.7	g
Potassium carbonate (anhydrous)	12.5	g
Potassium hydroxide	4	g
Distilled water	700 mℓ .	

Stock solution B

Sodium thiosulfate	100	g
Thiovanol (monothioglycerol)	10 mℓ .	

Prior to use, solution B substances are added to solution A and diluted to make one liter of working solution. The developing time is 5 minutes at room temperature.

In the above case one should use thiovanol (monothioglycerol) as a co-complexing agent with sodium thiosulfate, as this will result in clear holograms and little or no silvering of the processing vessel. Organic silver-complexing agents for photographic monobaths have been studied by *Haist* et al. [4.57]. They recommend monothioglycerol, 2-thiobarbituric acid and the β-mercaptoethylamines in a phenidone-ascorbic acid monobath developer. Since phenidone-ascorbic acid developers are frequently used in holography, in cases when a solvent action of a developer is of interest in a particular regime, these silver complexing agents could be worth trying.

Monobath processing is actually employed by various companies for automatic film processing machines and in holographic cameras. The Holomatic system from Laser Technology, Inc. [4.58] uses a proprietary developer called "Holo-bath", which processes a hologram in 20 to 60 seconds. In the automatic film processing system from Keystone Scientific Co. [4.59] the standard Eastman Kodak 448 monobath is recommended to be used which will produce a hologram in 30 seconds upon exposure.

The thermoplastic camera is probably the most frequently used commercial recording device in real-time holography. Since silver halides are not utilized in thermoplastic cameras, these will not be discussed here as a separate product. The main advantage of the thermoplastic technique is the fast and dry processing. Nevertheless, it might be interesting to mention a new, experimental silver-halide material in the processing of which a dry processing technique is employed [4.60]. In the new material, the dry, silver-halide emulsion contains a developer and stabilizers. The emulsion is processed with gases, such as ammonia or methylamine gas. The sensitivity of the material is high (about $1.2\,\mu J/cm^2$) and the processing fast (2 to 8 seconds). The diffraction efficiency of holograms recorded on the new material is higher than that of identical holograms recorded on Kodak 649-F plates. The signal-to-noise ratio is, however, lower. The material was primarily intended for the application in real-time holographic interferometry. No commercial equipment based on this technique has been developed and therefore this type of recording material is not produced.

Another fast technique for holographic plate processing based on a contact-diffusion method has been described by *Ermolaev* et al. [4.61]. A web containing three layers is applied to the holographic plates (Russian PE-2 emulsion). The first layer which is in contact with the emulsion consists of a fine-structure fiber; the second layer is a coarser fiber material impregnated with a monobath developer of the Russian GP-type and the third layer is made of polyethylene and imparts the required strength to the strip. A special apparatus described in the paper was employed for the contact-diffusion processing of the plates.

4.3.6 Solution-Physical Developers

The monobath developers just discussed are similar to developers used in *solution-physical development*. The main difference between the two is that the purpose of the monobath developer is to omit the fixing step without affecting the quality of the silver image, whereas in solution-physical development the quality of the silver image can be affected in such a way that the diffraction efficiency increases in the final hologram. Therefore the solution-physical developers are slightly different from monobath developers.

The solution-physical development is dependent on the original grain sizes of the emulsion. The technique gives best results in ultra-fine-grained emulsions, such as the emulsions of the Russian type. Some Western materials can be processed in this way as well, but normally it is more difficult to obtain consistent results on them as well as a high image quality. Solution-physical development is also important when processing fine-grained emulsions using regular chemical development. The rate of solution of silver-halide grains in a developer has been described by *James* and *Vanselow* [4.35]. For example, the standard D-19 developer has a silver-halide solvent rate that is about 30 times higher than a nonsolvent metol-ascorbic acid developer [4.8].

The first publication on processing holograms using solution-physical development is not surprisingly a paper from the former USSR [4.62]. The advantages of solution-physical development were, however, noticed and made use of for the processing of Lippmann photographs. *Denisyuk* applied this old type of developing technique (Lehmann's ammonia-pyrogallol developer [4.63] for his first holograms [4.10]. In the paper by *Andreeva* and *Sukhanov* [4.62] a modified version of the ammonia-pyrogallol developer is given as

Solution A			Solution B		
Pyrogallol	1	g	Potassium bromide	20	g
Distilled water	100	mℓ	25% ammonia	30	mℓ
			Distilled water	240	mℓ .

Mix 2.5 mℓ part A + 5 mℓ part B + 92.5 mℓ distilled water. Developing time: 9 to 12 minutes at 17° C.

This developer was aimed for an emulsion (thickness $7\mu m$) prepared according to *Valenta*'s recipe [4.33]. A diffraction efficiency of 20% was obtained for these colloidal silver holograms. The bandwidth was about 15 nm.

Usanov et al. [4.20] reported on an interesting investigation dealing with the size of a developed grain as a function of exposure and the resulting diffraction efficiency versus grain size. The investigated material was the Mikrat LOI-2 plates (grain size $26\mu m$), developed in the solution-physical developer GP-2. The following results were obtained:

- As the amount of exposure increased, the average diameter of the developed silver particles decreased.
- The variation in grain size around the average value decreased with increased developing time.
- For the particular material under investigation, amplitude holograms were created when the exposure was ≤ 24 $\mu J/cm^2$ and phase holograms

when the exposure was ≥ 76 $\mu J/cm^2$. In the first case the developed silver grains were large and possessed a high absorption. In the second case the silver grains (colloidal silver) were reduced in size after development, due to increased exposure. The grains were also capable of altering the refractive index of the emulsion containing them. As the size decreased, the absorption was also drastically reduced (as is Rayleigh scatter noise).

- The size of the colloidal silver grains varied between 15 and 50 nm in the phase holograms. For example, the grain-size variation in the interference fringes formed in a reflection hologram exposed with HeNe laser (exposure $210 \mu J/cm^2$) and developed for 12 minutes in GP-2 was between 15 nm and 30 nm.

The first Western publication on solution-physical development appeared in the form of a short note by *Benton* [4.64]. He developed an interesting, two-step processing method for making reflection holograms, in which the second step made use of solution-physical development. The idea behind all the solution-physical processing techniques is that by preserving the polarizable volume of the emulsion a higher diffraction efficiency can be obtained. Instead of the prevailing concept of silver atom removal, a new concept of silver-halide molecular redistribution within the emulsion was introduced by *Benton*. It consists of the fact that any small area of the processed emulsion will have the same average silver-halide molecular content during reconstruction as during exposure. Instead of being uniformly distributed during exposure, the silver-halide microcrystals will here be redistributed over very small distances into the grating structure constituting the holographic interference pattern in the emulsion. This *Intra-Emulsion-Diffusion-Transfer* (IEDT) method proceeds as follows:

- The usual chemical development step, using the *James* and *Vanselow* **MAA-3** developer [4.35] (contains no silver-halide solvents):

Metol   2.5 g
Ascorbic acid   10 g
Sodium carbonate 55.6 g
Distilled water   1 ℓ .

- A solution-physical development step, using the following developer [4.65]:

para-phenylenediamine dihydrochloride 18.1 g
Sodium sulfite (anhydrous)   50 g
Potassium bromide   1 g
Sodium thiocyanate   0.25 g
Distilled water   1 ℓ .

Adjust to pH = 8 with sodium metaborate. Development time: 32 minutes (not critical, the diffusion-transfer process self-terminates when the silver-halide supply has been exhausted).

As this process was intended for materials with slightly larger grains than what is really desirable for a solution-physical development, the plates were bleached to convert all the silver created in the process to a transparent silver salt using a rehalogenating bleach.

The advantage of this method is that a high diffraction efficiency can be obtained, which, combined with the fact that no relief image is created on the emulsion surface means that low-spatial-frequency intermodulation noise is reduced. The main drawback of the IEDT method when used with Western materials is that in solution-physical development excessive scatter noise is produced after bleaching. In another publication *Benton* [4.66] compared a nonsolvent chemical developer TMRA-3 which is based on tetramethyl reductic acid and potassium hydroxide possessing inert decomposition products with the IEDT process.

In general, the solution-physical development works best with ultra-fine-grained emulsions in which the sizes of the colloidal silver grains formed in the emulsion are so small that very little absorbtion occurs (they act more like phase holograms), and where Rayleigh scattering is low. If coarser-grained materials are processed in this way and bleaching is applied after development, strong scattering will often occur. The reason for this is that the process of the solution-physical development is difficult to control since, unfortunately, silver also tends to precipitate at random within the emulsion, producing "dichroic fog" or forming a scum on the emulsion surface. These phenomena are more pronounced in the coarser-grained materials than in the ultra-fine-grained ones. The extent of the above-mentioned effects depends also on the exposure time as well as the exact adjustment of the silver-solvent complex in the developer, which has to be carefully balanced for a given material.

As already mentioned, solution-physical development has been frequently applied in the former USSR [4.10]. A well-known solution-physical developer is the **GP-8** developer which consists of

Methylphenidone	0.2	g
Hydroquinone	5	g
Sodium sulfite (anhydrous)	100	g
Potassium hydroxide	10.6	g
Ammonium thiocyanate	24	g
Distilled water	1	ℓ .

Mix 60 mℓ of the developer with 400 mℓ of distilled water. Development time: 6 minutes at 20°C.

This developer works very well with the ultra-fine-grained materials from Russia. Nevertheless, it is the GP-2 developer which is most frequently used in that country. Other popular Russian developers are presented in Chap.10.

GP-2 stock solution:

Methylphenidone	0.2 g
Hydroquinone	5 g
Sodium sulfite (anhydrous)	100 g
Potassium hydroxide	5 g
Ammonium thiocyanate	12 g
Distilled water	1 ℓ .

Working solution: 15 mℓ stock solution + 400 mℓ distilled water. Developing time at 20° C is 12 minutes without agitation.

Several papers discussing the use of the GP-8 developer for the processing of Western materials, in particular the Agfa HD-materials, have been published [4.67-72]. In most of the investigations some modifications to the basic formulation for the developer were made.

Ruzek and *Fiala* [4.67] used the GP-8 developer strictly adhering to its basic formulation for Agfa 8E75 plates exposed with a HeNe laser. They only modified the development time (3 minutes at 20° C). The plates were exposed with the energy density of 300 $\mu J/cm^2$, which is about 10 times higher than when chemical development is applied. High-quality holograms with a diffraction efficiency of 25-30% were obtained.

Aliaga et al. [4.68] employed a slightly modified version of the GP-8 developer for Agfa 8E75 HD plates. They called that particular processing technique the **CPA-1** processing:

	CPA-1	GP-8
Phenidone	0.02 g	0.026
Hydroquinone	0.65 g	same
Sodium sulfite (anhydrous)	13 g	same
Potassium hydroxide	1.4 g	1.38
Ammonium thiocyanate	3.1 g	3.12
Distilled water	1 ℓ	same .

Development time: 2 minutes at 22° C (+/-1° C) (3s initial agitation).

Aliaga et al. emphasized the importance of preparing and diluting the developer with clean water, preferably with deionized water. The difference between the composition of CPA-1 and GP-8 is actually very small - the main difference being the *reduced development time*. This seems to be the only outstanding difference between the processing of Western materials as opposed to Russian materials with the GP-8 developer. The reduction of development time when processing Western materials is necessary for the reason that the Agfa emulsion is more active than the Russian materials. By reducing the time of development one is able to keep the dichroic fog at an acceptable level. The degree of dichroic fog depends on the developer activity, agitation, development time, temperature, and concentration. Exposure levels used in this investigation for 8E75 HD material were between 500 and 1000 $\mu J/cm^2$. The CPA-processed Agfa materials 8E75 HD and 10E75 show some interesting features described by *Aliaga* and *Chuaqui* [4.69]. Electron micrographs of the emulsions indicate a slight grain growth for the 8E75 HD emulsion when fixed after development in CPA-1. For the 10E75 emulsion an opposite effect is observed; a reduced grain size is obtained after fixing. In the same paper a micrograph of a CW-C2 developed emulsion shows the common filamentary structure. If such an emulsion is bleached (using PBQ-2) instead of being fixed, the grains are fairly regular in shape, not very different from the original grains. *Aliaga* et al. claimed in [4.70] that the solution physical developer CPA-1 followed by a fixing step produces holograms with high archival stability combined with a diffraction efficiency up to 26% for Agfa 8E75 HD materials. Another advantage is that an improved linearity is obtained with this process.

Spierings [4.71] discussed some practical aspects of the Western materials' processing with the GP-8 developer. He mentions the importance of reducing the development time when processing the Agfa HD-materials and, in addition, recommends to watch the following points as well:

- The activity of the developing part of the processing solution which is affected by dilution, PQ ratio and pH.
- The activity of the fixing part of the solution which seems to be of crucial importance. Fixing for too long and/or too quickly will result in the disappearance of the image. Agitation will increase the fixing rate. If the solvent action is too low, no physical development takes place and fog will be created, caused by the precipitation of silver thiocyanate in the emulsion.
- The temperature of the processing solution. Development rate increases faster with temperature than does the fixing rate.
- The processing time is critical. Normally, a rather short processing time produces better results, with the hologram showing less tendency to dichroic fog, provided the material has undergone a much higher exposure than what is normal in chemical development.

Bonmati et al. [4.72] investigated four different solution-physical developers for processing reflection holograms on specially made silver-halide emulsions containing various silver-halide concentrations. They found that the Russian developers GP-2 and GP-8 were the best. The optimum silver-halide concentration in an emulsion suitable for solution-physical development is about 1 g/m^2 of coated emulsion (thickness $10 \mu m$). *Fiala* et al. [4.73, 74] used the GP-2 developer to process their own silver-halide material from Czechoslovakia (SHE-ZE-3) applying the solution-physical development technique.

Leclère et al. [4.75, 76] investigated solution-physical development and optimized a developer consisting of pyrogallol, sodium hydroxide and ammonium thiosulfate for the Agfa 8E75 HD material. By varying these three constituents it is possible to obtain a hologram with either large bandwidth and low reflectivity or with a narrow bandwidth and high reflectivity. Also a developer for the Ilford SP 737 T emulsion has been formulated by *Renotte* et al. [4.77]. The following optimized developers for producing large bandwidth (70÷200nm) reconstruction on Agfa and Ilford materials are described in these three papers. No accurate information as regards the needed exposure energy is, however, provided in these papers. It is only mentioned that the Ilford emulsion needs three times higher exposure than the Agfa material:

For processing:	Agfa 8E75 HD	Ilford SP 737 T
Pyrogallol	3.0 g	3.0 g
Ammonium thiocyanate	5.5 g	4.4 g
Sodium hydroxide	2.3 g	2.5 g

| Distilled water | 1 ℓ | 1 ℓ |
| Development time at 20° C | 3 min | 3 min . |

The materials are normally fixed after the development. The publications also contain information concerning bleaching, fixation-free processing and comparisons with chemical development. For holograms processed in the solution-physical type of developer the MTF of the material was increased 20% over chemically developed holograms.

In China some papers have been published [4.78-85] on using diluted developers for the processing of the domestic holographic plates: HP-633P, Thianjin and Hep-F ui. All these plates are of the ultra-fine-grained type. For the development of the super micro-fiche holography storage system (SMH) at the Tianjing Radio Technique Institute a very-high-resolution recording material was required [4.78, 79]. A theoretical analysis (based on the Maxwell-Garnett effective medium theory) of the development of silver-halide materials with developers containing a diluted solvent has been presented [4.85]. The recipes for highly diluted Chinese developers are given in [4.80-84]. One of them is the N6 developer [4.84] presented here:

Metol	0.5 g
Sodium sulfite (anhydrous)	100 g
Hydroquinone	45 g
Sodium carbonate	30 g
Potassium thiocyanate	5 g
Potassium bromide	10 g
Distilled water	1 ℓ .

Mix 1 part developer + 8 parts distilled water. Development time not mentioned in [4.84].

Another developer mentioned in [4.83] is the Kodak D-76 to which 2.5 g ammonium thiocyanate per liter has been added. The working solution was then diluted 1:15 (development time 30 minutes). The diffraction efficiency obtained on the HP 633P plates using this developer was 50%.

It is worth mentioning that often it has been found that by overexposing the holographic material one can obtain higher-quality holograms. This is, in particular, true when using a developer containing silver-halide solvents. *Solman* [4.5] found for conventional fine-grained emulsions that if physical development also takes place in a chemical developing process the grain sizes will depend on the exposure. This was also confirmed, as already mentioned, by *Usanov* et al. [4.20] as well as *Bonmati* et al. [4.72] concerning holographic processing. When the recording material is exposed to a high level more grains are made chemically developable and thus there are fewer undevelopable grains left to provide silver ions for physical development. Hence, the filament dimensions decrease with increased exposure which was experimentally verified in *Solman*'s investigation. This is important to remember when developing holograms that are supposed to be bleached using a rehalogenating bleach process. If a developer containing silver-halide solvents is used the exposure of the material must be high to produce smallest possible silver grains after development. Therefore, often developers containing no silver-halide solvents are recommended for this type of processing.

4.3.7 Physical Development

There are very few reports on processing holograms using pure *physical development*. An early investigation demonstrated the possibility of using the old Agfa Scientia 10E70 materials for producing reflection holograms [4.86]. After exposure, the plates were fixed for one minute. After that they were washed for five minutes. The physical development took three minutes (in daylight). The following developer was used (after *Lumière* and *Seyemetz*):

Solution A		Solution B	
Sodium sulfite (anhydrous)	180 g	Sodium sulfite	20 g
Silver nitrate solution (1/10)	75 ml	Metol	20 g
Distilled water	1 l	Distilled water	1 l .

Mix 5 parts solution A + 1 part of solution B. Development time: 3 minutes.

The following findings were reported in [4.86]:

- The physically developed plates show less surface relief than the chemically developed plates.
- The exposure of the holographic material intended for physical development should be 2 to 5 times higher than that of the material intended for chemical development.
- An increased noise level was observed compared to chemically processed holograms (partly due to the fact that the old 10E70 material was rather coarse-grained).
- The grain size of the physically processed emulsion depends solely on the time of development. A three-minute development time was found optimum in this case.

Another investigation by the same group [4.87] reported on the possibility of using reflection holograms for microscopic recordings. This time the material used was the Agfa Scientia 10E75. A comparison was made between physically developed plates using the above-mentioned technique and conventional processing for amplitude holograms developed in Kodak D-19b, or phase holograms bleached in Kodak R-10. The main conclusions were that the efficiency of the physically developed plates is higher than for conventionally developed amplitude holograms, but not as high as for chemically processed phase holograms. The bandwidth, however, is narrower for the physically developed plate (23 nm) compared to the amplitude hologram (28 nm) and the bleached hologram (31 nm).

Two physical developers from the former USSR were used by *Verbovetskii* et al. [4.88] to make holograms with binary data on VRL holographic plates. One developer (F1) was mixed in the following way:

Amidol	4 g
Sodium sulfite (anhydrous)	30 g

Silver nitrate	3 g
Potassium bromide	2 g
Sodium thiosulfate	45 g
Distilled water	1 ℓ .

Developing time 8 minutes.

Using this developer, both physical and chemical development occur simultaneously. It is recommended to fix the material after development for 2 to 3 minutes.

The other developer (F2) was mixed as follows:

Metol	10 g
Sodium sulfite (anhydrous)	100 g
Silver nitrate	2 g
Potassium bromide	2 g
Sodium thiosulfate	30 g
Distilled water	1 ℓ .

Developing time 35 minutes.

When processing holograms in this developer the process of developing and fixing occurs simultaneously which means that no additional fixing is necessary after development.

It is mentioned in the paper that a much higher exposure is needed when using physical development than in the case of chemical development. The best quality (very high contrast) was obtained using the F2 developer.

4.4 Photographic Fixation

4.4.1 Stop Baths and Fixation Solutions

When the development has been completed a conventional photographic material must be treated in an acid stop bath or it must be rinsed in water, after which it is treated in a fixation bath. The fixation solution will dissolve the unexposed silver-halide crystals leaving only the silver grains in the gelatin. The stop bath is normally a diluted *acetic acid* (CH_3COOH) solution. The fixation step is either performed in a rapid or a slow fixer, with or without hardener. In principle, the fixation step can be expressed by the formula

$$AgBr + 3\ Na_2S_2O_3 \rightarrow Ag(S_2O_3)_3^{5-} + 6\ Na^+ + Br^- . \tag{4.4}$$

The silver thiosulfate ion in the formula is the most important one of the several possible silver thiosulfate complexes. It is easily soluble and will diffuse from the emulsion into the fixing bath. The remaining ions can be readily washed out from the emulsion in the subsequent wash. However, if the silver concentration in a fixing bath becomes excessive, less soluble complexes are formed which are difficult to wash out. If these remain in the emulsion they can cause yellow silver-sulfide stains. Therefore, the fixing bath should be replaced with a fresh one before the silver concentration becomes too high.

There are four different types of fixing baths used in photography:
(i)   Plain fixing baths.
(ii) Acid fixing baths.
(iii) Hardening fixing baths.
(iv) Rapid fixing baths.

4.4.2 Plain Fixing Baths

The advantage of a plain or "neutral" fixing solution is that it will not attack the very fine silver grains in the emulsion, which the acid fixer attacks. The disadvantage of using a neutral bath is that if some residual developing solution has been mixed with this type of fixing bath, there is a risk of creating dichroitic fog in the emulsion. However, if the material is properly washed this danger can be eliminated and the gentle performance of this fixing bath is exactly what is needed when processing fine-grained holographic emulsions. A standard "neutral" fixer (pH about 5.2) is based on *sodium thiosulfate (hypo)* $(Na_2S_2O_3 \cdot 5H_2O)$ and can be prepared in the following way:

Sodium thiosulfate (hypo)	200 g
Distilled water	$1\,\ell$.

4.4.3 Acid Fixing Baths

The most common fixing bath for photography is the acid fixing bath. Acid baths are based on *sodium thiosulfate* (like the previous fixing bath) combined with either *potassium metabisulfite* $(K_2S_2O_5)$, *sodium meta bisulfite* $(Na_2S_2O_5)$ or *sodium bisulfite* $(NaHSO_3)$ in order to obtain an acid solution with a pH of 4 to 5. In order to protect the pyro from the action of the acid carried over from a stop bath if required, a small quantity of *sodium sulfite* (Na_2SO_3) is also added to the bath. When a strong acid is added to a hypo solution, the hypo starts to decompose, forming a fine suspension of sulfur in the solution. The sulfur enters the emulsion creating an opalescent structure in it, the so-called *colloidal sulfur*. Sodium sulfite reacts with the sulfur, forming sodium thiosulfate and acting as a preservative and regenerator of the pyro in the fixing bath:

$$Na_2SO_3 + S \rightarrow Na_2S_2O_3 . \tag{4.5}$$

153

4.4.4 Hardening Fixing Baths

The hardening fixer is an acid fixing bath to which a hardening agent has been added, such as, e.g., formalin, alum; chrome alum $[Cr_2(SO_4)_3 \cdot K_2SO_4]$ or potassium alum $[Al_2(SO_4)_3 \cdot K_2SO_4]$. These are buffered to obtain a rather low pH, about 3.1 to 4.1. The hardening of an emulsion is done in order to prevent excessive swelling and softening during washing, with the consequent danger of mechanical damage. The hardening is also important if the emulsion is going to be dried at a high temperature. These baths are not recommended for holographic processing.

4.4.5 Rapid Fixing Baths

The rapid fixer is a fixing bath that will clear the photographic material much faster than the conventional acid fixer. It is based on *ammonium thiosulfate* $\{(NH_4)_2S_2O_3\}$ or *ammonium thiocyanate* (NH_4SCN). These types of rapid fixing solutions are not recommended for the processing of fine-grained emulsions. The solvent action of ammonium thiosulfate on fine silver grains is stronger than that of sodium thiosulfate. The fixing rate in general is determined by the following factors:

- Thiosulfate type.
- Fixer concentration.
- Temperature.
- Amount of silver already in the fixer.

An old rule of thumb in photography is that the total fixing time should be about twice the clearing time of the emulsion.

4.4.6 Stop Baths

The use of a stop bath in ordinary photography is common, but when processing holograms certain points should be kept in mind as regards the type of hologram to be processed. Amplitude holograms and certain types of phase holograms are also normally fixed before bleaching. It is important to avoid contamination resulting from the use of different processing solutions, which is why careful washing between active baths is necessary. It is also important to maintain a constant temperature of all processing solutions, including all washing baths.

If a developer contains sodium carbonate (like in Kodak D-19) which when mixed with the acetic acid stop bath can cause liberation of carbon dioxide, it will result in the emulsion being perforated with a multitude of tiny bubbles. The other consideration is the colloidal sulfur formation if a plain fixing bath is used. Therefore, it is safer to employ just a pure water bath followed by rinsing for at least 5 minutes before fixation. However, a stop bath will immediately arrest the development process, which will shorten the necessary rinsing time compared to when using only water baths. If a stop bath is applied, it should not be too strong. A suitable stop bath for holography may consist of

Acetic acid (glacial)   10 mℓ
Distilled water   1 ℓ .

The treatment time is between 15 to 30 seconds at 20° C.

An alternative for holography is to use a stop bath based on citric acid and mixed in the following way:

Citric acid   15 g
Distilled water   1 ℓ .

4.4.7 Fixation of Holograms

Fixing baths are used mainly for the processing of amplitude holograms. Several processing techniques for phase holograms require also a fixing step. As all the holographic silver-halide materials are of the fine-grained type, slower fixing solutions can be used, since even here the solvation of the small unexposed silver-halide grains will take a very short time only. Rapid fixers containing such fixing complexes as ammonium thiosulfate for example, are not recommended. A general rule in hologram processing is that the slower the action of a solution, the more gentle its action on the emulsion. A nonhardening fixer, such as the Kodak fixing bath F-24 (pH about 4.6 to 5.0) for example, can be used for holograms:

Sodium thiosulfate (hypo)   240 g
Sodium sulfite (anhydrous)   10 g
Sodium bisulfite   25 g
Distilled water   1 ℓ .

Treatment time double the visible clearing time at 20° C.

The previously mentioned plain fixing bath is best to use, on the condition that the holographic material has been carefully washed and no traces of the developer have been left in the emulsion. After the fixing step the emulsion must be carefully washed again since holograms which are going to be bleached must be free of all hypo stains. One can use conventional hypo clearing baths to decrease the washing time.

The use of a hardening fixer is not recommended, especially for holograms which are to be bleached. If hardening of the emulsion is necessary for satisfactory drying of the hologram, a separate hardening bath can be employed later on.

During the fixation step, unexposed silver salts are removed from the emulsion, which will result in a shrinkage of about 15 to 20% of the original thickness of the emulsion. *Wilkomerson* and *Bostwick* studied the influence of the fixation process on the reconstructed image [4.89]. If the shrinkage is not uniform, it can affect the diffraction efficiency; it will slightly change the reference angle at reconstruction; and it will also reduce the resolution of the holographic image. The way in which shrinkage af-

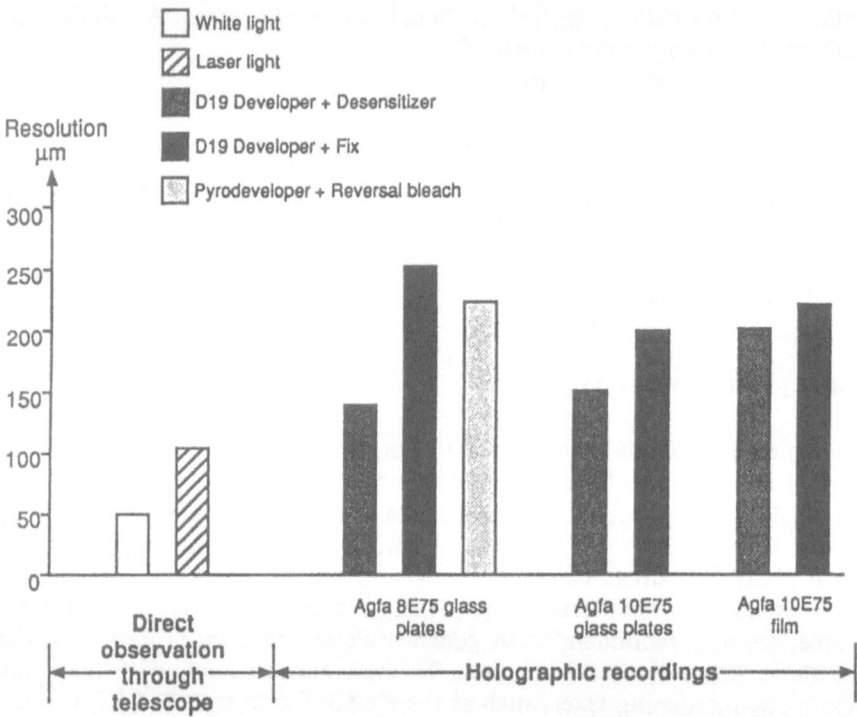

Fig.4.5. Holographic image resolution test. A test target positioned at a two-meter distance from the location of the holographic plate was observed through a telescope, using white-light and laser-light illumination. The reconstructed virtual image of the test target recorded in different holograms was also observed through the telescope. The influence on resolution due to different emulsions and substrates as well as different processing techniques is shown

fects the resolution of a holographic material in bubble chamber recordings has been studied by the present author [4.90] and some results are shown in Fig.4.5.

4.4.8 Desensitization

In order to avoid negative effects of shrinkage on the quality of the holographic image, it is sometimes best to avoid the fixing of the emulsion. Instead, the emulsion can be *desensitized*. Holographic materials can be treated in the following *desensitizing solution* - which procedure will produce stable amplitude holograms with a higher resolution than if they had been fixed:

Phenosafranine ($C_{18}H_{15}CIN_4$)   300 mg
Methanol   0.5 ℓ
Distilled water   0.5 ℓ .

Treatment time of about 3 minutes.

The experimental results of the investigation are summarized in Fig.4.5 which shows that the best image resolution is obtained on glass plates for amplitude holograms when the fixing step has been omitted. Here, the resolution is very close to "the best possible" which corresponds to that of the resolution observed in the test target illuminated with laser light and studied through a telescope (no hologram involved). The telescope placed in one and the same position was used to study different holograms recorded from the test target. The holograms were recorded on both glass plates and film, and processed with different techniques. The reconstruction was made in the same setup as the one used for the recording of the holograms and with the same wavelength (identical reference beam). In this way, the influence of the recording material and the processing regimes on the resolution of the holographic image could be investigated. And thus, the resolution observed directly (without holograms) through the telescope in laser light was 110 μm, a reconstructed hologram recorded on a desensitized glass plate without fixing gave a resolution of 140 μm, while a fixed plate gave a resolution of about 250 μm. When holograms are used for high-resolution imaging, the actual processing technique is very important for obtaining the optimal resolution. Sometimes it is recommended to employ a liquid gate when reconstructing holograms recorded on film [4.91].

Instead of using the desensitizing technique just discussed, a method of controlling the emulsion thickness is recommended [4.92]. If fixing has been performed (or a dissolving bleach has been used) the emulsion thickness can be permanently altered by imbibing a water-soluble monomer into the emulsion layer and polymerizing it in situ. For Kodak high-resolution plates and the Kodak 649-F plate a 70% aqueous solution of N-vinyl-2-pyrrolidone gave the same thickness of the emulsion after processing (development and fixing) as that during the recording. The treatment time was 30 minutes in the solution, after which the plate (still wet) was placed 10 cm from a 125 W mercury lamp for one hour where polymerization (and drying) occurred. For fine control of the profile it is necessary to calibrate the monomer concentration for each batch of plates, which applies also to materials other than the ones from Kodak.

In Chap.6 a paper by *Angell* [4.93] will be discussed. He used an organo-silane coupling agent in the fixing step, which makes it possible to maintain the emulsion thickness after the processing. The N-(2-aminoethyl)-3-amino-propyltrimetohoxy silane agent (Dow Corning Z-6020) [4.94] was recommendatd since it is compatible with the gelatin matrix. This new technique introduced by *Angell* makes it possible to obtain a permanent chemical way of controlling emulsion thickness in holographic emulsions. It is not adversely effecting the noise level of the hologram. Actually, it is improving the signal-to-noise ratio according to *Angell*. This method is dependent on emulsion type, silver-halide solids loading, exposure level, bleaching technique, etc. For the Kodak 649-F emulsion, *Angell* found that adding 2% to 4% to the fixer resulted in an emulsion thickness of 16 μm after drying which was equal to the original thickness. This percentage has to be adjusted in each and every case when other materials or processing methods are used.

Another way of reducing the shrinkage effect was proposed by *Hernandez* et al. [4.95]. They incoherently superimposed a second grating structure in the emulsion in addition to the holographic fringe pattern. The additional fringe structure was perpendicular to the holographic one, and it is mechanically strengthening the gelatin layer by increasing the mass of silver. It was demonstrated to be useful both for transmission and reflection holograms. The drawback of the method is that it reduces the diffraction efficiency of the hologram due to cross-coupling between the fringe systems.

5. Bleaching

Bleaching is an important phase in the recording of holograms on silver-halide materials, ensuring a high diffraction efficiency which is so important for holographic images. In this chapter the general theory for photographic bleaching is reviewed, followed by a detailed description of the holographic bleaching theory. The history of the holographic bleaching technique is summarized and common bleach baths used in holography today are carefully described. The printout problem of bleached holograms is treated at the end of the chapter, where also the stability of bleached holograms as well as methods of increasing the printout resistance are discussed.

5.1 Photographic Bleaching

Bleaching is the process of converting the silver image produced during the development into a water soluble or an insoluble silver compound. In conventional photography bleaching is used to reduce the density and to vary the contrast of black-and-white images. The processing solutions used for bleaching are therefore often referred to as *reducers*, a confusing term since from the chemical point of view bleaching is an *oxidation process* and not a reduction. The bleach itself becomes chemically reduced by silver during this process. The major photographic application of bleaching is in the process of removing the negative silver image from developed dye images in color photography. Information on the conventional bleaching and intensifying processing techniques can be looked up in the general photographic literature. In holography, bleaching represents a processing technique of vital importance and it will be comprehensively reviewed, placing emphasis on producing phase holograms.

Metallic silver is stable in aqueous solutions at all pH-levels. To affect the silver metal an *oxidizing agent* is needed. There are several agents that can be used in various bleaching baths. However, not all oxidizing agents are suitable for photographic purposes. Nitric acid, for example, is an oxidizer which readily dissolves silver to form a soluble salt, silver-nitrate, but it also attacks the gelatin, converting it into a soluble compound. Another example is sodium hypo chloride (household bleach) which affects the gelatin as well.

In the following, several oxidizing agents suitable for photographic applications will be presented. They can be grouped into three classes:

(i)   Metallic ion.
(ii) Metallic oxide.
(iii) Nonmetallic oxide.

Iron, copper, and cobolt belongs to metals that can be used as silver bleaching agents. Iron-based bleaches are very common (e.g., ferric EDTA solutions) because of their low toxicity. Copper bleaches are fast working and can be formulated as both acidic and alkaline versions. The type of metal contained in metallic ion bleaches is found in the transition metal region of the periodic system. The metal should be such that it can exist in multiple ionic states. Certain metal oxide ions are also good silver bleaches when in their highest oxidation states, e.g., dichromate and permanganate. The by-products of the chemical reactions in which they are involved have a staining effect on the gelatin, which can be advantageous in that it reduces scatter noise when producing holograms. Some elements of the periodic system, such as sulfur and boron, for example, have metallic properties without being metals. The important feature is their ability to form oxides in high oxidation states. Persulfate ion is an example of such a bleaching agent. Among the nonmetallic organic compounds quinone is a representative example.

5.1.1 Oxidizing Agents

This subsection gives a presentation of photographic oxidation agents in alphabetical order.

Bromine: Br_2. Bromine gas or bromine solutions can be used to oxidize the silver image to silver bromide. Bromate (BrO_3^-) is also a potential oxidizing agent.

Ceric: Ce^{4+}. The yellow ceric sulfate $Ce(SO_4)_2$ is a slowly acting agent. It dissolves readily in water, and is the most costly agent used for reversal bleaches. Sometimes ceric peroxide is used as a bleach solution.

Chlorine: Cl_2. Chlorine gas or chlorine solutions can be used to oxidize the silver image to silver chloride.

Cupric: Cu^{2+}. Cupric chloride ($CuCl_2$) or cupric bromide ($CuBr_2$) are used. Cupric bromide can be replaced by a mixture of cupric sulfate ($CuSO_4 \cdot 5H_2O$) and potassium bromide (KBr) which acts in the same way.

Dichromate: $Cr_2O_7^{2-}$. Potassium dichromate ($K_2Cr_2O_7$) has the advantage of a high valency (+6). In the presence of sulfuric acid, it readily gives up its excess oxygen to oxidize silver. This is a moderately rapid agent. Chromium salts, such as chromium sulfate have the property of combining with gelatin to form a complex, hardening the gelatin. The hardening effect is proportional to the amount of chromium sulfate, which makes hardening proportional to the optical density of the emulsion. This agent is therefore used when hardening is of interest, e.g., in the Carbro process. Employed with an acid, this oxidizer forms a popular reversal bleach bath frequently used in holography.

Ferric: Fe^{3+}. Iron can exist in the ferric form (valency +3) and a ferric salt can oxidize metallic silver. It is the most practical agent to use for rehalogenating bleach solutions. It must be employed with a chelate of high stability, such as EDTA for example, to prevent iron stains in the material.

160

Normally the iron III chelate of EDTA is utilized. Combined with bromide, it is slowly acting. The ferric-EDTA bleaches are very common, both in photographic color processing as well as in holography. However, there exist other chelates, such as disodium salts of Imino Diacetic Acid (IDA), Methyl Imino Diacetic Acid (MIDA) and Nitrilo Triacetic Acid (NTA). A fourth alternative is the dipotassium DiethyleneTriamine Pentaacetic Acid (DTPA) which Ilford is using in a holographic bleach. The ferric chloride-acid-oxalate-thiosulfate solution is known as the *Belitski reducer*.

Ferricyanide: $Fe(CN)_6^{3-}$. Potassium ferricyanide or hexacyanoferrate III {$K_3[Fe(CN)_6]$} (red prussiate of potash) is a complex salt containing the trivalent iron ion. In the presence of a soluble halide, potassium ferricyanide will convert silver to silver halide. This is a rapid agent which is most often used with bromide to slow down bleaching speed, which is why it is used only for rehalogenating bleaches. This agent was frequently applied to photographic color processing in the past since it does not attack the dye image, but because of toxic waste problems, it has currently been replaced by the Fe-EDTA bleaches. Ferricyanide solutions are decomposed by light, which is why they should be kept in darkness and used in dim light only. The photographic reducer called *Farmer's reducer* is based on this agent combined with thiosulfate. If combined with thiocyanate instead, the reducer is known as *Haddon's reducer*.

Hydrogen peroxide: H_2O_2. This agent tends to soften the gelatin in regions of high density during bleaching. It is often used in the etch bleach process combined with copper chloride to bleach the silver image and soften the gelatin so that these areas can then be washed out. It attacks badly most holographic gelatin emulsions.

Iodine: I_2. Iodine solutions can be used to oxidize the silver image to silver iodide. Most conveniently, iodine dissolved in potassium iodide should be used. Note that a mixture of hypo and iodine is impractical as the two react to form a complex which neither oxidizes nor fixes.

Mercuric: Hg^{2+}. Mercuric chloride ($HgCl_2$) is sometimes employed in photographic intensification processes. Mercuric chloride (corrosive sublimate) is a dangerous poison which must be handled with great care. It is always applied in a two step process since the silver chloride becomes white when bleached in mercuric chloride. The silver in the emulsion is converted to silver chloride (AgCl) and the mercuric chloride around each grain is converted to mercurious chloride (Hg_2Cl_2) (calomel). When the silver particles are very small, a possible reaction might be the forming of a double chloride (silver mercurochloride, $HgAgCl_2$). In photographic image intensifying, the white image is blackened using ammonia, sodium sulfite, or a sulfite-free developer. In addition to mercuric chloride, both mercuric bromide and mercuric iodide have sometimes been used.

Permanganate: MnO_4^-. Potassium permanganate ($KMnO_4$) is mainly used because of its high valency (+7). In the presence of sulfuric acid, it readily gives up its excess oxygen to oxidize silver. The crystals are of violet color and the dilute solution is bluish pink. The crystals must not be allowed to come into contact with concentrated sulfuric acid which decomposes them violently. This is a moderately rapid agent which combined with the acid forms a popular reversal bleach solution. Acid permanganate has little effect on the gelatin, which makes it suitable for photographic use. However, it stains the emulsion by the residual brown manganese dioxide. The acid permanganate reducer is known under the name of *Namia's reducer*.

Persulfate: $S_2O_8^{2-}$. Potassium persulfate ($K_2S_2O_8$), ammonium persulfate [$(NH_4)_2S_2O_8$], and sodium persulfate ($Na_2S_2O_8$) are oxidizing agents that can be used to form slow-acting bleach baths. Persulfate oxidizers are sensitive to salts of iron and copper, and to chlorides. The presence of chloride in a persulfate solution is revealed by the formation of a white precipitate in dense parts of the emulsion. Only persulfate of the highest purity should be used and solutions should be prepared using distilled water. Ammonium persulfate is obtained in colorless crystals which are very hygroscopic. When the crystals have absorbed moisture they become unstable. Ammonium persulfate is easily soluble in water, but the solution is not very stable. Potassium persulfate is less water soluble, but for this reason much easier to obtain in a pure form. Sodium persulfate is used in bleaches for color motion-picture film. Persulfate bleach removes residual stain.

Quinone: $C_6H_4O_2$. In conventional photography, quinone combined with thiosulfate has been used as a single-solution intensifier. *Para*-Benzo-Quinone (PBQ) has been introduced as a good bleaching agent for holograms. PBQ can cause eye irritation, and with longer exposure, damage to the cornea. To reduce the risks, in particular when mixing the solution, one can form *para*-benzoquinone in the bleach by oxidizing hydroquinone using potassium persulfate (Sect.5.2.3).

5.1.2 General Bleaching Theory

Bleaching is the reversed process of development. During the developing process, a silver ion is reduced to free silver whereas during the bleaching process, metallic silver is oxidized to a silver ion. It is interesting to consider the chemical equation for development being reversible. For example, the action of hydroquinone on silver bromide is:

$$C_6H_4(OH)_2 + 2AgBr + 2OH \leftrightarrow C_6H_4O_2 + 2Ag + 2Br + 2HOH . \quad (5.1)$$

The reversed action is the formation of hydroquinone and silver bromide caused by the oxidizing agent quinone acting on metallic silver. In practice, the reversal (oxidation) can only take place in an acid environment. The development (reduction) is only taking place in an alkaline solution. Actually, acid quinone bleaches are used in some cases.

There are various bleach solutions formulated on the basis of the different oxidizers just discussed. A concern in one area of conventional photography is to apply bleach baths which have different effects on the silver image depending on the subtractive, sub-proportional, proportional or super-proportional reducing effect. This is of no importance in holography where the entire silver image has to be converted to a silver compound of one type or another. Therefore, these properties of the bleach solutions will not be treated here.

Redox Reactions

Silver bleaching is an example of a reduction-oxidation reaction often shorten to a *redox reaction*. Oxidation can be described as de-electronation, and reduction as electronation. Both the oxidizer and reducing agent in such a reaction have their individual reaction voltages or redox potentials. Redox reactions are also known as *half-cell reactions* because of their importance in electrolytic cells. In chemistry handbooks electrochemical half-cell reactions and redox potentials ($E°$) can be found. The $E°$ value for the oxidation of silver metal to silver bromide is -0.07 V, found in an oxidation table. Using a bleaching agent, such as ferricyanide, the corresponding $E°$ value for the reduction of ferricyanide to ferrocyanide is 0.57 V, found in a reduction table. Since one electron is involved in both complexes, the obtained reference voltages have to be added. The net, balanced silver bleaching action is then $E°_{net} = 0.57 - 0.07 = 0.50$ V. The go or no-go character of a redox reaction is determined by the E^0_{net} value and the concentrations of the two reactants and byproducts. *Nernst's equation* takes all these factors into account. The equation expresses the driving force or ElectroMotive Force (EMF) for the reaction. For a fresh bleach the bleaching agent and alkali halide are at their peak values and byproducts are minimum. A starting voltage is then obtained. If the value is high enough the redox reaction starts. During bleaching, the reducing agent becomes more positive and the oxidizing agent more negative, and the amount of byproducts increases which reduces the voltage and the reaction rate decreases. Eventually, when the voltage is zero the reaction stops.

Conventional Photographic Bleaches

Dichromate Bleaches. In the presence of sulfuric acid, potassium dichromate reacts in the following way:

$$6Ag + K_2Cr_2O_7 + 7H_2SO_4$$
$$\rightarrow K_2SO_4 + Cr_2(S_2O_4)_3 + 3Ag_2SO_4 + 7H_2O .$$ (5.2)

Chromium sulfate that has been formed combines with gelatin to form a complex, which has a hardening effect on the gelatin. The degree of hardening is proportional to the amount of chromium sulfate, and therefore the hardening is proportional to the optical density of the emulsion. Dichromate is therefore used when a hardening effect is of interest, e.g., in the Carbro process. If alum is added to a dichromate bleach, the hardening effect can be increased. Employed with an acid, this oxidizer forms a popular reversal bleach bath. Dichromate bleaches are fairly corrosive and can cause dermatitis.

Permanganate Bleaches. Potassium permanganate with sufficient acidity to form a neutral sulfate acts on silver in the following way:

$$10\,Ag + 2\,KMnO_4 + 8\,H_2SO_4$$
$$\rightarrow K_2SO_4 + 2\,MnSO_4 + 5\,Ag_2SO_4 + 8\,H_2O \,, \qquad (5.3)$$

or ionically

$$(MnO_4)^- + 5\,Ag + 8\,H^+ \rightarrow 5\,Ag^+ + Mn^{2+} + 4\,H_2O \,. \qquad (5.4)$$

Acid permanganate has little effect on the gelatin, which makes this type of bleach suitable for many photographic processes.

In general, it is important that both dichromate and the permanganate bleach solutions are free from soluble halides (e.g., chlorides from tap water) as these will react with the silver sulfate to form silver chloride and stain the emulsion yellow. Therefore it is important to use distilled or de-ionized water for the preparation of bleach solutions.

Ceric Sulfate Bleaches. In this bleach ceric sulfate is converted to cerous sulfate, which is an example of a silver solution formed by a soluble salt of an element that changes its valency by a single unit during this process:

$$2\,Ce(SO_4)_2 + 2\,Ag \rightarrow Ag_2SO_4 + Ce_2(SO_4)_3 \,, \qquad (5.5)$$
$$Ce^{4+} + Ag \qquad \rightarrow Ag^+ + Ce^{3+} \,. \qquad (5.6)$$

A practical difficulty in the use of ceric salts is the ease with which they precipitate as basic compounds. Therefore, the presence of an additional, strong acid is usually required.

Copper Bleaches. One example of a copper bleach is the following reaction:

$$Ag + CuBr_2 \rightarrow AgBr + CuBr \,. \qquad (5.7)$$

Copper bleaches are used in photographic intensification processes.

Ferricyanide Bleaches. In the presence of a soluble halide, potassium ferricyanide will convert silver to silver halide, which occurs in two steps:

$$\text{(i)} \quad 4\,Ag + 4\,K_3Fe(CN)_6 \rightarrow 3\,K_4Fe(CN)_6 + Ag_4Fe(CN)_6 \,. \qquad (5.8)$$

Silver is oxidized to form silver ferrocyanide which, in turn, reacts with potassium bromide to form silver bromide and potassium ferrocyanide, i.e.,

$$\text{(ii)} \quad Ag_4Fe(CN)_6 + 4\,KBr \rightarrow 4\,AgBr + K_4Fe(CN)_6 \,. \qquad (5.9)$$

Potassium ferrocyanide is soluble in water and dissolves away.

In spite of its ability to oxidize silver, potassium ferricyanide is a mild oxidizer which, unlike dichromate and permanganate, can be combined with hypo in the same solution, as in the case of *Farmer's reducer*, for example.

If hypo and ferricyanide are combined, the following reaction takes place during the second step:

$$3Ag_4Fe(CN)_6 + 16Na_2S_2O_3 \rightarrow 4Na_5Ag_3(S_2O_3)_4 + 3Na_4Fe(CN)_6 . \qquad (5.10)$$

However, the mixed solution is unstable. An advantage of ferricyanide is that it does not attack the organic compounds, such as the dyes used in photographic color emulsions. One disadvantage is that ferricyanide decomposes in light and air to form cyanide ions, which causes a waste disposal problem in color processing laboratories. Nowadays, the ferricyanide bleaches employed in color processing are often replaced by ferric-EDTA bleaches which can be easily combined with hypo to form stable bleach-fixing solutions.

Mercurious Bleaches. Several photographic intensifiers are based on mercuric chloride. The silver image is converted to silver chloride, together with insoluble mercurous chloride:

$$Ag + HgCl_2 \rightarrow AgClHgCl . \qquad (5.11)$$

Persulfate bleaches. The reaction between ammonium persulfate and silver can be simply explained by the following formula:

$$2Ag + (NH_4)_2S_2O_8 \rightarrow (NH_4)_2SO_4 + Ag_2SO_4 . \qquad (5.12)$$

In actual fact, the reaction is more complicated since the solution becomes acid during use. Both ammonium sulfate and silver sulfate are water soluble and can be easily removed from the emulsion.

The reaction between potassium persulfate and silver is

$$2Ag + K_2S_2O_8 \rightarrow K_2SO_4 + Ag_2SO_4 . \qquad (5.13)$$

Persulfate bleaches which are least harmful from the environmental point of view are now replacing ferricyanide bleaches in motion-picture film processing. To increase the reaction rate, which is very slow in persulfate bleaches, reaction-rate accelerators are often added to persulfate bleaching solutions. Various onium compounds considerably accelerate the bleaching out of the silver. However, some of these components are toxic and can cause safety problems when used. The best accelerator seems to be an organic thiol compound which acts as an electron transfer agent, thereby increasing the reaction rate. The material must be treated in an accelerator before entering the bleach solution. Kodak has a special persulfate bleach accelerator on the market, the PBA-1, that is often applied in processing laboratories. This agent is adsorbed to the silver grain surface and facilitates electron transfer during the reduction process of metallic silver to its ionic state.

5.2 Holographic Bleaching

5.2.1 Background

The conversion of amplitude holograms recorded on silver-halide materials into *phase-contrast holograms*, commonly referred to simply as *phase holograms*, was first performed by *Rogers* [5.1] and *Denisyuk* [5.2]. As regards the laser-recorded holograms, the first paper that dealt with this area was written by *Cathey* [5.3]. After that publication, in the late 60's and the early 70's, a lot of research efforts went into the discovery of bleaching processes capable of producing both a high diffraction efficiency and a low scatter noise [5.4-113]. It was not until the mid 70's, however, when the holographic bleaching technique was better understood and became capable of getting more satisfactory results. This breakthrough was due to a great degree to the magnificent contributions made by such researchers (with co-workers) as *Hariharan* [5.44, 45, 47], *van Renesse* [5.32, 48, 50], *Graube* [5.49] and *Phillips* [5.51, 52]. At the same time Agfa introduced a new line of better recording materials possessing finer grains. The most important contributions will all be discussed in the following subsections, including a chronological survey of the progress made in the field of holographic bleaching. In the early days, the testing of various bleaching baths was often performed on diffraction gratings only, which meant that the influence of intermodulation noise caused by extended, diffuse objects was not always considered. Therefore, high diffraction efficiencies obtained for a given process could be rather misleading since the noise level for holograms of diffuse objects could be extremely high when applying that particular process.

Bleaching is a particularly important technique for holograms recorded on Western holographic materials since it constitutes the only way of producing holograms with high diffraction efficiency on these materials. Therefore, most papers discussing bleaching can be found in Western literature. The Russian and the East-European works concern mainly colloidal development in order to obtain high efficiency and low noise, which has already been discussed in Chap.4. Likewise, the general theory of dielectric, thin or thick holograms has already been given in brief in Chap.2. More details concerning this can be found in the references mentioned. This chapter concentrates on theories directly related to the *bleaching of silver-halide materials* for creating phase holograms.

5.2.2 Holographic Bleaching Theory

General Considerations

Bleaching of the interference fringes recorded in the emulsion will change the emulsion in such a way that the light used for reconstruction of the image will be modulated by *retardation of the wave front* instead of by attenuation as in the case of amplitude holograms. Theoretically, this can lead to a substantial increase of the diffraction efficiency (100%) of a hologram. Retardation of the wavefront is caused either by variations of the refractive index within the emulsion (inner image) or by local variations of the emulsion thickness (relief image). At high spatial frequencies, the inner image predominates, whereas at low spatial frequencies the relief image is more pronounced, shown by *Hannes* [5.6]. A characterization of bleached photographic materials has been given by *Lamberts* [5.40], where these mechanisms are treated. The first of the two mechanisms is mainly in current use for producing high-quality phase holograms on silver-halide materials. Concerning the pure relief phase hologram, a theoretical analysis has been provided by *Collins* [5.8]. The possibility to create relief images using silver-halide emulsions will be further discussed in Chap.6.

Most often the theories of bleached, silver-halide holograms assume that the hologram is a pure phase hologram of the index or the relief type, or else a combination of the two (*phase-only hologram*). As a matter of fact, a bleached gelatin emulsion is actually a combination of both an *amplitude and phase hologram* (a *complex hologram*). Several papers take up this subject as, for example, in the case of reflection volume holograms a theory has been formulated by *Alekseev-Popov* and *Gevelyuk* [5.70].

Lorber [5.18] extended the *Kelly's three-stage model* [5.19] of photographic information recording to include the formation of phase holograms. Briefly, *Kelly* described the photographic process in three stages, independent in time and space. The first stage consists in exposure where optical diffusion takes place within the emulsion and its backing. The second stage concerns the latent-image forming process - a sensitometric stage acting upon the silver-halide crystals. The last stage is the chemical-diffusion stage taking place during the development. *Lorber* has modified this model for exposure with coherent light and the chemical-diffusion stage is discussed here in terms of the developed silver density rather than the optical density. He also added a fourth stage, another chemical-diffusion stage, to account for granularity and bleaching or the etching of the emulsion. The first three stages were formulated in terms of nonstochastic functions. Effects of granularity were discussed in terms of the theory of nonhomogeneously filtered Poisson processes. The granularity theory was formulated in such a way as to make use of the valuable frequency-domain concepts of the stationary-process theory without forcing the analyst to assume a uniform, featureless exposure pattern. The four-stage model was introduced to define the signal-to-noise ratio for a bleached hologram and to determine the optimal frequency-domain characteristics of a grain cloud in the bleached emulsion.

Chang and *George* [5.20] analyzed the photographic emulsion as an artificial dielectric, which made it possible to relate the refractive change after bleaching to its prebleached density. They suggested that the Lorentz-Lorenz equation could be applied to predict the bulk refractive index of the emulsion if the respective concentrations of the gelatin and the silver halide were known. They established that a linear density vs exposure relationship is needed to produce a sinusoidally modulated index of refraction in a bleached emulsion. They also defined the phase-MTF function for the bleached (but partly lossy) case, which showed that the phase-MTF relation is equal to the modulation index m:

$$\Phi_{MTF} = m \ . \tag{5.14}$$

The thickness of a holographic emulsion was also analyzed by applying a modified Raman-Nath formalism to incorporate a loss. Their conclusion was that an emulsion thickness of about 20 μm is necessary to obtain high efficiency for photographic-recording materials. The *Chang* and *George* paper will be further discussed later on in this chapter.

Upatnieks and *Leonard* [5.21, 22] investigated theoretically and experimentally the conditions important for imaging holography, having a diffuse signal beam interacting with the reference beam, which is recorded in dielectric media. They considered only the noise generated by the intermodulation terms in their analysis. They also assumed a linear phase modulation as a function of exposure, the Rayleigh-probability amplitude distribution in the signal beam, and two-dimensional recording of the intermodulation terms. They found that the diffraction efficiency was limited to 22% for thin holograms (compared to 33.9% for thin, two-wave gratings) and to 64% for thick holograms (compared to 100% for thick, two-wave gratings).

Electrical Polarizability Theory

In the beginning, it was incorrectly believed that the refractive-index modulation increased with the increased refractive index of the silver-compound grains in the gelatin emulsion. Bleaching experiments performed by *van Renesse* and *van der Zwaal* [5.32] indicated that, to obtain the identical diffraction efficiency, holograms had to receive a higher exposure if converted to silver bromide than if converted to silver iodide. This fact is contradictory to the refractive-index theory since the refractive index is higher for silver bromide (n = 2.25) than for silver iodide (n = 2.20) (Table 5.1). This observation resulted in the *electrical-polarizability/molecular-volume theory* developed by *van Renesse* and *Bouts* [5.48].

The first important part to consider in the bleaching process is the density obtained in the holographic plate during development. The density D is proportional to the partial atomic concentration of silver in the emulsion N_s

$$D = aN_s d/\cos\theta \tag{5.15}$$

where d is the thickness of the emulsion and θ the angle between the normal and the incident direction of light. The factor a is normally constant and depends on the size, shape and weight of the silver grains. Being an absorption constant in nature, it can be calculated in the following way

$$a = M_s/PN \qquad (5.16)$$

where M_s is the atomic weight of silver, N Avogadro's number, and P the photometrical equivalent which is a quantity by weight of silver per unit of surface emulsion with unit density.

The molecular concentration of a silver compound during bleaching is considered proportional to the original density which is the concentration of silver atoms. Therefore, any variation in density (ΔD) per unit of emulsion surface will result in a difference of the number of silver-compound molecules ΔA_c per unit of emulsion surface. The relation between ΔD and ΔA_c uses (5.15)

$$\Delta D = a\,\frac{d_2 N_{c2} - d_1 N_{c1}}{q\cos\theta} = a\,\Delta A_c/(q\cos\theta) \qquad (5.17)$$

where N_{c2} and N_{c1} are the respective partial molecular concentrations of the silver salt in two units of the surface emulsion. The factor q indicates how many molecules of the silver compound are formed from one silver atom. In Fig.5.1 d_1 and d_2 are defined.

The theory is based on some additional assumptions. Firstly, that the silver in the emulsion is converted into one and only one compound (no other bleaching products are formed), contributing to the light modulation. Secondly, the ultra-fine-grained holographic emulsion is considered to be a homogeneous mixture of gelatin and a silver compound. Fine-grained holographic emulsions can be regarded as Rayleigh scatterers. If that is the case, then the *Lorentz-Lorenz equation* will give the refractive index of the mixture of substances, n, which is

$$n^2 = \frac{1 + \dfrac{8\pi}{3}\displaystyle\sum_i \alpha_i N_i}{1 - \dfrac{4\pi}{3}\displaystyle\sum_i \alpha_i N_i}, \qquad (5.18)$$

where α_i is the electrical polarizability of the molecule of type i, and N_i is the concentration of type i molecule. The electrical polarizability of a substance has the dimension of a volume and is given by the dipole moment of a molecule which is brought about by a unit electric field. The electrical polarizability increases with the increase of the molecular volume.

If the differences in emulsion thickness, Δd, are due only to the differences in concentration of the silver compound, then

$$\Delta d_t = \Delta A_c V_c \qquad (5.19)$$

where V_c is the molecular volume of the silver compound, namely

$$V_c = 1/N_c = M_c/(N\rho_c) \qquad (5.20)$$

where N_c is the molecular concentration of the compound which is a material constant, M_c is the molecular weight of the compound, N is Avogadro's number, and ρ_c is the specific gravity of the silver compound. The electrical polarizability of a compound can be calculated using (5.15 and 17), provided the refractive index, the molecular weight, and the specific gravity are known for the specific compound. The electrical polarizability and other parameters for some common silver compounds used in holographic bleaching are found in Table 5.1.

Table 5.1. Properties of silver compounds

Silver compound	Molecular weight	Refractive index	Electrical polarizability $[10^{-30}\ m^3]$	Molecular volume $[10^{-30}\ m^3]$
AgCl	143.32	2.07	5.3	42.7
AgBr	187.78	2.25	6.6	48.5
AgI	234.77	2.20	9.2	68.5
$Ag_4 Fe(CN)_6$	643.43	1.56	35.9	465
$AgHgCl_2$	379.36	1.82	12.4	120
Ag	107.87	–	–	–

There are other factors which influence the thickness of the emulsion, such as the tanning activity of solutions, the drying method, etc., but here only the differences in concentration will be considered. The thickness modulation of the emulsion is then

$$m_e = \Delta d/\Delta d_t = \Delta d/(V_c \Delta A_c) . \qquad (5.21)$$

From (5.17 and 21), Δd, representing the actual difference in thickness between two units of the emulsion surface, is found

$$\Delta d = (q V_c \Delta D m_e \cos\theta)/a . \qquad (5.22)$$

For a small change of the molecular concentration in the Lorentz-Lorenz equation (5.15), to a first-order approximation, a linear equation describing the refractive index n of the mixture, has been as

$$n = c \sum_i \alpha_i N_i + n_0 \,. \tag{5.23}$$

For $c = 8.7$ and $n_0 = 0.89$, (5.23) is valid with an accuracy better than 1% for $1.5 < n < 1.7$.

Phillips [5.98] introduced another approximation to the refractive index n derived from a quadratic fit and the regression method, which considers the influence of large change of N due to physical transfer, namely

$$n = 1 + 1.3904X + 1.0689X^2 \,, \tag{5.24}$$

where $X = (4\pi/3) \sum_i \alpha_i N_i$.

Phase Variations as a Function of Pre-Bleached Differences in Density

The phase of the light which has passed through the bleached emulsion is

$$\phi = \frac{2\pi nd}{\lambda\cos\theta} \tag{5.25}$$

where n is the refractive index of the emulsion, d is thickness, λ is wavelength, and θ the angle between the incidence direction and the normal to the emulsion.

According to Fig.5.1, the phase difference $\Delta\phi$ between the two indicated beams is

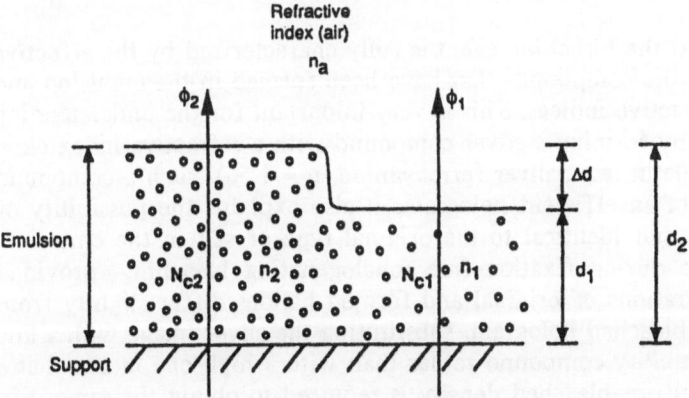

Fig.5.1. Phase difference between two beams passing through the emulsion surface with differences in thickness and refractive index, caused by differences in concentration [5.48]

$$\phi_2 - \phi_1 = \Delta\phi = \frac{2\pi(d_2 n_2 - d_1 n_1 - \Delta d n_a)}{\lambda \cos\theta} \tag{5.26}$$

where n_1 and n_2 are the refractive indices of two different parts of the emulsion traversed by the two beams, and n_a is the refractive index of the surrounding medium, usually air.

Combining (5.17, 23 and 26), the phase difference as a function of the original differences in density can be obtained

$$\Delta\phi/\Delta D = kcq\alpha_c - kqm_e V_c(n_a - n_0) \tag{5.27}$$

where $k = 2\pi/(\lambda a)$. Since k contains the absorption coefficient a, k is constant only when the fine-grained emulsions are processed in the same way. At increasing spatial frequencies, the modulation m_e caused by thickness variations has a tendency to decrease. Such is also the case when rehalogenating methods are used and minimum thickness variations occur ($m_e = 0$) and (5.27) takes on the simple form

$$\Delta\phi/\Delta D = kcq\alpha_c . \tag{5.28}$$

It is obvious that the phase change that will affect a light beam passing through the emulsion is dependent only on the number of polarizable molecules on the way through the emulsion.

The diffraction efficiency η is directly proportional to the square of the phase difference $\Delta\phi$, if $\Delta\phi < 1$,

$$\eta \propto (\Delta\phi)^2 . \tag{5.29}$$

Combining (5.28 and 29) we get

$$\eta \propto (kcqa_c)^2 \tag{5.30}$$

which means that the bleaching agent is fully characterized by the effective polarizability of the compounds that have been formed in the emulsion and not by their refractive indices. This is very important for the understanding of the fact that by forming a silver compound with a refractive index close to that of the gelatin (e.g., silver ferrocyanide, n = 1.56), such a compound can actually form an efficient hologram. It also explains the possibility of forming a silver salt identical to the original type of salt in the emulsion. This takes place during fixation-free rehalogenating bleaching, provided that the concentrations of original and formed halides differ slightly from each other. In a bleached hologram, substituting the silver image with a low electrical polarizability compound rather than with a high one means that a higher than usual pre-bleached density is required to obtain the same diffraction efficiency.

As regards variations in the emulsion thickness which often give rise to noise problems, the change in thickness is proportional to the molecular

172

volume of the substituting compound and is therefore also proportional to density variations. Generally, the electrical polarizability of an ion increases with its volume, which means that a compound of high polarizability will directly be associated with an increased noise level.

Van Renesse and *Bouts* [5.48] supported the theory with convincing experiments. Holographic emulsions were exposed through a grey scale with incoherent light, resulting in a continuously increasing density. The local change in density as a function of position within the emulsion on the plates was measured with high accuracy. After that the plates were bleached in various bleach baths. The phase changes within the emulsion of the test plates related to the previously measured density variations were measured with laser light in a Mach–Zehnder interferometer. Measurements were also performed on the plates submerged in an index-matching liquid to eliminate the influence of phase modulation due to emulsion thickness variations. A linear relation between the original density variations and the resulting phase modulation was revealed.

The relative diffraction efficiencies obtained for different common silver compounds related to silver chloride and having the lowest polarizability is presented in Table 5.2.

Table 5.2. Relative diffraction efficiency for different silver compounds

Silver compound	Relative diffraction efficiency
Silver chloride (AgCl)	1
Silver bromide (AgBr)	1.6
Silver ferrocyanide ($Ag_4Fe(CN)_6$)	2.9
Silver iodide (AgI)	3.0
Silver mercurochloride ($AgHgCl_2$)	5.5

Sirko and *Slaby* [5.67] used the approach applied by *van Renesse* and *Bouts* to find an expression for holograms bleached with the reversal technique. They introduced the quantity $m(\nu)$, which is a function of the spatial frequency ν. It is used to calculate the change in the emulsion thickness Δd

$$\Delta d = m(\nu) V_c \Delta N_c \qquad (5.31)$$

where ΔN_c is the variation in number of silver-compound molecules (N_c is the number of silver-compound molecules per unit area), and V_c is the molecular volume of the silver compound. For the reversal-bleaching method and for $1 < \nu \leq 500$, m is negative [5.40]. It means that the emulsion is thicker in places with fewer silver compounds. In the reverse process, the developed silver is removed from the emulsion whereas the undeveloped silver-halide grains remain in the emulsion. We get

$$N_c + N_s = N_0 \qquad (5.32)$$

where N_s is the number of silver atoms per unit area, and N_0 is the initial number of silver-halide molecules per unit area of the unprocessed emulsion. For reversal bleaching, $N_c = N_0 - N_s$, and $\Delta N_c = - \Delta N_s$. Assuming that during the processing the gelatin is neither washed out nor dislocated and using the same notation as in the previous subsection, the expression for a reversal hologram will be

$$\frac{\Delta \phi}{\Delta D} = \frac{2\pi}{\lambda a}[-c\alpha_c' + m(\nu) V_c'(n_a - n_0) + cp] \qquad (5.33)$$

where $p = N_g \Delta \alpha_g / \Delta N_s$. α_c' indicates the average polarizability of the virgin silver-halide molecules in the emulsion, and V_c' is the molecular volume of these molecules. In a practical situation when using reversal bleaching, the dichromate ions will affect the polarizability of the gelatin in a way different from what happens when applying the direct bleaching technique.

The corresponding expression for direct bleaching can also be evaluated. With $N_c = N_s$ and $\Delta N_c = \Delta N_s$ we obtain

$$\frac{\Delta \phi}{\Delta D} = \frac{2\pi}{\lambda a}[c\alpha_c - m(\nu) V_c(n_a - n_0)] . \qquad (5.34)$$

Sirko and *Slaby* discussed the displacement of gelatin during the drying process in the same paper.

The electrical-polarizability theory was studied by *Quintanilla* et al. [5.76]. Here, the variation of the emulsion thickness at zero frequency was of main interest. They modified (5.27) by *van Renesse* and *Bouts*, to consider the volume effect of salt,

$$\frac{\Delta \phi}{\Delta D} = kcq\alpha_c \left[1 - \frac{m_e V_c}{c\alpha_c}(n_a - n_0) \right] . \qquad (5.35)$$

Syms and *Solymar* [5.71] made important contributions to the understanding of silver-halide emulsions by investigating both theoretically and experimentally volume phase holograms formed in bleached photographic emulsions using Kogelnik's coupled-wave theory. They found that the theory works well close to the Bragg condition but away from the Bragg condition certain deviation between theory and experiments were found. This led to the formulation of a more accurate model based on the coupled-wave theory in which phase and absorption modulation, the second harmonic in the grating profile, and the appearance of some higher diffraction orders are taken into account. The agreement between the new model and

174

the experimental data seems to be much better. The model is also used to deduce the major characteristics of the photographic recording material, including saturation of the modulation with exposure. This theoretical model enabled *Syms* and *Solymar* to offer certain suggestions for the improvement of the recording material.

The previously discussed paper deals with diffraction gratings only, which is not sufficient to describe the behavior of a phase hologram of diffuse-wave recordings normally employed in, e.g., display holography. According to the coupled-wave theory it is possible to obtain 100% diffraction efficiency for phase gratings, provided the refractive-index modulation is high enough for a given thickness of the recording emulsion. If the material has a linear relation between the refractive-index modulation and the exposure modulation, the diffraction efficiency as a function of exposure will follow a $\sin^2$ form for a transmission grating and a $\tanh^2$ form for a reflection grating according to Kogelnik's theory. This is fine as long as the two collimated, interfering beams have the same intensity ($K = 1$). If one of these beams is diffuse and has a lower intensity level than the other ($K > 1$), the diffraction efficiency of the recorded hologram will be lower than for the case when $K = 1$. *Ward* and *Solymar* [5.94] presented a model concerning phase holograms, valid for different beam ratios and exposures. Initial experimental data yielded the efficiency of the recorded holograms where one beam consisted of light emitted from a diffusing screen. The following diffraction efficiencies as a function of K-values were obtained for holograms recorded on Agfa 8E56 HD (utilizing CW-C1 - PBQ-2 processing [5.77]):

K	η [%]
1.05	40
2.1	37
5.6	34
13	26
32	15
64	10 .

It is clear that the efficiency reaches a maximum for each K- value. Increased exposure at the higher K-values did not increase the efficiency. The following three reasons are mentioned as possible explanations:
(i) A material nonlinearity.
(ii) A competition with the noise grating.
(iii) A modulation nonlinearity.

As regards the material nonlinearity it was found that the optimum efficiency of holograms recorded at high K-values was obtained at exposure levels below the saturation level, i.e., the recording was within the linear part of the H-D curve. Therefore reason (i) does not constitute the main limitation. Reason (ii) was also ruled out since various experiments showed that noise gratings formed within the emulsion had very little influence on the modulation of the main grating. The main problem seems to come from the modulation nonlinearity during the bleaching process, which is actually not fully understood.

175

The intensity maxima and minima created by two interfering waves are

$$I = I_1 + I_2 \pm \sqrt{I_1 I_2} = (I_1/K)(\sqrt{K} \pm 1)^2 \qquad (5.36)$$

where

$$I_1/I_2 = K \quad (\text{or} = m, \text{ modulation}).$$

The mean intensity is

$$I_0 = I_1 + I_2 = (I_1/K)(K + 1) . \qquad (5.37)$$

The linear part of the H–D curve can be approximated by the relation

$$D = \gamma \log H + c . \qquad (5.38)$$

The optical density is directly proportional to the optical absorption (being proportional to the silver density in the emulsion, $S \propto D$). The silver-density modulation δS caused by the spatially varied exposure is

$$\delta S = \tfrac{1}{2}(S_{max} - S_{min}) \propto \gamma \log \left(\frac{\sqrt{K} + 1}{\sqrt{K} - 1} \right) . \qquad (5.39)$$

The relation between the silver modulation δS and the refractive-index modulation δn after bleaching was actually found to be (after experimentally confirming the linear relationship between δn and δS), too:

$$\delta n \propto \gamma \log \left(\frac{\sqrt{K} + 1}{\sqrt{K} - 1} \right) \qquad (5.40)$$

which means that the refractive-index modulation (and hence the diffraction efficiency) of a bleached hologram is independent of the exposure H for a fixed beam ratio and that the modulation is smaller for higher beam ratios (large K-values). The refractive-index modulation is directly proportional to the contrast γ. This equation is valid as long as the spatial maxima and minima of the exposure lie within the linear part of the δn-δS curve, which normally corresponds to the D–logH curve.

Recently, *Churaev* and *Artyomova* [5.106] investigated the phase shift as a function of the optical density for various materials from the former USSR, such as LOI-2, VRL, FP-R and IAE. They studied, in particular, the influence of the developer type and the developing time as well as the effect of various spatial frequencies. The main results are that for some developers the slope of the phase shift versus optical density curve is steeper for a short developing time compared to a longer time. Also the slope is steeper for high spatial frequencies ($>1000\ell p/mm$) compared to low frequencies ($<100\ell p/mm$). The spatial frequency effects will be further discussed later on.

As already mentioned, *Chang* and *George* [5.20] treated theoretically the bleached silver-halide emulsion. The developed and fixed emulsion consists of a suspension of discrete small silver particles in the gelatin layer. If these particles, by rehalogenating bleaching, are converted to a dielectric compound, e.g., silver bromide the effective dielectric constant ϵ of the emulsion can be obtained on the basis of Mie's theory. The polarizability α of a small ($k_1 a \ll 1$) absorbing sphere is

$$\alpha = a^3 \epsilon_1 \frac{\epsilon_2 - \epsilon_1}{\epsilon_2 + 2\epsilon_1} \tag{5.41}$$

which is substituted into the Lorentz-Lorenz formula to get ϵ, i.e.,

$$\frac{\epsilon}{\epsilon_1} = \frac{1 + 8\pi N\alpha/3}{1 - 4\pi N\alpha/3} . \tag{5.42}$$

Including losses, the complex dielectric constant for the emulsion is

$$\epsilon/\epsilon_1 = 1 + \frac{3f}{(\epsilon_2 + 2\epsilon_1)/(\epsilon_2 - \epsilon_1) - f} \tag{5.43}$$

where ϵ_1 and ϵ_2 are the dielectric constants of the gelatin and the lossy scatterer, respectively. The radius of the scatterer is a. The number of scattering centers per unit volume is N, and k_1 is the wave number. The filling factor f is defined by $f = 4\pi Na^3/3$. F is proportional to N and, for $f \ll 1$, (5.43) can be expanded using the refractive indices $n_1 = \sqrt{\epsilon_1}$ and $n = \sqrt{\epsilon}$ to give approximately

$$n = n_1 \left(1 + 1.5f \frac{\epsilon_2 - \epsilon_1}{\epsilon_2 + 2\epsilon_1} \right) . \tag{5.44}$$

The change in the index $f\Delta n_2$ is

$$f\Delta n_2 = 1.5fn_1 \frac{\epsilon_2 - \epsilon_1}{\epsilon_2 + 2\epsilon_1} . \tag{5.45}$$

If the real and imaginary parts of the effective index of refraction is used, we have

$$n = n_2' + ifn_2'' \tag{5.46}$$

where

$$n_2' = n_1 \left(1 + 1.5f \frac{\epsilon_2'^2 + \epsilon_2''^2 - 2\epsilon_1^2 + \epsilon_1\epsilon_2'}{(\epsilon_2' + 2\epsilon_1)^2 + \epsilon_2''^2} \right) \tag{5.47}$$

177

and

$$fn''_2 = \frac{4.5fn_1\epsilon_1\epsilon''_2}{(\epsilon'_2 + 2\epsilon_1)^2 + \epsilon''^2_2}. \tag{5.48}$$

If the emulsion is uniformly exposed and developed, the attenuation of a normally incident plane wave $[exp(-2k_0fn''_2d)]$ is given in density units (D_2) as

$$D_2 = 0.869k_0fn''_2d \tag{5.49}$$

for a thickness d and a free-space wave number k_0.

To find the index of refraction of the bleached emulsion, (5.44 and 49) are combined and if the subscript (3) denotes bleaching, i.e. $\epsilon_2 \rightarrow \epsilon_3$, we have

$$n_3 = n_1\left(1 + \frac{1.73\varsigma D_2}{n''_2k_0d}\frac{\epsilon_3 - \epsilon_1}{\epsilon_3 + 2\epsilon_1}\right) \tag{5.50}$$

where the expansion factor ς is the volume ratio of the bleached to the unbleached globule.

Diffraction Efficiency and Noise

The question about the emulsion's optimum thickness for holograms in general was discussed in Chap.2. Specifically, bleached phase-reflection hologram recording has been treated by *Hariharan* [5.57]. The volume regime is well valid if Q > 10. If two laser beams enter the recording material perpendicularly from opposite sides, the fringe separation is at the minimum and the Q-value is about 100 for an emulsion of an approximate thickness larger than 2 μm. The diffraction efficiency for the wavelength and the reconstruction geometry that satisfy the Bragg condition, is then

$$\eta = \tanh^2\psi \quad \text{with} \quad \psi = \pi\delta nd/\lambda_n. \tag{5.51}$$

The efficiency increases with an increase of ψ, first rapidly and then more slowly to finally reach 100%. For a given wavelength λ_0, the parameter ψ is proportional to the total phase modulation available in the emulsion and is determined by the product of the emulsion thickness d and the amplitude of the refractive index modulation. It is obvious that it is easy to obtain a high efficiency if the thickness is increased or else by increasing the refractive index modulation δn.

For a reflection hologram reconstructed with *white light*, a compromise between bandwidth and efficiency must be made. The more selective the hologram is (higher thickness) the less efficient it becomes. If the hologram is reconstructed with white light of the wavelength band $2\Delta\lambda_0$, as *Hariharan* [5.57] has shown, the luminance of the image is determined not only by the peak diffraction efficiency but also by the bandwidth diffracted by the hologram. For a relatively narrow bandwidth, the luminance is

$$L = G(\lambda_0) \int_{\lambda_0 - \Delta\lambda_0}^{\lambda_0 + \Delta\lambda_0} \eta_\lambda \, d\lambda \tag{5.52}$$

where $G(\lambda_0) = GE_\lambda K_\lambda$ for $\lambda = \lambda_0$. E_λ is the spectral irradiance of the source, K_λ is the spectral luminous efficacy of the radiation, and G is a parameter determined by the hologram recording geometry. The integral in (5.52) was numerically evaluated for the emulsion-thickness variation and the refractive-index modulation δn. For a possible refractive-index modulation of 0.025 (e.g., silver halide), the results reveal that the image luminance did not improve if the emulsion thickness was above 7 μm. What did improve then was the selectivity of the hologram, which means better sharpness for deeper objects. Unfortunately, the improvement in the hologram's selectivity implied also lower luminance of the image for a given white-light reconstruction irradiance.

Grain growth during rehalogenating bleaching contributes to light scattering in the emulsion after bleaching. *Joly* [5.73] investigated some bleaching agents and various concentrations of the rehalogenating agent (KBr). In particular, he studied closely the effects of using tanning developers during processing. The important factors which contribute to scattering are:
- Particle size.
- Particle composition.
- Environment (the degree of tanning).

The intensity of light scattered by a small particle is proportional to the square of its electrical polarizability α. The polarizability of a spherical particle (radius r) with a refractive index n_1 imbedded in a homogenous medium with a refractive index n_3 is given by

$$\alpha = \frac{4\pi r^3}{3} \frac{n_1^2 - n_3^2}{n_1 + 2n_3^2} \, .$$

This is the case when a *non-tanning developer* is used. The silver-bromide particles (n_1) are imbedded in the gelatin where the refractive index is constant (n_3).

If a *tanning developer* is used, the silver-halide particle is enveloped by a shell of tanned gelatin with a refractive index n_2 which differs from both n_1 and n_3. The polarizability of such a particle (radius: particle + shell = r, particle itself: fr, where $0 \leq f \leq 1$) is then

$$\alpha \simeq r^3 \frac{(n_2^2 - n_3^2)(n_1^2 + 2n_2^2) + f^3(2n_2^2 + n_3^2)(n_1^2 - n_2^2)}{(n_2^2 + 2n_3^2)(n_1^2 + 2n_2^2) + 2f^3(n_2^2 - n_3^2)(n_1^2 - n_2^2)} \, . \tag{5.54}$$

179

The refractive index for AgBr is 2.25. The refractive index of non-tanned gelatin is 1.5426 and that of tanned 1.5488, using values for dichromated gelatin from *Shankoff* [5.114]. Based on the formulas above, calculations were performed for the following cases:

(i)   No tanning occurs, particle size 50 nm.
(ii) A tanned shell is formed: size, including shell, 50 nm.
(iii) A tanned shell is formed: size, including shell, 70 nm.
(iv) The gelatin layer is uniformly tanned.

The results are presented in Tables 5.3 and 5.4.

In conclusion, it can be said that the use of a tanning developer creating a shell around the silver-bromide grain will increase the scattering by a factor of two at the very least. For the silver-halide grains that grow larger during the rehalogenating bleaching process, this difference will be even bigger. This means that if tanning development is used, to minimize wavelength-shift in white-light reflection holography, grain growth must be strictly controlled in order to avoid excessive scattering.

Joly discussed also some experimental results on the Agfa 8E56 and 10E56 film, developed in GP 62, fixed and bleached in PBQ, ferric nitrate, and potassium ferricyanide. The bleaching time depends on the speed with which silver ions are formed and removed. The formation of silver halide will be faster if the concentration of the halide ions is high. A strong oxidant will create silver ions faster, thus speeding up the bleaching process. As in emulsion making, grain growth depends on the KBr concentration. A high KBr concentration will yield larger grains. Ferric nitrate bleach gives smaller grains than the PBQ bleach due to the fact that Fe^{3+} is a stronger oxidant than PBQ. A high concentration of Ag^+ means that the relative concentration of Br^- will be lower even if a strong oxidant bleach and a weak oxidant bleach contain identical amounts of KBr. The ferric nitrate bleach has also a low pH value (about 2.35), which will reduce grain growth. The results indicate that the ferric nitrate bleach gives the smallest rehalogenated grains with a rather narrow grain-size distribution. If the concentration of KBr is increased in the bleach bath, both the grain size and their distribution increase, which means increased scattering (but also higher diffraction efficiency).

Staselko and *Churaev* [5.79] presented a theory describing how the diffusion of light in silver-halide emulsions affects the contrast of holographic images of diffusely scattering objects. Scattering from all the following sources was taken into account: the grainy structure of the emulsion before and after processing, the emulsion surface irregularities, speckle patterns produced by nonuniform object beams, the beam ratio and the angular size of the object. The theory has been verified experimentally by recording volume phase holograms of diffusely scattering objects on the Russian material PE-2.

The general phase characteristics of holographic recording media are discussed in a paper by *Staselko* and *Churaev* [5.74], where a physical model is constructed based on a function describing the phase characteristics ob-

Table 5.3. Grain-size distribution of the AgX formed during rehalogenation bleaching

Bleach solution		Bleaching time		Mean grain diameter $\overline{X}$ [nm]		Spread σ [nm]		$\sigma/\overline{X}$	
		8E56HD	10E56	8E56HD	10E56	8E56HD	10E56	8E56HD	10E56
PBQ	25 g KBr	105″	105″	46	57	11	17	0.25	0.30
PBQ	50 g KBr	105″	90″	61	69	15	16	0.25	0.23
PBQ	100g KBr	75″	60″	73	82	29	24	0.40	0.29
PBQ	32 g KCl	14′	12′	87	104	46	42	0.53	0.40
$Fe(NO_3)_3$	30 g KBr	40″	46″	34	38	8	11	0.25	0.28
$K_3Fe(CN)_6$	5 g[a]	39″	46″	37	35	12	12	0.33	0.34
Compare: Original AgBr in the emulsion				40	75	10	15	0.25	0.20

[a] The bleach solution contains 5 g of $K_3Fe(CN)_6$, the halogenating agent being $(Fe)CN_6^{4-}$

tained from the experiments performed on the most widely used Russian silver-halide materials for holography.

Qui and *Jiang* [5.96] published findings with a multi-factorial method of a fuzzy-set theory to predict the diffraction efficiency of bleached holograms. The factors considered were exposure, developer dilution, development time, and bleaching time. Experimental verification of the theory was performed using 33 specimens. The obtained diffraction efficiency versus the predicted one revealed a fitting rate of 85%.

Table 5.4. Influence of the grain size of AgBr formed after rehalogenation bleaching on the scattering ($n_1 = 2.25$, α is measured in 10^{-24} m³, and R stands for $r_{AgX+shell}$)

Radius of AgBr	Non-tanning developer $n_3 = 1.5426$ $n_2 = n_3$ $F = 1$		Tanning developer R = 50 nm $n_3 = 1.5426$ $n_2 \neq n_3$ $f \neq 1$				Tanning developer R = 70 nm $n_3 = 1.5426$ $n_2 \neq n_3$ $f \neq 1$				Uniformly tanning developer $n_3 = 1.5488$ $n_2 = n_3$ $f = 1$	
[nm]	α	α^2	f	n^2	α	α^2	f	n^2	α	α^2	α	α^2
26	3.8	14.4	0.52	1.549	6.3	39.6	0.37	1.5488	5.7	32.5	3.7	13.7
34.5	8.8	77.4	0.69	1.635	16.0	256	0.49	1.572	15.1	228.0	8.7	75.7
39	12.7	161.3	0.78	1.745	23.3	542.9	0.56	1595	22.7	515.3	12.6	158.8

5.2.3 The Holographic Bleaching Technique

The holographic bleaching techniques can be divided into three categories:
(i)   Conventional or direct (rehalogenating) bleaching.
(ii) Fixation-free rehalogenating bleaching.
(iii) Reversal (complementary) or solvent bleaching.

In *conventional bleaching*, the developed hologram containing the silver image is converted into a phase hologram by changing the silver image into a transparent silver halide. This is performed after fixation, i.e. after the unexposed silver-halide crystals have been removed. We say that the developed silver grains have been rehalogenated. *Fixation-free rehalogenating bleaching* means that the hologram is bleached directly after development, without fixing, leaving the unexposed silver-halide crystals in the emulsion (Fig.5.2a and b). Essentially, the same type of bleach bath is used for both the conventional and the fixation-free processes. A rehalogenating bleach bath consists of

- an oxidizing agent,
- an alkali halide (often KBr),
- a buffer (often H_2SO_4), and
- additives such as printout stabilizers (dyes).

Rehalogenating bleaches will be further discussed in Sect.5.2.3a.

In *reversal bleaching*, the developed silver image is converted into a soluble silver complex which is removed from the emulsion during bleaching, leaving the original, unexposed silver-halide grains in the emulsion (Fig.5.2c). These crystals modulate the light to reconstruct the holographic image. In reversal bleaching, the hologram is not fixed after the development, as otherwise all the silver-halide grains (exposed and unexposed) would disappear after bleaching. A reversal bleach bath consists of

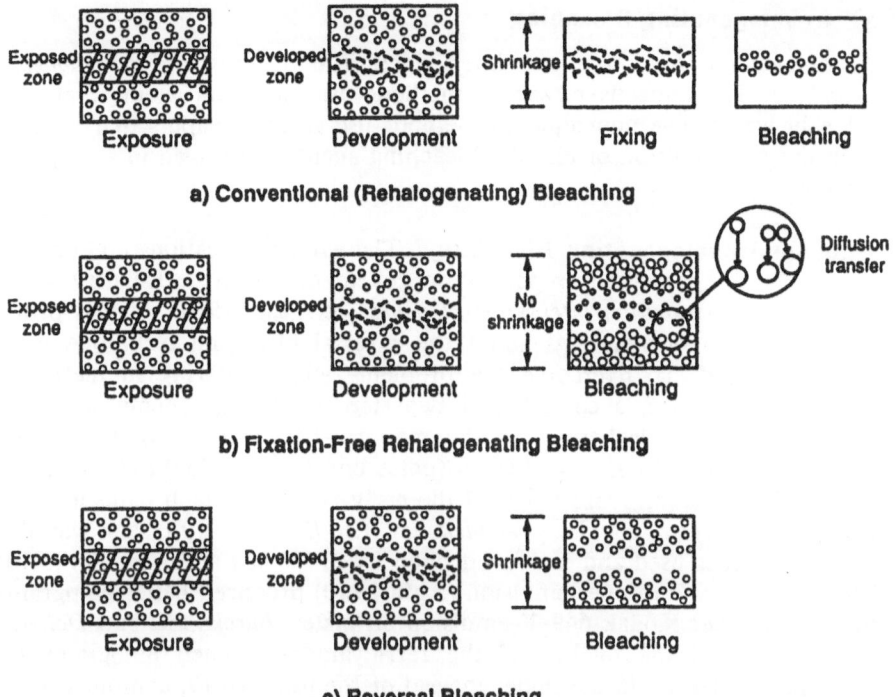

a) Conventional (Rehalogenating) Bleaching

b) Fixation-Free Rehalogenating Bleaching

c) Reversal Bleaching

Fig.5.2. Various bleaching techniques

- an oxidizing agent,
- a buffer (often H_2SO_4), and
- additives such as printout stabilizers (dyes).

Reversal bleaching will be further discussed in Sect.5.2.3b.

It should be mentioned, however, that certain special methods can be applied to remove all silver-halide crystals from the emulsion and thus to create modulation by a superficial effect, producing gelatin relief images. These very special techniques, which are normally not included among the conventional holographic bleaching procedures, will be treated in Chap.6.

The first bleaching techniques for laser-recorded holograms was the one introduced by *Cathey* [5.3]. It is based on a bleach from Kodak used in the chromium-intensifier process. Many rehalogenating-bleaching techniques (with or without intermediate fixing) have been suggested during the years, but only a few have produced sufficiently high quality to survive the tide of time. The latest holographic bleaching products and procedures will be treated in greater detail in the next section. Here, various important rehalogenating processes of earlier days will also briefly be discussed. In Chap.7 the reader will find complete processing schemes that have been successfully applied for different commercial silver-halide materials.

a) Rehalogenating Bleaching

Rehalogenating bleaching can be divided into *aqueous* and *non-aqueous bleaching*. The aqueous-bleaching methods use bleaching agents that dissolve in water. The non-aqueous methods use gases or vapors that interact with the dry emulsion or else the bleaching agent is dissolved in an organic solvent.

Aqueous Rehalogenating Bleaching. The early publications on conventional bleaching methods reported improvements of the efficiency when *mercuric and chromium intensifier bleaches* were used [5.9]. Various problems and misunderstandings about the reversal-bleaching technique were also mentioned. To overcome the problems with scattering in mercury-bleached holograms, a complicated two-step processing scheme was suggested, involving blackening of the mercury-bleached emulsion and re-bleaching it again in another bleach (potassium ferricyanide) [5.9]. *Upatnieks* and *Leonard* [5.11] investigated the early types of bleach baths and introduced *ferricyanide* and *copper bleaches*. The *Kodak R-10* bleach was also among the ones used and improved upon by *McMahon* and *Franklin* [5.13] They found that a thinner emulsion (Agfa 8E70) produced better holograms than the thicker Kodak 649-F emulsion on plates. *Burckhardt* and *Doherty* [5.16] improved the quality of the ferricyanide-bleached holograms by using the Kodak D-76 developer instead of the usual D-19, combined with a sophisticated drying procedure to minimize the scattering from the gelatin surface in the dry hologram. In an early investigation by *Pennington* and *Harper* [5.23] some of the important factors contributing to noise production in holographic bleaching techniques were addressed, as well as the importance of fidelity in the hologram recording process. Also the emulsion stress (due to manufacturing, fixation and drying) and the ways of reducing it were discussed. The main source of noise was found to be the reticulation of the emulsion surface after drying. Stress-relieved plates were therefore processed using nontanning developers and a new bleach - the *cupric halide bleach Kodak EB-2*. Before developing, the exposed plates were prehardened in the Kodak SH-5 prehardening solution, which reduced any tendency towards emulsion tanning. The drying was controlled by successive alcohol baths of increasing concentration. An obvious comment to this would be that the hydrogen peroxide used in the EB-2 bleach attacks the gelatin. This fact was also mentioned by *Pennington* and *Harper* who, however, claimed that prehardening eliminated most of that effect.

    Lehmann et al. [5.25] utilized a similar bleaching technique based on *cupric bromide*. *Lehmann* [5.26] recommended the copper-sulfate bleach (EB-2). He also provided another copper-sulfate bleach formulation intended for Agfa materials. *Lehmann*'s copper sulfate bleach

Potassium alum	20	g
Sodium sulfate	25	g
Copper sulfate	40	g
Potassium bromide	20	g

Sulfuric acid (conc.)   5 mℓ
Distilled water   1 ℓ .

Upatnieks and *Leonard* [5.22] suggested that the hardening of the emulsion should take place just before bleaching and not before development since that retards the speed of development and hampers development in depth. They used the Kodak SH-1 hardener instead of the SH-5.

Schmackpfeffer et al. [5.28] found that the EB-2 bleach combined with the Agfa 8E70 emulsion badly attacked the latter which is softer compared to the 649-F emulsion. The 8E70 material required a 25-minute prehardening in a SH-5 bath to prevent the EB-2 bleached emulsion from peeling away. (649-F only required a 10-minute prehardening). The article also contains SEM photographs of the EB-2 and the R-10 bleached emulsions. These photographs confirm that the R-10 bleach produces a relief structure with a periodically varying emulsion thickness due to the tanning action of the reaction products. They concluded that the Agfa materials require less-active bleaches.

The real improvement in the bleaching of transmission holograms came when *Phillips* and *Porter* [5.51] published their work on the low-noise technique which uses a mild-bleaching agent (*ferric nitrate*). This technique consists of a process which works well even for the argon-laser-recorded holograms (λ = 514nm). The development was performed in the concentrated *Neofin-blue developer* (already described in Chap.4) including the options of adding antifogging agents or increasing alkalinity of the developer. The prehardening of the emulsion was performed in an alcoholic bath (Drysonal) which had the advantage of simultaneously dissolving the antihalation dye from the emulsion. (Agfa used at that time a dye to suppress scattering in the green-sensitive emulsion. If not dissolved, the dye leaves a heavy residue of scatter in the finished emulsion).

After being prehardened and exposed, the plate was developed in

- concentrated Neofin blue for 5 minutes. (18° C, D $\simeq$ 2),
- 5-minute wash in running water,
- fix in Agfa G334 without hardener,
- pre-bleach clean (optional) (5-10 g/ℓ ferric nitrate),
- bleach in (Stock solution):

Ferric nitrate	150	g
Potassium bromide	33	g
Glycerol	20	g
Phenosafranine	300	mg
Isopropyl alcohol	500	mℓ
Distilled water	500	mℓ .

  Dilute: 1 part stock solution with 4 parts of water
- a wash (15 minutes) and bath in Drysonal (2 minutes), and thereafter
- drying.

In another paper *Phillips* and *Porter* [5.52] discussed the influence of organics in inorganic bleach baths. Phenosafranine, normally used as a desensitizing dye for silver halides, can also be employed for the acceleration of bleaching. Methylene blue has the same effect. The presence of this kind of dye in a bleach bath reduces ionic migration of silver in two ways:

(i)   By causing a reduced ion mobility which is due to the presence of large dye cations in the neighborhood, and

(ii) by shortening the ionic lifetime before precipitation of the halide due to the influence of the local anionic cloud.

These two processes have an image-sharpening effect since they prevent the growth of large crystalline halide structures by clumping.

Phillips et al. [5.59] further advanced the science of bleaching by introducing the organic oxidizing agent quinone or *Para-BenzoQuinone* (PBQ) bleach. The existing ferric-nitrate bleach from the same group was discussed in light of the experience gained, in which the main concern was the bleach's acidity and its tendency to soften the gelatin. In addition, the large amount of the oxidizing agent necessary made it also an expensive bleach bath.

The main feature of the PBQ bleach is that it does not destroy the oxidation products created during the pyrogallol development in which emulsion tanning occurs. The pyrogallol developer produces a blanket or a stain around the grain (the same applies to the catechol and hydroquinone developers) which is of the quinone type of structure. The idea behind the PBQ bleach was that a bleach based on an oxidizing agent such as quinone would probably not destroy the blanket around the grain. Normally, this blanket is destroyed in bleaches containing free bromine. *Para*-BenzoQuinone (PBQ) proved to posses the expected and desirable quality of not affecting the grain's blanket.

Although the PBQ bleach has proved successful from the point of hologram processing, it must be noted that this chemical is associated with toxic hazards, which is why severe precautions must be taken when using it. Quinone can cause eye irritation and, with longer exposure, damage to the cornea of the eye [5.60]. The upper limit of exposure to quinone vapor in room air has been set to 0.44 ppm in the USA. (Typically, quinone vapor just above an open tray is about 0.30ppm). The *PBQ bleach* (**PBQ-1**) was mixed in the following way:

p-Benzoquinone (PBQ)	2	g
Potassium bromide	30	g
Boric acid	1.5	g
Distilled water	1	ℓ .

The pH of this bleach is adjusted to about 5.

The importance of a buffer was mentioned. A correctly buffered solution in which, e.g., KOH and an acid were used is not as desirable as the formulation above which uses only a very small amount of acid. The reason for this is that the buffer constituents can affect both the gelatin-surface reticulation and the grain coagulation. Equally important is the KBr level, since excessive values of this substance can also cause coagulation which will increase scattering. For soft gelatin (Agfa) the upper level of KBr is about 30 g/ℓ. For harder gelatin (Kodak) the limit is about 50 g/ℓ. However, concerning KBr concentrations, see recent investigations by, e.g., *Hariharan* and *Chidley* [5.90] indicating a lower optimum KBr level, namely 16 g/ℓ (for Agfa materials).

Phillips et al. also recommend a *tanning version of the PBQ bleach* (PBQ-3) to be used with nonhardening developers (such as, e.g., Neofin blue):

p-Benzoquinone (PBQ)	2 g
Potassium dichromate	2 g
Potassium bromide	30 g
Borax	15 g
Distilled water	1 ℓ.

Note that the PBQ should be added just before use, and that its lifetime will then be about 15 minutes only. *Ackermann* et al. [5.97] have found that the lifetime of PBQ bleaches can be prolonged if, when used or stored, the solution is protected by a gas (e.g., CO_2).

Another interesting point mentioned by *Phillips* et al. [5.59] is the concept of the *fixation-free method* in rehalogenating processing – a rather new and for some readers confusing approach. It was, however, accidentally discovered and reported by *Hariharan* as early as in 1971 [5.36]. The present author has been using this method since the mid 70's. Holograms developed in HOLODEV 602 were bleached (without fixing) in the ferric sodium EDTA type bleach.

Eastes [5.56] made a careful investigation of six bleaching techniques using Kodak 649-F and 120-01 materials:
1) Cupric bromide/Hydrogen peroxide.
2) Bromine vapor.
3) Cupric bromide/Ferric chloride.
4) Potassium ferricyanide.
5) Iodine bleach.
6) Reversal R-9 type bleach.

In conclusion, *Eastes* stated that none of the investigated rehalogenating bleaches produced better results than the reversal bleach. The latter is considered to be the best because of its ability to produce phase holograms of high diffraction efficiency and with an acceptable signal-to-noise ratio over a wide range of exposures. This indicates that none of the "pre-Phillips" rehalogenating bleaches could really compete with reversal bleaching at that time.

Oliva et al. [5.68] and *Fimia* et al. [5.69] have noted that overexposing the material ($3 < D < 6$) will improve the signal-to-noise ratio of phase holograms processed with potassium-ferricyanide and ferric-nitrate rehalogenating bleaches. Using the high-contrast Kodak D-8 developer instead of D-19 improved the overall recording results additionally. They claim that the noise reduction was caused by the nonlinear recording of the intermodulation noise that will eventually occur at high exposure levels. The effect was also confirmed by microscopic studies of the emulsion surface as well as by performing index-matching investigations.

A significant contribution to the processing of phase reflection holograms was made by *Cooke* and *Ward* [5.77]. They presented another bleach of the PBQ-type (PBQ-2). A tanning catechol developer was used because of its lower staining effect (CW-C2, already discussed in Chap.4). The new PBQ-bleach was actually able to eliminate the slight stain obtained in catechol development. The acidity of the bleach was increased using citric acid instead of boric acid, which affected the thickness of the final emulsion. This was a completely opposite direction to take compared to the way which *Phillips* et al. [5.59] approached the problem. The main difference between the two approaches depended on the fact that *Phillips* et al. were mainly interested in display holography while *Cooke* and *Ward* were interested in diffraction gratings of the reflection type. As a matter of fact, the *Cooke and Ward* method (CW-C2 and PBQ-2 bleach, no fixing) has proved to be a successful technique for producing phase reflection display holograms as well. The bleach (**PBQ-2**) was mixed in the following way:

p-Benzoquinone (PBQ)	2 g
Citric acid	15 g
Potassium bromide	50 g
Distilled water	1 ℓ .

Bleaching time: 45 s to 2 min at 20° C, continuous agitation.

In this scheme, both the developer and the bleach will affect the emulsion in such a way that it will shrink, which can be useful for processing display holograms recorded with red-light lasers. However, immersing the processed holograms in propanol before drying (50% propanol for one minute and then 100% propanol for two minutes) increases the emulsion thickness (about 4.3%).

The main difference between PBQ-1 and PBQ-2 is their influence on the stain of the emulsion caused by the oxidation products in tanning developers. Mix the PBQ bleach with *boric acid* in order to *preserve the stain* generated during development. To *remove or reduce the stain*, mix instead the PBQ bleach with *citric acid*.

Fixation-free rehalogenating bleaching became the dominating technique for the processing of phase holograms after the publication by *Crespo* et al. [5.83]. Here, holograms developed in PAAP, ACC and CW-C2 developers are bleached in PBQ-1 and PBQ-2 bleach baths. They concluded that developing holograms in a chemical (nonsolvent) developer followed by a bleach in which the diffusion of silver and bromide ions is promoted inside the emulsion can actually result in holograms with a high diffraction efficiency and low scatter noise. This means that *the bleach should soften the gelatin slightly to promote the diffusion process*, which will render holograms with markedly increased diffraction efficiency. However, such a mechanism is also known to affect the position and the quality of the recorded interference fringes, thus reducing the image resolution.

Phillips [5.84] advocated the use of the **ferric-EDTA bleach** (Ferric-Ethylene Diamine Tetra acetic Acid) which he considers to be both safe in use and successful in the results it can achieve, trying to convince people to stop using the hazardous PBQ-based solutions. Ferric-EDTA can be obtained on the market or made by mixing di-sodium EDTA with a simple ferric compound such as ferric sulfate $Fe_2(SO_4)_3$:

Ferric sulfate	30 g
Di-sodium EDTA	30 g
Potassium bromide	30 g
Sulfuric acid (conc.)	10 ml
Distilled water	1 l .

Phillips maintained that in this bleach the complexing of silver ion by the sequestering agent EDTA assists the ionic migration of silver. It is also the present author's impression that there seems to be no doubt as to the exceptional ability of this mixture to promote physical transfer during the bleach process.

Concerning *copper bleaches Blyth* [5.85] has introduced the following rehalogenating bleach which produces, however, rather printout-sensitive holograms:

Copper sulfate (pentahydrate)	35 g
Potassium bromide	110 g
Acetic acid	10 ml
Distilled water	1 l .

The persulfate bleaches [5.86-88] have so far not attracted much attention among researchers in holography. However, an interesting conventional-photographic bleach bath of the rehalogenating type, based on a mixture of *persulfate and para-benzoquinone* has been formed by *Keiler* and *Pollakowski* [5.88]. The solution contains potassium peroxodisulfate as oxidizer and *p*-benzoquinone as activator. To avoid the hazard of mixing a bleach using pure PBQ, the compound can be formed in the bleach by a reaction between hydroquinone and persulfate. Potassium bromide acts, as usual, as a rehalogenizer. The solution is buffered with the acetic acid to a pH level between 3 and 4. Two, slightly different versions, are mixed in the following way:

	Version 1	Version 2
Sodium acetate (anhydrous)	3.5 g	4.5 g
Potassium peroxodisulfate	10 g	10 g
Potassium bromide	10 g	20 g
Hydroquinone	0.75 g	1 g
Copper sulfate, crystals	0.5 g	1.5 g
Acetic acid (conc.)	15 ml	15 ml
Distilled water	1 l	1 l .

Version 1 is for the Orwo color film, and Version 2 is for the Fuji/Eastman negative color film. In a persulfate-based solution, quinone tends to form humic acids, which will reduce the effect of PBQ as an activator. To reduce this effect a small quantity of an inhibitor is added in the form of copper sulfate crystals ($CuSO_4 \cdot 5H_2O$). It is important to note that this bleach must be mixed at least 6 hours before use in order to form a sufficient amount of PBQ.

The ideas of mixing a bleach by using an oxidation process between persulfate and common developing agents (e.g., ascorbic acid, amidol, metol, and hydroquinone) have been used to formulate a set of new rehalogenating bleach baths for holography by *Bjelkhagen* et al. [5.107]. These baths have very good performance concerning both high efficiency and low noise.

The best working formulas will be given in the following. The PBQ bleach for holography was slightly modified compared to the formula in the previously mentioned paper by *Keiler* and *Pollakowski* [5.88]. The **PBU-quinol** bleach is mixed in the following way:

Cupric bromide	1 g
Potassium persulfate	10 g
Citric acid	50 g
or: Sodium hydrogen sulfate	50 g
Potassium bromide	20 g
Distilled water	1 ℓ .

Add 1 g hydroquinone (quinol) after the other constituents are mixed.

This bleach must be mixed at least 6 hours before use in order to form a sufficient amount of PBQ. The potassium persulfate oxidizes the hydroquinone to form quinone which is revealed by the noxious vapor of this substance. However, when mixing this bleach the unoxidized form of quinone is used which means a much safer way of mixing a PBQ bleach. Once the quinone is formed, the bleach works like the normal PBQ bleach. In the bleach described by *Keiler* and *Pollakowski* [5.88], cupric sulfate was used to prevent polymerization (manifested by browning) of the quinone. To avoid too many different ions in the bleach above, cupric bromide was utilized instead of cupric sulfate. The effect of this compound makes this bleach very long lasting which means that it can be used over and over again.

Well-working bleaching solutions can also be based on other developing agents. After a systematic investigation of possible candidates the following two agents were found to be most successful, namely *amidol* and *metol*. The amidol version produces probably the cleanest results and shows the highest bleaching rate, but the bleach bath is not as stable as the metol-based version. Both bleach baths can be used after about half an hour after being mixed. They are mixed in the same way as the PBU-quinol bleach with the only difference being the developing agent used. The **PBU-amidol** or **PBU-metol** bleach is mixed by adding either 1 g of amidol or metol to the stock solution above.

If a rehalogenating bleach is desired in which an emulsion shrinkage of about 50 nm can be obtained after processing, the **PBU-ascorbic acid** bleach is worth trying. In this case 5 g of ascorbic acid is added to the stock solution. It is a slow bleach that can be employed 6 hours after mixing.

The amidol based bleach is port-wine colored, and the metol version is pale green when a sufficient amount of oxidizer is formed in the solution. The four just described bleaches perform very well and are much safer to mix and use than some of the previously introduced solutions for holography. These bleaches have been named PBU (Phillips-Bjelkhagen Ultimate) bleaches followed by the name of the developing agent on which they are based.

Most of the bleaches discussed so far entail the application of an intermediate oxidation step, which can affect the MTF due to the diffusion process involved. For this reason, it may be worthwhile to have a look at the rehalogenating solutions that work in a more direct way. One of such bleaches in popular use is the **bromine-water** bleach. It has been recommended by *Benton* [5.58] and it is claimed to produce clean, bright, and printout-stable holograms. It is also safer and easier to use than the bromine-vapor method that will be discussed in the next subsection. The bleach is mixed in the following way (*Benton's* recommendations [5.58]). A glass-stoppered glass bottle is filled with distilled water (about half a liter), after which a small amount of liquid bromine (about 5 mℓ) is poured into the bottle which is then closed. After about one day the bromine is dissolved and the bleach is ready to use. Under a fume hood, some of the solution is poured into a glass tray. The bromine will begin to slowly outgas, which is why it should be used rather quickly and the solution returned to the bottle as soon as the bleaching is finished. Bleaching takes about a minute or so, depending on the strength of the bleach. After bleaching, the plates should be washed and dried in the usual way. It is recommended to keep a bottle with a sodium-sulfite solution (50g/ℓ) handy to neutralize any spills or to clean stains in the trays. *Benton* recommended to use this bleach with the IEDT-processing method described in Chap.4. The bromine-water bleach is also often used for bleaching rainbow transmission holograms.

Non-Aqueous Rehalogenating Bleaching

A direct dry-bleaching method to convert silver into silver halide was first introduced by *Thiry* [5.42]. Here, bromine vapor was utilized to act upon the emulsion for half an hour, after being kept in vacuum for one hour. Higher diffraction efficiency combined with higher noise were obtained compared to those obtained for holograms bleached in a potassium ferricyanide solution.

Graube [5.49] has presented a systematic investigation on the dry-bleaching methods using elemental halogens. Since the refractive index for silver bromide is higher (n_{AgBr} = 2.25) than that for silver chloride (n_{AgCl} = 2.07) the paper suggests that this was the reason why bromine bleaching produced the best diffraction efficiency - the statement that *van Renesse* [5.50] has contested supporting his view with the electrical polarizability theory (already discussed in Sect.5.2.2b).

Graube did not find it necessary to apply a vacuum treatment to plates before bleaching. Transmission gratings recorded on Kodak 120-02 plates were developed in D-19 (density about 4) and fixed in a Kodak rapid fixer with a hardener added. The plates were then washed and dried, after which they were bleached in bromine vapor in a closed container partly filled with bromine liquid where they were held above the liquid surface for about 15 minutes. Bromine bleaching leaves unreacted bromine in the emulsion which appears in the form of a yellow stain. This stain can be removed by leaving the plates overnight in a fume hood providing good air circulation.

Graube found that a diffraction efficiency of about 70% could be obtained for both transmission and reflection diffraction gratings. Unfortunately, such high efficiency was always associated with a rather high noise level. One of the advantages of this method is that no physical transfer from the nearby grains takes place; the bleaching technique applied is a pure rehalogenating technique in which no grain movement within an emulsion occurs. In wet development combined with a fixing step prior to dry bleaching, the emulsion undergoes considerable changes. The main advantage here is that the bleached holograms are almost completely resistant to printout effects. The elemental bromine did not only react with the silver grains but it also oxidizes functional groups on the gelatin molecules that would have otherwise acted as reducers for the silver-bromide crystals. This resistance to printout is also found in holograms bleached in aqueous-bromine bleaches. However, long exposure to vapors produces considerable bromination of the gelatin molecules, causing associative hydrogen-bonding destruction and a breakdown of the three-dimensional structure of the emulsion. Other disadvantages include high scattering levels, destruction of developer stain by the free halogen and a high toxicity of the bleach chemical, considered deleterious to both the skin and the respiratory tract. The upper-limit value for exposure to bromine is 0.1 ppm, which is why the bleaching operation should be carried out in a fume hood with an air velocity of at least 0.8 m/s with protective goggles and rubber gloves.

The use of *non-aqueous solvents* such as methyl or ethyl alcohol to dissolve bromine has the advantage of being easier to dissolve. The use of iodine in a direct bleach requires an alcohol to dissolve the crystals. Sometimes a small amount of water is added to swell the emulsion and promote the reaction. *Benton* [5.58] formulated the following bleaching bath (**direct iodine** bleach) producing clean, printout-stable transmission holograms:

Iodine (crystals)	2÷5 g
Methyl or ethyl alcohol	750 mℓ
Distilled water	250 mℓ .

To remove the deep yellow stain after bleaching, *Benton* recommended that the bleached hologram is rinsed in an identical solution without iodine. This bleach has become very popular for the processing of rainbow holograms.

Popular Rehalogenating Bleach Baths

To summarize the section on rehalogenating bleaching a list of bleach baths currently in popular use for holographic processing is presented below:

- Ferric nitrate bleach (*Phillips*).
- PBQ-1 (*Phillips*).
- PBQ-2 (*Cooke-Ward*).
- Ferric EDTA (*Phillips*).

- PBU-(developing agent) (*Bjelkhagen-Phillips*).
- Bromine/iodine (*Benton*).

b) Reversal Bleaching

Only wet-processing schemes for reversal bleaching have been discussed in the holographic literature up till now. The first paper introducing this bleaching technique for holography was published by *Kiemle* and *Kreiner* [5.10]. They used potassium dichromate and sulfuric acid to bleach reflection diffraction gratings recorded with a HeNe laser on Agfa 8E70 plates developed in the Kodak D-19b developer. This technique is interesting in that it utilizes the original, unexposed silver-halide crystals to create the image at reconstruction of the hologram. This means that the size of the grains remains, in principle, unchanged after the processing, which causes the scattering to be low or at least determined by the original grain size in the emulsion. Unfortunately, by using the common reversal bleaching solutions the unexposed silver-halide grains will increase in size by a process similar to the Ostwald ripening. Normally, this phenomenon is not as pronounced as the grain growth which occurs in rehalogenating bleaching. Low scattering in reversal bleaching in which potassium dichromate was used is discussed in another publication by *Kiemle* [5.12]. During reversal bleaching, silver salt is removed from the emulsion which reduces the potential diffraction efficiency of the phase hologram in comparison with the fixation-free rehalogenating bleaching technique. The reversal bleaching technique shows, however, yet another advantage. After bleaching, the emulsion shrinks, which often gives rise to a desirable color change of the holographic image. Reflection holograms recorded with red-wavelength lasers will reconstruct in the yellow or green part of the spectrum depending on the pre-bleach density and the recording wavelength. The reversal bleaching technique is rather popular among holographers working in display holography.

About two years after the first publication dealing with the reversal bleaching process [5.10] for holography other researchers joined in the discussion, introducing small changes and modifications to the above mentioned. *Chang* and *George* [5.20] tried all the three existing photographic reversal bleaches, but finally decided on a bleach containing only a concentrated nitric acid which produced phase gratings with very low scattering noise on 649-F plates. Electron micrographs revealed that the original unexposed grains were reduced in size after bleaching. *Chang* and *George* had demonstrated the advantages of the reversal method, but warning, however, that it was undesirable to use highly volatile acids. *Lamberts* and *Kurtz* [5.29, 37] presented a reversal-processing scheme for Kodak materials. They also introduced the tanning pyrocatechol developer (Kodak SD-48) into their processing scheme (omitting potassium bromide) which seemed to reduce the surface relief effects due to low-spatial-frequency intermodulation noise in the object beam. *Verbovetskii* and *Fedorov* [5.33] found that chromium-bleached holograms were clearer than rehalogenated phase holograms which "took on a milky color". *Buschmann* [5.34] formulated a processing scheme for the new Agfa 8E75 materials, comparing it with the formulation given in the Kodak P-230 pamphlet [5.30] on reversal bleaching of the 649-F plates (based on the publication of *Lamberts* and *Kurtz* [5.29]). In it he used phenosafranine as a desensitizing dye for the bleached holograms. At the same time *Hariharan* published a paper [5.35] describing a more complicated reversal-bleaching technique involving a re-exposure of the holographic plate to white light after the first development. This method is especially well suited for the thick emulsion used in the 649-F plates. *Hariharan* claimed that the noise level in the holograms processed this way is very low. Soon thereafter *Hariharan* et al. [5.36] described a simple reversal technique in which the bleach contained potassium iodide to convert the unexposed, virgin silver-bromide grains to silver-iodide grains for better light stability of the phase hologram. Soon later, the same group [5.45] published a new reversal-processing scheme in which a different developing technique was introduced. By adding a silver solvent to the developer, controlled etching of the silver grains (the virgin, unexposed crystals) could take place during development. The bleached plates contained only slightly smaller grains than the original size. At the same time, the Agfa 8E75 emulsion had a grain size of 50 nm. The effect was lower scatter noise, without any substantial decrease of diffraction efficiency. To obtain such an etching effect, 0.5 g/ℓ of sodium thiosulfate was added to a D-19 developer for Agfa 8E75 materials and 1.0 g/ℓ for Kodak 649-F materials. The results are very similar to the ones obtained with the pyrogallol-based developer, HOLODEV 602, which has a high content of sodium sulfite (Chap.4). This is why the (as yet unpublished) pyrogallol developer used with the old type of Agfa materials produced very low noise holograms back in the times when the present author was experimenting with the Agfa materials. The HOLODEV 602 (with a reduced sodium sulfite content as compared to the first version) combined with a reversal bleach still represents a rather good way of processing CW-laser reflection holograms on the finer-grained Agfa materials of today.

Talking about the results obtained by the application of reversal bleaching it must be mentioned that the discussed results are slightly contradictory. The established view claims that a developer used for phase holograms of the reversal type *should not contain silver solvents* (i.e., a non-physical developer). This is certainly correct when an optimum diffraction efficiency is the main goal, as has been shown experimentally by *Lamberts* and *Kurtz* [5.37] as well as by *Slaby* and *Sirko* [5.66] and also by *Crespo* et al. [5.83]. If the content of the image forming silver salt in the emulsion is reduced, a reduction in diffraction efficiency is expected. On the other hand, the signal-to-noise ratio is often more important, in particular, for the overall quality of a hologram. In the earlier versions the Agfa materials had an average grain diameter of such a size that controlled etching of the virgin unexposed silver grains was beneficial to the signal-to-noise ratio of the image. It has been pointed out in *Hariharan*'s papers, for example, that the reduction in efficiency can be kept rather small and the overall image quality is improved. Even today, the effects of silver solvents for scatter reduction in the developer for reversal bleach processing should not be over-looked when applied to emulsions with grain sizes of 30 to 40 nm.

To reduce scatter noise in bleached holograms *Hariharan* et al. [5.44] used a diluted fixing bath to obtain a controlled etching effect on the silver grains. They tested Kodak 649-F plates containing normal AgBr grains as well as plates where the original grains had been converted to AgI. A normal non-hardening fixing solution was diluted 1:40 and acted upon the emulsion for 5 minutes. About 30% of the silver content was thus removed, which corresponds to a 50% reduction in scattering.

At the present author's laboratory a similar but simpler technique was used based on a sodium sulfite solution $(70 \div 100 g/\ell)$. It had two purposes: to reduce scattering and to introduce a color change for reflection phase holograms exposed with red laser light. The plates were treated in the sodium sulfite bath for 10 to 20 minutes depending on the desired shrink-age.

Common Reversal-Bleach Baths

The most common reversal bleaching technique is probably the "*pyrochrome processing*" technique [5.62,63]. Here, pyrogallol-developed holograms are bleached in the following bleach bath, which is actually a modified version of the Kodak R-9 bleach. The **Pyrochrome** bleach is mixed in the follow-ing way:

Stock solution:

Potassium dichromate	4	g
Sulfuric acid (conc.)	4	$m\ell$
Distilled water	1	ℓ .

Working solution: Use undiluted, or mix 1 part stock solution + 4 parts dis-tilled water. Bleaching time: 1 to 3 minutes. pH is about 2.4.

Some modifications of this bath are possible, concerning mainly the concentration of its constituents. A higher concentration (used undiluted) gives a faster bleaching action (sometimes desirable for mass production and machine processing). A lower concentration means a slower action but also a more gentle effect on the gelatin emulsion, which often results in higher quality holograms.

Blyth [5.82] recommended to use $20 \div 100$ g/ℓ sodium hydrogen sulfate (sodium bisulfate) instead of sulfuric acid as a safer method and an easier way to adjust acidity. The use of a solid acidifying agent avoids the decant-ing of strong liquid acids. For each milliliter of sulfuric acid, 2.6 grams of the bisulfate should be used. An even milder acid recommended by *Blyth* is the toluene-4-sulfonic acid (20g/ℓ). *Blyth* has also stressed the importance of using distilled water not only for mixing the bleach solution but also for washing the plates in it, directly before and after the bleach bath. This is to prevent chlorine (from tap water) or any other substance from interacting with the bleach bath, which produces a scum difficult to wash away. If the scum remains in the emulsion, it will increase the scattering of the bleached hologram. In addition to these precautions, *Buschmann* [5.34] recommended the use of a clearing bath after bleaching to dissolve remaining silver chromates ($Ag_2Cr_2O_7$ and Ag_2CrO_4) out of the emulsion. The clearing solution was mixed in the following way:

Sodium sulfite	50 g
Sodium hydroxide	1 g
Distilled water	1 ℓ .

Since the solubility of the silver chromates is so much higher than that of the silver bromide in the emulsion only a short processing time is necessary in this bath. About 1 to 2 minutes is quite sufficient and should not be exceeded in order not to dissolve too much of the silver bromide. After 2 minutes, less than 5% of the AgBr is dissolved, after 4 minutes, 10% and after 8 minutes, 20%, as shown in *Buschmann's* investigation.

Another reversal bleach used in holography is the permanganate based bath. *Benton* [5.58] recommended the **KP-4** recipe:

Potassium permanganate	3 g
Sulfuric acid (conc.)	10 mℓ
Distilled water	1 ℓ .

Phillips et al. [5.59] introduced a slightly different version of the permanganate bleach:

Potassium permanganate	0.5 g
Nitric acid	0.25 mℓ
Distilled water	1 ℓ .

He also recommended that the plate be cleared in the following solution:

Sodium metabisulfite	10 g
Distilled water	1 ℓ .

Thomas [5.95] investigated the reversal-bleach baths mentioned above combined with different holographic developers and found that the following combinations gave the best results as regards both the diffraction efficiency and the signal-to-noise ratio for *Denisyuk* reflection holograms exposed with the HeNe laser, using Agfa HD materials:

1) The CW-C2 developed holograms showed very good results when combined with both types of permanganate bleaches.
2) Holograms developed in HOLODEV 602 yielded similar results but only with the KP-4 permanganate bleach.
3) The HOLODEV 602- or CW-C2-developed holograms showed almost identical results when bleached in pyrochrome bleach - similar in quality to the ones obtained in point *1* above.

Combined with other developers, such as pure pyrogallol, Kodak D-19 or Agfa GP 62 the resulting quality was not equally good as that of the ones obtained above.

The *ceric sulfate reversal bleaches* have been employed by some holographers, but did not secure the same high-quality results as those achieved by the previously discussed baths. The investigation performed by *Thomas* [5.95] in which he tested the ceric sulfate bleach bath, supplies the evidence for the above statement. The investigated bleach bath was mixed in the following way:

Ceric sulfate (yellow)	10 g
Sulfuric acid	8 mℓ
Distilled water	1 ℓ .

To close the reversal-bleach section it is important to mention that a brand-new, well-working, and safe solvent-bleach bath for holography was recently (July 1992) introduced by N. Phillips. The extensive use of dichromates in bleach solutions became a point of concern to him, which is why he set off to find a reversal bleach that could replace the current ones. The new bleach **Phillips' Safe Solvent Bleach (PSSB)** is mixed in the following way:

Ferric sulfate *or* ferric nitrate	30 g
Potassium persulfate	20 g
Sodium hydrogen sulfate	30 g
Distilled water	1 ℓ .

The bleaching time is rather long, about five minutes or more.

The new bleach performs very well. It requires less exposure ($\frac{1}{4}$ to $\frac{1}{2}$, which means 50 to 100 μJ/cm^2) compared with the other reversal methods. Combined with the pure pyrogallol developer, this bleach produces bright holograms with a very high signal-to-noise ratio. The hardening effect of the pyrogalled developer is important in this processing scheme. This bleach does not attack the pyrogallol-produced stain in the gelatin, which the dichromate-based solution does.

If, instead, the hologram is developed in the CW-C2 developer (less stain and hardening), a slightly lower signal-to-noise ratio is obtained, combined with more shrinkage of the emulsion. Therefore, it seems that the best developer to use with this bleach is the pure pyrogallol developer, unless a rather large wavelength shift is of interest.

Popular Reversal-Bleach Baths

To summarize the subsection on reversal bleaching, a list oof bleach baths currently in popular use for holographic processing is presented below:

- Dichromate bleach (Kiemble & Kreiner, von Renesse, Spierings).
- Permanganate bleach (Benton).
- PSSB (Phillips).

5.2.4 Comparison Between Different Bleach Processes

Several experimental investigations have been performed with the view to compare different processing methods and to delineate the influence of the various parameters on the bleaching process. The way in which a developer affects a bleached hologram has been investigated by *Hariharan* and *Chidley* [5.89] as well as by *Hegedus* and *Hariharan* [5.100]. The main concern associated with the processing of holograms when using the rehalogenating method is not so much to obtain high efficiency, which is rather easy with many bleaches, but to reduce the noise in the final image. Noise is caused by both surface emulsion irregularities as well as by scattering from the silver-halide grains, other compounds, or voids in the emulsion. If surface irregularities are minimized by careful processing and drying, granularity will be the significant source of scattered light. The first question that can be posed then is how the developer will affect the noise in both reversal and rehalogenating bleaching. In reversal bleaching, the virgin unexposed grains are used, which, in principle, means that the problem of grain growth during rehalogenating bleaching is eliminated. However, in practice, a slight grain growth may occur also during reversal bleaching when applying common bleach baths. The effect of controlled etching by using silver solvents in the developer can control the size of the grains, which has already been discussed [5.44]. The main conclusion from the previously discussed tests is that for reversal bleaching the noise level can be reduced which, however, is always accompanied by a slight reduction in the diffraction efficiency. The overall quality (dynamic range) can be improved using controlled etching. Elimination of the sodium sulfite and the potassium bromide from the conventional types of metol-hydroquinone developers increases the diffraction efficiency without really increasing the noise level very much. In other words, *reducing the sulfite content increases the noise and the diffraction efficiency in a hologram*. This is valid for MQ developers but not necessarily for pyrogallol or catechol developers. The developer strongly recommended by *Hariharan* for reversal bleaching of CW-laser-exposed holograms is the *ascorbic acid based developer*.

Hariharan [5.101] made a comparison between the hardening non-solvent MQ developer and an ascorbic acid developer for reversal bleaching. The local hardening around the exposed grains caused by the MQ-developer causes some additional refractive-index modulation which is *180° out of phase* with the modulation of the unexposed grains. This explains the reason why the diffraction efficiency is lower when using hardening developers for processing holograms with the reversal method as compared to non-hardening developers. It is important to remember this fact when using the common, pure pyrogallol developer in the reversal-bleaching process (pyrochrome). It also explains why, developers such as CW-C2, HOLODEV 602, PAAP and SM-6 work so well (actually better than the sulfite-free pyrogallol developer) in the reversal bleaching process.

In *rehalogenating bleaching*, the conventional metol-hydroquinone developer containing sulfite and potassium bromide produces holograms with higher scattering levels than if sulfite and bromide were omitted. In other words, *reducing the sulfite content decreases the noise with only a slight loss of diffraction efficiency*. This is valid for MQ developers but does not necessarily apply to developers based on other agents. The attempts to use ascorbic acid-based developers for the rehalogenating bleaching of CW-laser-exposed holograms have not been very successful, according to *Hariharan*. An interesting aspect of processing rehalogenated holograms with MQ developers containing large amounts of sodium sulfite has been taken up by *Hariharan*. It is the fact by exposing the holograms to a high level, in order to obtain a high density, reduces the solution-physical development since the amount of the unexposed silver-halide crystals in the emulsion decreases. (Compare Sect.4.4.5). Concerning potassium bromide in the developer it seems that for rehalogenating bleaching the developer can contain KBr but not for reversal bleaching [5.69].

Hariharan and *Chidley* [5.90] have also investigated the influence of the halide type and its concentration on rehalogenating-bleach baths. A non-hardening ascorbic acid-metol developer was used here to avoid physical transfer during development. The bleach contained acidified potassium dichromate to which potassium bromide was gradually added to convert it from the reversal to the rehalogenating type. At first, the diffraction efficiency decreases, but later when a certain concentration level is reached, it starts rising again and reaches a maximum value that is higher than that for the zero concentration, i.e. for reversal bleaches. It was found that the KBr concentration of above 6 g/ℓ starts showing the effects of the rehalogenating process, namely an increase in diffraction efficiency. The optimum concentration was found to be 16 g/ℓ, above which the diffraction efficiency did not increase further.

In another experiment in the same testing procedure, potassium bromide was substituted with potassium iodide. This produced very different results. Small amounts of potassium iodide added to the bleach (0.2÷0.8 g/ℓ) resulted in an increase of the diffraction efficiency, which would drop when a concentration level of about 3.2 g/ℓ was reached. A continued increase of potassium-iodide concentration resulted in an increasing diffraction efficiency reaching its maximum at a concentration of about 12.8 g/ℓ. The rise of diffraction efficiency was usually accompanied by a large increase of scatter noise.

De Winne and *Phelan* [5.81] reported on an investigation into the silver-halide grain sizes as a function of KBr concentration in a PBQ bleach. The investigated material was the green-sensitive Agfa 8E56 HD emulsion. The results are listed in Table 5.5.

When 100 g/ℓ KBr was used the noise was increased by 50% due to the increased mean grain size and the grain-size spread of AgBr particles

Table 5.5. Silver-halide grain size as a function of KBr concentration

PBQ concentration [g/ℓ]	KBr concentration [g/ℓ]	Main grain diameter [nm]	Spread [nm]
2.5	25	46	11
2.5	50	61	15
2.5	100	73	29

formed during rehalogenating bleaching. It was also shown that the noise is increasing much faster than the brightness of the holographic image is with increased KBr concentration.

Another observation made was that it is not possible to use a ferricyanide bleach if the holographic material has been developed in a tanning developer, e.g., a pyrogallol developer. In spite of the fact that very small silver-halide grains (mean grain size 37 nm) were produced by the ferricyanide bleach, no detectable holographic image could be reconstructed from such a hologram. This is in good agreement with polarizability calculations performed by *Phillips* [5.72]. Under certain conditions, the modulation can vanish for fixation-free rehalogenating bleaching using ferricyanide bleaches.

Two other investigations performed by *Hariharan* and *Chidley* [5.91, 92] were aimed at showing the way in which *spatial frequency affects the diffraction efficiency in different bleaching processes*. In the first paper [5.91] the reversal- and the conventional rehalogenating-bleaching methods (including fixing) were compared when applied to transmission gratings with varied fringe spacing formed on Agfa 8E75 HD materials and exposed with the HeNe laser. In order to avoid physical transfer and hardening during development, a metol-ascorbic acid developer was used. In the rehalogenating mode of operation, potassium dichromate bleaches ($0.8\,\mathrm{g}/\ell$) were used with 4 g/ℓ potassium bromide. Both types of bleaching processes show a peak in efficiency at a fringe spacing of about 1 μm. A drop in efficiency starts at 0.8 μm and progresses rapidly, especially for conventional bleaches showing a 50% reduction at about 0.5 μm. As a way of comparison, it was shown that amplitude holograms are not affected by variations in spatial frequency to the same degree and that their diffraction efficiency is rather constant in the same fringe spacing region.

Hariharan and *Chidley* explained this phenomenon by the diffusion of the reaction products, which takes place in all types of bleaching processes. For the fixation-free rehalogenating processes the diffusion is the actual mechanism behind it, but even in other types of bleaching processes diffusion plays an important role. In *conventional rehalogenating bleaching*, the bleach converts developed silver into silver ions which then react with bromide ions (supplied by the bleaching solution), forming silver bromide in the exposed areas. Some silver ions, however, diffuse away to nearby unexposed grains, where they enter into a chemical reaction with bromide ions. The formation of silver bromide in these areas results in lower modulation and, consequently, a lower diffraction efficiency. The strength of this effect depends, obviously, on the *diffusion length* over which the diffusion occurs, explaining the spatial-frequency dependency.

In *reversal bleaches* diffusion is caused by formation of bromide and silver ions coming from the unexposed silver-bromide grains and entering the bleaching solution. Some of the bromide ions formed this way diffuse away from the unexposed areas to exposed areas, forming silver bromide with the silver ions also diffused from the unexposed grains. This process affects the diffraction efficiency to the extent which depends on the fringe separation in the affected area, i.e. the area in which diffusion takes place. From this we can see that *diffusion length* is as important in reversal bleaches as it is in conventional rehalogenating bleaches.

These findings initiated yet another investigation by *Hariharan* and *Chidley* [5.92] who now studied the *fixation-free rehalogenating bleaching process*, where the diffusion is believed to be the mechanism behind the process. The same testing technique as the one described in the previous paper was used, except for the fact that here 8 g/ℓ potassium bromide was used in both bleaches, as compared to the previous experiment in which the second rehalogenating bleach contained ferric sulfate (30g/ℓ). What is interesting in the results obtained is that both types of bleaches show much higher efficiency at a fringe spacing of about 0.5 μm (and below) than they did in the previous test on conventional bleaching. The drop in efficiency occurs at higher fringe spacings. The diffraction efficiency for the dichromate bleach (whose overall efficiency is higher than that of ferric sulfate bleach) drops rapidly for fringe spacings of above 1.2 μm. For a ferric sulfate bleach the diffraction efficiency falls already at a fringe spacing of about 0.6 μm. The diffusion process occurring here can be explained in the same way as in the previous paper. The emulsion contains both the unexposed silver bromide and silver. A part of the dissolved silver ions diffuse away to create silver bromide in the unexposed silver-bromide grains. The remaining silver ions are redeposited in situ on the exposed and developed grains. The material transformation process taking place in the emulsion occurs without any significant loss of silver salts, which means that the emulsion shrinkage is minimized. The diffusion process can only take place over a certain distance (the diffusion length of the silver ion) which now becomes an important characteristic for a fixation-free rehalogenating bleach. In the investigation of *Hariharan* and *Chidley* [5.91] on bromide concentration in the bleach discussed earlier in this section it was found that at the fringe spacing of 0.83 μm the efficiency drops down to zero if the concentration of potassium bromide is below 1 g/ℓ. This means that the diffusion length at that concentration is below 0.83 μm. The highest efficiency was obtained at the concentration of 16 g/ℓ, indicating that the diffusion length was then >0.83 μm. *Hariharan* and *Chidley* also mentioned the fact that by changing the fringe spacing in the gratings, a transition from the thick to the thin recording regime will eventually occur (for an emulsion thickness of 6μm). This transition takes place at a fringe separation of somewhere above 4 μm. From the above follows that the observed drop in efficiency is not really affected by regime transition. One of the side-effects of the rehalogenating process contributing to the increase of diffraction efficiency at smaller fringe separations is the progressive improvement of the MTF, caused by various adjacent effects during development. This does not affect the overall process to any great degree, however, which is why *Hariharan* and *Chidley* concluded that the rehalogenating-bleaching process hinges on the diffusion length of the silver ion, since above a certain value the material transfer process in the emulsion stops and the efficiency is drastically reduced. The paper suggests that a certain drop in diffraction efficiency at a certain fringe spacing should be used for determining the diffusion length of the silver ion for a given bleach. In the present investigation, the diffusion length is 0.6 μm for the ferric sulfate bleach and 1.6 μm for the dichromate bleach. The larger diffusion length found for the dichromate bleach is probably due to its higher activity.

Phase-contrast microscopy investigations for studying these effects have been described by *Kostuk* and *Goodman* [5.108]. This technique will be discussed in Sect.6.9.3. Their results support the theory that a redistribution of silver-halide grains occurs during the rehalogenating bleaching process and that this mechanism is spatial-frequency dependent.

Hariharan [5.101] investigated the mechanism of material transfer during fixation-free rehalogenating bleaching. He suggested that a process similar to the Ostwald ripening is involved in this type of bleaching. Tests were made with a bleach containing ferric sulfate as the bleaching agent and potassium bromide as the alkali halide. The quantities of these compounds in the bleach were varied for different experiments. The diffraction efficiency and scattering of recorded holograms were measured. The best results were obtained when the molarity of potassium bromide was slightly higher than that of ferric sulfate. If the oxidizing agent in the bleach is further increased the diffraction efficiency drops off. If the alkali halide is increased above the optimum level only an increase in scattering is obtained. *Hariharan* explained the process in the following way: First the developed silver is converted by the oxidizing agent into ionic silver which goes into solution. In the vicinity of the exposed areas the silver ions react with the bromide ions and form silver-bromide nuclei. Then material transfer takes place from the exposed areas to nearby unexposed areas, which is similar to Ostwald ripening. The formed silver-bromide nuclei is dissolved caused by excess bromide present in the bleach solution. The solubility of the newly formed small silver-bromide crystals is much higher than that of the bulk material. Material transfer by diffusion then takes place and silver bromide is deposited on larger silver-bromide crystals in the unexposed areas. Thus, this process results in a grain growth of original crystals in the emulsion. Grain growth can also occur by coalescence of crystals which were originally separated. A large excess of alkali halide speeds up this process. *Hariharan* has reviewed all the just discussed investigations in [5.102].

Jeong et al. [5.103] studied the diffusion transfer mechanism by re-developing holographic gratings which had originally been rehalogenated without fixing. The obtained density of low and high spatial frequency gratings was measured. By comparing the difference of optical density for these gratings, relative amount of rehalogenated silver halide and the diffusion transfer for high-spatial-frequency gratings could be estimated. In this study, it was found that a KBr-bleach concentration of $16 \div 18$ g/ℓ is the optimum, which is in good agreement with *Hariharan's* results.

From the above, the general conclusion that may be drawn is that fixation-free rehalogenating bleaches with optimum diffusion lengths can attain both a high diffraction efficiency and low noise, provided that the grain

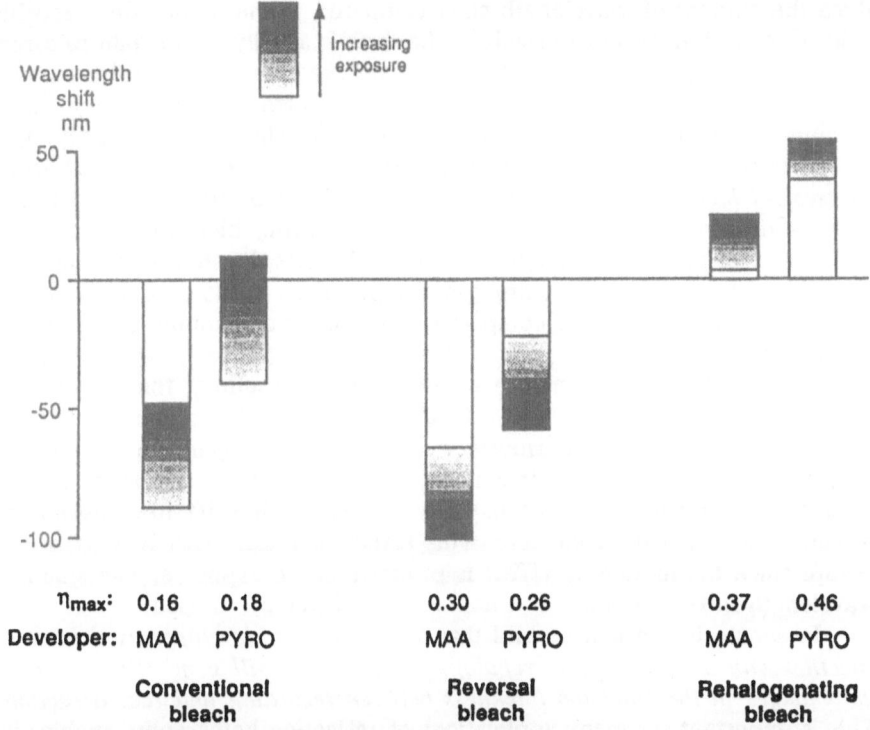

Fig.5.3. Wavelength shifts (shaded areas) and variation with increasing exposure (increased darkness) obtained with reflection gratings over an exposure range of 4:1 using a nontanning Metol-Ascorbic Acid (MAA) developer and a tanning pyrogallol (PYRO) developer with three different bleaches [5.93]

growth in the emulsion is controlled in some way. The intermodulation noise remains in the low-frequency range in which the fixation-free rehalogenating bleaches show a low response, as we have just seen in the above investigations. That, combined with the fact that emulsion thickness is hardly affected in this processing scheme, makes it possible to record phase holograms of high quality. Continued research is, however, needed for further studies of the diffusion mechanism as a function of emulsion and its processing parameters.

Hariharan and *Chidley* [5.93] have also compared different bleaching processes for reflection holography, where the effect on color shift is of interest. Holographic reflection gratings developed in non-tanning metol-ascorbic acid or strongly tanning pyrogallol were bleached in a potassium dichromate rehalogenating bleach (with or without fixing) or a reversal dichromate bleach. The wavelength shifts for the different bleach-processing schemes are shown in Fig.5.3.

As could be expected, *in both the conventional and the reversal bleaching a shift to shorter wavelengths* occurs because of the removal of some material from the emulsion. The tanning pyrodeveloper will, however, reduce the amount of wavelength shift compared to the nontanning ascorbic acid developer. In reversal bleaching the shift is usually larger than in direct bleaching. *Increased exposure in direct bleaching* causes more grains to convert into silver than in reversal bleaching, which is why only a small amount of silver halide is left in the emulsion. The fixation process will therefore remove very little material, *causing only a small wavelength shift.* In *reversal bleaching, increased exposure* has the opposite effect - a larger amount of silver is developed and removed during bleaching, which, in turn, causes a *larger wavelength shift.* In both cases the shift will, however, be reduced by the use of a tanning developer, such as the pyrogallol developer for example, as when compared to when a non-tanning developer is used.

In the case of fixation-free rehalogenating bleaching, the wavelength shift that occurs here is very limited due to the fact that no material is removed from the emulsion. However, a tanning developer can be used to harden the wet emulsion so that it will not revert to its original thickness after the processing. This means that a wavelength shift towards longer wavelengths is experienced here. This effect increases with increased exposure since the hardening effect is proportional to exposure. The smallest wavelength shift is obtained if a non-tanning developer is used.

In conclusion, it can be said that *a non-tanning developer used in conjunction with a fixation-free rehalogenating bleach will generally minimize the changes in the emulsion thickness between recording and reconstruction.* This is important for many applications of reflection holography, such as in HOEs, in reflection masters used for contact copying, etc. The only other efficient processing technique able to compete with the above-mentioned and to minimize the emulsion changes, is the colloidal development of ultra-fine-grained emulsions, in which the fixing step is omitted. The last method of phase-hologram processing is also the one that shows probably the highest fidelity in recording the interference fringes in the emulsion and therefore yield the highest possible image resolution.

For amplitude-hologram processing without shrinkage the highest image resolution has been obtained in holograms which were desensitized directly after development, omitting the fixing step (Chap.4). This method appears, however, to be rather impractical for amplitude-reflection holograms which show low diffraction efficiency.

Variations in the emulsion thickness are actually quite useful for varying the colors of a hologram. Shrinkage occurring in reversal bleaching is often used as a means of obtaining a more pleasant color in reflection holograms. Special swelling agents can be introduced in the emulsion during manufacturing like, for example, the technique used by Ilford (Chap.3) for the red-sensitive emulsion, so that fixation-free rehalogenating bleaching could be applied in which a wavelength shift occurs. In pseudo-color holography, artists have been using various swelling methods before exposure to bring about emulsion shrinkage in order to produce different colors in their holograms. By applying multiple exposures combined with different degrees of swelling in-between exposures, it is actually possible to obtain several colors in a single hologram when reconstructed in white light. We shall again return to these special methods in Chap.9.

Various reconstruction properties of a hologram will be affected by varying the emulsion thickness with the help of different processing methods. In Chaps.2 and 4 some of these phenomena have already been discussed. For bleached silver-halide emulsions *Syms* and *Solymar* [5.75] have studied the effect of emulsion swelling in transmission holograms. Both dry holograms and water-swollen holograms were reconstructed and different parameters, such as Bragg sensitivity, transmission, diffraction efficiency, etc., were measured and compared. It was found that the thickness of the Agfa 8E56 material exposed with a 514.5 nm wavelength, immersed in water increased from 4.7 to 12.8 μm, which constitutes an approximate increase of 170%. At the same time the selectivity of the hologram increased as expected, whereas both the transmission and the diffraction efficiency were reduced due to the fact that the average refractive index in the emulsion was lowered when water was used as the swelling agent. Using the definitions introduced in Sect.5.2.2b, the change in n = $\sqrt{\epsilon'_0}$ has slightly increased the Bragg angle in the emulsion. Since no additional modulating material is introduced by swelling, the volumetric filling factor f is inversely proportional to emulsion thickness d, which means that $\epsilon'_1 \propto 1/d$ and $\epsilon'_1 d$ is more or less constant. The important parameter in the Bragg diffraction process is $\kappa'_1 d/\cos\theta_0$ where both $\kappa'_1 d$ and θ_0, are slightly increased by the drop in $\sqrt{\epsilon'_0}$. This effect is less significant if the refractive index of the swelling liquid is closer to that of gelatin, e.g. glycerol. According to Kogelnik's theory on the efficiency of volume diffraction gratings replayed at Bragg incidence, the overall efficiency of the water-swollen hologram decreases. One advantage of the swollen hologram is that the response to spurious scatter gratings is reduced, which means a better separation between spurious scatter gratings and the desired main grating. The above-mentioned paper also provides the reader with numerical examples of how Kogelnik's theory should be applied in practice, which is rarely the case in other publications.

Another, vital factor in hologram processing is the *drying procedure*, discussed in some detail by *Burckhardt* and *Doherty* [5.16] where it was found that the drying method applied affected the quality of a hologram. A very important point concerning washing and drying of bleached holograms has been raised by *Hariharan* [5.80] in an investigation comparing the drying process for holograms bleached in conventional and reversal potassium dichromate based bleaches. The difference in washing and drying consisted in the fact that some of the bleached plates were immersed in various alcohol solutions before being dried at room temperature, whereas other plates, exposed and processed the same way, were just washed and dried at room temperature. The results were striking. The *plates bleached in the conventional rehalogenating bleach showed improved efficiency after treatment in an alcohol bath* before drying. The *plates bleached in the reversal bleach and treated in alcohol showed a dramatic reduction in diffraction efficiency*, compared to the holograms that had been just washed and dried. The results obtained depended to a large degree on the use of *a tanning bleach*, potassium dichromate, which is a tanning bleaching agent. During conventional bleaching tanning takes place at the exposed sites of the emulsion. The chromium ions formed during the bleaching process harden the gelatin molecules by crosslinking them. This takes place in the exposed areas, i.e., the areas where silver had been formed during the development. As a result, the local hardening in the exposed areas, combined with the usual refractive index modulation which occurs after bleaching, will increase modulation and, therefore, also diffraction efficiency if, prior to drying it, the emulsion is treated in an alcohol bath - very much in the same way that dichromated gelatin holograms are processed. The two mechanisms for modulation will cooperate to increase the efficiency of the hologram.

Similarly to the conventional rehalogenating bleach, when a tanning bleach is used with a *reversal bleach*, tanning takes place at the exposed sites. Phase modulation is normally obtained from the unexposed silver-halide grains located in the areas which have not undergone hardening. In other words, the *alcohol method secures modulation in the hardened sites, which, in turn, counteracts modulation caused by the unexposed grains. The resulting effect is a dramatic reduction in diffraction efficiency.* This applies only to cases when *tanning bleach baths* are employed in the processing of phase holograms. Since the dichromate reversal bleaching technique is in common use, it is important to be aware of the effects just discussed.

A comparison between reversal (pyrochrome) and rehalogenating (CW-C2 - PBQ-2) bleaching on Agfa 8E75 HD materials was made by *Joly* and *Jacobs* [5.99]. They found that the a prebleached optical density of 2.5 yields the best efficiency for both processes. However, the reversal bleaching technique shows a more narrow range of exposures about the optimum than the rehalogenating method. The results are shown in Fig.5.4.

The pyrochrome bleached holograms exposed at 633 nm reconstructed at 625 nm with a bandwidth of 30 nm and the PBQ-bleached holograms at 615 nm with a bandwidth of 50 nm. These results are interesting. Although in the pyrochrome process material is removed the thickness of the emulsion is better retained than with the CW-C2 developed and rehalogenated holograms. This means that the tanning action of the pyrogallol outweighs the removal of material. The pyrogallol-developed holograms also show a more narrow-band reconstruction.

They also recommend that the developer should yield the highest possible local gradient in the working point and develop silver with fairly low covering power.

The pyrochrome process has its merits (high signal-to-noise ratio) although it does not yield the highest diffraction efficiency. The use of rehalogenating bleaches will give higher efficiency, but the noise level in these holograms is always difficult to control.

Kumar and *Singh* [5.109-112] have investigated various bleaching processes intended for Agfa and Kodak materials, including fast holographic materials, such as Agfa 10E75 and Kodak SO-253. In addition, blue

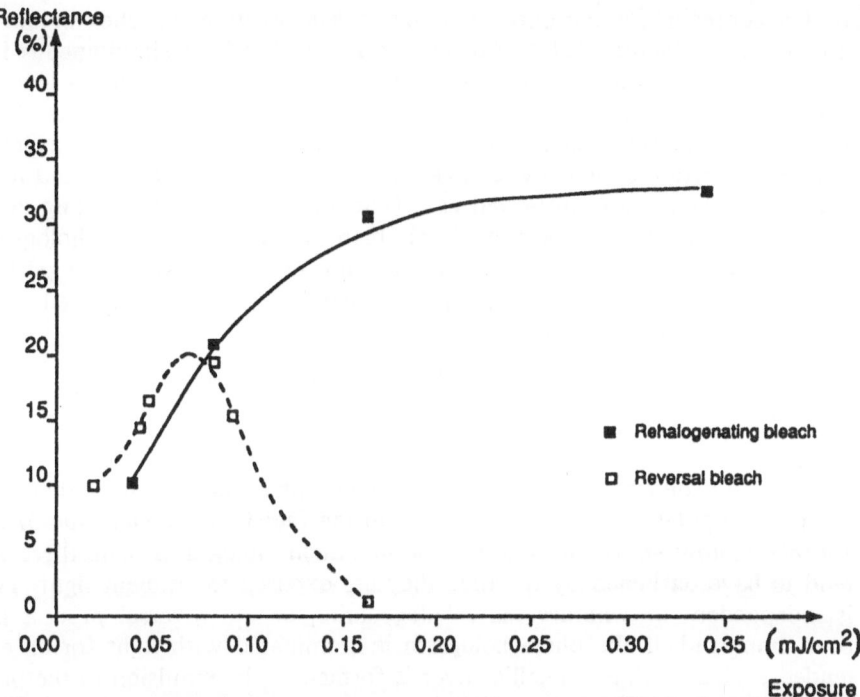

Fig.5.4. Reflectance as a function of exposure for a reversal bleaching method (pyrochrome processing) and a rehalogenating bleaching method (developer CW-C2 and PBQ-2 bleach). The reversal method shows a narrow range in which the optimum reflectance is reached. The rehalogenating method yields higher reflectances over a wider range of exposures [5.99]

recording laser light (442nm) was utilized in some of the tests [5.111, 112]. For high-resolution materials from Agfa and Kodak, *Kumar* and *Singh* found that the rehalogenating potassium ferricyanide bleach works very well. For holograms recorded on the faster materials, which are difficult to convert to phase holograms, the potassium iodine and iodine bleaches are recommended [5.110]. For holograms recorded at 633 nm and reconstructed at a different wavelength (442nm), the chromium intensifier bleach seems to be the best choice [5.112]. Only diffraction gratings were generated in all the tests by *Kumar* and *Singh*, which means that the processes recommended have not been tried for recording diffuse objects.

As already mentioned, *Hariharan* and *Chidley* [5.91, 92] investigated the diffraction efficiency as a function of spatial frequency for bleached holograms. In a recent publication by *Beléndez* et al. [5.113], it was shown that the diffraction efficiency is independent of the slant angle of the interference fringes in the emulsion and depends exclusively on the spatial frequency.

Finally, in order to study the effects of various bleach constituents and their concentrations, *Kostuk* [5.104, 105] recommended that a factorial design approach for the optimization of rehalogenating bleaches should be applied. An advantage of this technique is that it allows simultaneous improvement to a number of hologram performance characteristics as a function of several input parameters. Interactive effects of combinations of input parameters can also be examined. Specifically, he shows an example where diffraction efficiency, average light loss, thickness change, and noise grating efficiency are optimized as a function of the rehalogenating agent and acid concentration of the bleach bath Kodak R-10. The holograms were developed in an ascorbic acid-catechol developer. Both the performance of Agfa and Ilford red-sensitive materials were investigated using the factorial design approach.

5.2.5 Stability of Bleached Holograms

One of the problems connected with bleached phase holograms is that they are rather light-sensitive, which is due to the fact that the *emulsion of the finished hologram contains silver halide*. Such holograms will therefore tend to have darkened by the time they are exposed to ambient light. This is independent of whether the rehalogenating or the reversal process has been employed. If the phase hologram is illuminated with light for an extended period of time, metallic silver is formed in the emulsion in the process of *photolysis*. This process is otherwise referred to as the *printout effect* in the professional jargon. There are several ways of reducing or eliminating the printout effect [5.115-130]. It should be pointed out that the printout effect should not always be viewed as an undesirable, negative process. For example, in display holography which uses high-resolution materials, darkening of a phase hologram reaches a maximum after which no more printout takes place. Actually, a slight reduction of density occurs on dark storage of holograms suffering from printout exposure (printout relaxation). Such hologram darkening reduces the diffraction efficiency to a certain degree, but at the same time it also decreases scattering in the material. The combined effect often results in an overall improvement of the image quality. As a matter of fact some holographers expose their freshly bleached holograms to light, as it is often desirable to obtain a certain degree of printout-noise reduction. In particular, if the exposure is performed when the material is still wet, darkening will be faster. It is generally known that the photolytic effect increases in an aqueous environment, also shown in an investigation by *Locker* [5.120]. The influence of humidity on bleached holograms was tested by *Chenoweth* [5.116] who found that the diffraction efficiency decreased if the emulsion was kept in high humidity (>75%). This is caused by two factors - a decrease in overall diffraction efficiency and a change in the spacing of the grating structure due to emulsion-thickness variation as a function of humidity.

　To reduce the printout effect and thus make the holograms more light-stable, several means are at one's disposal. The first thing to remember here is that the bleached emulsion must be very thoroughly washed so that no chemical compounds that might trigger off the printout process are left in the emulsion. If a swelling agent, such as, for example, triethanolamine is used with a view of increasing the emulsion thickness in the finished product, the photolytic process also increases dramatically, which is why it is not recommendable to employ this agent for permanent post-processing swelling.

Chang and *George* [5.20] introduced safranine to desensitize phase gratings bleached in the chromium intensifier process. *McMahon* and *Maloney* [5.115] investigated several different bleaching schemes and found that the stability of bleached holograms increases as the silver halide changes from chloride into bromide and into iodide. *Hariharan* and *Ramanathan* [5.118] found also that silver iodide is the most stable of the three silver halides. They suggested that holograms should be bleached directly to form silver iodide, or alternatively, that already bleached holograms should be converted to this silver compound. However, sometimes rather poor results are obtained (with a high noise level) when using KI in a bleach bath. The R-10 bleach frequently used at the time and containing in its original iodide version 128 g/ℓ of KI can be drastically reduced to only 2 g/ℓ of KI for holographic purposes, in the opinion of *Hariharan* and *Ramanathan*. To demonstrate the stability of holograms which contain silver iodide, *van Renesse* and *van der Zwaal* [5.32] reported that the Agfa Neutol S developer for photographic paper did not develop any silver in the exposed, silver-iodide-bleached emulsion in ambient light.

Laming et al. [5.117] suggested a hardening procedure for bleached holograms which can be accomplished in two steps. First a chemical hardening solution is used (ammonium dichromate and sulfuric acid), which is followed by a heat treatment (30 minutes under vacuum in an oven at 200° C). This process increases substantially the lifetime of bleached holograms. *Norman* [5.119] investigated various desensitizing dyes which could reduce the printout effect. The following were found useful: phenosafranine, rhodamine B, malachite green, and methylene blue. Picrin acid (2,4,6-trinitrophenol) was found to produce superior stability against printout exposure at the green argon wavelength (514.5 nm).

Graube [5.49] found that bleaching done with the dry bromine method produced very stable holograms. This is due to the fact that the mobility of the silver ions within the emulsion is inhibited and that halogen remains in the emulsion to recombine with any photolytically formed silver. During bleaching, the bromine does not only react with the silver but it also oxidizes functional groups on the gelatin molecules that could otherwise act as reducers for the silver-bromide crystals, which is an additional factor improving the printout stability. The advantage of the high stability of bromine-vapor-bleached holograms could eventually be carried over to other phase holograms bleached by different techniques. *Phillips* and *Porter* [5.51] recommend that a plate bleached in ferric nitrate should be washed in isopropyl alcohol and then immersed into bromine vapor for a few seconds, whilst it is still wet with alcohol. This will create a stable hologram, though at the expense of increased scatter and a possible gelatin structural damage.

The idea of using bromine in the bleach as a method of reducing printout was recently suggested by *Weiss* et al. [5.129]. They formulated some new bleaches based on ammonium dichromate and ferric sulfate, in which bromine was either chemically synthesized in-situ or simply added or dissolved in the solution. Ammonium persulfate was utilized for the oxidative, in-vitro synthesis of bromium from potassium bromide. As expected, holograms treated in these bleaches possessed a high photolytic stability.

A completely different approach was taken by *Nishida* [5.121] who suggested that silver halide should be converted to a nonsilver, less light-sensitive compound. After investigating many different compounds he recommended two of them: choose from lead ferrocyanide $[Pb_2Fe(CN)_6]$ or nickel ferrocyanide $[Ni_2Fe(CN)_6]$. After exposure, development and fixation, holograms on Agfa 8E75 were bleached for 5 minutes in the following solutions, depending on whether lead- or nickel ferrocyanide was desired.

Bleach converting silver to lead ferrocyanide

Solution A

Lead nitrate	7 g
Distilled water	100 mℓ

Solution B

Potassium ferricyanide	1.4 g
Distilled water	100 mℓ

Solution C

Acetic acid (glacial)

Mix 50 parts of solution A with 60 parts of solution B and one part of solution C just before use.

Bleach converting silver to nickel ferrocyanide

Solution A

Nickel nitrate	5 g
Potassium citrate	15 g
Distilled water	100 mℓ

Solution B

Potassium ferricyanide	4 g
Distilled water	100 mℓ

Solution C

10% nitric acid

Mix 20 parts of solution A with 10 parts of solution B and one part of solution C just before use.

The results reported were rather good. Of the two, lead ferrocyanide showed higher stability than nickel ferrocyanide. Both were better than silver ferrocyanide.

Inayaki et al. [5.122] found that almost printout-stable holograms could be obtained if the material is treated in a cleaning bath followed by a short rinse in a potassium-iodide solution (20g/ℓ) after being bleached in the EB-2 bleach process. The cleaning bath in which the holograms are treated for three minutes was mixed in the following way:

Solution A
   Potassium permanganate   5 g
   Distilled water   1 ℓ
Solution B
   Sulfur acid (conc.)   10 mℓ
   Potassium bromide   40 g
   Distilled water   1 ℓ
Mix solution A with solution B just before use.

Phillips [5.84] recommended to treat the material after bleaching in 1% acetic acid solution as the final rinse. This treatment will dissolve the remaining sensitizers and therefore largely diminish the printout tendency. The adsorption of desensitizing bromide ions to the silver halide in place of gelatin is favored by increased acidity, which will reduce the printout.

Holograms with colloidal silver are regarded to be very stable. This processing technique has mainly been employed for ultra-fine-grained materials so far, such as the Russian materials. The method of converting bleached holograms recorded on Western materials to the colloidal type, will be presented in full in Chap.7. This new technique is actually a good solution to the printout problem. The method contains a step in which the bleached material is exposed to incoherent white light. *Hüttmann* [5.123] described a way of reducing the printout tendency *by a long exposure to intense white light* after the hologram has been bleached, washed, and dried. After that, the hologram is rebleached in a reversal bleach, washed and dried again. During the exposure to light, photolytic silver is produced, which is then converted to a soluble silver complex in the bleach bath. The silver-halide grains previously containing photolytic silver will now be reduced in size when the silver is dissolved. The smaller silver-halide grains left in the emulsion will now be much less sensitive to light than they were before because of their reduced size. However, this will also affect the resulting diffraction efficiency which will be somewhat lower after this treatment.

Vila and *Wesly* [5.124] investigated the printout of the Agfa 8E75 HD emulsion on both glass and film using both low- and high-intensity illumination. They tested three, common processing regimes:
(i)   Developing, fixing and rehalogenating bleaching.
(ii) Developing and reversal bleaching.
(iii) Developing and rehalogenating bleaching without fixing.

In addition, raw film as well as its resistance to printout were tested, using anti-printout treatment solutions:
- Phenosafranine desensitization (0.3g phenosafranine in a 50-50 mix of water and isopropyl alcohol).
- Conversion of silver bromide to silver iodide (2.5g/ℓ solution of potassium iodide).
- Acid bath (10mℓ/ℓ acetic acid solution).

For the low-intensity test a 100 W mercury vapor lamp was used at a distance of one meter from the material, giving a flux of 10 mW/cm^2. The low-intensity test samples received continuous exposure for four months, totalling over 2500 hours. This corresponds to the total accumulated energy density of over 90,000 J/cm^2.

For the high-intensity tests an argon ion laser was used at 488 nm, giving a flux of 800 mW/cm^2. As a comparison, sunlight is about 100 mW/cm^2. The testing period here was only twenty minutes, corresponding to the accumulated energy density about 9600 J/cm^2. The main results of these test are as follows:

- Bleached holograms using a solvent bleach or a rehalogenating bleach plus fix had less printout than rehalogenated holograms without fixing.
- In most cases initial, quick darkening could be observed, after which a reversal lightening and an apparent stabilization occurred.
- The acid anti-printout treatment gave the best protection and the cleanest looking holograms.
- The reversed-solarization effect was not present in the high-intensity test.
- The printout rate is a function of the intensity of the irradiation source.

The fact that printout occurs very quickly at high-intensity illumination may cause problems when using bleached silver-halide masters for photoresist copying with the short-wavelength, high-intensity argon-ion laser light.

Vorzobova et al. [5.125] recommended the use of methyl viologen for increasing the photostability of holograms. In their experiment processed holograms were exposed using a filament lamp of 200 lux illumination for 400 hours without any change in diffraction efficiency of the holograms.

Vakhtangova et al. [5.126] successfully used water solutions of two-quaternary salts of 4,4-dipiridilium to stabilize phase holograms.

Recently, *Kumar* and *Singh* [5.127] described the influence of developer composition on the stability against printout. They found that holograms which were beached in potassium-iodide and iodide solutions show an improved stability if urea (50g/ℓ) was added to a metol-hydroquinone developer containing sulfite (100g/ℓ). However, a developer without sulfite did not indicate an improved stability by adding urea. In addition, the papper mentions that ferric EDTA bleaches do not dissolve the sensitizing dyes in the emulsion which means that a rapid printout will occur. To prevent this the EDTA-bleached holograms must be treated in a solution to dissolve the dyes, e.g., the acetic-acid treatment.

In another recent investigation by *Jacobson* and *Baxter* [5.128] tests of the printout stability for transmission and reflection holograms were performed using artificial daylight exposure (765W/m^2 at 270÷800nm and up to 15 hours). The tests were similar to the experiments by *Vila* and *Wesly* [5.124] which were just described. The main conclusions from the *Jacobson* and *Baxter* investigation were:

(i) Re-bleaching printout-affected holograms make them more stable.

(ii) Conversion to silver-iodide inhibited printout. However, the image contrast was reduced in this process.

(iii) Desenitizers can reduce the printout but often at the expense of reduced image contrast.

As described in the beginning of this section, the printout process is not always a negative factor to be considered when making holograms. It is actually possible to take advantage of this effect, not only for noise reduction, but also for some special imaging purposes, as described by *Jonathan* and *Kinany* [5.130]. They found for printout caused by strong illumination due to linearly polarized light, the resulting plate is optically anisotropic. They actually suggested an application of this birefringence effect for the storage of two images and their separate restitution without contrast inversion.

However, in addition to printout problems, there are several other factors that can affect the long-time stability of holograms. These factors will be discussed in Chap. 8.

6. Special Techniques

Many different problems are encountered in the application of silver-halide materials recorded with various types of lasers. For example, reciprocity failure and latent-image fading are the phenomena which become more pronounced in holographic applications than in conventional photography. Therefore, the first part of this chapter is devoted to these problems.

Increasing the sensitivity of holographic fine-grained materials is sometimes desirable, which is why various hypersensitization and latensification mathods are carefully examined.

Index matching is a technique frequently employed in the holographic recording and reconstruction processes and the following chapter devotes a section to this particular matter, including a list of commonly used chemical products for index matching.

The chapter also describes techniques for obtaining surface-relief holograms on silver-halide materials as well as processing methods for silver-halide sensitized gelatin holograms. They are sometimes used as an alternative to dichromated gelatin holograms.

Finally, methods employed for the investigation of emulsion, such as electron microscopy and other techniques, are examined.

6.1 Problems Due to Short or Long Exposure

6.1.1 Early Pulsed Holography

Pulsed lasers are being used on an ever increasing scale in holography, mainly because of less severe demands on stability during the recording process and a growth in the variety of objects that can be holographed. *Brooks* et al. [6.1] reported the first use of a pulsed ruby laser for recording holograms. They utilized both Q-switched and free-running operations for recordings made on Kodak 649-F plates. Dynamic events, such as a bullet in flight or a water jet, were captured. Other early publications on pulsed holography are the following papers: *Jacobson* and *McClung* [6.2] employed various Kodak materials (649-F, V-F and High Contrast Microfilm) to record off-axis holograms of 35 mm transparencies with 60 mJ pulses. *Siebert* [6.3], one of the real pioneers in this field, made several front-lighted holograms of moving objects, e.g., a fan rotor, jet spray, hand pouring milk from a beaker, and a smoke stream. A hologram of a hand, holding a ballpoint pen, was also recorded. Later, he made the first pulsed hologram of a human subject, a portrait of himself documented on October 31, 1967 [6.4]. *Zech* and *Siebert* [6.5] made the first pulsed reflection portrait. *Ansley* [6.6] described various aspects of producing holograms of people.

The quality of solid-state lasers (e.g., ruby or Nd:YAG) has improved considerably since the early days of pulsed holography so that nowadays they are as good as CW gas lasers, both with respect to temporal and spatial coherence. For holographic purposes solid-state lasers are operated mainly in the Q-switched mode (20÷30ns pulse length). Free-lasing operation (~1ms) is also possible for some imaging techniques. For a ruby laser made for bubble-chamber holography a special pulse stretching technique has been developed to obtain pulses from 1 to 100 μs, which brought about an improved coherence length (10 meters and more) [6.7]. However, from the photographic point of view these exposure times are all "short".

6.1.2 Reciprocity Failure

For the Q-switched operation laser energy is released during that very short time producing very high output peak power. For a 10 J pulse (20ns) the peak power is 500 MW. Such high power is certainly desirable when exposing holographic materials of relatively low sensitivity. However, short exposure times are associated with a certain problem - the *reciprocity failure* or, more precisely, *failure of the reciprocity law*. In general, the exposure H of the photographic material is: $H = E \cdot t$, as discussed in Chap.2. The reciprocity law was originally formulated by *Bunsen* and *Roscoe* [6.8], stating that a given exposure H is independent of the two factors separately. This is not true for the extreme values of E or t; the phenomenon can also affect hologram recordings of even very long exposure times at low light levels (the *Schwarzschild effect*). In particular, using a CW laser with a weak output, which is often the case especially in small holographic laboratories, can sometimes create problems for some recording materials.

Curves showing the reciprocity law (or its failure) are often plotted as $\log E \cdot t$ vs $\log E$ for a fixed optical density. A typical reciprocity-law failure curve is shown in Fig.6.1. Between any two points on a horizontal part of the curve there is no reciprocity failure. If the two points are not on a horizontal part, then the reciprocity law would not hold between the corresponding exposure times.

In reality, the exposure necessary for obtaining a certain density in the developed material is not constant but depends on time t. For very short exposures at high intensities E, as well as for very long exposures at low intensities, H has to be strongly increased to get the same density as the one required for the optimal values of E and t. These effects are called *High-Intensity Reciprocity Failure* (HIRF) and *Low-Intensity Reciprocity Failure* (LIRF), respectively. The HIRF becomes of importance even for pulses much longer than those from a conventional Q-switched laser and the effect is roughly constant for times $< 10^{-5}$ s in the case of conventional photographic materials.

In order to better understand these phenomena and the effect of different ways of increasing the sensitivity of the holographic material it is advisable to go back to Chap.2, where the formation of the latent image is explained.

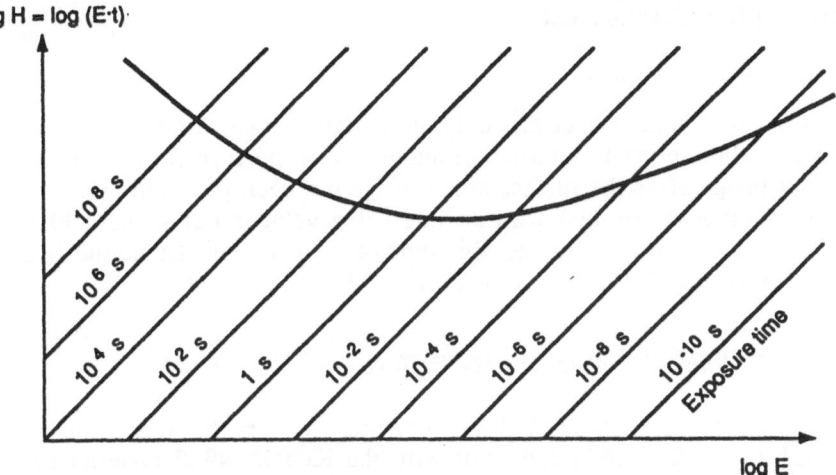

log H = log (E·t)

10⁸ s — **10^8 s**
10⁶ s — **10^6 s**
10⁴ s — **10^4 s**
10² s — **10^2 s**
1 s
10⁻² s — **10^{-2} s**
10⁻⁴ s — **10^{-4} s**
10⁻⁶ s — **10^{-6} s**
10⁻⁸ s — **10^{-8} s**
10⁻¹⁰ s — **10^{-10} s**
Exposure time

log E

Fig.6.1. Typical reciprocity-law failure curve. Curves showing the reciprocity law (or its failure) are often plotted as logEt vs logE for a fixed optical density. A material without reciprocity failure is a horizontal curve for all exposure times. If the two points of the curve are not within a horizontal part of the curve, then the reciprocity law would not hold between the corresponding exposure times

At HIRF, as the intensity increases, more absorbed photons are required per grain to produce the same density in the developed material as when compared with exposures at lower intensity levels. HIRF is caused by the silver ions' motion and their concentration in the emulsion. The exposure time of about one second is sufficient for a mobile, interstitial silver ion to neutralize a trapped electron before another arrives. At high intensities, electrons are produced at such a rate that there is not enough time for the mobile silver ions to neutralize the trapped electrons. Because of electrostatic repulsion the second electron is not trapped at the same site as the first electron. Recombination of holes and electrons may occur instead at an increasing rate or trapping of electrons may take place at some other site. In each case the process of latent-image formation becomes inefficient.

LIRF, on the other hand, depends on the thermal stability of an isolated silver atom which, if not stabilized by combining with another silver atom within its lifetime (~2s), will decompose into an electron and a silver ion again. This, of course, makes that a long exposure at a low light level becomes a very inefficient process of latent-image formation.

Another result of short exposure is the localization of the latent image within the silver-halide crystal. At exposure times longer than 10^{-2} s the latent image is almost entirely localized on the surface of the silver-halide crystals. At shorter exposure times it is also formed inside grains. Therefore, in order to develop the internal latent-image specks as well, it is necessary to use the right developing technique. The development methods for silver-halide materials can be divided into three groups:

217

(i)   Surface development.
(ii) Internal development.
(iii) Total development.

A surface developer acts only on the latent image on the surface of the silver-halide crystal. Internal development is performed on the internal latent image after the surface latent image has been bleached off. Total development is performed with an internal developer but without bleaching off the latent surface image. This type of development technique is definitely preferable for holograms exposed with a pulsed laser.

6.1.3 Holographic Reciprocity Failure

The first publication on HIRF concerning holography was published by *Hercher* and *Ruff* [6.9] and dealt with the Kodak 649-F material exposed with a Q-switched ruby laser. They found that the effects of reciprocity failure were of significant magnitude. They also studied the effects of post-exposure, and increased development time and temperature on the reduction of the reciprocity failure. *Nassenstein* et al. [6.10] found that for all Agfa materials exposed with a Q-switched pulse from a ruby laser, the exposure had to be increased by a factor 2 to 4, as compared to CW-laser recordings, in order to obtain equal densities. Reciprocity failure has also been observed in conventional photographic materials when exposed to ruby laser pulses, which was reported by *Rice* and *Macomber* [6.11].

Hüttmann [6.12] showed that exposing Agfa 8E56 HD with a 8 ns pulse compared to a 10 s exposure result in lowering the slope of the T_a-log exposure curve after development. Since the diffraction efficiency grows proportionally to the square of this slope, the efficiency of a pulsed hologram is often lower than that of a CW-laser-recorded hologram receiving the same total exposure.

Vorzobova and *Staselko* [6.13, 14] noted that there was a reduction in the γ-value of holographic materials exposed with pulsed lasers. The diffraction efficiency for a Q-switched hologram fell to 10% compared with a hologram recorded in the free-lasing mode. They conclude that the observed drop in diffraction efficiency of pulsed holograms is due to a change in the optical characteristics of the photographic layer when the illumination time is reduced. In another investigation, *Benken* and *Staselko* [6.15] studied the latent-image-formation process in Russian holographic materials (LOI, PE-2 and IAE). They were particularly interested in the influence of the exposure time on obtainable diffraction efficiencies. A dramatic difference in the diffraction efficiency of the PE-2 material was found when the material was exposed to a 20 ns pulse (Q-switched) as compared to a 300 μs pulse (free lasing): the longer pulse produced a 100 times higher diffraction efficiency. Using a special scattering technique, *Benken* and *Staselko* could measure the diffraction efficiency of the dynamic latent-image grating as well as the efficiency of the static grating (the developed grating). The former was divided by the latter and the ratios compared. The ratio that was obtained was about 10 to 20 for free-lasing pulses, whereas for Q-switched pulses it could sometimes be as high as 1000. This difference is due to some extent to *latent-image fading* which will be discussed later in Sect.6.3. These examples may shed some light on the problems one has to contend with in pulsed holography.

Recently, *Pantcheva* et al. [6.16] has discussed emulsion-manufacturing methods to reduce HIRF for silver-halide materials in pulsed holography. HIRF can be reduced by

- Creating hole traps by introducing reducing agents.
- Increasing electron lifetime by introducing shallow traps for electrons using, e.g., metal ions.
- Formation of a few stable and active sensitivity-specks on the microcrystal's surface by chemical sensitization.

The different methods listed above were tested with emulsions having a grain size of about 30 nm. The best reducing agent appears to have been *ascorbic acid*, which means that a pretreatment of a holographic emulsion in an ascorbic acid solution before exposure can improve pulsed-hologram recordings. Introducing metal ions in the emulsion at the preparatory stage increases the latent-image specks' lifetime. The best results were obtained using lead ions, but also cadmium ions gave fairly good results. Gold sensitization with, e.g. $HAuCl_4$, also reduces HIRF, but any combination with sulfur or sulfur used alone, e.g. $Na_2S_2O_3$, increases HIRF and should therefore be avoided. This confirms similar results obtained for conventional photographic emulsions [6.17-20]. However, as mentioned in these papers, HIRF in sulfur-sensitized emulsions can be reduced by gold latensification (Sect.6.2.3a). The very best results were achieved with a *combination of lead and gold* (8 mol% $Pb(NO_3)_2$ + gold) which gave an improvement of a factor of about 13.

There are very few publications on holographic LIRF. However, a recent investigation by *Binfeld* et al. [6.21] treated this problem. Exposure times between 1/8 of a second and 8 seconds at two nominally constant exposure levels (40 and $200\mu m/cm^2$) were studied. Agfa 8E56 HD material was exposed with an argon-ion laser beam at 514 nm wavelength. After development of the samples, the developers were applied for the processing: Agfa GP61, GP62, Ilford Autophen, a holographic pyrogallol-ascorbic acid developer, Neofin Blue as well as J. Webster's Marchwood developer. The last-mentioned developer is actually a developer for pulsed holograms, where HIRF is the problem. Between the extreme exposure times, variations up to 50% in optical density was observed, when applying strong developers for the processing. Slower developers exhibit less reciprocity failure than faster ones. However, the slower ones give an overall lower optical density.

Johnson et al. [6.22] discussed a slightly different aspect of reciprocity failure. It concerns the decrease in diffraction efficiency of reconstructed images in multiple-exposure holograms recorded with equal energy per exposure. They called this phenomenon *Holographic Reciprocity Law Failure* (HRLF). Here, Monte-Carlo simulations of the multiple recording process are presented with experimental verifications. The HRLF effect is dependent on the time Δt between the exposures, and is directly related to the lifetime of individual silver atoms in the latent-image prespecks. Due to the dissociation of single silver atoms the first exposure is favored. Subsequent exposures reinforce the signal of the previously recorded images, increasing the noise in later recordings. For a $\Delta t = 0.5$ s, there is hardly any reciprocity failure, but as soon as $\Delta t > 2$ s the HRLF effect can be seen. Experiments performed with the time delay $\Delta t = 30$ s between exposures support the presented theory.

To compensate for the HRLF effect a first-order correction technique can be utilized when recording multiple images in the same material. The diffraction efficiency is dependent on Δt which is expressed in the correction expression for the diffraction efficiency of the nth image in a multiple-exposure hologram as

$$\eta(\Delta t) = \frac{[\alpha_n(\Delta t) \, H_b \, M_i \, M(\nu)]^2}{4} \, . \tag{6.1}$$

If the modulation M_i and the modulation transfer function $M(\nu)$ are both constant, the decrease in $\eta_n(\Delta t)$ is caused by the decrease in $[\alpha_n(\Delta t)]^2$ which was demonstrated in both computer simulations and experiments. If M_i and the $M(\nu)$ are constant, then an improvement in the uniformity of the diffraction efficiency in multiple-exposure holograms involves increasing the bias exposure H_n of the nth recording for a given Δt according to

$$H_n = \frac{H_b}{\sqrt{\eta'_n}}; \quad \text{with} \quad \eta'_n = \frac{\eta_n(\Delta t)}{\eta_1} \quad n = 1, 2, ..., N \tag{6.2}$$

where H_n is the exposure increase for recording n.

These just described reciprocity effects are important for many holographic applications where multiple exposures are used to superimpose holograms in the same emulsion. It is also of importance for understanding the effect of pre- and post-exposure of the holographic material. The demand for high-resolution recording materials means that the sensitivity often is rather low. In addition, sometimes only low-power lasers are available for the recording [6.23]. These problems were addressed by many researchers in the early days of holography. *Caulfield* et al. [6.24] described how pre- and post-biasing differ from simultaneous biasing in holography. They found that pre- or post-exposure of the material sometimes is necessary when the total energy from the light source during the required exposure time is not sufficient. They also succeeded in recording a 4000-point multiple exposure hologram by using a pre-biased plate. *Nishida* and *Sakaguchi* [6.25] found that the optical density obtained for exposures recorded earlier was higher than those of recordings made later. They realized that this effect was caused by the latent-image-formation process in the silver-halide emulsion, as previously described. Also, *Akahori* and *Sakurai*

[6.26, 27] reported difficulties in obtaining uniform diffraction efficiency for multiply exposed holographic plates.

The pre- and post-exposing technique was successfully used by *Beesley* and *Castledine* [6.28] to reduce the exposure time of gratings recorded in photoresist materials. *Vikram* and *Sirohi* [6.29] discussed in great detail the effect of pre- and post-exposure for any reference-to-object beam ratio.

Landry and *Phipps* [6.30] investigated single- and multiple-exposure holograms using the Agfa 10E75 emulsion for the recordings. They studied not only the influence of multiple exposures on the diffraction efficiency, but also how signal-to-noise ratio and image resolution were affected.

Couture and *Lessard* [6.31, 32] have investigated the reciprocity-law failure for incoherently superimposed holograms (up to twenty) recorded on Kodak 649F plates with green laser light (514.5 nm).

Phipps et al. [6.33] have described techniques to reprocess holograms having a nonoptimum optical density caused by an incorrect exposure energy in the first place.

The possibility of increasing the sensitivity of silver-halide holographic materials is often important. Therefore, this subject will extensively be treated in the following section.

6.2 Increasing Sensitivity by Hypersensitization and Latensification

A photographic material can be treated in different ways before exposure in order to increase its sensitivity. Increasing the material's sensitivity is referred to as *hypersensitization*. If the treatment is performed after exposure but before development, it is called *latensification* (*latent*-image *intensification*).

Some of the methods commonly used in astronomy for increasing the sensitivity of photographic materials and described by *Hoag* and *Miller* [6.34] have been reported to increase the sensitivity of holographic materials as well [6.35, 36]. In particular, these methods are in frequent use in Russia, as the ultra-fine-grained silver-halide holographic emulsions have very low sensitivity [6.37].

The total increase of the holographic material's sensitivity that can be obtained with the help of these methods depends on the material used, the manufacturing method, the ripening, the finishing, etc. Only very few investigations have been performed with the Western holographic materials. The results obtained for the Russian materials cannot be applied directly to Agfa, Ilford or Kodak materials. The reason for this is the difference in the manufacturing methods.

6.2.1 Hypersensitization

For practical purposes hypersensitizing methods can be divided into two major groups:
(a) Wet methods.
(b) Dry methods.

a) Wet Methods

A wet method means that the material is placed in a solution for a period of time. Then, it is washed and dried before the actual exposure takes place. The following solutions have been in use:
- Water.
- TEA solution.
- Sodium-sulfite solution.
- Ammonia solution.
- Silver-salt solution.
- Gold-salt solution.
- Developing (reducing agents) solution.
- EDTA solution.

Water. Soaking the material in distilled water to which a few drops of a wetting agent have been added, will remove excessive bromide and increase the concentration of silver ions. This, in turn, increases the material's sensitivity.

The bath temperature should be $10°$ to $12°$ C and drying should take place in a low ambient temperature ($13°$ to $15°$ C). The durability achieved by this method is low and the material must therefore be exposed directly after treatment - otherwise the fog level will increase. Placing the material in a low-temperature surrounding of about $-18°$ C directly after treatment will increase its stability. This method has been tested for Agfa's holographic materials by *Biedermann* [6.35] where the increase in sensitivity can be expressed by a factor of $1.5 \div 2$.

Water Solution of TEA. Triethanolamine (TEA) [$(HOCH_2 CH_2)_3 N$] has been used extensively in holography to increase the sensitivity of recording materials [6.35-38]. The treatment of the material is performed in a bath with a TEA concentration of 0.7% to 2%, which provides an increased sensitivity factor of about 2. Higher concentrations (up to 10%) are recommended for materials intended for pulsed-laser exposures according to Russian investigations. It is also recommended to use a bath at a temperature of $10°$ to $15°$ C and not to dry the material in hot air. This method produces quite stable results but it is advisable to expose the material soon after the treatment to keep the fog level low. Storing the material at low temperatures ($-18°$ C) ensures better stability than storing it at room temperature. *Kirillov* [6.37] demonstrated that the grain size in the emulsion is slightly reduced during the TEA treatment, resulting in a holographic image of a higher quality.

Sodium-Sulfite Solution. This method in which 0.5 to 1% sodium sulfite concentration is used, is similar to water treatment. Sodium sulfite has also a slight solvent effect upon the silver-halide grains. The factor of sensitivity increase is similar to that for a water treatment. *Crespo* et al. [6.39] employed this method and reported an increase in sensitivity by a factor of about 2 for thus-treated 8E75 HD plates.

Ammonia Treatment. The ammonia treatment is a quite an effective method except that it is difficult to obtain uniform results with it. In this method one may use either a bath containing 10 mℓ ammonia/ℓ of distilled water (ammonia 25%, specific gravity 0.91) at 20° C for 3 minutes or a bath of 4% ammonia (regent ammonium hydroxide, 28%) for 4 to 6 minutes at 5° C. The most uniform results are obtained when the action of the ammonia solution is quenched with a 1% solution of acetic acid. The drying of the material should be rapid but should not be done in hot air. To speed up the drying process one may immerse the material in a bath of alcohol and water directly after treatment before drying. Ammoniated materials are very unstable and should be used as soon as possible after the treatment. Gain factors of 2÷10 have been reported for conventional photographic materials. For holographic materials a factor of about 3 has been obtained for Agfa plates in the investigation by *Biedermann* [6.35]. Ammonia has been combined with silver chloride, but for the Agfa 10E material the best results were found in an ammonia solution without silver chloride. Hypersensitization in a water solution of ammonia depends primarily on the increase in silver-ion concentration, although an increased pH-value has a certain effect too.

Silver-Salt Solution. Another interesting way for increasing sensitivity is the application of a silver-salt solution:

Silver nitrate	5 g
Acetic acid	5 mℓ
Ethyl alcohol or water	1 ℓ .

After one minute in this solution the material's sensitivity has been reported to increase as much as 50 times in Russian holographic materials [6.33]. This value seems to be very high and depends probably also on the processing techniques applied during the manufacture of the material, when little or no sensitization takes place. For the already highly-sensitized Western materials one cannot possibly obtain such a high gain factor.

Gold-Salt Solution. Instead of silver salts, gold salts can be used to increase the material's sensitivity. High values increase have been reported from Russian work concerning holographic emulsions [6.37]. Here, gold chloride is utilized alone or in combination with an alkali thiosulfate. Solutions of gold-complex salts, such as aurous thiocyanate or aurous thiosulfate have also been employed. Like a mercury-vapor treatment, gold increases the sensitivity in the red end of the spectrum, but in general, gold treatment produces more stable results than when applying dry methods.

Developing Solution. The use of a developer as a hypersensitizing solution for holograms has been reported by *Biedermann* and *Molin* [6.40] and also by *Spiering*'s laboratory [6.41, 42]. The technique is to soak the plate in the developer that will be employed for the processing for a few minutes before the exposure and then let it dry. This has an effect on the latent-image fading and the reciprocity failure since introducing a reducing agent to the emulsion before the exposure creates an effective hole trap, preventing the recombination of latent-image subspecks. Developers containing phenidone or ascorbic acid seem to be particularly effective in this respect [6.16].

Another interesting aspect of this technique is that since the developer has been introduced to the emulsion before the development starts, one can eventually get a more uniform processing within the emulsion. (Compare processing of nuclear emulsions, Chap.4).

EDTA Solution. *Todorova* et al. [6.43] have shown that Pb(II)-EDTA (a complex of Pb(II) and ethylenediaminetetraacetic acid) can be used as both a sensitizer and a hypersensitizer. Mixing a 0.1 M solution of $Pb(NO_3)_2$ and a 0.1 M solution of disodium salt of EDTA resulted in a 0.05 M solution of Pb(II)-EDTA which was then further diluted when used. A concentration of $2 \cdot 10^{-4}$ M was found optimal for hypersensitizing. Treatment times were 15 to 60 minutes.

Todorova and *Stankova* [6.44] have found that Tl(I)-EDTA (a complex of monovalent thallium with ethylenediaminetetraacetic acid) a high hypersensitizing effect combined with low fog can be obtained for photographic emulsions if this complex is introduced at concentrations from $8 \cdot 10^{-4}$ M to $1.6 \cdot 10^{-3}$ M. It has been shown that the less sensitive the photographic emulsion, the higher the hypersensitizing effect of this treatment.

b) Dry Methods

There exist several dry methods that can be taken advantage of to increase the sensitivity of the material:

- Light.
- Heat.
- Vapor.

These methods can be used separately or in combinations as well as in conjunction with wet methods.

Light. "Light" means that the emulsion is exposed to light before the actual image-information exposure takes place. Pre-exposure is generally made with a rather short light pulse (a flash) resulting in a density of about 0.15. *Chang* and *George* [6.45] showed that the dynamic range was increased and the H&D curve was linearized at the lower portion after such treatment. Although this method produces some fog, it is useful in reducing the effect of LIRF. The contrast will be reduced by using pre-exposure. The light method should not be applied if HIRF is the problem, i.e. when short image-information exposures are utilized.

Heat. Ripening occurs during the manufacture of photographic emulsions. The initial heat treatment of an emulsion is the so-called *Ostwald ripening*, during which the grains grow in size. At this point the potential light sensitivity of the material is determined. The following heat treatment (the after-ripening) causes no additional grain growth. The second ripening takes place in the presence of sensitizing dyes and the sensitivity of grains undergoes important changes.

Emulsion heat treatment for the purpose of hypersensitizing can be regarded as a second after-ripening which can affect the sensitivity of the material. Baking must be done in an environment free from contaminating vapors or in an inert atmosphere, e.g., in nitrogen, and in the temperature ranges of 50° to 80° C for several hours. The baking time decreases rapidly with the increase in temperature. For example, a 24 hour treatment at 50° C is equivalent to 8 hours at 65° C. Emulsion damage can occur if the temperature exceeds 80° C.

Hypersensitized holographic materials from Russia are known to have been baked for 30 to 40 hours at 70° C where the gain factor obtained was 2 to 4 times [6.35, 46, 47]. The fog level increases with the increase in the duration of baking. However, the increase of sensitivity is stable and the material can be exposed several months after the treatment. There are different opinions as to the effects of such a treatment. In general, this method is satisfactory when LIRF is the problem and it also works for reducing the effects of HIRF. Conventional, red- and IR-sensitive emulsions do not respond well to baking, but contrary to this statement some Russian investigations concerning the red-sensitive holographic materials show that the response is good. *Yaroslavskaya* et al. [6.46] mention that if a material has been heat-treated to its maximum sensitivity, additional hypersensitization in TEA has absolutely no effect. However, *Kosobokova* and *Usanov* [6.47] have shown that if TEA treatment is performed before the heat treatment, the baking time necessary to reach the maximum sensitivity will be shortened.

Vapor. At least four kinds of vapor or gases are known to increase the sensitivity of a photographic material. In vapor treatment it seems to be important to start removing the oxygen and the moisture from the emulsion by vacuum degassing. The material must be kept in vacuum for a considerable time.

Mercury vapor and *sulfuric dioxide* have been used in photography from the early days.

Mercury vapor will increase the material's sensitivity especially in the red part of the spectrum with a gain factor of $1.5 \div 2.5$, provided the material has been exposed to a mercury atmosphere during its treatment, which may be from a few hours up to many days. The effects are very unstable unless the material is kept cold (-18° C) after the treatment. Mercury vapor is hazardous and is not recommended to be used indiscriminately.

A treatment with sulfuric dioxide gives a similar gain factor. During the manufacturing the emulsion sulfur is often used for sensitizing (sulfur-sensitized emulsions). Such emulsions have a high HIRF, hypersensitization with sulfur dioxide will probably have little effect on holographic materials used with pulsed lasers.

Hydrogen-gas treatment has been employed to increase the sensitivity of conventional photographic materials [6.48-50]. After degassing the emulsion in vacuum for 16 hours at $5 \cdot 10^{-7}$ torr, dry hydrogen gas of one atmosphere was allowed to act upon the material at room temperature for 3 to 6 hours. Later, it was found that the use of heated hydrogen results in an improvement of sensitivity [6.49]. The time of degassing was reduced to about 30 minutes at $1 \cdot 10^{-2}$ torr. The material was exposed to heated hydrogen of 65° C for several hours. It was established that the sensitization level depends on the duration of the hydrogen treatment, the upper limit being about 6 hours.

The first degassing step to remove oxygen and moisture is very important. Already after this step an increased sensitivity is obtained, especially for LIRF, but with no considerable effect on HIRF. After hydrogen treatment the gain factor is $3 \div 4$ and HIRF is almost eliminated. It is not fully understood how the process works in detail, but the resulting gain factor could be due to the following:

- Reduction of silver ions by hydrogen to form silver, yielding a very effective type of reduction sensitization.
- Absorption of hydrogen at the silver-halide surface, possibly at preferential sites, providing a very efficient hole trap.
- Hydrogen reaction with some reducible species in the gelatin which, in turn, reacts with the silver halide.
- Displacement of residual oxygen.

In the presence of normal chemical sensitization (with sulfur, reduction plus sulfur, or sulfur plus gold) a hydrogen treatment increases the sensitivity, most probably by adding new silver centers or by increasing the size of the existing chemical sensitivity centers.

This method has been tested at CERN for holographic Agfa materials by *Bjelkhagen* [6.36]. The experiment was carried out in the following way: Film and plates (Agfa Holotest 10E75) were put in a cylindrical tank where the pressure and the temperature inside the tank could be measured. The tank was then connected to a vacuum pump and, depending on the time of pumping, different vacua were obtained. A vacuum of $2 \cdot 10^{-5}$ torr obtained after 5 hours was found to be optimal. Hydrogen gas was then introduced to the tank and different treatment times were tested at room temperature. The effects of a temperature treatment were also investigated by immersing the hydrogen container into a larger water tank with heated water. To study the treated material's stability, the material was exposed directly after the completion of the hydrogen treatment, as well as after a certain lapse of time during which the material was kept in nitrogen. The test procedure is depicted in Fig.6.2, and the results are presented in Figs.6.3 and 4.

Hydrogen hypersensitization made it possible to increase the sensitivity by a factor of 3 at the most. No considerable reduction of HIRF was found in the tests performed with a pulsed laser. The fog level increases rapidly

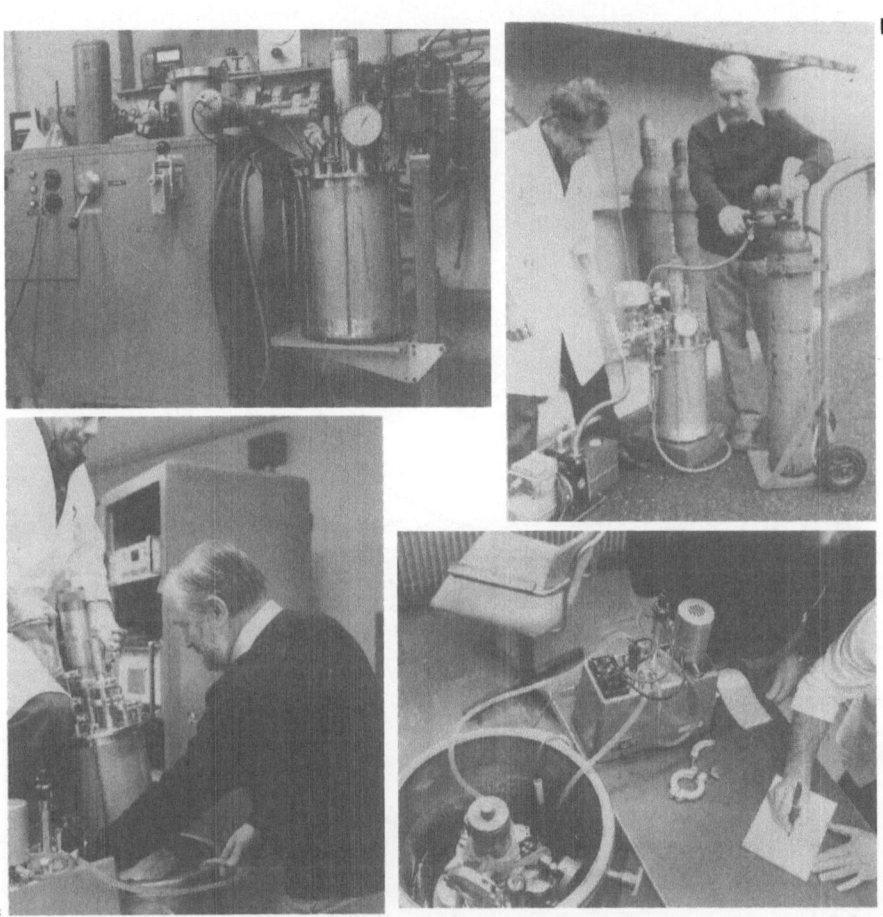

Fig.6.2. Hydrogen hypersensitization test procedure. Holographic film and plates were put in a cylindrical tank where the pressure and the temperature inside the tank could be controlled and measured. The tank was connected to a vacuum pump to obtain different vacua (a). Hydrogen gas was then introduced to the tank at room temperature (b). The effects of increased temperature were tested by immersing the hydrogen container into a larger tank with heated water (c and d)

with the increase in the hydrogen treatment time, which is why it should be kept short (15 to 30 minutes). Hydrogen gas has been used here at 40° to 50° C. The fog level increases also with time if the treated, unexposed material is stored in air. A few days after the treatment the fog level is still acceptable, but after a few weeks the material becomes useless. However, if the material is stored in nitrogen, there is hardly any increase in fog level (Fig.6.4).

It should be mentioned that pure hydrogen is highly explosive and flammable when vented into air or oxygen. Therefore, it is recommended to utilize *forming gas* instead of pure hydrogen if the safety aspect of the hy-

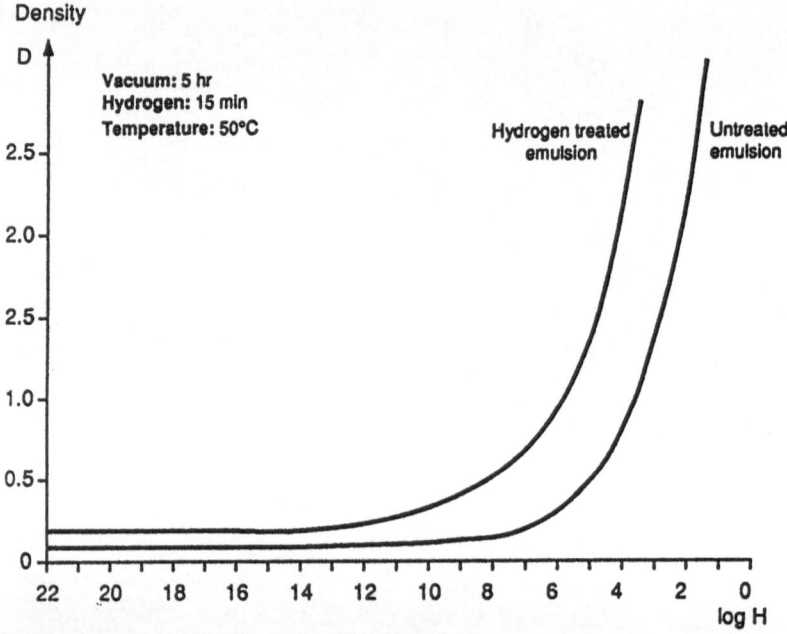

Fig.6.3. Increased sensitivity obtained after hydrogen treatment (vacuum 5 hours, hydrogen 15 minutes at 50° C) of the Agfa 10E75 material

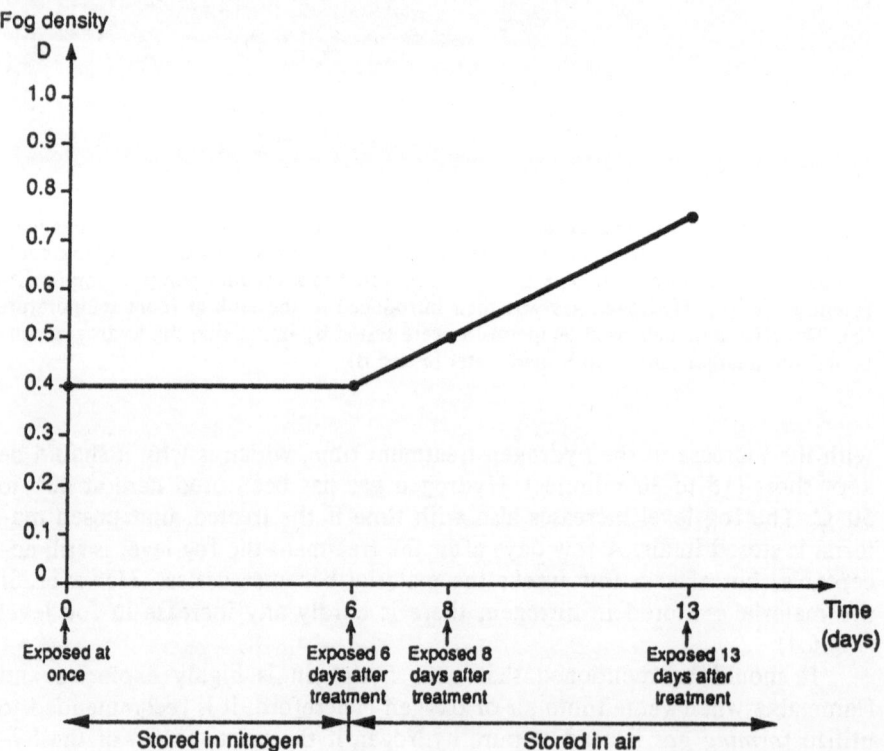

drogen treatment is not completely under control. The forming gas is a mixture of hydrogen and nitrogen which is nonflammable or nonexplosive. The forming gas mixture can vary from 2% (2% hydrogen and 98% nitrogen) to 8% (8% hydrogen gas and 92% nitrogen). This mixture has been used for hypersensitization of photographic materials for astronomy described by *Sliva* [6.51, 52]. A 5% mixture has been successfully employed for treating Kodak spectroscopic plates by *Scott* et al. [6.53].

Another interesting technique uses the *vapor from the Ammonium Ethylene-Diamine-Tetra-Acetat* complex ion (EDTA·2NH$_4$). The material is put in an oven at 65°C for 3 hours with a tray containing the above-mentioned liquid inside the oven. The method has been tested at CERN for Agfa Holotest 10E75 materials and the gain factor obtained was about 2.5 with a low fog level [6.36].

Finally, to close the discussion about the hypersensitization methods a statement made by *Biedermann* should be brought to the reader's attention: "Hypersensitization sometimes changes the modulation so that a change of beam ratio is necessary. The gain in sensitometric speed results in lower diffraction efficiency if not compensated. This compensation, however, makes the effect of hypersensitization less pronounced" [6.35].

6.2.2 Increasing Sensitivity During Exposure

During the actual exposure of the material some procedures may be applied to increase the material's sensitivity. The sensitivity is temperature dependent since the mobility of the silver ions required to build up the latent-image centers depends on the temperature of the material during exposure [6.34]. Elevated temperatures increase the ion mobility and thus improve the material's response to short exposures of high intensity. Low temperatures inhibit the regression of sublatent-image centers that occur during long exposures at low intensity, thereby improving the response at low light levels. Increased sensitivity has been reported for exposures at low and extremely low temperatures where LIRF is a problem. The LIRF effect almost ceases to exist at a temperature below -50°C. In astrophotography special cameras are sometimes used in which the film is cooled down to about -78°C using dry ice.

For problems connected with HIRF, on the other hand, the temperature during exposure must be raised above room temperature. Unfortunately, gelatin starts softening at about 80°C which is why an elevated temperature of at most 50°C may be used.

For holographic materials a higher temperature could lead to the movement of silver-halide crystals between exposure (as the gelatin becomes softer) and after processing. This would show up in a poor quality of the image. Therefore this method is not favored.

Fig. 6.4. Fog density obtained in hydrogen-treated Agfa 10E75 material as a function of time after treatment. The samples were stored first in nitrogen and then in air

Another way for increasing the sensitivity is to apply an electric field across the material during exposure, as described by *Rothstein* [6.54]. This method is not very easy to perform. It has only been used in cases where the emulsion was coated on a special, transparent electrode. The applied electric field was about $2 \cdot 10^{-6}$ V/cm across the emulsion. The gain in gamma and contrast was high. The idea behind the method is that, since the release of the electrons in the emulsion represents an important step in the photographic process, this process might ensure high sensitivity if the charge density of the electrons in the emulsion could be increased. By applying an intensive electric field, this becomes possible. If the electric field could be somehow applied to the film holder instead of using transparent electrodes on which the emulsion is coated, the method would be even more efficient in practice.

6.2.3 Latensification

Generally speaking, latensification leads to the acceleration of the development process giving an apparent speed increase at short development times only [6.55]. At long development times the effect is less pronounced. There are, however, exceptions where a true increase in sensitivity can be obtained. The methods used for latensification are very similar to the ones used for hypersensitization. Both wet and dry methods are in use.

a) Wet Methods

One of the most effective methods for conventional photography is bathing the latent image in gold-salt solutions, which usually requires a few minutes only. Aurous thiocyanate solution is the one most commonly used here. Gold latensification increases the number of countable and developable specks but does not change their distribution.

James et al. [6.56] have published a recipe for a gold latensification bath:

Potassium thiocyanate	0.5 g
Potassium chloraurate (0.1% solution)	40 mℓ
Potassium bromide	0.6 g
Distilled water	1 ℓ.

Treatment time 5 minutes at 20°C. After that the material should be washed for 30 minutes. Potassium thiocyanate is added to the potassium chlorate solution which is heated to the boiling point. After that potassium bromide is added and the solution is diluted with water.

The effect of HIRF is reduced after gold treatment, in particular for sulfur-sensitized emulsions [6.19, 20]. It has also been reported that gold treatment is applicable for hydrogen-hypersensitized materials [6.48]. The fog level can, however, increase during this treatment.

In addition to gold-salt solutions, silver salts can also be employed for the same purpose. The following solution, which actually represents a silver intensification method, can give a gain factor of $1.5 \div 3$:

Solution A

    Silver nitrate solution (10%)   70 mℓ

    Sodium sulfite (anhydrous)   90 g

    Distilled water   1 ℓ

Solution B

    Sodium sulfite (anhydrous)   20 g

    Hydroquinone   6 g

    Distilled water   1 ℓ .

Mix solution A with solution B before use.

Electron-injection methods with developing agents for latensification of the internal image have been reported by *James* [6.57]. Some developing agents, such as, e.g., *phenidone* (1-phenyl-3-pyrazolidone) can latensify the internal latent image. (Refer to Chap.4 where developers are discussed). It has been suggested that latensification depends here on the initial formation of isolated silver atoms which subsequently lose electrons to the conduction band of the crystal. Conduction electrons formed in this way act to build up latent-subimage centers in the same way as photoelectrons formed by the exposure. If "action at a distance" occurs between the developer and the latent subcenters or the very small latent-image centers, as suggested, this action should also lead to latensification of the internal image.

This effect can be obtained by adding 1 to 2 g/ℓ phenidone to a metol-hydroquinone developer. The reason why developers containing phenidone work so well for holograms exposed with pulsed lasers is that internal image latensification takes place in combination with the superadditive effect due to another developing agent. Figures 6.5 and 6.6 show the results of a test on Agfa 10E75 materials exposed to a 13-ns pulse at 694 nm and developed with the Kodak D-19 developer with and without phenidone [6.36]. The test reveals that the sensitivity increases by a factor of about 3.

Latensification taking place in the developer is not directly recognized as a separate method for hologram treatment. Therefore, more detailed information on this matter is given in Sect.6.4, where special developers for holograms recorded with pulsed lasers are discussed.

b) Dry Methods

Essentially, two dry methods are used for latensification:
- Light.
- Vapor.

Light. Latensification using low-intensity light offers a possibility for true speed increase by actually using the LIRF mechanism. Post-exposure can be made at a suitable wavelength depending on the spectral sensitivity of the material. A very low light intensity must be applied for a long time. The latensification exposure takes between 15 minutes and two hours for normal photographic materials.

Light latensification has been tested for Agfa 10E75 materials [6.36]. Here, an underexposed image was created using coherent light, after which

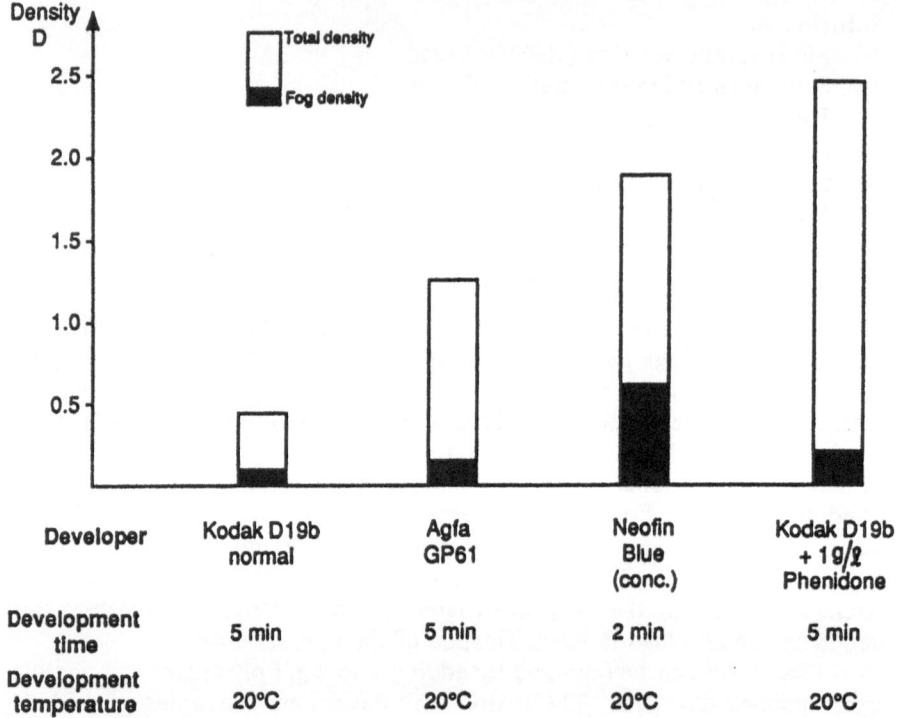

Fig.6.5. Density obtain on Agfa 10E75 material for a fixed short exposure (13ns pulse at $\lambda = 694\,\text{nm}$) using different developers (with or without phenidone)

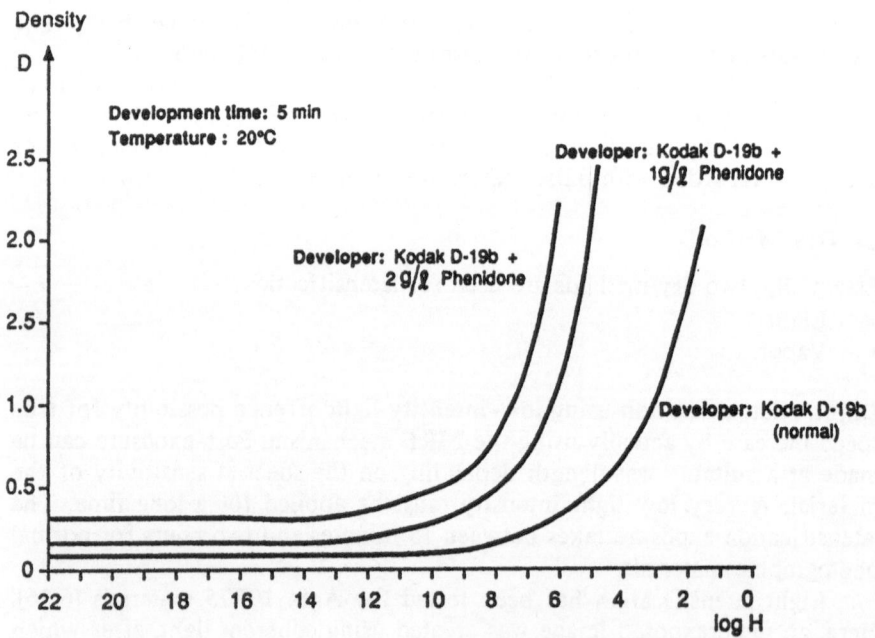

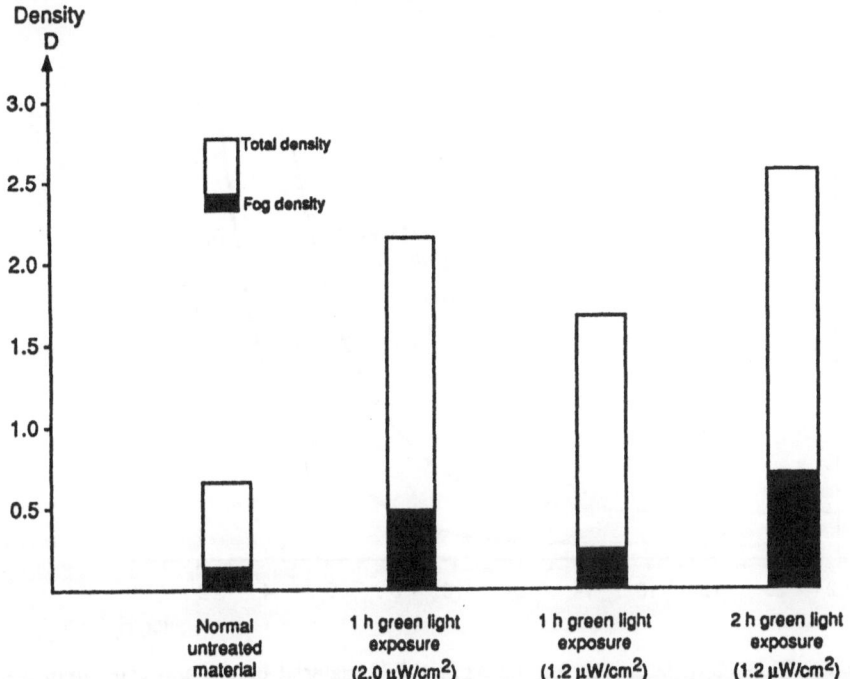

Fig.6.7. Density differences obtained before and after light latensification of Agfa 10E75 material. The latensification was performed using weak green light exposures (1.2 and 2.0 μW/cm^2) for one or two hours. The material was developed in Kodak D-19b (normal, 2.5 minutes at 27° C)

an additional exposure of the material was performed with incadesent light. The incoherent light employed for latensification was in that part of the spectrum in which the material has low sensitivity. An ordinary, safe-light lamp with a 25-W light bulb and a dark green filter (Kodak Wratten No.3) was used. The material was exposed at two different distances from the lamp (110 and 165cm), corresponding to the power-density level at the film of 2.0 and 1.2 μW/cm^2, respectively. The exposure time varied between 30 minutes and two hours. At optimal conditions, an increased sensitivity of about 2.5 was obtained. Figures 6.7 and 8 display results of the latensification investigation.

Vapor. The vapor or gas used for latensification is absorbed by the small latent-image specks which grow and become developable during the treatment. Mercury and sulfur dioxide have been utilized for conventional photographic materials, the treatment-time being between a few hours and sev-

---

Fig.6.6. Characteristic curves for the Agfa 10E75 material developed in Kodak D-19b (normal) as well as with 1 g/ℓ and 2 g/ℓ phenidone added. The material was given a fixed short exposure (13ns pulse at λ = 694nm)

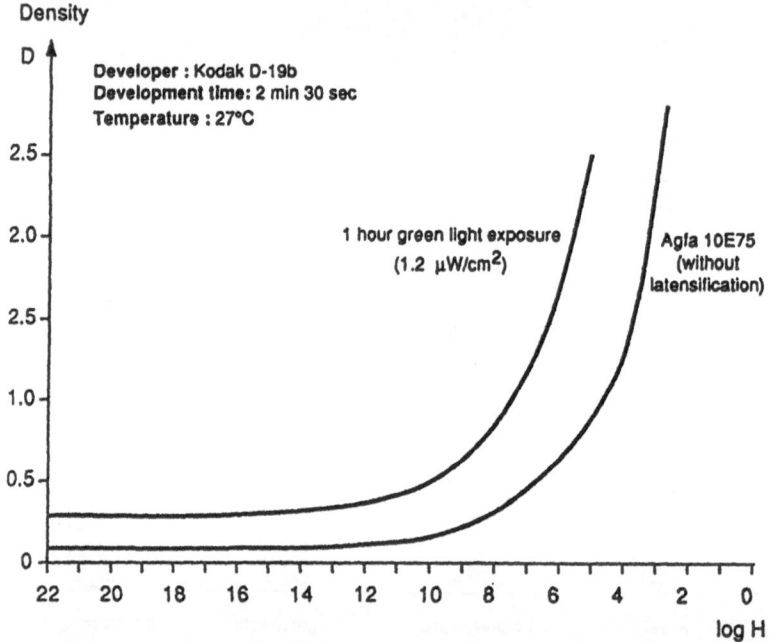

Fig.6.8. Characteristic curves for the Agfa 10E75 material before and after light latensification. The latensification was performed using weak green light exposures (1.2 μW/cm² for one hour). The material was developed in Kodak D-19b (normal, 2.5 minutes at 27° C)

eral days. Here, gain factors of 1.5÷2 were obtained. Acetic acid vapor and hydrogen gas are also known to latensify, and are especially effective in materials suffering from HIRF.

6.2.4 Autoradiography

Autoradiography is a *post-processing* intensification method which is not very common [6.58]. In this case intensification takes place *after the exposure and processing* of the material. The term autoradiography derives from the fact that the radiation used for the exposure comes from the subject itself rather than from an external source. The autoradiographs for post-processing image intensification are made by exposing the photographic emulsion that has been activated by nuclear or chemical methods to the radiation from the silver contained in the emulsion. This intensification method finds obvious applications in cases where photographs have been underexposed or underdeveloped.

Standard development procedures yield about 10^9 silver atoms for each stable silver atom in the latent image of the original exposure. In autoradiography, part of the 10^9 atoms are made radioactive through a chemical combination or exchange with radioactive atoms or through neutron bombardment. Each of these activated atoms is then a potential source for amplification of the original latent image. A contact print of the original underexposed emulsion is made on an autoradiographic emulsion. The radioactive exposure takes several hours. A safe method using Thiourea-S^{35} (developed at Marshall Space Flight Center in the USA) can be applied in photographic laboratories. A basic aqueous solution of thiourea (H_2 NCS NH_2) labeled with S^{35}, will react directly with the emulsion silver to produce radioactive levels appropriate for autoradiography. The radiation from S^{35} is pure β of 0.167 MeV maximum energy. Five mCi will activate up to 5000 cm^2 of the emulsion area. The contact-exposure times vary from two to 20 hours. The speed increases 8 to 13 times and the contrast, 4 to 18 times.

It is not clear whether this method would be useful for holographic high-resolution emulsions. In any case, it could only be considered for thin, transmission holograms. As a contact print is made from the original plate there is certainly some loss of the resolution. For tested, conventional photographic materials the resolution was reduced to half of what it was in the original photograph. Very-low-density images in the original material cannot be enhanced without a corresponding enhancement of fog. It is claimed that new solutions to photographic reciprocity failure may be found by the correct application of this technique. It is worth mentioning here that the technique is time-consuming and involves certain safety hazards. It may be considered to be used for saving, e.g., very underexposed recordings of special value.

6.3 Latent-Image Fading

6.3.1 General Latent-Image Fading

One of the problems in hologram recording is the *fading of the latent image*. If the holographic material is not developed immediately after the exposure a certain amount of latent-image fading will occur, i.e., the optical density will be lower than it would have been had the plate been developed right after the exposure. The fading rate varies depending on the exposure time and the type of material used. Fading is much more pronounced in holograms exposed with a Q-switched laser, for example.

Latent-image fading has been investigated for conventional photographic materials [6.59-61]. It is believed that it is caused by both the chemical processes resulting from the contact of the material with the air surrounding the emulsion (in particular, the oxygen), as well as from the physical (thermal) process, such as ejection of electrons from the silver specks of the latent image. However, it is the chemical processes that are responsible for 90% of fading. Latent-image fading is also linked with the grain size and the initial density (exposure) of the film. It has been observed that latent-image fading increases with a rise of the *temperature*, *moisture, the size of silver-speck centers in the emulsion,* as well as *with the decrease in grain size.* Therefore, by storing the film under favorable and appropriate conditions between its exposure and processing fading can be minimized.

6.3.2 Holographic Latent–Image Fading

The effects of latent-image fading in holographic materials were tested at Fermilab for bubble-chamber holograms recorded on Agfa 10E75 film by *Akbari* and *Bjelkhagen* [6.62]. The holographic plates were exposed to 4 μs laser pulses. Each plate was then cut into small pieces, some of which were developed immediately, whereas others were stored under different conditions, e.g., at a low temperature of about 3° C or in a vacuum of about 10^{-3} torr, and developed a week later. The reference parts were kept in open air at room temperature of about 18° C. All plates were developed in a D-19-Phenidone developer (4 min, 20.5° C). The results reveal that the storage environment is vitally important for the fading rate and that most of the fading occurs within the first few minutes, which is why holograms exposed with a short pulse should be developed almost immediately (within seconds) after the exposure. If this is not possible, one has to compensate for the estimated fading that will occur by overexposing the film. The fading rate of the latent image can be predicted according to the following formula which has been experimentally verified:

$$(D_0 - D_t)/D_0 = 1 - \exp(-ct^n) \tag{6.3}$$

where D_0 and D_t are densities at the time 0 and t, and c and n are constants depending on the exposure, temperature and humidity.

Using the data obtained from the fading test on holographic materials fitted to the fading curve above gave: c = 0.21 and n = 0.16 for the fading of the Agfa 10E75 material stored under normal conditions. When the material was stored at low temperatures (3° C) these constants were c = 0.0013 and n = 0.55.

6.4 Pulsed-Hologram Development

As described in the previous sections there are many factors to consider when processing holograms recorded with extremely short exposure times. The difficulties connected with this have been solved to some extent by formulating special developers to be employed for this type of processing. However, before recording a pulsed-laser hologram, a presoaking of the unexposed material in a water solution of ascorbic acid (20g/ℓ) can be very effective to increase the sensitivity of the material.

Based on the considerations already described, the *SM-6* developer has been formulated and has proved very successful in this type of development. SM-6 is the pulse version of the *Benton PAAP developer* [6.63], where the phenidone content has been drastically increased, and its combination with ascorbic acid has made it especially suited for pulsed-hologram processing. The **SM-6** developer is mixed in the following way:

Ascorbic acid	18 g
Phenidone	6 g
Sodium hydroxide	12 g
Sodium phosphate (dibasic)	28.4 g
Distilled water	1 ℓ .

The development time is 2÷2.5 minutes at 20°C to 24°C. The required exposure is between 5 to 100 $\mu J/cm^2$ depending on the material, the recording wavelength, and hologram type. The SM-6 developer requires a very low exposure compared with other developers, e.g., the standard Kodak D-19. It works for both transmission holograms and reflection holograms. In particular, Denisyuk single-beam holograms of high quality can be produced using this developer. The whole process will be described in detail in Chap. 7. The *Neofin blue developer* introduced by *Phillips* and *Porter* [6.64] is another choice for pulsed-hologram development, as is the Kodak D-19 to which 1.5 g/ℓ phenidone is added for the Agfa 10E75 materials, or 2.0 g/ℓ for the Agfa 8E materials and Ilford emulsions [6.36]. Figure 6.5 illustrates how adding phenidone to the D-19 developer affects the developed density for a given exposure. The D-19-phenidone developer can actually produce a higher density compared to concentrated Neofin blue for the same exposure. Compared to these two, the performance of SM-6 is even better. Slightly different formulas have also been published for pulsed hologram development, e.g., the Royal College of Art (RCA) developer by *Phillips* [6.65] (which is essentially the D-19-phenidone without metol and KBr) and the **Quephe** developer by *Unterseher* [6.66] which is a PQ developer to which ascorbic acid is added:

Ascorbic acid	10 g
Hydroquinone	8 g
Phenidone	2 g
Sodium sulfite (anhydrous)	30 g
Sodium carbonate (anhydrous)	60 g
Distilled water	1 ℓ .

The development time is 2÷4 minutes at 19°C.

Unterseher recommends that the ascorbic acid is added to the developer just before it is ready to be used. The development time is 2 to 4 minutes at 19°C. Hydroquinone is not really needed in a developer where the superadditive effect is obtained by phenidone and ascorbic acid alone. However, in this case the amount of ascorbic acid has to be increased.

Boone [6.67-69] formulated a developer for recording primarily pulsed Denisyuk reflection holograms. The developer (**Boone's Pet**) is mixed the following way:

Ascorbic acid	10	g
Metol	2.5	g
Sodium carbonate (anhydrous)	45	g
Sodium EDTA	2	g
Potassium bromide	0.5	g
Distilled water	1	ℓ .

Development time 4 minutes at $20°C \pm 1°C$. Constant agitation of the holographic material is recommended. About 15 sheets of film ($30 \times 40\,cm^2$) can be developed in one liter of this developer.

J. Webster has formulated a proprietary developer, the *Marchwood developer* when working at the Marchwood Laboratories in the UK. It is believed to contain potassium nitrate and formulated to counteract HIRF.

Also in the former USSR pulsed-hologram processing techniques have been investigated. Ruby lasers have been used for the recordings, but also frequency doubled Nd:YAG lasers are common. The use of a free-running ruby laser for recording large pulsed holograms was reported by *Nikolaev* and *Starobogatov* [6.70]. A yellow pulse (580nm) generated from a pulsed dye laser was also used to record holograms, as described by *Aristov* et al. [6.71]. Concerning pulsed holograms recorded on Russian materials *Kliot-Dashinskaya* et al. [6.72-74] recommend to apply either the LOI- emulsions or the PE-2 material with the *GP-2 developer* (30-min development time). The LOI-3 plates have a flatter diffraction efficiency-exposure curve than the PE-2 material. The sensitivity of the LOI-3 material for the free-running ruby laser is 500 $\mu J/cm^2$. This material has a silver content of about 10 g/ℓ emulsion. The LOI-2 (630nm) emulsion was the best material for recording holograms at the yellow wavelength. An exposure of only 50 $\mu J/cm^2$ was necessary provided that the material had been pretreated in a TEA solution as well as in sensitizing dyes, such as, e.g., rhodamine B. Also in this case the GP-2 developer produced the best results.

Concerning the remaining steps of pulsed hologram processing the procedure is very much like any other type of processing technique described in previous chapters. However, in Chap.7 a few complete processing schemes for pulsed holograms will be described.

6.5 Index Matching

6.5.1 Index Matching in General

Index matching can be of special interest for holography in some cases. Firstly, when a sheet of unexposed film is attached to a glass plate for exposure with a CW laser. Secondly, when contact copies are made on film from a glass master. When a sheet of film is placed on a glass plate, the optical contact between the two is not perfect because of a layer of air that remains between the plate and the film. Since the refractive index of the atmospheric air is much lower than that of the film and glass, unwanted re-

238

flections can cause interference that will affect the quality of the image. In order to avoid this effect, a liquid with a refractive index as close as possible to that of the film can be applied between the glass and the film. This is often called *index matching*. The use of a liquid gate for recording and/or reconstructing a hologram is another example where index matching is applied in holography.

Index matching has successfully been employed in the printing of motion-picture films over many years and information as regards liquids used for this purpose is available in a publication by *Delwiche* et al. [6.75].

There is a wide choice of liquids that can be utilized to improve index matching. The choice of a particular liquid depends on the task to be performed and is guarded by the following general principles: The liquid must not affect the emulsion or the support material; the liquid must be easy to remove; and finally it must have a good index match for the particular film. It is also important that it is safe to use.

The main object for index matching liquids in motion-picture film copying is to reduce the effect of scratches, which can also be important in hologram contact-copying. It is important to know the refractive index of the support material. For example, the refractive index for film material on triacetate base is 1.485; a film polyester base has a refractive index of about 1.6; the refractive index of pure gelatin is about 1.52 to 1.54. *Sklar* [6.76] determined an average value of 1.5384±0.0001 for the Kodak SO-243 film emulsion measured at 546.1 nm. If the gelatin is wet (swollen with water) the index is 1.36. Gelatin containing silver bromide has a refractive index of about 1.62. Recently, *Kostuk* and *Goodman* [6.77] measured the refractive index of the emulsion and glass support for the Agfa 8E75 HD plates. They found the following values (glass: 1.51, unprocessed emulsion: 1.64, fixed emulsion: 1.54, and fixation-free rehalogenated emulsion: 1.64). This last value was independent of exposure and prebleached density. For index-matching considerations one must remember that in most cases there is a protective pure gelatin coating of the support material to avoid curling of the film.

In Table 6.1 several index-matching liquids, each with its index of refraction, are presented.

According to the publications concerning the motion-picture film copying technique, it can be stated that from the point of scratch elimination an index difference of 0.10 or more from the support is not acceptable. A match of about ±0.02 is necessary.

In order to obtain a desired value of the refractive index, two different liquids can be mixed, giving an index calculated according to

$$\frac{P_1}{100} = \frac{n_m - n_2}{n_1 - n_2} \tag{6.4}$$

where P_1 denotes the volume per cent of component 1, n_1 is the refractive index of component 1, n_2 that of component 2, and n_m that of the mixture.

Table 6.1. Index-matching liquids. (The values for the refractive index have been measured at 20° C)

Product	Refr. index
Methyl alcohol	1.328
Freon-113	1.358
1-Butanol, 3 methyl	1.405
Kodak dispersant MX-1320	1.420
Stoddard solvent	1.435
Methyl chloroform	1.438
Kerosene	1.460
Carbon tetrachloride	1.461
Decalin solvent	1.475
Glycerin	1.475
Mineral oil	1.475
Trichloroethylene	1.478
Tetrachloroethane	1.494
Diethylbenzene	1.496
Toluene	1.496
p-Xylene	1.496
Di-n-butyl phthalate	1.497
Xylene (commercial)	1.499
Tetrachloroethylene	1.504
o-Xylene	1.506
Pyridine	1.509
Dimethylphthalate	1.515
Benzyl ether	1.517
β-Ionone	1.520
Ethyl benzoil acetate	1.523
Chloro benzene	1.524
Methyl salicylate	1.536
Benzyl benzoate	1.570
Bromo naphthalene	1.658

This expression is approximate as it assumes volume additivity, which is why small errors are possible if the two liquids differ greatly in physical properties.

The index of refraction is temperature dependent and decreases at the rate of about 0.0005 per degree centigrade as the temperature increases, which is normally quite insignificant for practical applications.

The liquids for holographic purposes must not affect the emulsion in any way, which is why neither water nor any aqueous solutions can be used. Some liquids, such as acetone or ethyl acetate, act as solvents for the triacetate support, which is why they cannot be considered for a film on an acetate base. Other liquids, such as alcohol, e.g., are also unsuitable as they have a tendency to react with antihalation layers and dyes in the emulsion.

Other aspects to consider when choosing the liquid are the toxicity of it and the simplicity of removing the liquid after it has been used. If the liquid is sufficiently volatile it will evaporate very quickly, but it may become a safety problem if it is employed frequently. If it is less volatile, it has to be removed from the film before the film can be processed.

6.5.2 Index Matching for Holograms

The consideration of index matching in holography is different from that in conventional photography, where the reduction of visibility of scratches is the main purpose. In holography, index matching is taking advantage of both for reducing Fresnel reflections as well as for reducing the effect of phase distortions in film materials [6.78,79]. Random phase distortions introduced by the substratum which the emulsion is coated on will affect the recorded hologram at reconstruction, introducing image aberrations. Phase noise in photographic film materials is produced by lack in their optical homogeneity, i.e. random variations in thickness and the refractive index. Variations in substrate thickness constitute usually the main phase-noise contributing factor [6.80]. Typically, a 140 μm triacetate film can show a variation of 3 to 10 μm and an 80 μm film of 1 to 2 μm.

In accordance with aberration-free imaging in holography, the optical path difference between rays passing through different portions of the emulsion should not exceed one-quarter of the wavelength (Rayleigh limit), and thus the thickness variation Δh must be

$$\Delta h = (n - 1) \leq \lambda/4 \tag{6.5}$$

where n is the refractive index of the emulsion and λ the wavelength of the laser light. If n = 1.5 and λ = 500 nm, the variations indicated above far exceed the maximum thickness variation permitted in this case ($\frac{1}{2}\mu$m). To eliminate or reduce the effect of thickness variations, an immersion liquid with an index equal or close to that of the film has to be applied. Since emulsions (dry gelatin) and film substrates have different refractive indices, it is not possible to obtain perfect index matching. The refractive index of the immersion liquid will always differ from that of the layers constituting the film. Knowing that the refractive index of a liquid differs slightly from that of the immersed material we can calculate the remaining phase difference from a defect in the material. Assume that there is a depressed spot in the film base (refractive index n_b) of thickness Δh, which is filled in by the liquid (refractice index n_ℓ). The remaining phase difference ΔL between this point on the film and a point without defect will then be

$$\Delta L = \Delta h(n_b - n_\ell) . \tag{6.6}$$

If the thickness variation of the substrate is known, this formula can be used to find the largest difference in refractive index permitted for index matching with an $\lambda/4$ or $\lambda/8$ phase criteria.

a) Index-Matching Liquids for Dry Holograms

Dry, pure gelatin has a refractive index of about 1.54. A holographic raw emulsion has a higher refractive index on average since it also contains silver bromide. *Bryngdahl* [6.81] reported that the Kodak 649-F emulsion undergoes a change from 1.61 to 1.56, and the Agfa 8E75 emulsion from 1.63 to 1.53 during (amplitude) processing, both measured at 633 nm. *Syms* and *Solymar* [6.82] found that the Agfa 8E56 emulsion had a refractive index of 1.62 before processing and 1.59 after bleaching, a drop of 2%. These values are the real values of the index. The imaginary parts are about three orders of magnitude smaller and can be neglected. Normally, the refractive indexes of the emulsion and the substrate differ, which is why it is not possible to obtain perfect index matching.

A commonly used index-matching liquid is *xylene*, but it is not to be recommended because of its toxicity. Another commonly used by holographers and a safer liquid is a paint thinner or *white spirit* (turps substitute) called *Thin-X* in the USA [6.83]. The advantage of this liquid is that, in addition to its index-matching property, it can also act as an adhesive between the film and a glass plate. Another popular index-matching fluid is *kerosene* (Naphtha), which is, however, more toxic than Thin-X. The best liquid would probably be the *decalin solvent* (decahydronaphthalene, $C_{10}H_{18}$) as it is both easy and rather safe to use. Also, Kodak's product: Kodak dispersant MX-1320, can be used with good results. Trichloroethylene, tetrachloroethylene and mixtures of Freon-113 or methyl chloroform with tetrachloroethylene, or with toluene are other possibilities, if a very good index matching is required. *Quintanilla* et al. [6.84] reported the use of a mixture (80% benzylic alcohol and 20% triacetate glycerin) with a refractive index of 1.5142 for the 632.8 nm wavelength.

In general, toxic liquids, such as orthoxylene and toluene, work very well. Kerosene and dimethylphthalate give quite acceptable results, while mineral oil and glycerine give lower quality results because of their high viscosity. N. Phillips, who has been very concerned about safety, recommends *baby oil* as a good index-matching liquid for holograms. Another safe solution is sugar water (50 % solution, n = 1.42) suggested by *Wuest* and *Lakes* [6.85]. This liquid, containing water, is not used in contact with the emulsion. The sugar-water solution is placed between the back of the holographic glass plate and another glass plate coated with an AH layer.

b) Index-Matching Liquids for Wet Holograms

In some cases index matching would be of interest for holograms which are still wet from processing, e.g., when liquid gates are used. As the wet emulsion has a refractive index of 1.36 and the base may be 1.48, a somewhat lower-index liquid would seem necessary. However, when a film is used, phase distortions are mainly caused by the base material and not by the swollen gelatin. The use of most organic immersion liquids is not recommended, as they do not dissolve in water. Here, the application of aqueous inorganic salts is recommended. In saturated solutions only a few inorganic salts have a refractive index larger than that of the solution itself. Good results have been reported for greatly undersaturated aqueous solutions of potassium and ammonium thiocyanate. For wet holograms on glass plates the application of water (n = 1.33) or alcohol (n = 1.36) produces good results.

c) Practical Methods for Large-Format Film Holograms

Finally, two well-working methods for combined index matching and film mounting to glass plates will be mentioned. At the Atelier Holographique de Paris, a technique is used to laminate a piece of film to a glass plate using an index-matching silicone sealant, applied during both exposure and processing. After the film is dry it can be removed from the glass plate. This technique works best with Agfa materials which have a suitable gelatin coating on the back of the base substrate. Another similar technique developed by J. Kaufman in California is to employ gelatin for the same purpose. The film is laminated to the glass plate with a solution of gelatin. It is also processed while still attached to the glass plate. After the processing has been finished the film can be removed from the glass simply by soaking the hologram in warm water. Actually, large sheets of film are often more easy to process in trays when they are attached to a rigid support. Therefore, these methods are convenient both for exposing and processing large-format film holograms, where both index matching and stability are achieved at the same time.

6.6 Surface-Relief Holograms

The generation of a surface relief on a silver-halide emulsion can have some applications in mass production, e.g., of HOEs and display holograms. The embossing technique for mass replication of holograms utilizes the transfer of a relief pattern. The dominating recording material for producing master holograms for embossing is the photoresist material. This material works very well but has the disadvantage of having a rather low sensitivity and being only UV or blue sensitive. Powerful argon-ion lasers are required for the recording. The possibility of using other materials for masters for embossed holography is therefore of interest. A silver-halide emulsion has the advantage of high sensitivity and strong sensibility in any part of the visible spectrum. Large plates for embossing can therefore be produced with cheap low-power lasers, such as the HeNe laser. Actually, silver-halide masters were manufactured by one European holographic company for producing embossed holograms. The quality of the embossed holograms from such masters is very similar to the photoresist-produced holograms. The important question here is how to create the surface-relief pattern on the silver-halide emulsion.

6.6.1 Surface-Relief Emulsion Structures

The generation of a surface-relief hologram is one way for obtaining transmission phase holograms already discussed in Chap.5. However, concerning silver-halide materials the effect is normally limited to low spatial frequencies. Also sometimes an unwanted surface-relief pattern is created as a result of certain bleaching processes. Such a pattern is then only considered a source of noise. We shall therefore devote some time to discussing the possibilities of enhancing and extending the relief pattern to higher spatial frequencies.

The fact that a surface-relief image can be obtained in a photographic silver-halide emulsion has been known since the 19th century. The Koppmann process, for example, uses gelatin relief images for dye printing. In photography such images have also been used for color reproduction.

The main procedure to obtain a relief structure on photographic materials is to use either a tanning developer or a tanning bleach, or else a combination of the two. The gelatin of the emulsion must be soft (very little or no hardening of the unexposed material). Pyrogallol or pyrocatechol developers are often used. After development the emulsion is washed in hot water to remove the gelatin (from unexposed areas) which has not been hardened during the development. This gives a surface-relief image which is, however, obtained mainly at low spatial frequencies. A tanning bleach, such as Kodak R-10, can also be employed. Concerning the tanning development of photographic dye-transfer images a good review by *Tull* has been published [6.86]. The following developing agents have a tanning effect on gelatin according to *Tull*: pyrogallol, pyrocatechol, amidol, and hydroquinone. All these tan the gelatin at various degrees when mixed with suitable alkali. No sodium sulfite is allowed in the developer. One interesting remark mentions that the developer promotes a better relief image at a lower temperature (12° C) than when used at higher temperatures (20° C). Another important finding was that a supplementary oxidation step applied before the hot-water etching takes place, improves the relief structure. The oxidizing agents tested were potassium ferricyanide, dichromate, and ferric nitrate, which all enhanced the relief image. After the tanning development the material is washed, fixed in a nonhardening fixer, washed and then soaked in the oxidizer for about one minute. After that the material is washed again and then etched in hot water (80° C). The oxidizing step can be substituted by an alkali treatment (carbonate). The paper contains a lot of other useful information that could be important for potential holographic applications as well. The above-described technique has only been tested for low-spatial-frequency information recording.

An investigation on relief images on photographic materials (Kodak Minicard film 6451 and 649-GH) was made by *Smith* [6.87]. Both tanning developers and bleaches were tested. Only spatial frequencies between zero and 200 lines/mm were studied. The main conclusions are:

- The height of the relief image is dependent on the degree of tanning which is proportional to the amount of silver in the image (proportional to the optical density).
- The maximum relief is obtained for a certain spatial frequency. This maximum depends only on the thickness of the emulsion and varies inversely with the thickness.
- The material processed in a tanning bleach has a higher relief at higher spatial frequencies than the material processed with a tanning developer.
- The drying method (slow or fast, drying in high or low humidity) does not affect the height of the relief image.

Later, *Smith* [6.88] discussed the production of relief images with an arbitrary profile mainly for low-spatial-frequency applications, such as lenticular lenses and screens.

Altman [6.89] investigated Kodak high-resolution photographic plates and the 649-GH film. He established a relation between the height of the relief image h_μ and the optical density D. A linear relationship is valid for densities between 0 and 2.5. In the linear range, the relation is

$$D = 5.9h_\mu \tag{6.7}$$

where h_μ is the height in μm. This is valid for images developed in the Kodak HRP developer for 5 minutes at 20° C. This is also an upper limit for the highest density depending on the recorded line width of the object.

6.6.2 Surface-Relief Holograms

Most publications concerning holographic relief images on silver-halide materials came from the former USSR. However, an early paper on this topic was a Western publication by *Altman* [6.90]. He used the Kodak 649-F emulsion, which is a thick and rather hard emulsion, for producing pure relief images. After development (Kodak HRP developer), the plate was bleached in Kodak R-9 bleach. A surface-relief of about 1 μm was observed when the emulsion was exposed to an optical density of 4. If a tanning bleach (Kodak R-10) and a fixing step were applied instead, a relief of 3 μm was obtained. It was verified that an index variation had no or very little influence on the phase modulation, which indicates that the relief pattern alone is responsible for the diffraction. The spatial-frequency response was also discussed and it was speculated that a relief image was probably impossible to achieve at high spatial frequencies.

To prove that the surface relief alone was responsible for the reconstruction of the holographic image, *Rigler* [6.91] aluminized the emulsion side of a 649-F plate. An excellent image of high brightness was obtained in reflection illumination. He assumed that the relief pattern was a fraction of a wavelength in depth. This was confirmed by *Cathey* [6.92] who measured the emulsion-surface profile using a Tayler Hobson Talysurf. Gratings with different spatial frequencies recorded on 649-F plates were investigated. *Cathey* found that at 65 lines/mm the height of the fringes varied from 30 nm to 140 nm by increased exposure. At 275 lines/mm, the depth was only 10 nm. Similar measurements were made by *Hannes* [6.93].

Russo and *Sottini* [6.94] reported that phase holograms were obtained on 649-F plates at a spatial frequency of 1200 lines/mm, using a reduced amount of silver chloride (1/10 of the original) in the Kodak R-10 bleach. Unfortunately, their findings were rather misleading since the observed effect was actually due to volume/refractive index phase modulation mainly. They could not make a successful hologram recording if they first bleached the hologram and then fixed it, which means that no relief image was actually produced. When the two processing steps were reversed, phase holograms were obtained because of index modulation within the emulsion.

Brandt and *Rigler* [6.95] demonstrated that an aluminized emulsion in which a focused-image hologram had been recorded could be reconstructed in white light.

All the early investigations indicate that it is difficult to obtain good surface-relief holograms at high spatial frequencies. Therefore, by the end of the 60's this particular field of research had been abandoned by almost all researchers.

One country that did not follow this trend was the former USSR in which the interest was kept alive and papers describing improved surface-relief holograms continued to appear [6.96-103]. Naturally, much of the Russian work has been carried out on the ultra-fine-grained emulsions, such as, the PE-2 material, for example; but also Agfa materials have been employed. *Butusov* and *Ioffe* [6.96] investigated the parameters of holographic periodic structures recorded in silver-halide materials. They compared the height of the relief pattern before and after bleaching for the Mikrat-VRL and Agfa 8E75 materials. Bleaching increased the relief pattern and the Agfa emulsion performed slightly better than the Russian material. They also found a dependance of the height of the surface relief on the spatial frequency, which was explained in the following way: In the gelatin there is a competition between the intermolecular attractive forces, generated by drying, and the surface tension forces. When the spatial frequency of the relief structure becomes higher, the curvature of the gelatin surface becomes larger, which means increased surface-tension forces which create a reduced height of the surface relief. They also investigated the various profile types. Three types were considered, sinusoidal, triangular and rectangular. For the case of diffraction by transmission the modulation index α is

$$\alpha = \pi\lambda^{-1}(n - 1)h \tag{6.8}$$

where n is the refractive index of gelatin and h the height of the profile.

For $\lambda = 633$ nm and $n = 1.53$ the optimum height is for the three profiles: $\alpha_{sin} = 577$ nm, $\alpha_{tri} = 687$ nm and $\alpha_{rect} = 404$ nm. For diffraction by reflection the surface-relief height is lower and for the same conditions we find: $\alpha_{sin} = 145$ nm, $\alpha_{tri} = 184$ nm and $\alpha_{rect} = 105$ nm. In this case the efficiency can also be increased by metallization of the relief pattern.

Beinarovich et al. [6.97] described the possibility of making copies based on *Rigler'*s [6.91] aluminized reflection-relief hologram. They used Mikrat-VR and Mikrat-300 materials. No details concerning processing of the materials are given in the paper which is devoted mainly to the description of the method to make copies from silver-halide relief emulsions. This method will be described in Sect.6.6.3.

It was not until the mid 80's when some real improvements had been obtained in the former USSR. *Galpern* et al. [6.98] presented then a processing method for recording surface-relief holograms with spatial frequencies extended to 1600 lines/mm using the PE-2 emulsion. The experience from the 60's, documented in the references discussed above suggested to use the D-19 developer and the Kodak R-10 bleach for obtaining relief structures. *Galpern* et al. formulated a new developer for the PE-2 material mixed in the following way:

Relief developer, No.1 (stock solution)

Metol	2.2 g
Hydroquinone	8.0 g
Sodium sulfite (anhydrous)	50.0 g
Sodium carbonate (anhydrous)	30.0 g
Potassium thiocyanate	4.0 g
Sodium hydroxide	4.0 g
Potassium bromide	5.0 g
Distilled water	1 ℓ .

Dilute: 1 part stock solution + 4 parts distilled water.

The following processing steps were performed:
1) Development 10 minutes at 20° C.
2) Rinsing in distilled water for 30 seconds.
3) Fixing in a sodium metabisulfite fixer for 15 minutes.
4) Rinsing in flowing water for 10 minutes at 16° to 18° C.
5) Bleaching in Kodak R-10 until clear.
6) Rinsing in running water form 5 minutes at 16° to 18° C.
7) Drying the hologram.

Gratings produced in this process were then aluminized and the diffraction efficiency of the masters and copies was measured. Using the following method, the corresponding depth values could be calculated:

If a sinusoidal character of the relief structure is assumed, the diffraction efficiency η is given by

$$\eta = J_q^2(\tfrac{1}{2} m) \tag{6.9}$$

where J_q is the Bessel function of q order and where $\tfrac{1}{2}$ m determines the phase difference between the maximum and the minimum of the grating.

By measuring the diffraction efficiency η, it is possible to calculate the value of $\frac{1}{2}$ m by means of a table for the Bessel functions. Then, the depth h of the relief pattern is

$$\frac{1}{2}m = \frac{\pi h(n - n_m)}{\lambda} \tag{6.10}$$

where n is the refractive index of the grating material, n_m is the refractive index of the surrounding medium, and λ is the wavelength. If a reflecting coating is applied to the transparent copy of a grating, then (6.10) assumes the form

$$\frac{1}{2}m = 2\pi h/\lambda \tag{6.11}$$

which means that the phase change increases by a factor of four in comparison with that of a transmission grating. An even greater increase of the phase change can be obtained by applying a transparent coating with refractive index $n_r > 1$. In this case, (6.11) becomes

$$\frac{1}{2}m = 2\pi n_r h/\lambda .$$

The results of measurements and calculations are listed in Table 6.2.

Table 6.2. Diffraction efficience (Diff. eff.) of relief holograms

Angle between recording beams	Spatial frequency lines/mm	Diff. eff. of bleached samples [%]	Diff. eff. of aluminized samples [%]	Relief depth [nm]
20°	550	24	33	210
40°	1080	50	14	90
50°	1340	30	4.2	60
60°	1580	22	2.4	50

The above method was also applied to the production of relief holograms of diffuse objects. Mass replication of the L. Cross 360° multiplex holograms was of prime interest here, using such relief masters for embossing. However, the paper mentions that the technique needs further improvements. In two other papers by *Galpern* et al. [6.99, 100] various processing methods were investigated. Here, holographic gratings of spatial frequencies of 1080 and 1340 lines/mm were tested. The investigation confirmed the earlier findings that *nonhardening development* followed by *hardening bleaching* produced the highest relief images for both spatial frequencies. A new developer composition was devised:

Relief developer, No.2:

Metol	4 g
Hydroquinone	8 g
Ascorbic acid	26 g
Sodium carbonate (anhydrous)	40 g
Potassium thiocyanate	4 g
Ammonium bromide	2 g
Distilled water	1 ℓ .

This developer produces the highest relief images on silver-halide emulsions even at rather high spatial frequencies. When used with Western materials, the potassium thiocyanate content must be adjusted (reduced to $2g/\ell$ or less) to suit the particular emulsion.

Galpern et al. also mentioned the reason why the other method (tanning development) does not work so well. It appears to be due to the fact that the hardening effect of the oxidation products of the developer, e.g., pyrogallol, extends beyond the boundaries of the developed image lines. This, in turn, leads to the leveling of the degree of gelatin hardening in the developed and the undeveloped regions of the emulsion. This effect is, of course, more pronounced at high spatial frequencies.

Concerning the hardening bleach bath, best results are obtained with the standard Kodak R-10 bleach. To increase the hardening effect the material was treated in a solution of ammonium dichromate after bleaching and then directly (without washing) soaked in a solution of sulfuric acid (pH 3). The surface-relief processing steps are the following [6.101]:

1) Preheat the unexposed material for 1 hour at 90° C.
2) Expose the plate to obtain a high density.
3) Develop in Relief developer No.2.
4) Wash.
5) Bleach in Kodak R-10 until clear. (No wash after this step).
6) Soak in ammonium dichromate solution (5 g/ℓ) for 5 minutes. (No wash after this step).
7) Dip in a sulfuric acid solution of pH 3.
8) Wash in water at 20° C.
9) Dry at an elevated temperature of 70° C.

Using this procedure it was possible to obtain about 30% diffraction efficiency of aluminized gratings recorded on PE-2 at 1080 lines/mm and 20% at 1340 lines/mm.

The influence of the thickness of the PE-2 emulsion layer was investigated by *Brui* and *Koreshev* [6.102]. They studied layers whose thickness varied from 0.1 to 10.5 μm. The processing was different from the one just described. *Brui* and *Koreshev* used the PRG developer, followed by the R-10 bleach and a fixing step, to remove the silver halide from the emulsion. The material was dried in isopropyl alcohol. Only holograms with low spatial frequencies (<300 lines/mm) were tested. An interesting way of enhancing the relief pattern was a superprotional intensification process (silver intensifier) directly after the development. The mass of silver is increased during this process which means that the degree of tanning is raised in the subsequent bleach bath.

It was found that thinner layers work better and that it is actually possible to obtain a relief depth equal to twice the original thickness of the emulsion for a layer thickness of about 0.1 μm. With a thin layer it is also possible to achieve a better uniformity of the relief height over a wider range of spatial frequencies. *Koreshev* and *Gil* [6.103] studied the thickness of PE-2 layers at spatial frequencies below 300 lines/mm, but only for layers in a thinner region (0.4 to 1.5μm). They discussed the profile shape of the relief phase holograms. It was shown that the profile shape is nearly trapezoidal in the thickness region studied. When the spatial frequency increases, the shape approaches a triangular shape.

The methods used in Russia can also be used quite successfully for Western materials, provided that the highest resolution materials with the softest emulsions are used. Here it is recommended to slightly modify the developer, as already mentioned. In addition, the emulsion can be further etched after the hardening bleaching procedure in a solution of hydrogen peroxide (H_2O_2) or a household bleach. This etches away part of the un-hardened gelatin and enhances the relief pattern. Another possibility is the copper etch solution described by *Sklar* [6.76]. This solution removes the gelatin associated with silver at a much faster rate than the gelatin not associated with silver because of the catalytic action by the silver. The reaction starts with surface gelatin and works down. The copper etch bleach is mixed in the following way:

Solution A

Acetic acid	150 mℓ
Water	360 mℓ
Cupric nitrate	200 g
Potassium bromide	10 g
Water to make	1 ℓ .

Solution B: 3% H_2O_2

Working solution: Solution A plus solution B in equal proportions.

Finally, a technique developed in Europe was recently published by *Ahlhorn* and *Kreye* [6.104]. The material is developed and fixed, and the surface relief pattern is enhanced using a biochemical etching process. For this purpose an enzyme is utilized which has a stronger etching effect on pure gelatin compared to gelatin containing silver. However, neither the enzyme applied is mentioned nor is the developer recipe revealed in the paper. The enzyme used is probably *trypsin* or *chymotrypsin*. The material can be developed in a nonhardening developer, such as, e.g. the PAAP developer. Their replication method will be described in the next section.

The techniques just described have been tested by the present author and the results indicate that the method from the former USSR [6.99-101]

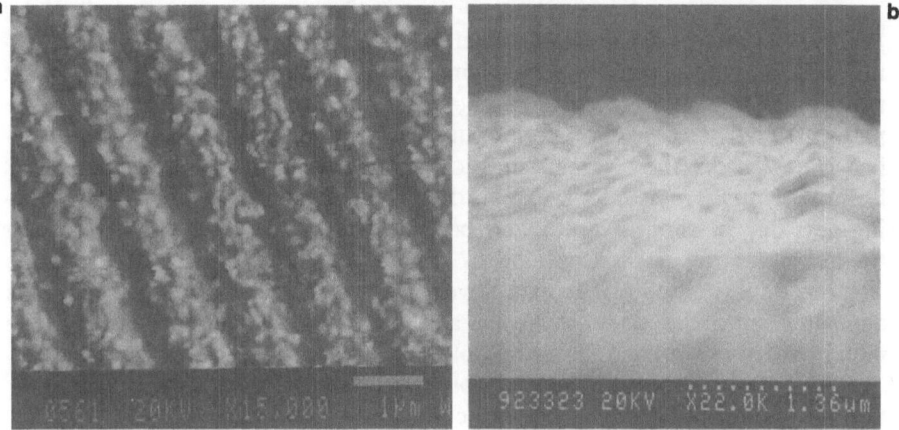

Fig.6.9. SEM pictures of the relief pattern created by the Russian method on the Agfa 8E56 HD emulsion. A fringe spacing of ~1 μm is revealed [Mag.15,000] (a). The relief depth is ~180 nm, seen at the edge of the emulsion [Mag.22,000] (b)

works very well but also the enzyme method produces good relief patterns. Holographic plates (Agfa 8E56 HD) were used for the recording of diffraction gratings with a fringe separation of 1 μm. The samples were gold coated and investigated with the SEM technique. In Fig.6.9 a relief depth of about 180 nm with the expected grating period of 1 μm is revealed in the SEM pictures.

6.6.3 Matrix Fabrication from Silver-Halide Relief Masters

Gale and *Knop* [6.105] published a book on relief images and copying techniques including photoresist materials and the embossing technique. The procedure based on photoresist materials is well established, which is why only the way concerning the preparation of a matrix from a silver-halide relief will be presented here.

As already mentioned, *Beinarovich* et al. [6.97] described a technique for making copies from silver-halide relief emulsions. In their experiment, the emulsion layer was filled with a cold-hardening synthetic liquid resin (e.g., polyester) and covered by a glass plate. After the resin (a liquid at room temperature) had hardened, the matrix was separated from the backing with the resin and the relief pattern remaining on the glass. This was achieved by the initial treatment of the backing with an adhesive substance composed of an alcohol solution of methacrylic methyl triethylane. The separation of the original from the copy was obtained by the treatment of the matrix surface with an anti-adhesive, such as diethyl dichlorosilane. As was expected, the copies obtained had a higher efficiency at low spatial frequencies than at higher frequencies.

Several different techniques of matrix fabrication were described by *Farberov* et al. [6.106], some of which will be reviewed here. A rather complicated method consists of applying a one-component sealant (silicone rubber) on the relief pattern, a catalyst under vacuum, and allowing the pattern to harden at room temperature for 48 hours. Then the silicone replica is peeled off the original emulsion and a thin layer of a solution of polyisobutyl methacrylate in ethylene dichloride is poured over it. This coating gets dry in about 3 hours and becomes the polymer hologram copy. A tin oxide coating is then applied to the polymer copy. After that metallic silver is deposited onto it from a silver-ammonium complex. Then nickel is deposited onto the silver layer by an electroplating method. Finally, the metal matrix is removed from the polymer copy.

In a simpler method thermal polymerization of a monomeric composition is used for the first polymer copy. It is important that the polymer copy is carefully degreased before beginning of the whole process. Then a thin layer of conducting copper is deposited by vacuum evaporation $(5.3 \cdot 10^{-3} \, \text{Pa})$ onto the surface of the polymer copy. This process takes about one minute and produces a copper layer of about 0.02 μm. After that an electrolytic deposition of nickel is performed. When the thickness of the nickel layer is about 0.5 mm the electroplating process is finished. The metallic matrix is separated from the polymer copy. Before the matrix can be used for embossing, the top copper layer covering the nickel is chemically removed.

Yet another method to make a copy of the relief pattern would be to use techniques established for making impressions for electron microscopy. The surface-relief is covered with a synthetic material, e.g., a zapon lacquer. When dry, a coating of epoxy cement is applied. After hardening the epoxy copy is removed from the silver-halide emulsion. This method was employed by *Ahlhorn* and *Kreye* [6.104]. The molding is silvered by immersing it into silver-nitrate reduction solution based on ammonia and glucose. On top of this silver layer a nickel layer is deposited in a sulfamate electrolyte solution. In a next step the resin layer is dissolved and another nickel coating is deposited on the back of the silver layer, previously covered by the resin. Before this step is performed, the silver layer is passivated in potassium dichromate so that the final nickel layer (thickness about 300 nm) can be used as a master shim. Such a shim can be utilized for embossing about 5000 holograms according to the paper.

A review of making relief-phase holograms and methods for mass replication has been given in a publication by *Galpern* and *Smaev* [6.107].

6.7 Silver-Halide Sensitized Gelatin Holograms

Silver-Halide Sensitized Gelatin (SHSG) holograms are similar to Dichromated Gelatin Holograms (DCG). Holographic Optical Elements (HOEs) are often recorded in dichromated gelatin which offers a high diffraction efficiency combined with low noise. The drawback of a DCG is its low sensitivity and limited spectral response. Therefore, a lot of interest has been directed at silver-halide materials processed in such a way that the final hologram will have properties like a DCG hologram. This can be performed using some special processing steps. The results are rather good, which means that holograms of high efficiency and low scattering can be produced. In addition, the SHSG hologram is free from the printout effect. However, *during the recording* of the hologram on a silver-halide emulsion scattering will take place, which differs from the recording of a hologram on dichromated gelatin where very little or no scattering occurs. Therefore, only the fine-grained silver-halide holographic emulsions should be considered for SHSG holograms.

Briefly, the technique for generating an SHSG hologram is to expose a silver-halide emulsion and then, using tanning developers or bleaches, process it in such a way that local tanning will occur within the emulsion. It seems that a tanning dichromate bleach gives best results. The material is fixed in order to remove all the silver halides from the emulsion, leaving only gelatin. The last step of the processing is to dehydrate the material using isopropyl alcohol in the same way in which dichromated gelatin holograms are processed. A dry SHSG hologram must be sealed to prevent moisture penetrating into the emulsion which would destroy the image. The difference between the SHSG process and the DCG process is that in the DCG process hexavalent chromium is *photolytically reduced* to the trivalent state, while in the SHSG process the same is done *chemically*. The developed silver in the SHSG emulsion reduces the chromium during bleaching in the dichromate bleach solution

$$6Ag^0 + Cr_2O_7^{2-} + 14H^+ \rightarrow 6Ag^+ + 2Cr^{3+} + 7H_2O \ . \qquad (6.13)$$

Maximum tanning occurs in the regions of maximum bleaching. A local hardening of the gelatin is caused by cross-link bonds between neighboring gelatin molecule strands.

The SHSG process has been known since the beginning of the 70's. *Pennington* et al. [6.108] used a complicated, 21-step chemical process to create the first SHSG holograms. However, first in the early 80's did the process become more manageable in practice. A simpler process was presented by *Gladden* and *Eastes* [6.109]. They took advantage of Kodak 649-F plates, a D-19 developer, the Kodak R-9 bleach and a nonhardening fixer (Kodak F-24). They found that a slightly acid rinse bath (pH about 5) applied after the dichromate bleach improved the efficiency. Their method was described in more detail in a paper by *Graver* et al. [6.110]. In particular, the importance of the acidity of the washing bath following the R-9 bleach was pointed out here. They claimed that it is possible that only the reduction of the hexavalent to trivalent chromium takes place in the bleach bath. The actual tanning will not occur until the material is immersed in a wash bath with a pH of about 5, which differs from the acidity of the bleach (pH about 1.7). Therefore, the rinse bath is very critical in this process.

Chang and *Winick* [6.111,112] used a slightly different processing method to create SHSG holograms on 649-F plates. The development was performed in the D-19 developer. Instead of the reversal Kodak R-9 bleach, they used a modified version of the Kodak R-10 bleach which is a rehalogenating dichromated bleach bath. An additional step was introduced by soaking the material in an 0.5% ammonium dichromate solution in order to increase tanning before the fixing step. Before exposing the material the plates were pretreated in a weak fixer in order to dissolve some of the dyes as well as to reduce the size of the silver-halide crystals. The absorption of the plates was then reduced and the diffraction efficiency increased compared to the untreated plates. Impressive diffraction efficiencies were obtained, demonstrating the potential of this new processing method.

SHSG hologram recordings at 633 nm on Agfa 8E75 HD plates were reported by *Fimia* et al. [6.113]. The Agfa emulsion is more difficult to use for regular DCG holograms because it is thinner and has a different gelatin type than the Kodak 649-F emulsion. For the SHSG process, *Fimia* et al. employed the Kodak D-8 developer, a modified Kodak R-10 bleach, and the nonhardening F-24 fixer. By using a washing bath of a slightly higher than room temperature (30° C) as well as a dehydration bath at an elevated temperature (60° C), the diffraction efficiency was increased compared to that obtained for room-temperature solutions. However, the signal-to-noise ratio decreased with temperature. In three other publications from the same group, *Boj* et al. [6.114] and *Fimia* et al. [6.115,116] revealed some interesting results. The emulsion on the Agfa 8E75 HD material is normally too hard to be used directly for recording SHSG holograms. The plates should therefore be soaked in hot water to soften the gelatin before the exposure. The influence of the bleach-bath temperature (Kodak R-10) was also investigated. The diffraction efficiency obtained for a transmission grating was higher at 50° C ($\eta = 80\%$) than at 20° C ($\eta = 40\%$). As regards reflection holograms, the non-solvent PAAP developer [6.117] was recommended instead of Kodak D-19 or D-8. With the PAAP developer better resolution can be obtained. The performance at high spatial frequencies is improved which is important for reflection SHSG holograms. *Ferrante* [6.118] investigated the spatial frequency response of the Agfa 8E75 HD emulsion for SHSG transmission holograms. Using the D-19 developer and essentially the same technique as *Graver* et al. [6.110], he obtained experimental data points which fitted a theoretical MTF function for an emulsion with an apparent grain size of about 70 nm. The particular processing technique which utilizes the D-19 developer seems to give larger grain growth than if the PAAP developer were used. This indicates the fact that good reflection SHSG holograms can be produced with the PAAP developer. This developer is then, of course, also a better choice for transmission holograms at higher spatial frequencies. In general, overexposing the material will reduce noise according to *Boj* et al. [6.114]. The exposure needed for SHSG holograms developed in D-19 is about 80 and 200 $\mu J/cm^2$ for PAAP-developed holograms. Another interesting finding shows that the diffraction efficiency could be increased by interrupting the process after the bleaching step (without washing) and letting the plates slowly dry (in darkness) in a high humidity atmosphere (~80%). After 24 hours (under high-humidity conditions) the processing continued with washing, fixing and dehydration. An increased efficiency was found in this case. A possible explanation is that during the remaining time, the trivalent chromium could form a large number of cross-links. High humidity during the storage is important, since at lower humidity levels lower diffraction efficiencies were achieved.

The influence of the temperature on the processing solutions was also investigated by *Hariharan* [6.119]. In addition, he compared a tanning bleach (dichromate) and a nontanning bleach (ferricyanide). He wanted to prove that tanning can actually be performed during development instead of during bleaching. Moreover, using a developer which is normally classified as a nontanning developer can actually produce a hardening effect if the material is subsequently brought in contact with an oxidizer (even nontanning). An MQ developer containing a large amount of sodium sulfite will not harden the gelatin. During development, oxidized developer products are bound to the gelatin immediately surrounding the developed silver. No cross-linking will take place during development, but the oxidized developer molecules constitute available bridges whose free ends can react if they come in contact with an oxidizer. This effect described by *Tull* [6.86] has already been discussed in Sect.6.6.2 for relief-pattern holograms. *Hariharan's* experiment confirmed that SHSG holograms could be obtained on such grounds. Concerning the temperature of the washing bath after the fixing step, he found that 55° C was the best for the tanning bleach, and 35° C for the nontanning bleach.

Angell [6.120] introduced a 13-step processing scheme intended for the Kodak 649-F material. The process, slightly modified later [6.121], is revealed in Table 6.3. Commercial processing products (mainly Kodak) are applied in this process.

Although rather a complicated and time-consuming procedure, the processing scheme introduces some very interesting aspects of SHSG processing. So far, the main tanning bleaches used have been the dichromate bleaches (Kodak R-9 or R-10). *Angell* claimed that *potassium chlorochromate* (Peligouts salt) used as a bleaching agent is an improvement. This is solution A of the Kodak chromium intensifier (Kodak CIA). He also stated that a *fixer with a hardener* can improve the dynamic range and the signal-to-noise ratio but it will also reduce the sensitivity of the material.

The above-mentioned processing scheme contains some new details, such as the stabilization and emulsion protection, for example. In particular, the use of an organo-silane coupling agent (Dow Corning Z-6020) in the fixing step makes it possible to maintain a certain emulsion thickness after the processing, which is important for many HOE applications. This emulsion-thickness control technique will also apply to other types of holograms. The organo-silane coupling agent N-(2-aminoethyl)-3-amino-propyltrimetohoxy silane (Dow Corning Z-6020) [6.122] is recommended since it is compatible with the gelatin matrix. The chemical formula for the agent is of the following type: $RCH_2CH_2CH_2Si(OCH_3)_3$, where R is any organic

Table 6.3. Angell's SHSG processing [6.121]

Step	Process	Formulation	Time
1	Develop	Kodak D-19	4 min
2	Rinse	Water	0.25 min
3	Stop	Indicating stop	0.5 min
4	Rinse	Water	3 min
5	Bleach	Kodak CIA	5 min
6	Rinse	Water	5 min
7	Fix	Hypo (25%)	3 min
8	Fix	Rapid fixer w. hardener +2÷4% Dow Corning Z-6020	5 min
9	Clear	Hypo-clearing agent	3 min
10	Rinse	Water	10 min
11	Dehydrate	50% Isopropanol	5 min
12	Dehydrate	100% Isopropanol	5 min
13	Dry	In oven at 49° C	20 hours

molecule forming a functional group which is compatible with the emulsion matrix. The diamino functional group which exhibits a high degree of gelatin solubility is suitable in this case.

This new technique introduced by *Angell* makes it possible to obtain a permanent chemical way of controlling the emulsion thickness in holographic emulsions. This does not effect the noise level of the hologram negatively in any way, but it actually improves the signal-to-noise ratio according to *Angell*. The actual method depends on the emulsion type, the silver-halide solids loading, the exposure level, the bleaching technique, etc. For the Kodak 649-F emulsion, *Angell* found that adding 2% to 4% of the coupling agent to the fixer resulted in an emulsion thickness of 16 μm after drying, which was equal to the original thickness. This percentage has to be adjusted in each and every case when other materials or processing methods are used.

Weiss and *Millul* [6.123], and *Weiss* et al. [6.124] have recently worked out an alternative simpler SHSG processing method for the 649-F emulsion. Based on *Hariharan's* [6.119] experiments on tanning developers, *Weiss* and his co-workers introduced the CW-C2 developer to the SHSG process. This developer gives very good results when used in regular silver-halide processing. They tested both reversal and rehalogenating dichromate bleach solutions. Their best working SHSG processing scheme was performed according to Table 6.4. In Table 6.5 a method is described where all processing solutions are kept at room temperature.

256

Table 6.4. Weiss-Millul's warm SHSG processing [6.123]

Step	Process	Formulation	Time
1	Develop	CW–C2	5 min
2	Stop/Rinse	Distilled water	3 min
3	Bleach	RB5–2 (reversal)	5 min
4	Rinse	Distilled water	3 min
5	Fix	Rapid fixer, no hardener	5 min
6	Rinse	Dist. water (50–80° C) + 10% TEA solution	7 min
7	Dehydrate	50% Isopropanol	5 min
8	Dehydrate	100% Isopropanol	3 min

Table 6.5. Weiss-Millul's room temperature SHSG processing [6.123]

Step	Process	Formulation	Time
1	Develop	CW–C2	5 min
2	Stop/Rinse	Distilled water	3 min
3	Bleach	RB–6 (rehalogen)	5 min
4	Rinse	Distilled water	3 min
5	Fix	Rapid fixer, no hardener	5 min
6	Rinse	Distilled water (20° C)	3 min
7	Dehydrate	50% Isopropanol	5 min
8	Dehydrate	100% Isopropanol	3 min

The **RB5–2** bleach is a reversal bleach, and **RB–6** is a rehalogenating bleach mixed in the following way:

	RB5–2	RB–6
Ammonium dichromate	5.0 g	2.5 g
Sulfuric acid (conc.)	2.0 mℓ	-
Citric acid	-	15.0 g
Potassium bromide	-	42.5 g
Distilled water	1 ℓ	1 ℓ
pH	1.5	2.3

Weiss and *Millul* compared the D-19 developer with the CW-C2 developer and found that CW-C2 gave significant higher signal-to-noise ratios. Actually, the D-19 developer can give a higher diffraction efficiency for relatively low spatial frequencies. This is in agreement with *Ferrante's* [6.118] investigation in which an apparent grain growth was noticed for D-19 developed gratings, reducing the MTF at higher spatial frequencies. The CW-C2 developer is therefore often a better choice, in particular for reflection holograms. The PAAP developer performs also very well as already discussed.

As regards the bleaching part of the process, *Weiss* and *Millul* [6.123] tested a variety of bleaching agents and found that only the reversal or the rehalogenating ammonium dichromate bleach could produce high diffraction efficiency SHSG holograms. They presented two possibilities here. The first is to use their reversal RB5-2 bleach, followed by fixing in Kodak rapid fixer without hardener and a *warm* (50° -80° C) *rinsing bath*. To this bath *Weiss* et al. [6.124] added a 10% TEA solution which resulted in an increased diffraction efficiency of about 10%. They also tested sorbitol (another swelling agent) which, however, had no influence on the diffraction efficiency. The reason why only TEA gave an improved diffraction efficiency was explained in the following way. In the reversal bleach the silver is bleached and dissolved, leaving voids in the emulsion. These voids are amplified during the warm rinsing step where the gelatin swells. If these voids are filled with air ($n = 1$), water ($n = 1.33$), sorbitol ($n = 1.33$) or isopropanol ($n = 1.38$), they will have a lower refractive index than the gelatin ($n = 1.52$), which will result in a lower diffraction efficiency. If they are filled with triethanolamine ($n = 1.47$) instead, index matching can increase the diffraction efficiency. However, this method works only for SHSG holograms. If the voids are filled with TEA in a regular silver-halide emulsion bleached in a reversal bath, no improvements are observed. This is because the voids are expected to be in phase with the low-index regions of the emulsion.

The other alternative presented by *Weiss* and *Millul* is to use the rehalogenating bleach RB-6, followed by fixing in the Kodak rapid fixer without hardener and a *rinsing bath at room temperature* (20° C). It is not exactly clear from the paper if this second method has any advantages over the first, warm TEA method based on the reversal bleach.

A new method for producing SHSG holograms was described in three publications from the former USSR [6.125-127]. The new principle is based on the formation of a Micro-Cavity (MC) structure. The gelatin in a photographic emulsion is absorbed on the silver-halide grains. In fact, only a part of the gelatin molecules is absorbed. The molecule chains are also linked with the gelatin mass of the emulsion. The thickness of the absorbed layer in a dry emulsion is 2.5 to 4 nm. Each silver-halide grain is surrounded by gelatin molecules linked at different points by active groups able to form complex compounds with silver grains produced during the development. The new method is based on the fact that these absorbed layers are less active and will be more difficult to harden than the surrounding gelatin mass. Variations in hardening between exposed and unexposed areas will therefore occur. After removing silver and silver-halide grains from the emulsion and dehydrating the hologram, micro-cavities will remain which will be responsible for the refractive-index variations. The processing is performed in the following way: After the material has been exposed, developed and fixed it is hardened in a potassium dichromate solution or in formaldehyde. The treatment in the hardening solution takes one hour or more in a potassium dichromate bath. The duration of treatment is six times longer in a formaldehyde solution. After that the material is bleached, fixed and finally dehydrated. Another possibility is to bleach the material before hardening it; after that it is fixed and dehydrated. A third possibility is to use a reversal bleach after development. Then the material is hardened and the final steps are fixing and dehydration. The second processing method results in holograms with better spectral selectivity than holograms processed according to the third method. The new SHSG technique was tested with the Russian material PFG-03. High-quality holograms were produced with diffraction efficiencies between 70% and 90% at 458 to 647 nm. In the third publication [6.127] it is mentioned that the D-19 developer which contains a large amount of potassium bromide ($20g/\ell$) is a better alternative for processing reflection SHSGs in the PFG-03 emulsion than the GP-2 developer, previously used.

An important contribution was very recently made (February 1993) by *Phillips* et al. [6.132]. The traditional view stating that the DCG system hardens gelatin thus preventing solubilization of the material must be balanced against other observations such as that of the reduction of the bulk index of the gelatin layer and the appearance of gelatin in the processing solutions. The researchers proposed that the large values of index modulation in DCG holograms are caused by *gelatin hydrolysis* in the nodal parts of the image structure. Based on the new ideas about the mechanism behind the DCG process, two new processing procedures for SHSG holograms were formulated. The first one is a reversal process which is presented in Table 6.6. The second one (Table 6.7) is a refined process based on high-speed development and rehalogenation beaching. *Fimia* et al. [6.129-131] have described the advantages of using rehalogenating of the silver image as an important feature of the SHSG process. Since silver is not removed from the developed site when rehalogenation of the silver takes place, the formation of hardened shells around the bleached silver site is enhanced by non-removal of the contents of each shell, according to *Phillips* et al. [6.132].

The reversal SS-1 and SS-2 bleaches and the rehalogenating SRH-2 bleach are mixed in the following way:

	SS-1	SS-2	SRH-2
Ammonium dichromate	-	10 g	10 g
Ferric nitrate (or sulfate)	100 g	-	-
Potassium persulfate	20 g	20 g	20 g
Potassium (or sodium) hydrogen sulfate	-	2 g	2 g
Potassium bromide	-	-	20 g
Distilled water	1 ℓ	1 ℓ	1 ℓ .

Table 6.6. Reversal SHSG process after *Phillips* et al. [6.132]

Step	Process	Formulation	Time
1	Develop	HC 110 (diluted 1 = 30) or Agfa G284c (conc.)[a]	2 min
2	Rinse	Water	5 min
3	Bleach	SS-2 (solvent)	until clear
4	Rinse	Water	5 min
5	Reactivate	Solution RA[b]	2 min
6	Rinse	Water	2 min
7	Expose to white light		–
8	Develop to completion	High-energy developer (HC110)	–
9	Rinse	Water	5 min
10	Bleach	SS-1 (solvent)	until clear
11	Dehydrate	Hot or graded isopropanol	–

[a] The Agfa G284c (lithography developer for Agfa's Millimask products)
[b] RA: Sodium sulfite (anhydrous) 50 g
Sodium hexametaphosphate 1 g
Distilled water 1 ℓ

Table 6.7. Rehalogenating SHSG process after *Phillips* et al. [6.132]

Step	Process	Formulation	Time
1	Develop	Agfa G284c (conc.)	2 min
2	Rinse	Water	5 min
3	Bleach	SRH-2 (rehalogenating)	until clear
4	Rinse	Water	5 min
5	Fix	Hardening fixer	5 min
6	Rinse	Water	5 min
7	Soak	Warm water (30° C)	–
8	Dehydrate	50% isopropanol (30° C)	–
9	Dehydrate	100% isopropanol (30° C)	–
10	Dry	Bank of infrared tungsten lamps	

These two new SHSG procedures outlined in Tables 6.7a and 8 have already proved to work very well and the reader is advised to follow the current work of *Phillips'* research group.

Comparing the results from all the publications discussed in this section, the SHSG processing technique can be summarized as follows:

A 70% diffraction efficiency for *transmission gratings* recorded in the Kodak 694-F emulsion at 514.5 nm (30° beam separation, K = 1) was reported by *Weiss* and *Millul* [6.123] who used the CW-C2 developer, a reversal dichromated bleach, and a warm rinsing bath to which a 10% TEA solution had been added. In addition, a high signal-to-noise ratio was obtained with this method.

Similar diffraction efficiencies (of about 70%) were also obtained with *Angell's* [6.120] more complicated process for *transmission holograms* recorded in Kodak 649-F emulsion at 632.8 nm (at high spatial frequencies, >1000 lines/mm).

Almost the same diffraction efficiency (>70%) for *transmission gratings* recorded in the thinner Agfa 8E75 HD emulsion at 632.8 nm (spatial frequencies between 1000 and 2000 lines/mm) was reported by *Fimia* et al. [6.115,116] who utilized the PAAP developer and a rehalogenating dichromate bleach. The process was interrupted after the bleaching step and taken up again 24 hours later with fixing, rinsing and drying. An efficiency of 80% was reached at a spatial frequency of about 1000 lines/mm.

However, a very high diffraction efficiency (80%) for *transmission holograms* recorded in the Kodak 694-F emulsion at 632.8 nm (no spatial frequency mentioned) was reported in one of the first SHSG papers by *Chang* and *Winick* [6.111,112] who applied *pretreatment* of the unexposed material in a weak fixing solution. The rest of the process was a D-19 - rehalogenating dichromate bleaching process. Probably, the spatial frequency was rather low here, which can give a high efficiency with the D-19 developer. The pretreatment, already described, was important in order to reach this high efficiency.

Recently *Fimia* et al. [6.130] investigated how to optimize the SHSG process. By measuring the diffraction efficiency at different steps during the process it was possible to find, for example, the optimum potassium-bromide concentrattion in the R-10 bleach. A diffraction efficiency of 80% for gratings (1000 lines/mm) recorded on Agfa 8E56 HD emulsion at 514-nm wavelength was obtained. The material was processed in an ascorbic-acid/sodium-carbonate developer (ACC), bleached in the optimum R-10 bleach, fixed in the Kodak F-24 fix, and dehydrated in isopropanol in the usual way. This result is another confirmation of the fact that a developer which avoids the influence of oxidation products during development is to prefer for the SHSG process.

As regards *reflection gratings*, high efficiency can be best obtained using either the PAAP or the CW-C2 developers. The D-19 developer for reflection gratings results in very poor holograms.

In general, it seems that the use of the CW-C2 or the PAAP developers combined with either a reversal or a rehalogenating ammonium dichromate bleach represents so far the best processing method for both transmission and reflection holograms. The temperature of both the bleach bath and the rinse bath can play an important role. As regards the material used, the Kodak 649-F plate is often preferred, although the Agfa 8E75 HD emulsion can also give good results. The pretreatment (with a weak fixer) as well as softening the emulsion in a warm-water bath should also be considered. For SHSG holograms produced with a pulsed laser, the SM-6 developer will probably work very well since it is the pulse version of the PAAP developer.

Finally, a method for increasing the sensitivity of regular dichromated gelatin (DCG) holograms was presented by *Mazakova* et al. [6.133,134]. This new method is based on very small silver-halide grains added to the emulsion. They sensitized the gelatin emulsion by producing super-fine silver-halide grains in an already coated gelatin layer by bathing the material in a silver-nitrate solution. Then, after washing in distilled water, the plate is immersed in an alkali-halide solution. The plate is then washed again and sensitized in an optical sensitizer (1,2-dimethyl-3,3-methylpiromedinidene-ketylidine). The 20-μm emulsion contained super-fine silver-halide grains of a concentration which is about 10 times smaller than in a normal holographic emulsion. An exposure of 1 J/cm^2 produced a density of only 0.15 in such an emulsion developed in D-19. However, this emulsion is about 10 times faster than the ordinary DCG emulsion. Both transmission and reflection gratings were produced at the 488-nm argon-ion laser wavelength. The processing was performed with a 0.5% ammonium-dichromate bath, followed by a fixing step, washing, and dehydration in isopropanol baths of increasing concentrations.

6.8 Dye Substitution for Silver in Holograms

In Chap.5 one method of improving the light stability in bleached holograms has been presented, suggesting lead or nickel ferrocyanide as substitutes for the light-sensitive silver halides in the emulsion [6.135]. Dyes can be another choice for the replacement of silver in processed holograms. The dyes tested for this purpose are the color couplers used in photographic color processing. This technique enables a holographer to obtain a wavelength-selective reconstruction combined with a high diffraction efficiency by phase modulation. The method was first described by *Nassenstein* and *Eggers* [6.136]. Thin holograms can be processed to act as wavelength-selective phase holograms or as amplitude holograms with reduced noise caused by phase variations. Which type of hologram is received will depend on the absorption band of the particular dye and the difference between its refractive index and the refractive index of the gelatin in the emulsion.

In the first case *dispersive refraction* is used, which means selective absorption outside the spectral absorption interval of the dye. The calculations performed by *Lashkov* and *Sukanov* [6.137] demonstrated that due to the presence of the absorption band the refractive index falls off with the distance from the center of this absorption band much more slowly than the absorption coefficient does. It is thus possible to effectively change the refractive index in a given spectral interval by adding a substance which is essentially transparent in the working spectral interval.

Amplitude absorption is based on pure absorption which means that the dye's absorption band center is peaked at the reconstruction wavelength. The refractive index of the dye should be equal (or almost equal) to that of the gelatin.

6.8.1 Chromogen Development

If a hologram is developed with the conventional color development technique, the silver grains in the emulsion are replaced by dye molecules. It is possible to convert the silver-grain distribution of a conventional amplitude hologram into a corresponding distribution of dye molecules. The steps to achieve this are the following:

- Exposed silver-halide grains are reduced chemically to silver with the usual developing method. At the same time, however, the developed silver grains are superimposed on the dye molecules formed as a result of an oxidative coupling reaction of the added color coupler in the developer.
- In the second step the developed silver grains are rehalogenized with a bleach solution.
- In the third step (which can be often combined with the bleaching step) the rehalogenized silver grains and the unexposed silver-halide grains are removed from the emulsion, so that only dye molecules are left in the emulsion.

Röhler et al. [6.138] introduced slight modifications to these processing steps to improve the performance of dyed holograms. They used the Agfa 10E75 emulsion and the following chromogen-processing procedure (Modified Color Coupler "MCC" processing) to make holograms based on a blue dye:

1) Conventional development in Agfa G3p developer.
2) Fixing in Agfa G334.
3) Bleaching in a special color bleach bath: N II.
4) Re-exposure of the material to uniform light of high intensity.
5) Chromogen development using blue color developer: NPS I.
6) Bleaching.
7) Fixing.

Röhler et al. claimed that the advantage of this rather complicated process is that the diffusibility of dye molecules increases when the unexposed silver-halide grains are removed before the chromogen development takes place. Also, the already processed amplitude holograms can be converted (in daylight) to dye holograms, using steps 3 to 7 above.

A comparison between conventional amplitude holograms and dye holograms of the amplitude type indicates that a similar diffraction efficiency is obtained up to a spatial frequency of about 1500 lines/mm. Beyond this frequency, the efficiency decreases, due to the relatively large size of the blue molecules. The noise level of dye holograms is rather low for a wide range of spatial frequencies, but because of the drop in efficiency at a spatial frequency of above 1500 lines/mm, the signal-to-noise ratio is lower for dye holograms than for conventional amplitude holograms.

Sukhanov et al. [6.139] discussed the possibility of producing efficient dye holograms of the phase-type, using dispersive refraction. Different dyes were tested for the processing of the Russian material LOI-2-633. For the best performing dye (the type was not mentioned in the paper) the theoretical diffraction efficiency calculated for a given recording case was 28%. They measured 22% for that particular dye in the emulsion. In this case, the diffraction efficiency of the dye-phase hologram was much higher than that of the corresponding amplitude hologram which had a diffraction efficiency of about 4% only.

The mechanism of dye holograms is similar to the underlying principle of the behavior of the new organic recording material Reoxan, developed in the former USSR [6.140] and briefly described in Sect. 1.2.2b.

6.9 Microheterogeneous Recording Media

A very interesting development project concerning a new volume-recording material for holography is currently taking place in Russia [6.140-143]. The tested product is a heterogeneous material with a capillary structure and the typical recording layers' thickness of between 100 and 1000 μm. It consists of a rigid porous frame and a light-sensitive substructure firmly connected to the walls of the inner frame structure. A fine-grid structure of the inner frame cavities remains open during its processing, enabling the processing solution to quickly reach the entire volume of the material. The pore radii of these channels range from 10 to 200 nm. The penetration ability of the processing solution is enhanced by the presence of the capillary pressure. The material's permeability to various processing solutions was tested and it was found that the average rate of passage through the structure was about $3.3 \cdot 10^{-4}$ cm/s. Due to its rigid frame, the material is shrink proof. The rigid materials must be transparent over the entire visible spectrum and contain a large free-volume of interconnected inner cavities. Porous glass meets these requirements. It consists of 90% to 99% silicon dioxide. Figure 6.10 depicts a structural model of the capillary recording material.

The inner walls of the channels must undergo special treatment to ensure that the light-sensitive substratum is rigidly attached to the pore surfaces. The light-sensitive coating can consist of a number of common photosensitive materials. The following section discusses only the silver-halide capillary structure holograms, however. In this case we are dealing with a latent-image material that needs post-exposural processing. The new recording material is called *Focar*, which is a Russian acronym for phase, volume, capillary and reinforced material. When the silver-halide grains act as a light-sensitive substratum, the material is called *Focar-S*. To reduce

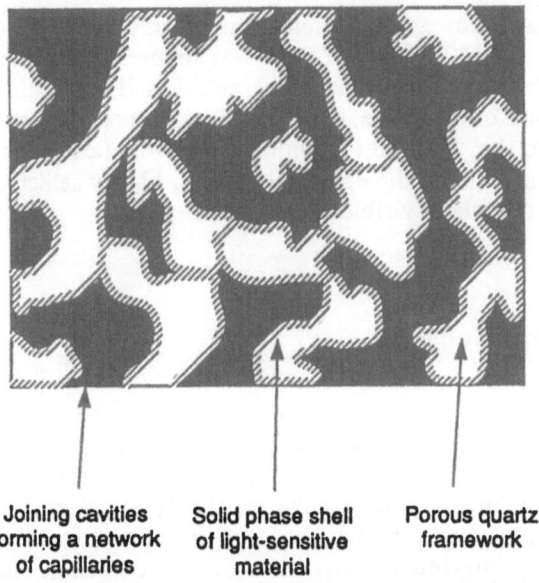

Joining cavities
forming a network
of capillaries

Solid phase shell
of light-sensitive
material

Porous quartz
framework

Fig.6.10. Structural model of the Focar-type capillary recording material [6.140]

light scattering caused by its capillary structure, the inner cavities can be filled with a liquid that has a refractive index similar to that of the frame material. Index matching has to be performed at both the recording stage and at the reconstruction of the hologram, which means that this material is suitable for the production of low-noise holograms.

6.9.1 The Focar-S Recording Material

The Focar-S volume medium consists of a porous, quartz-like frame, (porous glass) with 15% to 30% of the walls of its inner cavities covered with an ultra-fine silver-halide suspension in gelatin. The synthesis of silver-halide grains takes place inside the structure, which gives very fine silver-halide grains (15÷20nm). Currently, the sensitivity of this material is about 100 mJ/cm^2, but it is believed that it can be increased by one to two orders of magnitude. Focar-S materials with a thickness of 2 mm have been produced.

The development is performed with a slow-acting, diluted colloidal developer. In such thick materials the colloidal silver particles will attenuate the reconstructed waves, which affects the diffraction efficiency. This effect is more pronounced at shorter wavelengths. For example, at 633 nm a diffraction efficiency of 90% could be obtained for a mixed (absorption-dielectric or phase-amplitude) hologram. For a wavelength of 488 nm a diffraction efficiency of only 7% could be obtained for the same type of hologram. For this reason, such holograms are most often bleached after the

265

colloidal development process. For a phase hologram of this type with a thickness of 2 mm a diffraction efficiency of 96% was reported. The same efficiency could also be obtained for Focar holograms based on dichro-mated gelatin (*Focar-G*) but only at the expense of a higher exposure. Focar-S seems therefore to be the most promising microheterogeneous volume recording material, producing dimensionally stable, highly selective, high-efficiency HOEs over the whole visible spectrum.

6.10 Investigation of Emulsions

It is often important to understand how holographic recording, e.g., in the silver-halide emulsion, really takes place, in order to better understand the effect that various processing methods may have on the final product. It may interest the reader that already the researchers working with Lippmann photography investigated the interference structure of the emulsion, not only for the purpose of improving the color-recording technique but also to prove W. Zenker's photochromic theory applied to Lippmann photography. This theory had already been verified experimentally by O. Wiener who de-monstrated stationary light waves. However, using optical microscopy to study the periodic silver-grain structure in the emulsion is pushing this technique to its very limit. Nevertheless, in the late 1890's, *Neuhauss* [6.144] succeeded in microscopically imaging the interference structure of red light recorded in a Lippmann emulsion. He used a long-wavelength recording where the interference layers are most widely separated. In air, the distance between the interference layers is 330 to 380 nm; in the emulsion it is closer and depends on the refractive index of the emulsion. A fringe separation of 220 to 250 nm must be microscopically resolvable in order to see the re-corded structure in the emulsion. *Neuhauss* used different techniques for embedding the emulsion samples to be studied, such as paraffin, Canada balsam, and glycerin, which slightly increased the separation of the layers. Thin emulsion layers (about $2\,\mu$m in thickness) were prepared at the Ana-tomic Institute in Berlin using a very fine microtome. The quality of the microscope lens was considered very important, too, which is why the oil-immersion technique and a high-quality apochromat lens (numerical aper-ture 1.40) from Zeiss were applied. Short-wavelength illumination will also increase the resolution. *Neuhauss* utilized sunlight passing through a dark-blue dye to get a wavelength peaked at about 450 nm. With the above-men-tioned technique, *Neuhauss* succeeded in recording photographs of the emulsion through a microscope, receiving a linear magnification of about 4000 times. The separation of interference layers was clearly visible and the distance between them could be measured. The results of the experiment were in good correlation with the theory. Thereafter, optical microscopy was often used by Lippmann-photography scientists to study the influence of various processing methods (see, for example, the work by *Ives* [6.145]).

Today, of course, the electron microscope is a more powerful instrument in the hand of a holographic researcher who studies cross sections of the holographic emulsion. In the following, some of the results of such studies will be discussed.

6.10.1 Electron-Microscopic Investigation of Holographic Emulsions

An early publication by *Akagi* et al. [6.146] gives photographs of the Kodak 649-F emulsion recorded in a Transmission Electron Microscope (TEM). The investigated specimen was prepared by embedding the holographic material in an epoxy adhesive mixture and cutting it with an ultramicrotome. The thickness of the emulsion section was 150÷200 nm. *Akagi* et al. made transmission gratings using a HeNe laser. Two collimated interfering beams of equal intensity (K = 1) were employed for some of the tests. Other tests were made with the object beam containing an on-off bit pattern and the value of K = 2.5 in this case. The plates were developed, fixed and some were bleached in a potassium ferricyanide bleach. It was found that the interference pattern had been recorded throughout the whole thickness of the emulsion layer. It was also found that the grain size in bleached holograms was about 1.5 times larger than in similar amplitude holograms. The recording containing object information showed some expected irregularities, due to the modulation of the signal beam by the information of the digital image.

In addition to the 649-F emulsion, some tests were performed on the high-sensitive, coarsely grained Agfa Scientia 14C75 emulsion. These revealed almost no periodic structure of the layers of the silver grains, due to the rather large grains of this recording material. Reconstruction from such an emulsion is, of course, rather poor.

At Agfa, *Buschmann* [6.147] investigated the bleaching process applied to the 8E75 emulsion, using electron microscopy. Also, *Joly* and *Vanhorebeck* [6.148] from the same company took advantage of this powerful technique to study the influence of various developers on the quality of the holographic images recorded on Agfa 8E56 HD plates. The holograms were bleached in a PBQ solution. The main findings from the electron-microscope images were the following. The developed silver structure is filamentary. The pyrogallol developed structure contains longer filaments of silver than filaments developed by a metol-hydroquinone developer, but not as compact. Another important finding shows that the rehalogenated silver grains become spherical after bleaching. The interesting thing is that the mean grain size of silver-halide particles formed in the emulsion is more or less uniform, which indicates that it is not dependent on the developer employed in the particular case.

Aliaga and *Chuaqui* [6.149] used 50 nm thick emulsion samples for TEM investigations. They compared the Agfa 10E75 and 8E75 HD emulsions which were processed with a solution-physical development (CPA-1 processing). This type of processing forms spherical silver particles. The

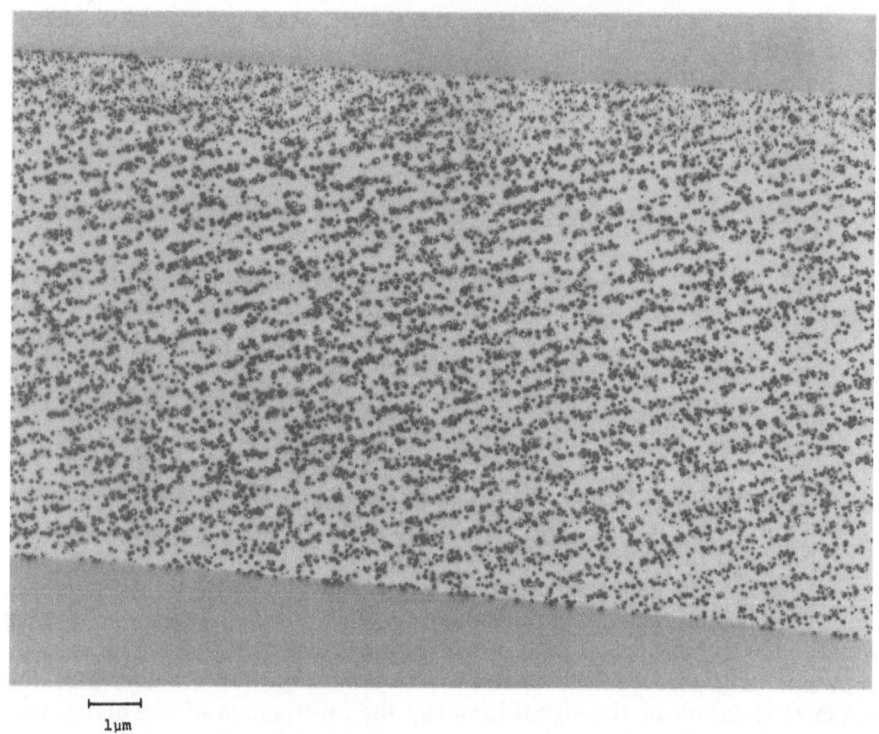

1μm

Fig.6.11. Electron microscope picture of an emulsion cross section from a reflection hologram recorded on the Agfa 8E75 HD material and processed with the CW-C2 developer and the PBQ-2 bleach. The hologram recording geometry used produced a period of 0.20 μm and a 9° inclincation of the fringes in the emulsion. The thickness of the sample is 80 nm. Due to deformations during the preparation of the cross section, the fringe separation measured is 0.33 μm and the inclination angle 12°. The grain size in this sample is about 50 nm [6.150]

size of these particles was more or less the same (50 nm) for the two emulsions after fixing (not before) and the only noticeable difference was grain density. In the same investigation holograms processed with the CW-C2 developer and fixed or bleached in the PBQ-2 bath were also examined. The fixed samples showed the familiar filamentary structure. The PBQ bleached samples revealed spherical particles similar in size to the original grain sizes of the emulsion.

Kubota [6.150] published an electron microscope picture of the emulsion of a reflection hologram recorded on the Agfa 8E75 HD material. One collimated beam from a HeNe laser entered the plate at the emulsion side at normal incidence and the other beam entered at the glass side at an angle of 30° (in air) to the plate normal. This recording geometry produces a period of 0.20 μm and a 9° inclination of the fringes in the emulsion, assuming an emulsion refractive index of 1.62. Three different processing methods were investigated:

268

(i)   CW-C2 developer - PBQ-2 bleach.

(ii) D-19 developer - fixing in Kodak rapid fixer.

(iii) D-19 developer - 3% bromine water bleach.

The thickness of *Kubota*'s samples was about 80 nm. Figure 6.11 shows the CW-C2 - PBQ-2 processed emulsion.

The measured values are slightly different from those calculated above. The fringe separation was 0.33 μm and the inclination angle 12°. The discrepancies can be explained by the fact that the emulsion samples were cut aslant to the surface and that, in addition, some water was absorbed in the emulsion during the preparation of the cross section. The grain size in this sample was about 50 nm and it was relatively uniform. The amplitude hologram developed in D-19 revealed a filamentary structure which confirms the *Joly* and *Vanhorebeck* investigation [6.148]. When such a hologram is bleached the rehalogenized grains become spherical again, which has also been confirmed in Kubota's test on holograms bleached with bromine water.

6.10.2 Phase-Contrast Microscopic Investigation of Holographic Emulsions

At the beginning of this section conventional optical microscopy was discussed. Although it is not as powerful a tool for emulsion studies as electron microscopy, it has some advantages over the last mentioned. The possibility to use phase-contrast microscopy was described by *Kostuk* and *Goodman* [6.77]. This technique offers a more economical alternative to electron microscopy and also allows for larger areas of the emulsion to be studied. In addition, there is no need of preparing the ultra-thin emulsion samples for these investigations. The phase-contrast microscope is applied in conjunction with a Fresnel-zone plate formed in the film under test. Using this technique *Kostuk* and *Goodman* were able to demonstrate the spatial-frequency-dependent diffusion process of silver-halide grains during the fixation-free rehalogenating bleaching (Fig.6.12). The diffusion process, which is an important part of this bleaching mechanism, does not take effect until the spatial frequency has reached a certain value. (Refer to Sect.5.2.4).

6.10.3 X-Ray Fluorescence Analysis of Emulsions

X-ray fluorescence for the analysis of holographic emulsions has been discussed by *Eichler* et al. [6.151] In this method a radioactive source is used to generate a photon beam to which the holographic emulsion is exposed. Photoelectric absorption of photons from the incident beam leads to the creation of K-shell vacancies which are subsequently filled up by electrons from higher shells. Characteristic K_α and K_β X-rays are then emitted from the emulsion. A detector is used to record the emitted X-ray intensities which are proportional to the atomic concentrations of silver and silver halide in the emulsion.

Fig.6.12. Example of a holographic emulsion bleached with the fixation-free rehalogenation process and investigated with the phase-contrast microscopy technique showing the spatial frequency dependent diffusion process: 1 cycle/70μm (a), 1 cycle/35μm (b), 1 cycle/15μm (c), 1 cycle/10μm (d). The diffusion process does not set off until the spatial frequency has reached a certain value. Compare a and d in the figure [6.77]

Eichler et al. investigated various processing methods for the Agfa 8E75 HD emulsion. The results are summarized in Table 6.8. They represent an example of the possibilities which this method offers to the investigator. The rehalogenating bleach was of the PBQ-type and the reversal bleach of the pyrochrome type. As can be seen from the table, the silverhalide concentrations after bleaching, depend strongly on the exposure and the optical density obtained after development.

The paper also discusses the relationship between the concentration of the absorbing silver x (in mg/cm²) and the optical density D. The attenuation of light by the emulsion is given by

$$E = E_0 e^{-\mu x} \tag{6.14}$$

where E_0 and E are the incident and the outgoing light intensities, respectively. The absorption coefficient μ is then given in cm^2/mg. The optical density is expressed by

$$D = \log(E_0/E) \tag{6.15}$$

Table 6.8. Concentrations of emulsion constituents

Element	Unexposed emulsion (rehalogenated) [mg/cm²]	Bleached emulsion (reversal) [mg/cm²]	Bleached emulsion (reversal with KI) [mg/cm²]	Bleached emulsion [mg/cm²]
Ag	0.296	0.288	0.222	0.117
Br	0.220	0.216	0.167	0.080
I	0.011	0.010	0.010	0.024

which means that

$$x = 2.303D/\mu .\tag{6.16}$$

Assuming a linear relationship and considering the Agfa 8E75 HD emulsion and the Kodak D-19 developer, the absorption coefficient μ was found to be 50.16 cm²/mg.

Eichler et al. claimed that the combination of the X-ray fluorescence method with microstructure emulsion analysis leads to a better understanding of the way in which the silver-halide emulsion works. This can help us in the optimization of the performance for holographic purposes.

7. Processing Schemes

This chapter presents a wide collection of complete processing schemes for different types of holograms. The presentation is based on the various aspects of hologram recording discussed in detail in the previous chapters. Many of the well-working techniques, tested and used for quite some time in holographic processing, have appeared in various scientific journals. They may, however, be difficult to locate just when they are needed. The idea of this chapter is to provide the reader with easy but extensive reference to the existing processing schemes and techniques published so far and help holographers to further improve their processing skills as regards silver-halide materials. It is just as well to remember though that even as this book is being written the existing techniques are undergoing modifications and that new processing techniques might have appeared by the time this book is finished. The reader may also be glad to know that after some time in the field he or she will have worked out his/her own modifications, which, even though they may be small, may have a considerable influence on the final results. Finally, attention should be drawn to the fact that only processing schemes for commercial Western materials are discussed in this chapter.

7.1 Recording Considerations

The processing technique to be chosen is dependent, to some extent, on the recording technique applied. As regards the choice of a suitable recording wavelength for a given hologram type, the following considerations should be kept in mind:

• For transmission holograms a long recording wavelength is highly preferable since light scattering during the recording will be the lowest possible for a given silver-halide material. At times, however, other considerations can be more important, such as the reduction of aberrations when a short wavelength is to be used at the reconstruction of the hologram. One such case is, for example, when recording silver-halide transmission masters for photoresist copying. One must also remember that the highest image resolution is obtained at a short wavelength.

• A long recording wavelength is preferable also for reflection holograms from the scattering point of view. However, for both HOEs and display holography the peak of the reconstructed image bandwidth is important and therefore a decision as to the most suitable recording wavelength must be carefully weighted. It will also depend on whether a pretreatment, such as emulsion preswelling, will or will not be performed.

- Availability of various recording wavelengths obtainable from commercial lasers must be considered. Actually, the choices are rather limited, with less than ten wavelengths in practice. The most widely used CW laser for red recordings is the HeNe laser (633nm). A more powerful alternative is the krypton-ion laser (647nm) which is common in commercial laboratories. Newcomers in the area are red diode lasers. The most popular CW laser for green recordings is the argon-ion laser (514nm) which is especially common in large-scale production holographic laboratories. This is the most powerful CW laser employed in holography today with more than 10-W output power possible in a single line (with an etalon). The argon-ion laser emits also several powerful blue lines (476nm, 488nm). Newcomers in the green recording area are the green HeNe laser (543nm) and the CW frequency-doubled diode-pumped Nd:YAG laser. One blue wavelength frequently used is the 442 nm emission from the HeCd gas laser. The most commonly used laser for holographic pulsed recordings is the red ruby laser (694nm). The main green pulsed laser source is the frequency-doubled Nd:YAG laser (532nm). Sometimes, there is no other alternative for recording a hologram but a pulsed laser, which means a choice between these two wavelengths (694nm and 532nm).

- Various scientific or industrial applications may require a very specific recording wavelength. In such a case the recording material is of less importance. Display holography is mainly concerned with the best possible image quality which is usually determined by the combined effect of the recording wavelength, the recording material and the processing method. In display color holography, the recording wavelengths and their combination are equally important, as will be discussed in Chap.9.

- Economical considerations must not be overlooked. The cheapest lasers are the red HeNe lasers which are employed in a variety of holographic applications. However, a mass production of holograms as well as the recording stability demands can justify investing in more expensive, high-power CW lasers, such as ion lasers or even pulsed lasers.

The recording geometry, beam ratios, polarization control, recording stability, etc. will not be discussed here. These factors and their influence on the image quality have been closely discussed in other books on holography. However, with regard to the influence of the state of polarization, *Phillips* [7.1] has made an important contribution to the understanding of interference structures recorded in holographic emulsions. The conflict between interference symmetry to obtain the highest fringe contrast and polarization control to avoid internal reflections in the plate results in reduced image quality and increased fog. The popular Brewster-reference-angle method with *p*-polarization (Sect.2.2.4) to suppress internal reflections during recording of holograms is not generating maximum fringe contrast in the emulsion. Therefore, in order to obtain the highest possible image quality it is instead recommended to use *s*-polarization and to dump the internally reflected light within the material, employing, for instance, the index-matching technique.

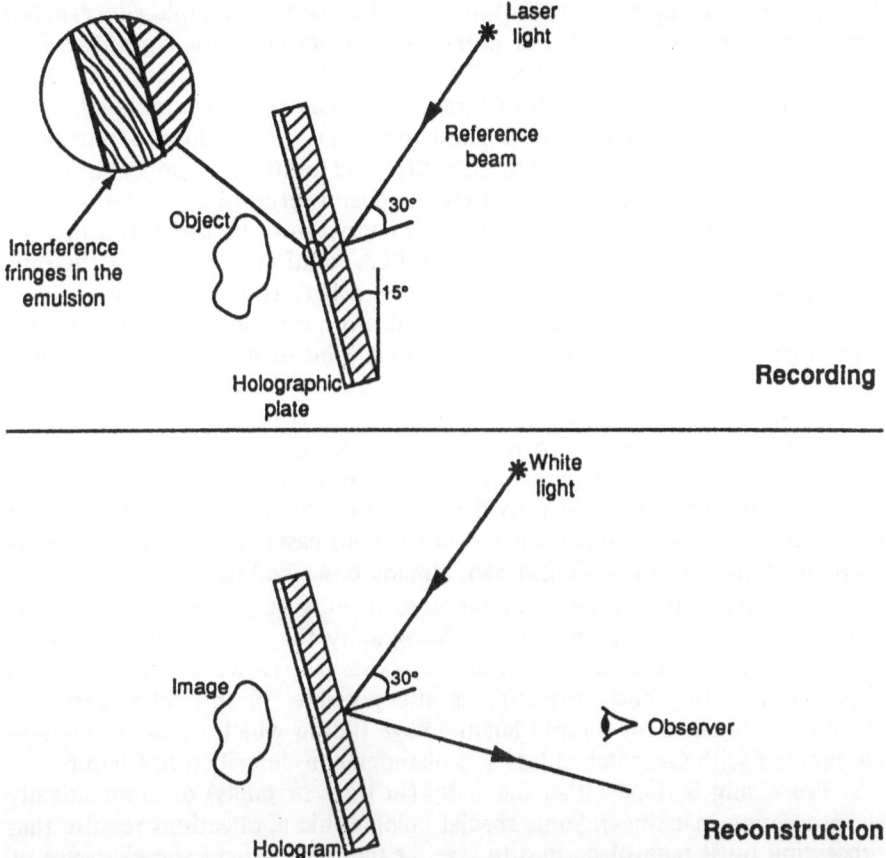

Fig.7.1. Optimal geometry for recording white-light reflection holograms with minimum image blurring. For the recording of Denisyuk single-beam holograms with a reference angle of 45° from above, the top of the plate should be tilted backwards 15°. One should try to arrange the setup in such a way that the recorded interference fringes run as parallel as possible to the surface of the emulsion

Another detail, often overlooked in other publications, concerns the recording of white-light reflection holograms, and particularly ways of reducing image blurring. *Ward* et al. [7.2], and *Ward* and *Solymar* [7.3] have extensively treated this particular problem. Briefly summarized, they recommended that for the recording of Denisyuk single-beam holograms with a reference angle of 45° from above, the top of the plate should be tilted backwards 15° (Fig.7.1). This is the optimum recording geometry for reducing dispersion blurring in white-light reflection holograms. One should try to arrange the setup in such a way that the recorded interference fringes run as parallel as possible to the surface of the emulsion. In addition, *Kubota* [7.4] discussed various methods for image sharpening of Lippmann reflection holograms.

Before a holographic recording can take place, a suitable silver-halide emulsion must be selected. For a choice of materials from Agfa, Ilford or Kodak the reader is referred to Chap.3. The three companies offer also a wide choice of materials with different sensitivity. For display holography, only the materials with the highest resolving power/lowest sensitivity should really be considered. For scientific and industrial applications very often the faster materials are required and even preferred.

In some applications, the recording material must be presoaked in various solutions before exposure, e.g., in a TEA solution or water for the purpose of hypersensitizing, swelling or stress relief. However, in most cases the material can be used directly without any pretreatment. The exposure takes place with the material mounted in a film or a plate holder. Sometimes, the material is exposed in a liquid gate or the film is attached to a glass plate, taking advantage of an index-matching liquid. In holography, it is common that the exposure time is extremely long or short, e.g., a long exposure time during which a weak laser beam illuminates the emulsion, or a very short exposure time provided by a Q-switched laser pulse of very short duration and a high-peak power. In both cases the effects of the reciprocity failure must be considered, already described in Chap.6.

After the material has been exposed, it must be processed in order to obtain the recorded information. In holography the material is quite often processed almost immediately after the exposure. However, a long delay between recording and processing is also possible. In this latter case one should be prepared for a rapid latent-image fading which occurs in materials exposed with Q-switched lasers, a phenomenon described in Chap.6.

Processing is done either manually (in trays or tanks) or automatically (in processing machines). Some special holographic applications require that processing must take place in-situ (i.e., at the place where the exposure of the recording material was done). The following sections provide a detailed description of the necessary processing steps. Generally, it can be said that machine processing offers a high degree of uniformity which is rather difficult to achieve when the material is processed manually.

7.2 Processing of In-Line Transmission Holograms

7.2.1 In-Line Amplitude Holograms

The in-line holograms of the Gabor type normally present few problems in the development process. If the holograms have been recorded on fine-grained, high-resolution holographic materials, the processing is quite straightforward and any high-contrast MQ developer (such as, e.g., Kodak D-19) will give good-quality amplitude holograms. These holograms are very often recorded with Q-switched lasers, where the pulsed holography processing technique applies (compare Chap.6).

For an in-line type of hologram, the holographed object must be either transparent or consist of small particles only. This means that the object signal will be low and the K-value high. This, in turn, means that the intermodulation noise will be low and the holograms will have a relatively good signal-to-noise ratio. The main problem with this type of hologram is the background reference light, which, if it is not spatially filtered at the reconstruction stage, will affect the image contrast. The in-line hologram is normally used only for scientific applications (spatial filters, holographic microscopy, bubble-chamber recordings, etc.), which is why amplitude holograms are often sufficient. In general, as explained by *Vikram* and *Billet* [7.5], an amplitude, in-line hologram gives a higher signal-to-noise ratio if the density is high (low amplitude transmission) than a normal amplitude off-axis hologram. Optical densities from 1.5 to 2.0 give usually very good results. *Bexon* et al. [7.6] claimed that an optical density of 1.7, in combination with a γ-value between 6 to 9, will produce optimum quality. *Dunn* and *Walls* described processing of in-line holograms and the ways of obtaining high-quality amplitude holograms as well as producing good phase holograms [7.7, 8]. Their technique uses the *Neofin Blue developer* and high-resolution materials, such as Agfa 8E75 HD. The general rule for holography by which to obtain high-quality holograms applies, also, here: the finer the grain size, the better the signal-to-noise ratio of the hologram. As this type of hologram can sometimes be recorded on faster, coarser-grained materials, the noise level will increase considerably. The processing of in-line holograms is done along lines indicated by *Dunn* and *Walls* [7.7]:

(i) Predevelopment in a bath of 200 mg/ℓ Benzotriazole for 2-10 seconds.
(ii) Develop in concentrated Neofin Blue, for 1.5 minutes.
(iii) Wash and fix in Agfa G 334 for 2 minutes.
(iv) Before drying, soak the plate in a 90% solution of methanol.

The density level of the in-line amplitude holograms should be about 1.2 according to the paper.

As fixing is applied, the image will be affected, so that various aberrations and distortions may appear [7.9]. To minimize these aberrations, visible only on a microscopic scale, the fixing phase should be omitted. Instead, *desensitize* the plate in a phenosafranine ($C_{18}H_{15}ClN_4$) solution. This will produce a stable amplitude hologram with a higher image resolution than if it had been fixed:

Phenosafranine   300 mg
Methanol   0.5 ℓ
Distilled water   0.5 ℓ .
Treatment time about 3 minutes.

In another investigation, astigmatism was reported in phase holograms developed in undiluted Neofin Blue, which was not the case of holograms developed in Agfa GP 62 [7.10]. However, this was noticed after the holograms had been bleached. To really minimize aberrations at the reconstruction it is often necessary to use a liquid gate [7.11]. For holograms recorded with a Q-switched pulse, the SM-6 developer is an alternative to Neofin Blue.

7.2.2 In-Line Phase Holograms

As mentioned in the previous subsection, for most applications of in-line holography, amplitude holograms are quite satisfactory, mainly because of their better signal-to-noise ratio. However, if a phase hologram is really needed, bleaching should be done. *Dunn* and *Walls* [7.8] have also developed a method for obtaining in-line phase holograms. They employed the same developing technique as the one described for amplitude holograms. The developed density level required for holograms intended for bleaching is higher (>2), however. After predevelopment and the 1.5 minute development in concentrated Neofin Blue, the plate is washed for 2 minutes. No fixing is done and the following bleach is recommended:

Ferric nitrate	75 g
Phenosafranine	150 mg
Distilled water	1 ℓ.

Bleaching time: Until clear + 1 minute. Washing and drying as for an amplitude hologram.

Dunn and *Walls* claimed that this process gives rather noise-free reconstructions with a good edge definition, although contrast and signal-to-noise ratios are reduced when compared to an identical amplitude hologram. The diffraction efficiency is, of course, higher. They do not discuss their method's influence on the image resolution. In many applications of in-line holography, the image resolution is often the most important feature.

One investigation on the bubble-chamber holography resolution [7.9] shows that *fixing* or *bleaching* (using a solvent bleach) both resulted in a reduction of the image resolution of about 1.6, compared with a desensitized amplitude hologram. In both cases bulk material was removed from the emulsion, which affected the image quality. One may suspect that a fixation-free rehalogenating bleaching might have given better results. In general, the fixation-free rehalogenating technique is not working well at low spatial frequencies, as discussed in Chap.5, making it a bad alternative for in-line holograms. Tests performed using rehalogenating bleaching at high spatial frequencies (reflection phase holograms) where the technique works well, did show a slightly worse resolution compared to a hologram made using a solvent bleach. The reason for this was found in another investigation in which contact Denisyuk reflection holograms recorded on the Agfa 8E75 HD emulsion were made [7.12]. The surprising result is caused by the fact that in rehalogenated emulsions high scattering occurs, reducing the image contrast and thus affecting the resolution on a microscopic scale. Even if the emulsion thickness remains the same at both the recording and the reconstruction, this fact is not enough to guarantee that a higher resolution will be obtained. The diffusion process involved in the fixation-free rehalogenating bleaching is probably also a contributing factor affecting resolution. It may be interesting to know that the best image resolution in a phase in-line hologram recorded on the Agfa 8E56 HD plates was obtained with the direct bleaching technique (bromine-methanol), where no ion diffusion takes place during processing. This result was obtained in an investigation by *Fusek* et al. [7.13]. However, this bleaching technique can be used only after a fixing step, which, as we know, affects the image resolution. *Watson* and *Britton* [7.10] employed a rehaloganating PBQ-bleach (Agfa G 432) after the Agfa 8E56 HD plates had been developed in the GP 62 developer. As already mentioned, they found that this processing technique gave lower astigmatism than when the Neofin Blue developer was used in the same process. However, the Neofin Blue developer and the G 432 bleach are generally not a good combination; moreover, no comparison with desensitized amplitude holograms was made in the investigation.

It seems that at the present moment the desensitized colloidal development technique (without fixing) is the most superior processing method as regards the achievement of the highest possible image resolution. In the contact Denisyuk-hologram recording test [7.12] it was found that the very best resolution was obtained on Bulgarian HP-650 plates developed in the Russian GP-3 developer. The resolution achieved here was actually better than the image resolution obtained from a dichromated gelatin emulsion which requires drastic dehydration. The colloidal processing technique ensures accurate transfer of the intensity distribution in the emulsion during exposure to form concentration variations of ultra-fine spherical silver grains after processing. As mentioned earlier, the desensitized amplitude hologram offers the highest image resolution. The colloidal-silver hologram is the best choice for the combination of high-resolution/high-diffraction efficiency because of the phase-type modulation character of these holograms. Unfortunately, such holograms cannot easily be produced on Western commercial holographic materials.

Summary: In-line transmission hologram processing.
The highest possible quality (including image resolution) is achieved in an *in-line transmission* hologram on a glass substrate by
- using a recording material (glass plate) with the highest possible resolving power;
- presoaking the plate for emulsion stress relief;
- recording an amplitude hologram, using a high-contrast developer yielding an optical density of $1.2 \div 1.5$;
- desensitizing the emulsion (instead of fixing);
- carefully washing and slowly air-drying;
- replaying the hologram with the recording wavelength; and
- optionally, one may use a liquid gate at recording and/or reconstruction.

7.3 Processing of Off-Axis, Transmission Holograms

7.3.1 Off-Axis Amplitude Holograms

Off-axis amplitude holograms are both used in display holography and scientific/industrial applications. Amplitude holograms are characterized by a very favorable signal-to-noise ratio, which makes them suitable for a number of scientific applications. Because of a rather low diffraction efficiency, they are seldom employed in display applications, with the exception of plates used as masters for hologram copies. In display holography, only commercial materials with the highest resolving power (such as Agfa 8E75 HD/8E56 HD, Ilford SP696T/Hotec/SP695T) are recommended. Scientific and industrial applications sometimes require faster materials (such as Agfa 10E75 or Kodak 131 or SO-253) which have to be processed for highest possible sensitivity. Sometimes, e.g., for HNDT applications, there is also a time factor to contend with - the processing must be very fast. In this case a monobath developer can be helpful (Sect.4.3.4). The other extreme would be when we want to achieve highest possible image resolution for a scientific-industrial application (e.g., holographic inspection of nuclear power fuel elements [7.14]).

a) Off-Axis Transmission Holograms Recorded with CW Lasers

Processing techniques for high-quality holograms without any time constraints will be described here. Normally, the processing is performed on the highest-quality recording emulsions. Except for semi-physical development and some bleaching techniques, faster emulsions will in most cases produce acceptable results. In principle, there is no substantial difference between processing glass plates or film, or even holograms recorded with different wavelengths. It must be remembered, however, that a holographic emulsion coated on film substrates is often faster than the same emulsion coated on glass. A glass plate is normally the only choice if the highest possible image resolution is desired. In this case, desensitized amplitude holograms are preferred.

Chemical development of a high-resolution holographic emulsion is performed by exposing the material so that it receives its optimal exposure, which will result in an optical density of about 0.5 corresponding to an amplitude transmission of about 0.6 assuming also that the reference/object beam ratio (K-value) is between 5 and 10. Depending on the kind of recording material, Kodak D-19, Neofin Blue, or a special holographic developer is used.

(1) For amplitude holograms on Agfa 10E- and 8E-emulsions as well as on Ilford and Kodak materials, Kodak D-19 or similar holographic MQ developers give good results if followed normally by fixing or desensitization, when a minimum image distortion is desirable.

(2a) For phase holograms on Agfa 10E-materials, a tanning developer, such as pure pyrogallol, is preferred, since the stain will reduce the unavoidable scattering noise in the bleached emulsion. Only reversal bleaching should be applied here in order to avoid increased scattering due to grain growth during rehalogenating bleaching. In general, the 10E-material is not really the emulsion one would recommend in the first place for phase holograms. However, *Kumar* and *Singh* [7.15] claim that high-efficiency diffraction gratings can be obtained on the 10E75 material applying a potassium iodide and iodine bleach process.

(2b) A general, low-noise developer for transmission phase holograms on both Agfa 8E and Ilford materials is the *adurol developer* introduced by *Phillips* [7.16]. Overexposed materials developed in ascorbic acid show a tendency to image solution in rehalogenating bleaches. The clorohydroquinone overcomes this problem, which makes this developer a useful all-purpose developer for high-efficiency transmission phase holograms recorded with CW lasers and treated with fixation-free rehalogenating bleaching. Adurol and potassium bromide contained in this developer indicate that it is a depth developer, which means that it can also be used in the fixation-free rehalogenating bleaching process of reflection holograms. The developer is mixed in the following way:

Ascorbic acid	10 g
Clorohydroquinone (adurol)	2 g
Sodium sulfite (anhydrous)	30 g
Sodium metaborate (Kodalk)	10 g
Potassium bromide	5 g
Sodium carbonate (anhydrous)	60 g
Distilled water	1 ℓ .

Use at 21° to 23° C.

For the develop-bleach processing technique, *Phillips* [7.16] recommended specifically the following developers for Agfa and Ilford materials:

Solution A	Agfa	Ilford
Catechol	20 g	20 g
Hydroquinone	10 g	30 g
Sodium sulfite	60 g	60 g
Potassium bromide	-	10 g
Distilled water	1 ℓ	1 ℓ .
Solution B		
Sodium metaborate	-	20 g
Sodium carbonate	120 g	120 g
Distilled water	1 ℓ	1 ℓ .

Mix solutions A + B and use at 21° to 23° C.

The potassium bromide and sodium metaborate additives control the rapid induction tendency of Ilford materials.

As regards bleaching, the most popular formula is currently the ferric EDTA bleach for fixation-free rehalogenating bleaching. The PBQ-2 bleach is also frequently used. A PBQ-based bleach is often giving the highest diffraction efficiency, but is avoided because of its toxic properties. However, the present author recommends the safer-to-mix PBU-quinol bleach instead of the normal PBQ-2 bleach if a PBQ-type of bleach is really desired. The PBU-quinol is obtained in the following way:

Cupric bromide	1 g
Potassium persulfate	10 g
Citric acid	50 g (or 50 g sodium hydrogen sulfate)
Potassium bromide	20 g
Distilled water	1 ℓ .

Add 1 g hydroquinone (quinol) after the other constituents have been mixed.

This bleach must be mixed for at least 6 hours before use in order to form a sufficient amount of PBQ. Instead of the PBU-quinol bleach the PBU-amidol bleach is worth trying.

The CW-C2 developer - PBQ bleach combination is an alternative fixation-free rehalogenating technique for transmission holograms on Agfa 8E materials. When processing for the highest possible *signal-to-noise* ratio, a pyrogallol developer or the CW-C2 developer must be selected, followed by a reversal bleach, e.g., the pyrochrome bleach. This process will, however, result in holograms with a lower diffraction efficiency. The application of the reversal-bleach process also leads to image distortions, as already described in Sect. 7.2.2.

As regards phase holograms on Ilford materials, Ilford's chemistry works well and is particularly recommended, since their chemicals come in concentrated solutions to which only distilled water is added. In most cases, Hotec chemicals produce satisfactory results. However, sometimes, when manually processing the material in a tray, one is occasionally faced with some fog problems. Therefore, a developer formulated by *Phillips* [7.16] is worth trying. The rapid induction tendency of other holographic developers used for processing Ilford materials can cause splash marks during the development which get out of hand in the bleach. It should also be mentioned that *Phillips* [7.17] has formulated a ferric EDTA bleach particularly intended for the Ilford green emulsion which is mixed in the following way:

Ferric sulfate	12 g
Disodium EDTA	12 g
Potassium bromide	30 g
Sodium bisulfate	50 g
Distilled water	1 ℓ .

Bleaching time should be in excess of 6 minutes without agitation.

Finally, *phase* holograms on Kodak materials, such as the Type 120 or SO173 emulsions, are best processed with the holographic version of the Kodak SD-48 developer followed by the reversal pyrochrome bleach according to E. Wesly. The GP 431 ferric nitrate bleach is an alternative for fixed phase holograms. As regards display holography, it seems that Kodak materials are mainly used for the recording of Benton transmission rainbow holograms. Such holograms are best processed by development in D-19, fixing and bleaching in bromine/methanol bleach or, if more printout stability is required in a iodine/methanol bleach according to *Berkhout* [7.18]:

Iodide	5 g
Methyl alcohol	900 mℓ
Distilled water	100 mℓ .

After bleaching, the material is rinsed in ethyl alcohol/water (9:1) for 2 minutes.

Kodak gelatin emulsions can withstand the action of direct bleach solutions, which is why bromine and iodine methanol/water bleaches produce satisfactory results, meaning very clean and stable holograms explained by *Benton* [7.19]. The printout stability is of the greatest importance in white-light display transmission holograms, since from an aesthetic point of view even a slight darkening of a transmission hologram, which does not really affect the efficiency in any serious degree, may be disturbing to the observer.

Summary: Off-axis CW transmission hologram processing.

(I) The highest possible quality (including image resolution) in an *off-axis amplitude transmission* hologram recorded on a glass substrate, exposed with a *CW laser* is achieved by

- using a recording material with the highest possible resolving power;
- presoaking the plate for emulsion stress relief (and removing the swelling agent from the Ilford material);
- creating an amplitude hologram using a high contrast developer (e.g., Kodak D-19 or Adurol developer) with an optical density of $0.5 \div 0.7$;
- desensitizing the emulsion (instead of fixing);
- carefully washing and slowly air-drying;
- replaying the hologram with the recording wavelength; or
- optionally, one may use a liquid gate at the recording and/or the reconstruction.

(II) The highest possible quality in an *off-axis phase transmission* hologram on a glass substrate exposed with a *CW laser* is obtained by

(1) using a recording material with the highest possible resolving power;
(2) developing in a high contrast developer (e.g., the Adurol developer) to obtain an optical density of about 2;
(3a) bleaching for *highest signal-to-noise ratio* in a reversal bleach;
(3b) bleaching for *highest diffraction efficiency* in a fixation-free process using, Fe-EDTA or the PBU-type of bleach;
(4) carefully washing and slowly air-drying; and
(5) replaying the hologram with the recording wavelength.

Phase holograms recorded on Kodak 120-01/SO-173 are developed in D-19 and bleached in a direct iodine bleach.

b) Off-Axis Transmission Holograms Recorded with Pulsed Lasers

Amplitude holograms recorded on any fast holographic material (Kodak, Agfa 10E75) are best developed in D-19 to which 1.5 grams of phenidone per liter of D-19 are added. Instead of D-19, the Neofin Blue developer can be utilized.

Amplitude holograms recorded on high-resolution materials can be developed in SM-6, Neofin Blue, or D-19 to which 2 grams of phenidone per liter are added. Holograms intended for high-resolution imaging should not be fixed, but desensitized instead. If high image resolution is of little or no importance, the material can be fixed and washed in the usual way.

Phase holograms recorded on high-resolution materials from Agfa and Ilford should be developed in the SM-6 developer, followed by a fixing step and bleaching in a ferric-nitrate bleach (Phillips' ferric nitrate I). This will yield low-noise, high-quality holograms. By omitting the fixing step and using a rehalogenating bleach, higher efficiency can be accomplished, but the signal-to-noise ratio often decreases slightly.

Phase holograms recorded on fast holographic materials (e.g., Agfa 10E75) are best developed in the pyrogallol developer with phenidone (pyro-plus), followed by reversal bleaching (pyrochrome) in order to suppress scattering noise as much as possible.

Summary: Off-axis pulsed-transmission hologram processing.

(I) The highest possible quality (including image resolution) in an *off-axis amplitude transmission* hologram recorded on a glass substrate, exposed with a *pulsed laser* is achieved by
- using a recording material with the highest possible resolving power;
- presoaking the plate for emulsion-stress relief (and removing the swelling agent from Ilford materials);
- creating an amplitude hologram, using a high-contrast pulse developer (e.g., SM-6) giving an optical density of $0.5 \div 0.7$;
- desensitizing the emulsion (instead of fixing);
- carefully washing and slowly air-drying;
- replaying the hologram with the recording wavelength; or
- optionally, one may use a liquid gate at the reconstruction.

(II) The highest possible quality in an *off-axis phase transmission* hologram on a glass substrate, exposed with a *pulsed laser* is achieved by
(1) using a recording material with the highest possible resolving power;
(2) developing in a fast, high-contrast pulse developer (e.g., SM-6), giving an optical density of $2 \div 3$;
(3a) bleaching for the highest signal-to-noise ratio after fixing in a ferric nitrate bleach (*Phillips'* ferric nitrate I), or, without fixing, in a reversal bleach (pyrobleach);
(3b) bleaching for the highest diffraction efficiency in a fixation-free process, using the Fe-EDTA bleach or the PBU type of bleach;
(4) carefully washing and slowly air-drying; and
(5) replaying the hologram at the recording wavelength.

7.4 In-Situ Processing

In-situ processing means that the holographic material is processed at the exact place in which it was exposed. This is of importance for some real-time holographic interferometry investigations to guarantee that no repositioning error of the plate will be committed. In addition, in-situ processing is of interest only in the processing of transmission holograms. This technique was first proposed by *Bolstad* [7.20]. The plate is held in an overhead holder in its exposure-position during the processing. This was performed by elevating beakers containing the appropriate solutions around the plate, which made it possible to bring the liquid in contact with the emulsion. *Casler* and *Pruett* [7.21] soaked an unexposed 649-F plate in a D-19 developer, after which the plate, still saturated with the developer, was exposed. The reconstructed image could be viewed during the exposure which terminated when a satisfactory image was produced. The technique was further developed by *Biedermann* and *Molin* [7.22]. They used a liquid-gate plate holder. The holographic plates were exposed while immersed in the developer which was then replaced by a fixing solution for reconstruction and evaluation. Their article presents the recipes for a developer and a fixing solution with identical refractive indices and indicates ways of controlling emulsion swelling and of maintaining the recording conditions at the reconstruction. The following developer and fixing baths intended for the old version of Agfa materials are recommended:

Rapid developer:

Sodium sulfite (anhydrous)	140	g
Hydroquinone	30	g
Phenidone	0.5	g
Sodium metaborate	40	g
Sodium hydroxide	15	g
Potassium bromide	2.5	g
Wetting agent	1	mℓ
Distilled water	1	ℓ .

pH $\simeq$ 11.5, n = 1.3628

Stop bath: 1.5 % acetic acid (glacial) solution.

Fixing bath:

Sodium thiosulfate	225 g
Potassium pyrosulfite	15 g
Distilled water	1 ℓ .

n = 1.3627

Emulsion swelling is adjusted in the fixing solution by keeping the correct proportion of potassium pyrosulfite to sodium thiosulfate. Swelling increases with pyrosulfite and decreases with the thiosulfate concentrations:

Procedure:
1. Fill liquid gate with developer, wait 2.5 minutes.
2. Expose the plate.
3. Develop for 1-2 minutes under flow agitation.
4. Drain developer.
5. Rinse with stop bath for 15 seconds.
6. Fill gate with fixing solution.
7. After 15 seconds, the reference beam can reconstruct the hologram.

In holographic recording devices such as Holomatic [7.23] and Keystone [7.24] the in-situ processing is utilized, nowadays, to reduce the time between exposure and reconstruction. To further speed up the process, monobath developers are used. For example, the Eastman Kodak 448 monobath developer can be employed with the Keystone equipment. (Refer to Sec.4.3.4 for more on monobath developers for holography).

7.5 Processing of Reflection Holograms

Reflection holograms intended for white-light or laser-light reconstructions are more difficult to process than both the laser-reconstructed transmission holograms and the white-light rainbow transmission holograms. The main problem connected with the processing of the first named is to obtain a high diffraction efficiency combined with a low noise level. This problem has already been discussed in Chap.5 at quite some length, which is why in this section only some of the more promising processing procedures will be taken up. The scientific applications of reflection holograms can be found in, for example, HOEs and "piggy back" interferometric hologram investigations [7.25, 26]. In white-light display holography, three main types of reflection holograms are produced:

- Denisyuk reflection holograms, single-beam or with additional object lightening.
- Image-plane reflection holograms transferred from transmission masters.
- Reflection holograms, contact-copied from reflection masters.

The most important recent contribution to the field of hologram processing and, in particular, in the area of reflection-hologram processing was made by *Phillips* [7.27, 28] with a *three-step processing method*. The new scheme for the processing of reflection holograms makes it possible to produce holograms on Western materials with a noise level previously obtained only on ultra-fine-grained Russian materials. In the three-step method the material is exposed to regular light (white or UV) as part of the process. This is not an entirely new idea. Many holographers have already noticed that the quality of a reflection hologram will improve for a certain period of time immediately after bleaching if it is exposed to light. The reason for this is that the silver-halide grains will darken slowly due to the photolytic printout process taking place. This reduces the scatter in the emulsion and improves the signal-to-noise ratio. The diffraction efficiency is slightly affected, but the overall increase in the quality of the hologram compensates this small reduction in efficiency. Actually, some laboratories expose their reflection holograms to regularly strong white light or UV light after the processing to obtain a noise reduction. *Blyth* [7.29] described a method for introducing a shrinkage (combined with noise reduction) by exposing the hologram to white light after the normal bleaching step, and then redeveloping and bleaching the plate in a solvent bleach. In this process a developer of regular strength is used and the produced silver is then dissolved in the reversal bleach, causing shrinkage. The exposure to white light is carefully adjusted in such a way that only part of the silver-halide grains are dissolved, which will give the desired reduction in emulsion thickness. The noise reduction is unavoidably accompanied by a reduction in diffraction efficiency in this method. As mentioned in Chap.5, this technique was also applied to reduce the printout tendency of transmission holograms, described by *Hüttmann* [7.30].

In most holographic processing schemes *Phillips'* three-step process can be utilized. It works both for holograms treated in rehalogenating as well as in reversal bleaches. The first two steps consist in developing and bleaching the material the usual way. Originally *Phillips* recommended a rehalogenating bleach based on zinc chloride, because the silver chloride produced in the process is photolytically very sensitive and is also more suitable for colloidal development. In this process, the bleaching is only an intermediate step:

Disodium EDTA	20 g
Ferric sulfate	20 g
Zinc chloride	20 g
Sodium hydrogen sulfate	50 g
Distilled water	1 ℓ .

The new PBU-amidol bleach has also been tested for this process and was found to work very well.

After the development and bleaching the material is carefully washed and soaked in distilled water. A very weak colloidal developer (pH-3) mixed in the following way is used in the third step:

Ascorbic acid	10 g
Distilled water	1 ℓ .

The plate is immersed in this developer for 30 s after which a spotlight situated above the tray is switched on to illuminate the plate during the remaining part of the development. For the processing of a 20 by 25 cm^2 plate, a 50-W halogen spotlight, positioned at about one meter above the processing tray is sufficient. The developing time is about 5 minutes. At this stage the plate turns into a light, brownish-red color which is created by the colloidal silver. After the colloidal development, the plate must be washed and dried in the usual way. The hologram will show no printout tendency and the emulsion will have received a slight increase in thickness at the completion of this step.

The three-step process works well for all types of phase holograms, both transmission and reflection, but it is in the white-light reconstructed reflection holograms when a really dramatic difference in their quality can be noticed. The process works also equally well for both Agfa and Ilford materials.

7.5.1 Reflection Holograms Recorded with CW Lasers

a) Red Recordings

Only the processing of phase holograms will be discussed here, since the low efficiency of the reflection-amplitude holograms makes them usually of no practical interest. Almost all reflection holograms are recorded on high-resolution Agfa or Ilford materials. Kodak has not really any good alternative to offer here. There are two distinct ways of processing phase reflection holograms:

(i) The first method is applied when the *reconstruction must be done at the same wavelength as the recording wavelength*. Examples where this applies are HOEs and masters for contact copying. In this case the hardening or the tanning processing scheme will apply. A *cathecol developer followed by a rehalogenating bleach* of the PBQ type (PBU-quinol) is recommended. Cathecol developers CW-C1 or CW-C2 can be used. In Ilford materials there is the BIPS factor to be considered, which sometimes means that the material must be presoaked in water and dried before exposure. The use of the three-step processing technique is not really necessary here if the laser reconstructions to be made are in the red part of the spectrum only.

For Ilford materials *Phillips* [7.16] suggests the following processing scheme for a developer which controls the induction of the Ilford material:

Solution A

Pyrogallol	10 g
Potassium bromide	10 g
Distilled water	1 ℓ .

Solution B

Sodium metaborate	20 g
Sodium carbonate (anhydrous)	120 g
Distilled water	1 ℓ .

Mix solutions A + B and add 0.3 g/ℓ of benzotriazole to the working solution. Use at 21° to 23° C. The use of the new bleach bath, the PBU–amidol, is producing very good phase holograms on Ilford materials.

Saxby [7.31] recommended the following ascorbate-metol developer for reflection masters which are later treated in a rehalogenating bleach:

Ascorbic acid	20 g
Metol	5 g
Potassium bromide	1÷2 g
Sodium carbonate (anhydrous)	20 g
Sodium hydroxide	6.5 g
Distilled water	1 ℓ .

Development time: 12 minutes at 20° C or 6 minutes at 30° C.

(ii) The other way of processing phase-reflection holograms is to bring about *emulsion shrinkage*, by using a reversal bleach. This means that the holograms recorded in red will reconstruct in green, yellow or orange/yellow, depending on the holographer's taste and the degree of shrinkage. In this case the following processing technique is frequently applied:

• For Agfa 8E75 HD with little shrinkage, a pure pyrogallol developer, followed by the pyrochrome bleach, gives good results.

Saxby [7.31] has formulated a modified pyrogallol-metol developer for Agfa materials as a result of recent changes in the Agfa emulsions:

Solution A

Pyrogallol	15 g
Metol	5 g

| Potassium bromide | 1÷2 g |
| Distilled water | 1 ℓ. |

Solution B

| Sodium hydroxide | 6.5 g |
| Distilled water | 1 ℓ. |

Mix solutions A + B immediately before use, and apply once only.
Development time: 6 minutes at 20° C or 3 minutes at 30° C.

• For Agfa 8E75 HD with more shrinkage, the CW-C2 cathecol developer, followed by the pyrochrome bleach gives very good results.

As regards the Agfa 8E emulsions used for phase holograms with a high diffraction efficiency, an alternative processing method results from the CW-C2 developer, followed by a rehalogenating bleach (the PBU-type). In this case, the material must be preswelled in a TEA solution, where the concentration of TEA will determine the color of the finished hologram. *Phillips'* aldurol developer can also be applied for reflection holograms, which gives results similar to the CW-C2 developer, consisting of

Solution A

Catechol	20 g
Ascorbic acid	10 g
Sodium sulfite (anhydrous)	10 g
Urea	100 g
Distilled water	1 ℓ.

Solution B

| Sodium carbonate (anhydrous) | 60 g |
| Distilled water | 1 ℓ. |

Mix solutions A + B. Developing time: 2 min at 20°C (continuous agitation).

It is sometimes advantageous to apply the divided development technique described in Sect.4.3.2. One possible method is to *first soak the material in solution A of the developer* for about one minute and then immerse it in the *mixed solution* (A+B) for the development. A more uniform development throughout the depth of the emulsion is then obtained. Holodev 602 is an alternative two-part pyrogallol developer containing sulfite for color control and noise suppression. A divided development is, particularly, recommended when Holodev 602 is used since the duration of time in which the plate is soaked in solution A will determine the color to be obtained.

Pyrochrome reversal bleach stock solution:

Potassium dichromate	4 g
Sulfuric acid (conc.)	4 mℓ
Distilled water	1 ℓ.

Working solution: Use undiluted, or mix 1 part of the stock solution + 4 parts of distilled water. Bleaching time: 1 to 3 minutes.

Instead of using sulfuric acid the safer sodium hydrogen sulfate (15 g) can be considered. Wash the material in distilled water *both before and after* the bleaching to prevent chlorine (contained in tap water) or any other substance from interacting with the bleach bath, which produces a scum difficult to wash away. With a solvent bleach, it is advisable to bleach the plate *upside down*, i.e., with the emulsion facing the bottom of the tray but without the emulsion coming into physical contact with it.

For people who feel concerned to work with dichromate-based bleaches the new PSSB bleach bath (Sect.5.2.3b) is recommended as a substitution for the pyrochrome beach.

Alternatively, the reversal permanganate bleach KP-4 can be applied which gives similar results.

As an option, clear the material for 1 minute in

Sodium sulfite (anhydrous)	50 g
Sodium hydroxide	1 g
Distilled water	1 ℓ .

For Ilford Hotec materials, the Ilford chemistry gives good results. Here, orange or yellow holograms are obtained. If more shrinkage is required, the Hotec bleach can be replaced by a reversal bleach, such as the pyrochrome bleach. An alternative here is the chemistry recommended by *Phillips* [7.17]. The BIPS factor of the Ilford materials must not be overlooked as it will make a larger color shift than when the same process is applied to Agfa materials.

The third colloidal developing step of the three-step processing technique can always be incorporated into any of the above-mentioned processing schemes.

b) Green Recordings

Almost all reflection holograms are made on high-resolution Agfa or Ilford materials. The green emulsion from Ilford is both more sensitive and less scattering than the Agfa 8E56 HD. As regards phase hologram processing there is really only one alternative, unless blue reconstructions are of interest. Normally, the reconstruction of the hologram should be made at the same wavelength as the recording (or even slightly longer, if possible). Examples here are HOEs and masters for contact copying where the hardening processing scheme will apply: a *hardening developer, followed by a rehalogenating bleach*. (A PBQ bleach, such as the PBU-quinol is the standard way to go about here). Cathecol developers CW-C1 or CW-C2 can be used – the former has a slightly higher hardening effect on the emulsion. *Heaton* and *Solymar* [7.32] used CW-C1 and PBQ-2 for processing HOEs recorded on the 8E56 HD emulsion. High-quality reflection HOEs were obtained with this processing method and Agfa materials. The results were in good agreement with the theoretical model of HOEs recorded in silver-halide emulsions.

The Ilford emulsion is definitely recommended for green and blue recordings despite the BIPS factor, which sometimes means that the material must be presoaked in water and dried before exposure. After the soaking, the sensitivity of the emulsion increases and its stresses are released. The

290

three-step processing treatment can be performed here if very low-noise reconstructions are of interest. Isopropyl alcohol treatment, applied after processing but before drying, can actually increase the emulsion thickness, as shown by *Cooke* and *Ward* [7.33]. The material is first soaked in a 50% propanol bath for one minute and then in a 100% propanol bath for two minutes. This increases the emulsion thickness about 4.3%, which result was obtained on the Agfa 8E56 HD emulsion in the CW-C2 - PBQ-2 processing regime.

Summary: CW reflection phase hologram processing.

The highest possible quality of a *phase-reflection* hologram recorded on a glass substrate and exposed with a *CW laser* is achieved by

(1) using a recording material with the highest possible resolving power: a) to obtain shrinkage or b) shrinkage-free processing. An option for the b-procedure below and the red Agfa recordings: preswell the emulsion in TEA solution and dry before exposure in order to obtain a wavelength shift. Ilford materials are already preswelled by 8.8%,

(2a) developing in a developer with or without sulfite (e.g., CW-C2) giving an optical density of about 3,

(3a) bleaching for the *highest signal-to-noise ratio* in a reversal bleach,

(2b) developing in a sulfite-free developer (e.g., CW-C2), giving a density of <3,

(3b) bleaching for the *highest diffraction efficiency* in a fixation-free process, using the PBU-type bleach,

(4) both processes: redeveloping the plate under white light illumination (*Phillips'* third step), and

(5) carefully washing and slowly air-drying.

7.5.2 Reflection Holograms Recorded with Pulsed Lasers

a) Red Recordings

In general, when using a short exposure time (less than about 1 ms) for the recording of holograms the processing schemes presented in the following sections are applicable. When using the red wavelength from a ruby laser for recording display holograms (Denisyuk or image-plane holograms) a green, yellow or orange/yellow color is often suitable. The following processing scheme which is a modification of *Phillips'* three-step process gives good results on both Agfa and Ilford materials. Since it demands the use of a reversal bleach, the color can be controlled by exposure. Higher exposure results in more shrinkage and a larger color shift in the final hologram. Note that when using Ilford materials the BIPS factor will cause a larger color shift, compared to Agfa materials exposed to the same prebleached density.

If the preswelling technique (using, e.g., TEA) for color control is applied, or if a minimal color shift is desired when using the ruby laser, then the processing scheme for the green pulsed laser can be followed:

Give the hologram an exposure of about 50 μJ/cm^2 or more, depending on the desired color shift.

(i) Develop in SM-6 at 22÷22° C for 2 minutes

Ascorbic acid	18 g
Phenidone	6 g
Sodium hydroxide	12 g
Sodium phosphate dibasic	28.4 g
Distilled water	1 ℓ.

(ii) Bleach in pyrochrome bleach

Potassium dichromate	4 g
Sulfuric acid (conc.)	4 mℓ
Distilled water	1 ℓ.

(iii) Develop in a colloidal developer under the illumination from a white halogen light.

Ascorbic acid	10 g
Distilled water	1 ℓ.

Develop 5 min or more.

The light should be switched on 30 s after the plate was placed in the developer.

b) Green Recordings

The following processing scheme is formulated for holograms recorded with short green pulses. Give the hologram an exposure of about 50 μJ/cm^2:

- Develop in SM-6 at 22÷22° C for 2 minutes
- Bleach in PBU-amidol bleach
- Develop in a colloidal developer (Phillips' third step) under illumination from white halogen light (5 minutes).

The light should be switched on 30 s after the plate has been put in the developer.

Summary: Pulsed reflection-phase hologram processing.

The highest possible quality of a *phase-reflection* hologram recorded on a glass substrate and exposed with a *pulsed laser* is achieved by

(1) using a recording material with the highest possible resolving power: a) to obtain shrinkage or b) shrinkage-free processing. An option for the *b* procedure below and the red Agfa recordings: preswell the emulsion in TEA solution and dry before exposure in order to obtain a wavelength shift. Ilford materials are already preswelled by 8.8%,

(2) developing in a nontanning pulse developer (e.g., the SM-6 developer), giving an optical density of about 3÷4,

(3a) bleaching for the *highest signal-to-noise ratio* in a reversal bleach (pyrochrome bleach),

(3b) bleaching for the *highest diffraction efficiency* without emulsion shrinkage in a fixation-free process, using PBU-amidol bleach,

(4) both processes: redeveloping the plate under white light illumination (*Phillips'* third step), and

(5) carefully washing and slowly air-drying.

For more detailed information on the developers and the bleach baths mentioned in this chapter, the reader is referred to the previous chapters.

7.6 Processing of Holograms Produced by Scanning

The scanning technique has become popular in hologram copying. The scanning method is described in Chap.8. As regards the processing of such holograms the best results have been obtained on Ilford red materials, processed with the help of the new chemistry from Ilford. Holograms copied with an HeNe (or krypton) laser beam will reconstruct as bright orange images, pleasant to the eye thanks to the BIPS factor, provided the fixation-free rehalogenating bleaching technique is applied. The Phillips' third step will increase the image quality of such copies but it also means an additional bath. The whole procedure will thus become more complicated and the scanned copied holograms more expensive in this otherwise inexpensive mass-production method. An alternative to the Ilford bleach is the PBU-amidol rehalogenating bleach which gives very low scattering without the necessity of using the third step. Ilford chemicals in a tray can occasionally cause problems due to the short incubation time. Therefore, sometimes, different processing chemistry may be required according to *Phillips* [7.17].

7.7 Secondary Effects in Hologram Processing

The final quality of a silver-halide hologram is a result of a great number of factors interrelated. Therefore, using identical recording materials and processing schemes on two different occasions is no guarantee of identical results. The following will contribute to the final quality of the product:

• Quality variations of the recording material from batch to batch.
• Variations in temperature, humidity and safelight in the laboratory.
• Processing variations, such as temperature and freshness of the solutions, pH, water quality used for the mixing solutions, processing time, agitation, washing in-between active solutions, final wash, type and amount of the wetting agent used, drying procedure, etc.

It is obvious that it is rather difficult to always maintain perfect control of all these factors. *Boone* [7.34, 35] has investigated the influence on the reflection hologram quality caused by some of these factors. The results are briefly summarized in Table 7.1.

In addition to the results presented in the table, *Boone* has also investigated the influence of the following sources: emulsion batches, safelight illumination, prebleach rinse water quality, and machine processing.

Table 7.1. Processing factors affecting the hologram quality. (D: density, λ: wavelength, $\Delta\lambda$: bandwidth, η: diffraction eficiency)

Solution	Quantity	Temperature	Freshness
Developer (MA type)	*Small -* lower D	*Low (20° C) -* narrow $\Delta\lambda$ high η *High (30° C) -* narrow $\Delta\lambda$	*New -* shorter replay λ higher η (const. D $\rightarrow$ time = t)
	Large - higher D	higher η *Very high -* broad $\Delta\lambda$ low η	*Old -* longer replay λ lower η (const. D $\rightarrow$ time > t)
Bleach (R-9 type)			*New -* shorter replay λ narrow $\Delta\lambda$
			Old - longer replay λ large $\Delta\lambda$

7.8 Final Remarks

Using the processing schemes described in this chapter, holograms of good quality can be obtained on a regular basis. In general, it often takes some practice to achieve these results even if one strictly follows the instructions and recipes. However, if the hologram quality is not satisfactory for a particular application when using any of the processes described here, some modifications may be necessary. Information provided in previous chapters may assist the reader in what directions to go in order to achieve the desired improvements.

8. After-Treatment

In this chapter some of the processing steps that are usually applied after the material has been fixed or bleached will be discussed. These steps are often important if the best possible quality of the final hologram is to be achieved. In addition, some other items concerning holography will be discussed here, e.g., hologram copying and archival properties of holograms.

8.1 Additonal Treatment After the Main Processing

8.1.1 Washing and Drying Holograms

As regards the final wash of the processed holograms the following considerations should be taken into account. It is necessary to remove the chemicals introduced during the previous processing steps from the emulsion if a long-term stability of the hologram is desired. Residual chemicals in the layer can also cause additional scattering and should, for that reason, be removed as well. A rinse in filtered tapwater is normally sufficient, provided that the material is soaked in distilled water before drying. For a bleached hologram an addition of a small quantity of acetic acid to the bath can reduce the printout effect, as recommended by *Phillips* [8.1]. If the emulsion contains residual, sensitizing dyes which are difficult to remove in water, *Coblitz* and *Carney* [8.2] recommend soaking the hologram in a 75% ethyl alcohol - 25% water solution as part of the final washing procedure.

A water rinse should last for at least 10 minutes, and for more important holograms even longer, with a guaranteed water flow across the emulsion. The temperature should be kept constant on the same level as the one used during the whole process in previous baths. A temperature of 20° C is often recommended. Cold water taken directly from the tap in the laboratory is not recommended, as it prolongs the rinsing time and can affect the gelatin, in particular if the material is suddenly introduced into a cold-wash solution.

When introduced into water and processing solutions, gelatin has a tendency to swell. This presents a problem in holography since for many holographic applications it is important that the emulsion thickness remains constant at both recording and reconstruction. In Chap.2 some of these problems were discussed. In Sect.8.1.2 ways to compensate for emulsion shrinkage due to processing will be treated. *Green* and *Levenson* [8.3,4] investigated the swelling of thin layers of gelatin in water. They found that the swelling is largest in the beginning and decreases as the equilibrium level is approached. They also formulated a mathematical expression to describe gelatin swelling, namely

$$\frac{S}{Z - S} = kt^n \, , \tag{8.1}$$

where S is the swell at times t, Z is the equilibrium swell at saturation, and k and n are constants. For various gelatin layers n may range from 0.5 to 2.5 and n, if greater than unity, implies the initial phase of acceleration in the rate of swelling. *Biedermann* and *Molin* [8.5] studied emulsion thickness variation for a holographic silver-halide material throughout the complete processing sequence.

After a thorough wash, the holographic material is soaked in a tray with distilled water to which a wetting agent is added. Normally, this lasts for one to two minutes, after which the material is quickly taken out of the bath so that no dust particles get stuck to the emulsion surface. This final washing solution should be frequently changed to avoid accumulation of dust particles in it. Particles remaining on the surface of the emulsion will attract water and will dry later than the area around, which can cause drying marks, particularly disturbing in reflection holograms. These small spots are slightly thicker than the surrounding areas and will therefore re-construct images in different colors. A good method of reducing the possi-bility of such marks is to shower the emulsion with clean distilled water (with a wetting agent added to it) from a bottle, holding the hologram in an almost vertical position above the tray. After that the holographic material can be hung to slowly dry in a dust-free drying cabinet. For this purpose, warm air as well as blowing air can be used, provided the air stream con-tains no dust particles that might get stuck to the emulsion. Some hologra-phers prefer to squeegee the holographic plate or film before drying. The advantage of such a procedure is that the plate will dry faster and that any dust particles that might be there will be removed from the emulsion sur-face. However, some practice is needed here since it is easy to leave streaks or even to damage the emulsion. Straight, high-quality car windshield wiper blades mounted onto some rigid holder are suitable for this purpose. These blades are made of rubber or polyurethane and can be used one at a time or, for optimum efficiency, in pairs. The application of squeegees in pro-cessing machines was described by *Perkins* [8.6] and in another publication, *Edgcomb* and *Zankowski* [8.7] discuss squeegee blades, too.

A more sophisticated procedure is letting the hologram move past an air knife which will squeegee the photographic material without touching the emulsion. Air knives are slotted tubes that direct a stream of oil- and dust-free air against one or both sides of the material to strip the water and surface particles from it.

The use of special photographic dryers can be advantageous for a ho-lographic film. Such dryers must be carefully tested before actually using them for holograms since the holographic emulsion is much more sensitive to, e.g., roller marks, than the conventional photographic film is. Reflection holograms are particularly difficult to dry without any marks. Holograms processed in a machine normally pass through a drying section as the final step.

If a hologram has a drying mark, it can be rewashed and sometimes the mark will disappear after being dried the second time. However, sometimes the gelatin is affected, e.g., by local hardening around the mark or else the particle could have damaged the gelatin layer, in such a way that the mark cannot be removed by rewashing even though the dust particle causing the mark in the first place has been removed. Normally, the sooner the hologram is rewashed the more likely it is that it will be recovered. Therefore, one should carefully inspect the wet emulsion layer and, if some dust particles are visible, rewash it immediately before it dries. Occasionally one may encounter dust particles which got embedded in the gelatin during manufacturing. Such particles (if they are small) do not usually cause any problems since they were already in the emulsion during the recording of the hologram. Only particles added after the exposure (during processing) can cause problems.

Drying a hologram constitutes a critical moment in hologram after-treatment. In many cases the emulsion is supposed to return to its original thickness (thickness during the exposure) when the hologram is dry. Emulsion stress relief by, e.g., presoaking the unexposed material in water can help in this respect, as described in more detail in Chap.2. The final rinse procedure can also be helpful, as explained by *Hariharan* [8.8]. He studied the way in which an alcohol bath may influence a hologram's diffraction efficiency if applied before drying and found that bleached phase holograms processed with the tanning *rehalogenating* technique showed an improved efficiency after such treatment (one ethyl alcohol bath and two isopropyl alcohol baths) prior to drying. As regards bleached phase holograms processed with the tanning *reversal* technique, a dramatic drop in efficiency was observed if an alcohol bath was used. In *Hariharan*'s opinion the hardening effect of gelatin is the contributing factor here. Alcohol treatment will create a second volume phase grating in the emulsion which works either in phase or out of phase with the ordinary phase grating. The second phase grating is similar to the SHG technique described in Chap.6. In the first case, maximum hardening occurs at the same sites at which silver bromide has been deposited by the bleach and the resultant refractive-index modulation is at its maximum. In the second case, using the reversal bleach, the silver-halide content of the emulsion layer after processing is at its minimum at the points at which maximum hardening occurs, which means that the net refractive-index modulation will be at its minimum.

8.1.2 Correction of Emulsion Thickness

The TriEthanolAmine (TEA) shrinkage-compensation method presented by *Denisyuk* and *Protas* [8.9] has been recommended by *Lin* and *LoBianco* [8.10] for color holography for the purpose of compensating for the shrinkage introduced by fixing a holographic emulsion. Fixing causes 15% to 20% of the emulsion shrinkage. The exact amount depends on the optical density of the developed hologram. *Lin* and *LoBianco* claimed that for a density of

2, a 6% TEA-solution treatment will compensate for the shrinkage in a fixed Kodak 649-F emulsion. This method works well when applied to amplitude holograms, but is not recommended for bleached holograms due to the fact that TEA sensitizes the silver-halide grains and increases the printout tendency. *Nishida* [8.11] studied the long-term effect (over 100 days) of TEA-treated amplitude holograms. He found that the diffraction efficiency remains unchanged provided that the humidity is kept constant. *Hariharan* [8.12] recommended *sorbitol* as an alternative to TEA for bleached holograms. A method of permanently controlling the thickness of a processed emulsion was presented by *Young* [8.13]. By imbibing aqueous solutions of N-Vinyl-2-Pyrrolidone (N.V.P.) into the emulsion layer, which is polymerized in situ by ultraviolet light, the emulsion thickness can be accurately and permanently adjusted. The material is soaked in the N.V.P. solution of a certain concentration for 30 minutes. The plate is then removed from the bath and the excess solution is swashed from the surface with acetone. A 125-W mercury lamp is used for the polymerization and drying of the emulsion. At a distance of 10 cm between the lamp and the material this process takes one hour. It was found that a N.V.P. concentration of 15% restored the original thickness of a fixed Kodak 694-F emulsion.

If the shrinkage is not compensated for, the diffraction efficiency of the hologram can decrease, which has been pointed out in many publications, for example, by *Dzyubenko* et al. [8.14,15]. However, as shown by *Kusakabe* et al. [8.16] it is also possible to increase the diffraction efficiency of finished reflection holograms by baking them. The material is repeatedly baked (in air) at a temperature of 130° C for periods of 10 minutes each time. According to *Kusakabe* et al. the increased diffraction efficiency is caused by an actual increase of the refractive index modulation in the emulsion after the baking process and is not caused by emulsion shrinkage. The effect was demonstrated with the Agfa 8E75 HD material developed in D-19 and bleached in a PBQ bleach (Agfa GP 432). Increased diffraction efficiency was also obtained by *Guo* and *Cai* [8.17] via a post-processing heat treatment of amplitude holograms which had been developed and fixed. A Chinese holographic silver-halide material (TJ-1) was baked at a temperature of 400° C until the gray or black color in the hologram becomes orange-yellow and more transparent. The increased diffraction efficiency is due to the splitting of silver particles in the emulsion during the heat treatment which transforms the amplitude hologram into a refractive-index-modulated phase hologram according to *Guo* and *Cai*.

Recently, *Wuest* and *Lakes* [8.18] utilized humidity to control the emulsion thickness. Even when using "nonshrinkage" processing a small emulsion-thickness change may take place after processing. By controlling the ambient humidity, both at the recording and the reconstruction of the hologram, it is possible to compensate for the small shrinkage of the emulsion due to the processing. The effect, which, of course, depends on the material, the developer, and bleach used, is in the order of 1-nm wavelength shift per one percent of humidity variation.

Serov et al. [8.19] described a technique allowing for drastic changes of the emulsion thickness between recording and reconstruction. With this technique, it is possible to change a reflection hologram to a transmission hologram, and vice versa. To convert a reflection hologram to a transmission type, glycerin swelling is applied. To perform the opposite transition, the hologram is exposed in a swollen condition (using distilled water). After processing and drying, the hologram behaves like a reflection hologram. As regards possible applications of this technique, the UV or IR reconstruction of holograms recorded in visible light was mentioned.

8.1.3 Protection of Transmission Holograms

When a transmission hologram has been dried, it can be reconstructed directly without the necessity of any further treatment. However, to reduce noise caused by bleaching-induced emulsion reticulation, reconstruction in a liquid gate containing an index-matching liquid may be considered. To avoid the liquid gate, *Buschmann* [8.20] recommended to coat the emulsion using a plastic spray (Kontakt Chemie 70) as the first step and then applying a light protecting lacquer from Agfa (Agfacolor Lichtschutzlack). After that the emulsion is treated in an 18.8% solution of cyclo-caoutchove in xylol and, finally, in a solution of 5% gelatin and 0.3% Formalin (30%).

Many of the bleaching processes will produce low-noise holograms for which no index matching would be necessary. However, to protect the emulsion it can be advisable to laminate a film hologram with a transparent, protective plastic layer, e.g. MACtac PB2113 [8.21]. A glass hologram can be protected by a clean glass plate cemented to the emulsion of the hologram, using optical cement or epoxy. *Berkhout* [8.22] who has a long experience in sealing rainbow transmission holograms recommended the following technique for this purpose: When the plate is still wet, 5 to 6 mm of the emulsion is completely scraped off around the edges. This can be done in the wash between development and bleaching. The purpose here is to prevent moisture from penetrating through the thin gelatin layer around the edges when it is sealed to a cover glass. The dry hologram is then laminated to a clean glass cover with a two-component epoxy, Epo-Tek 310 from Epoxy Technology, Inc. [8.23]. Epoxy takes 36 hours to set. It is important when coating the emulsion that the glue contains no air bubbles. Bubbles appearing from mixing the two components of the epoxy, can be removed before coating the hologram by using a vacuum or a centrifuge. *Berkhout* mentioned that using another product for sealing holograms, such as Norland UV cured optical adhesive, for example, causes considerable printout problems over time.

Phillips and *Porter* [8.24] presented a rather interesting way of protecting transmission holograms. A holographic plate of the same size as the hologram to be protected was fixed and used as a cover. This was done by attaching the still wet hologram to the wet gelatin emulsion of the protecting plate. These were allowed to dry together (at 40° C) which meant that they became sealed. The sealing was enhanced by applying cyanoacrylate around the edges of the plates. In addition to protecting the emulsion, this procedure also serves as an index-matching technique which eliminates any emulsion surface reticulation that could otherwise cause scattering noise in the hologram. A similar technique for reflection holograms was published by *Janowska* and *Szydlowska* [8.25]. Their method provides also a technique for emulsion shrinkage control. The emulsion is soaked in a solution of Canada balsam and ethyl alcohol in volume ratio 1:1. After that the emulsion is covered with a glass plate. The layer of Canada balsam fills the irregularities of the surface and thereby reduces the noise caused by emulsion surface irregularities.

8.1.4 Blackening of Reflection Holograms

Often it is desirable to blacken the backside of the ready-made phase reflection hologram. It is usually the emulsion side of the hologram. The most common way of doing this is to cover it with black paint. There are various spray paints that can be used for this purpose, but caution must be exercised when choosing a particular paint for a given material as some paints will react with the emulsion. And thus, a paint that will work with, for example, Agfa materials will not necessarily work with other materials. A flat black spray paint from Krylon, produced in the USA, works very well with Agfa emulsions. *Hariharan* [8.26] recommended a black PVC protective coating paint with a nearly emulsion-matching refractive index manufactured in England by DCMC Industrial Aerosols Ltd. The long-term effect of using paint on holographic emulsions is not completely clear. The safest to use are lacquers which contain chemical components that cannot penetrate the gelatin or produce peroxides. Suitable solvents here are hydrocarbons and halogenated hydrocarbons. Peroxides can be formed from oil-based paints, which is why such lacquers should not be used for emulsion protection. Instead of using spray paint, one can employ the silk screen coating technique (acrylic screen printing ink AMP-710 [8.27]) which gives a very uniform and thick protective layer.

Yet another way of coating the emulsion has been suggested by *Blyth* [8.28]. It is a chemical technique used for blackening of the back of the emulsion by immersing the dry plate, after normal processing in a bath of 0.5 g/ℓ sodium borohydride ($NaBH_4$), in a mixture of water and methanol. The bath acts as a reducing agent on both the exposed and the unexposed silver-halide crystals, creating a colloidal silver layer in the upper part of the emulsion mainly. The plate becomes reddish after the treatment and the ratio of signal-to-noise level is superior to that of any spray paint, according to *Blyth*. The following recipe is used for preparing this bath:

Sodium borohydride	0.5	g
Methanol	650	mℓ
Distilled water	350	mℓ .

300

Lamination of a clear reflection hologram in order to support the film material by, e.g., black plexiglass with a clear adhesive, such as MACtac PB2113, has been recommended by Ilford. This is a convenient way of combining blackening and mounting of film holograms. For the production of large quantities of film holograms, soft lamination with, e.g. MACtac MACal 9800 series or B2978, is often a convenient way of blackening the flexible substrate. An alterative laminating material is the Holopaque film from Imagys International, Ltd. [8.29]. The company has also a new, clear, non-birefringent material, Holoclear, which can be used to laminate film to glass plates even before the recording takes place.

8.1.5 Noise Suppression

Noise suppression refers to the ways of reducing the noise level in a processed hologram (transmission or reflection) by using dye staining techniques. One method by *Cantos* [8.30] is to use a 0.1% (1 gram per liter) nigrosin solution in which the hologram can be treated for two minutes. Nigrosin is a biological dye which stains some of the organic materials black. Here, nigrosin molecules get attached to the gelatin, absorbing the light scattered by the silver-halide crystals and thereby reducing the noise level.

Another dye for microscopy and optical filters that can be used for noise suppression in reflection holograms, is Orange II from Merck. This dye is suitable for holographic images in the yellow or the orange-red parts of the spectrum. The dye can be added to the final wash-bath of distilled water-wetting agent. When a hologram treated in this way is illuminated with white light, the blue scattered part of the spectrum is reduced, which ensures better contrast. Red or green dyes can be used in a similar way if the reconstructed hologram is to reproduce the image in the red or the green color. However, the three-step processing technique presented in Chap. 7 makes the noise suppression methods mentioned here less important.

8.1.6 Protection of Reflection Holograms

Reflection holograms on glass can be protected in the same way as transmission holograms, which was discussed in Sect. 8.1.3. In addition, some other methods can also be used. Black backing can be combined with laminating a black-painted glass plate onto the emulsion side of the hologram. The index-matching laminating adhesive makes it possible to avoid contact between the emulsion and the lacquer, since the other side of the emulsion of the protecting glass can carry the black paint. Using optical epoxy eliminates color changes caused by humidity variations and is recommended for holograms exhibited in hot and high-humidity environment. Another method, not as efficient as the previous one, is to seal the black coated hologram with a glass plate that is slightly larger than the hologram (extending about 4 to 5mm around the edges of the hologram). A string of

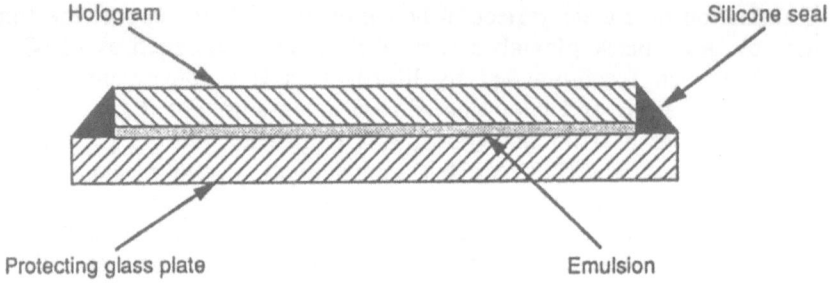

Hologram

Silicone seal

Protecting glass plate

Emulsion

Fig.8.1. Sealing a hologram using a silicone sealant. The black-coated emulsion of the hologram is sealed with a glass plate that is slightly larger than the holographic plate (extending about 4 to 5mm around the edges of the hologram). A string of silicone sealant is applied on the extending parts of the protecting glass plate as well as on the glass sides of the hologram

silicone sealant is applied on the extending parts of the protecting glass plate extending beyond the holographic plate as well as on the glass sides of the hologram (Fig.8.1). Silicone will not give a one hundred percent protection to the emulsion against humidity variations, but it will considerably delay emulsion changes. The advantage here is that this seal is not permanent and the protecting plate can be removed if necessary.

Reflection holograms on film cannot be protected against humidity variations since the substrate itself is hygroscopic. The only way to provide protection would be to put the film between two pieces of glass and seal them together. However, it is often sufficient to laminate the hologram onto a piece of black plexiglass, using a transparent index-matching adhesive. To protect the substrate surface of the hologram a transparent protective laminating film can be attached to it as it is more scratch resistant than the holographic film material itself. However, lamination is not recommended for reflection holograms which are supposed to last very long. It has happened that laminated film copies illuminated for extended periods of time have developed colored crack patterns over the entire surface (Fig.8.4c). This has occured with Ilford film laminated to black plexiglass. If, instead, the film was laminated to black lexan, these problems have not been observed, according to L. Lieberman.

The *Kiraly* method [8.31] of embedding Ilford Cibachrome photographic prints for archival protection could also be an alternative for film hologram protection. In this method, the photographic emulsion is surrounded by an acrylic microcrystal molecular layer. A partly polymerized acrylic resin is used by adding a catalyst of applying UV light to activate the polymerization. This forms a colorless, nearly perfect gas barrier which is impermeable by water and also acts as a filter for UV radiation. Actual tests of photographs embedded in this manner show no visible deterioration after being exposed to harsh weather conditions for over three years.

8.2 Preparation of Processing Solutions

The processing of holographic materials must be carefully controlled if the desired results are to be obtained. The following section provides the reader with some general rules and various practical hints that may help the reader to avoid some of the mistakes and common pitfalls awaiting an unexperienced person when working with the preparation of processing solutions for holography.

For mixing processing baths distilled or deionized water should be used. Recipes for baths are usually published with the chemicals listed in the order in which they are to be dissolved. In this book the recipes are listed with their chemical constituents in groups to which they belong, so that, for example, recipes for developers start with all the developing agents grouped together.

As regards the dissolution of chemicals, it is advisable to start dissolving sodium sulfite before developing agents are dissolved. There is one exception to this rule and that concerns metol which must be dissolved before sulfite is added. Phenidone is best dissolved after the alkali compound has been added.

The general rule for mixing solutions is that each chemical should be dissolved completely before the next one is added to the solution. Most processing baths can be mixed at a temperature of about 50° C. One starts with a volume of water which is equal to three-quarters of the final volume. When all constituents have been mixed well, cold water is added to make the required volume (often 1 liter). It is recommended to take advantage of a stirring apparatus for the mixing of solutions. A magnetic stirring bar is put at the bottom of the beaker which is then filled with water. Since the top of the stirring apparatus can be heated, it is easy to perform mixing, using this type of equipment.

When weighing chemicals, scales or balances adapted to the quantities required should be used. A new, clean sheet of paper should be placed on the scale pan for each chemical and the paper discarded after the chemical is weighed. With the exception of very unstable solutions, it is a good practice to mix a processing bath the day before it is to be used so that it achieves a uniform constituency. Sometimes a solution has to be filtered before it can be used. The necessity of labelling the containers with chemical solutions very clearly and immediately after they are bottled, cannot be stated too emphatically.

In this book the metric system is utilized for indicating weights and volumes. Temperatures are given in the Celsius grade system only. Chapter 10 provides tables and formulas for converting metric and Celsius units to corresponding avoirdupois, U.S. liquid, and Fahrenheit.

In addition to scales, measuring graduates, and a stirring table, it is also necessary to have access to a very accurate thermometer and a pH-meter. Conventional dark-room equipment, such as sinks with filtered tap water, trays, safelights, etc. are not treated in this monograph. The reader is referred to books on photography describing darkroom equipment.

8.3 Safety Aspects of Hologram Production

The hazards of working in a holographic laboratory can hardly be exaggerated and must be kept in mind at all times. The use of powerful lasers is one potential area of danger: another one results from the chemicals used for the processing of holograms. Some of them are highly toxic, e.g., *para*-benzoquinone, bromine vapor, ammonium dichromate, and mercuric chloride. If a person is regularly exposed to these products over extended periods of time, he/she must take certain precautions to eliminate the risks involved with their handling. And thus a holographic laboratory must comply with the standard conditions required for a normal chemical laboratory, which means that the processing facility should be equipped with sufficient ventilation and, in addition, a fume hood for handling the more toxic chemicals. A safety shower, an eye-wash bottle, a fire extinguisher, personal protecting devices, such as safety goggles, rubber gloves, respirators, etc., belong to the standard equipment that must be provided. In addition, large-scale holographic production must also conform with the waste regulations concerning disposal of used processing solutions.

Holographic film and plates should be fastened in special holders so that the operator does not have to put his hands into the processing solutions. If that is not possible, he must wear protective gloves. Figure 8.2 shows a holder and a developing tank for four 30×40 cm² plates. The tank contains about 12 liters of a processing solution.

More information on the safety hazards connected with various chemicals can be found in reference books, such as *The Hazardous Chemicals*

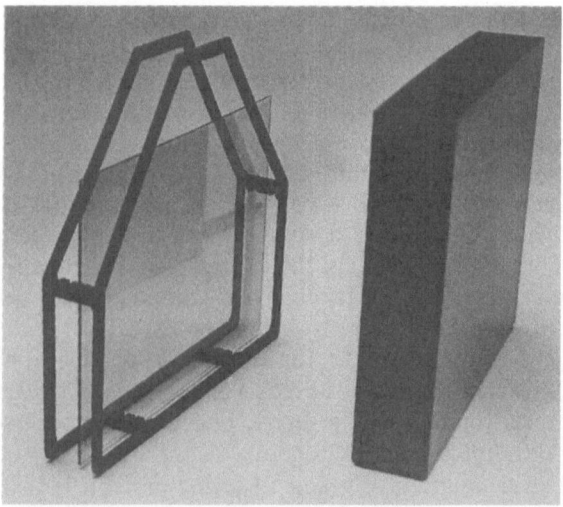

Fig.8.2. Developing system for 30×40 cm² plates. Plate holder and a developing tank for four glass plates. The tank contains about 12 liters of a processing solution. (Holovision AB, Stockholm, Sweden)

Data Book [8.32]. One can also approach the companies from which the chemicals are obtained for safety data sheets for their products. In addition, the reader may want to refer to *Cheung* [8.33] and *Crenshaw* [8.34], for information on the general safety aspects of holographic processing as well as a more comprehensive description of health hazards connected with the use of common chemicals in a holographic laboratory.

8.4 Hologram Copying

There are many reasons why hologram copies may be necessary to obtain and therefore various copying methods have been developed over the years [8.35-73]. Applications where hologram copies are of special interest include, e.g. copying to produce display holograms and holographic optical elements. In holography, the term copying refers to both the recording of a replica hologram as well as to the process of *transferring* one holographic recording to another type of hologram, e.g. generating an image-plane reflection copy from an off-axis transmission master hologram. In general, copying methods can be divided into *mechanical* and *optical*. Mechanical methods are used for copying relief structures into plastic materials. In holography this method is used for mass production of holograms using *embossing* techniques. The embossing technique is not treated in this book. Chapter 6 discusses, however, surface-relief holograms on silver-halide emulsions and gives a few details as regards copying such holograms. Optical methods can be divided into two fundamental types: *direct contact-printing* methods and *holographic interferometric copying* methods, described in the following section. Copying can take place in either the Fresnel-diffraction zone of the original hologram or in the Fraunhofer-diffraction zone, depending on the distance between the emulsions.

Normally, the master hologram (the original hologram to be copied) is called H1 and the hologram copy obtained from H1 is called H2. Three different copying schemes have been proposed for optical hologram copying:

(I) Direct contact printing of the interference pattern in the emulsion (H1-emulsion is in very close contact with the H2 emulsion). This method is identical with contact printing of photographic negatives. It can only be used for copying thin transmission holograms. Lasers, filtered narrow-band lamps or even broad-band light sources can be used.

(II) Contact copying of the reconstructed wavefront from H1. The H1 emulsion does not have to be in close contact with the emulsion of H2, but often enough they are placed rather close together. The same reference beam is used for both the reconstruction of H1 and as a reference beam for H2. In contact copying the use of lasers prevails. Contact copying can be used for copying both transmission and reflection holograms. This copying technique is also used for scanning (line scanning) with a laser beam.

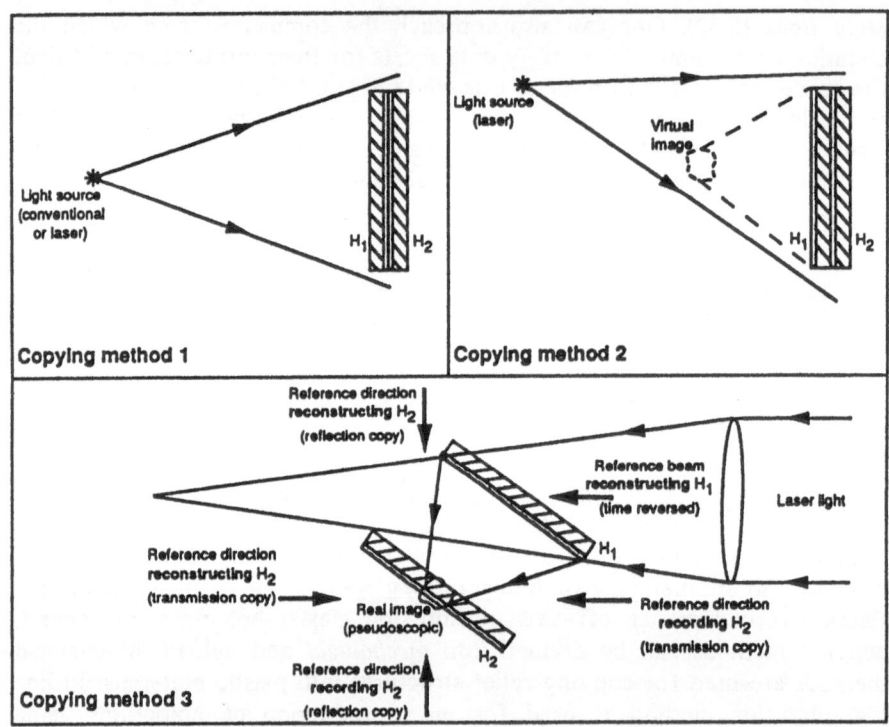

Fig.8.3. Hologram copying methods. H_1 is the master hologram to be copied. H_2 is the copy being recorded in the setup. *Method 1.* Direct contact printing of the interference pattern in the emulsion (H_1-emulsion is in very close contact with H_2-emulsion). *Method 2.* Contact copying of the reconstructed wavefront from H_1. (H_1-emulsion does not have to be in close contact with the emulsion of H_2). *Method 3.* Copying of the image wavefront reconstructed from H_1, recorded at a certain distance from H_1. The reference beam for H_2 is normally added as a separate beam to produce either reflection or transmission image-plane holograms

(III) Copying of the image wavefront reconstructed from H1, recorded at a certain distance from H1. The reference beam for H2 is normally added as a separate beam generated from the laser in the setup. Depending on whether the beam is introduced from the same side as the image wavefront or from the opposite side a transmission copy or a reflection copy is obtained. Most often than not the H1 is a transmission hologram. In this kind of copying lasers are used exclusively. This setup is frequently used for the generation of image-plane holograms and rainbow holograms. These three methods are illustrated in Fig.8.3.

306

8.4.1 Optical Copying of Holograms

The first publication on hologram copying appeared actually before laser-recorded holograms were made by E. Leith and J. Upatnieks. *Rogers* [8.35] produced a copy of an in-line hologram where the separation between the plates was 45 cm (Method II above). The hologram to be copied reconstructed both the virtual and the real image. The hologram copy was exposed to the two image wavefronts generated from the master hologram, as well as to the reconstructing reference beam which acted as a reference beam for the copy plate. Using a thin recording material, four images were obtained in the copy (H2). Later, *Horman* [8.36] described a similar technique but this time the first-order diffraction image was separated from the zero-order reference beam for the master (H1) hologram. The plate for the copy hologram (H2) was located at a large distance from the original hologram and had an in-line reference beam superimposed on the first-diffraction-order image wavefront created from the master.

Rotz and *Friesem* [8.37] were the first to introduce the type (III) of our discussed copying methods, showing the possibility to obtain a real, non-pseudoscopic image from the H2 hologram.

The first type of the copying techniques was actually the last one to appear on paper. *Harris* et al. [8.38] revealed that they had copied transmission holograms using three different lamps: a tungsten lamp, a high-pressure mercury lamp, and a zirconium concentrated-arc lamp positioned at one meter's distance from the hologram. The exposure times were typically 5 to 10 minutes, using Kodak 649-GH film developed in Kodak D-19 or D-8. The distance between the emulsions was 30 μm and the master hologram as well as the film copy were mounted in a vacuum printing frame. *Landry* [8.39] investigated the use of (HeNe) laser for this type of copying technique. He found that the highest quality copies could be obtained using laser light as compared to either filtered white light (0.3nm-wide filter centered at 694.3nm) or unfiltered white light. The 649-F plate and a vacuum-contact printer were used for copying. *Rose* [8.40] claimed that a copy generated in this way will always be noisier than the master even if the copy and master were in perfect contact.

Brumm [8.41] returned to method (II), pointing out the advantage of not having to preserve close contact between the copy and the H1 hologram; also the fact that a vacuum contact printing apparatus was not necessary here. An HeNe laser and 649-F plates were used. *Belvaux* [8.42] mentioned that this technique could also be employed for copying reflection holograms. This fact was also mentioned later by *Kurtz* [8.43]. *Landry* [8.44] was the first to draw attention to the fact that the reconstruction of the H1 hologram with this method should be made at the same angle that was used for the H1 recording. That was not obvious and clear in the previous papers.

The in-line copying principle introduced by *Rogers* [8.35] was mathematically treated by *Sherman* [8.45, 46]. The problem of creating four images, mentioned by *Rogers*, could be eliminated by using volume holograms. *Brumm* [8.47] discussed exclusively the double-image problem.

307

The application of a ruby laser for making transmission hologram copies in a setup according to method (II) was described by *Rieck* [8.48]. He used an experimental version of the Agfa 8E75 material which required less than 100 $\mu J/cm^2$ when exposed to 50 ns ruby pulses.

Nassenstein [8.49, 50] pointed out that the "woodgrain" interference structures often present in the H1 hologram, would not be copied if the H1 was bleached and the direct copying method (I) was applied with regular light. The possibility of improving the hologram's diffraction efficiency by copying was discussed by *Palais* and *Wise* [8.51]. The quality of hologram copies was studied by *Suhara* et al. [8.52] where the MTF concept was applied to the copying process. Practical applications of *Belvaux*'s technique [8.42] in making reflection copies using method (II) was presented by *Zemtsova* and *Lyakhovskaya* [8.53]. The copying technique was tested here by making holographic recordings of an aluminized mirror. The master hologram was recorded on the LOI-2-633 material and developed in the GP-2 developer. In the former USSR it was a common practice to hypersensitize the plates in a TEA solution prior to exposure. To avoid emulsion shrinkage after processing, the plates were washed for 30 minutes after the TEA treatment. This removes all the TEA from the emulsion without reducing the overall effect of the treatment. To demonstrate problems associated with using this technique for copying, the following experiment was performed. The master hologram was recorded in contact with the mirror. Since the mirror reflects about 90% of the light illuminating it, the beam ratio for this recording was 1:0.9. The processed master hologram had a diffraction efficiency of about 40%. This master was then used for making the hologram copy. Since the master hologram of the mirror reflects only 40% of the light illuminating it, the beam ratio is reduced so that the recording beam ratio for the copy hologram was 1:0.4. This means a reduction by a factor of two compared with the master recording, which is why the diffraction efficiency of the copy is lower than for the master, provided that the same recording material and the same processing technique are used for the copy. It was found that the efficiency of the copy was a little less than 25%. This shows clearly that the application of this technique for copying holograms of diffuse objects requires masters with a very high diffraction efficiency, combined with a large signal-to-noise ratio.

In the same paper, *Zemtsova* and *Lyakhovskaya* also described a method for making copies of pseudoscopic images from reflection masters using a collimated beam. In this case the plates are separated a certain distance. The obtained copy is then an image plane reflection hologram containing an orthoscopic image. The advantage of this method as compared to an image-plane reflection hologram recorded from a transmission master is the increased viewing angle over which the image can be seen. One weakness of this method is that high diffraction efficiency is difficult to obtain here. As regards the (II) copying method, which is more common, it should be mentioned that the size of the transmission master hologram (H1) affects the diffraction efficiency of the copied hologram (H2). This interdependence has been described by *Vanin* [8.54]. By increasing the size of the transmission master hologram, a reduction in the efficiency of the reflection copy is perceived but the viewing angle increases. In simple words it can be stated that a limited amount of light is available for the observers of the hologram. This light is either concentrated to a very limited region in front of the hologram, in which case the image will be very bright, or else it is distributed over a large area in front of the hologram, with the result that the image observed at a particular point within this area will not be as bright as in the first case. This explains also the difference in luminance of the image between, e.g., a rainbow hologram and a normal Denisyuk hologram. A better way of comparing diffraction efficiencies of various hologram types of diffuse objects would be to apply the following criteria. The integrated irradiance over the whole reconstructed field of the first-order diffracted image light divided by the whole area of the reconstruction field at the distance from the hologram where the irradiance is measured. This value is a measure of the average reconstructed image-power density per unit area (e.g., $\mu W/cm^2$). The average reference-power density at the plane of the film is also measured. Compensations for reflection losses may perhaps be taken into consideration as well. The ratio between these two values will then be the holographic diffraction efficiency for holograms of diffuse objects.

To avoid problems with emulsion shrinkage associated with reflection masters, *Vlasov* et al. [8.55] suggested that the Russian SP-4 developer be used. Unfortunately, they have not revealed the recipe for it. The pseudoscopic image reflection copying method was also used by *Bjelkhagen* [8.56] for producing large-format glass reflection copies.

Hologram copying using transferring techniques other than already discussed will not be further treated here. Details applied specifically in some of these copying setups, such as, e.g., the fringe locking technique and the polarization control, are described in other books on holography. Normally, exposure and processing of the recording materials here will not differ much from how it is done in other holographic techniques. However, there are some special features that should be noted and these will be briefly discussed in the following.

8.4.2 Practical Hologram Copying Methods

The direct contact printing technique (Method I) is rather seldom used for actual hologram copying. The contact copying technique (Method II) is the one in common use. Here, many variations are found. Although, so far, mainly lasers have been used here, other light sources, including the sun [8.57], have been suggested. *Phillip's* considerations [8.58,59] about using incoherent light, discussed in Chap.1, must be looked into more carefully when developing new techniques based on this method. *Chamberlin* and *McCartney* [8.60] investigated the possibility of contact copying of the volume phase transmission gratings recorded in silver-halide emulsions into silver-halide or dichromated gelatin emulsions.

A single-step copying process for multiplexed volume HOEs was recently presented by *Piazzolla* et al. [8.61]. The technique utilizes a set of coherent but mutually incohernt optical sources. *Pascual* et al. [8.62] have designed a copy system based on partially coherent light for HOEs recorded in dichromated gelatin and photoresist.

a) Copying with a Pulsed Laser

As regards the use of lasers for copying, two approaches can be taken here, both of them providing less severe stability demands on the copying system. The first is the more obvious one and is based on the use of a pulsed laser. Employing a pulsed laser makes such a system rather expensive and its use can be justified only for massproduction of holograms. Applied Holographics, Plc., introduced the "Holocopier" machine, utilizing a ruby laser and a holographic film from Ilford. The recording film used in Holocopier is based on thin polyester and uses a reversal bleach to introduce shrinkage necessary to obtain the desired hue in the reflection copy. The above system (no longer manufactured) described by *Brown* [8.63,64], was intended for both security applications and mass production of display holograms using the contact copying method (II). The properties of the Ilford film have been described in Chap.3. Using a pulsed laser for hologram recording seems to be very useful in avoiding recording problems due to, e.g., vibrations. However, maintaining the laser beam clean from a very folded beam path inside the copier is difficult. Maintaining a clean beam is necessary for the recording of holograms of high quality over a long period of time. The processing of pulsed holograms is also rather tricky as described in Chap.6, considering problems connected with latent-image fading and others. Problems of both technical and commercial nature have prevented this system from becoming more successful.

Pulsed lasers can be useful, however, in the making of contact copies of master holograms and even other types of copies. They can be also used in making holographic stereograms, as described by *Towers* et al. [8.65] or in the transfer step of large rainbow holograms [8.66]. Both the master and the rainbow copies can be made by pulses from a ruby laser. The advantage of using a pulsed laser for the rainbow copy produced from a pulsed master is that the same wavelength is used, which is why no wavelength distortions are introduced. Another advantage is the simplicity of the copy setup with hardly any stability requirements. No index matching of the film is necessary when exposing the copy which means that film copies can be produced much faster than using a CW laser transfer setup. The film is stretched flat in some sort of suitable film holder.

b) Copying with a Scanning Laser Beam

Another system used for contact copying holograms uses a scanning laser beam. *Palais* [8.68] proposed that a hologram could be made by using a method which he called *scanned beam holography*. The demonstrated experiment did not use a moving beam during the recording. The reference beam was positioned at different locations over the film in sixteen discrete steps. The main advantage was the reduced exposure time at these individual small regions of the holographic recording material, which reduces the stability demand of the recording setup. In a more recent publication *Imedadze* and *Kakichashvili* [8.69] demonstrated the technique of recording a Denisyuk reflection hologram by scanning a thin laser beam across both the plate and the object space behind it. The authors were mainly interested in recording large format dichromated gelatin holograms using this method. *Qu* et al. [8.70] described the basic principle of making scanning holograms including some examples that could be produced with a 50-mW HeNe laser such as, e.g., a hologram of a hand. However, the main application of the scanning method is for making hologram film copies. A vertical laser line is scanned across a holographic film which is in contact with the master plate. The laser line will both reconstruct the image (locally) and expose the film copy when it propagates across the plate. This technique has been described by *Cvetkovich* [8.71, 72], where he discusses the influence of the slit width on the quality of copies, assuming a given output power of the laser. A very narrow slit reduces the momentary exposure time, which can allow faster scans. This, however, requires a very accurate scanner and extremely stable output power from the laser. Wider slit widths require a slower scanning speed in which the motion of the film can destroy the recording. On the other hand, inaccurate scanning performance and laser power variations will be averaged out in this case. A compromise must be made between these factors when determining the slit width. Typically, using a 25 mW HeNe laser, the Ilford Hotec recording film, and a slit width of 5 mm the exposure time is about 5 s for a 30×40 cm² hologram. The scanning speed is then about 6 cm/s at the film plane. Compared to a full beam copying technique, this represents a more relaxed situation as regards the stability requirements since the momentary exposure time for the individual parts of the film is short. To obtain clean film copies the demand on the scanner is high and the intensity stability of the laser must be high as well. There are two different ways of copying the hologram. In both cases the copying is performed at the Brewster angle if possible:

1) The copying emulsion (H2) is in contact with the emulsion of the master hologram (H1). In this case no index matching is necessary. The problem here is that the emulsion of the master may get scratched after some time in use. This problem does not appear if a hologram master with a hard emulsion is used. The processing requirements for obtaining reconstruction at the HeNe laser wavelength favor the application of tanning processing, resulting in a hard emulsion surface. The reduced handling time makes this method very popular.

2) The copying emulsion (H2) is separated from the master hologram emulsion at the distance corresponding to the thickness of the master hologram's glass plate. The emulsion of the master hologram can be sealed in this case to protect it mechanically and to prevent it from humidity variations. An index-matching liquid between the film and the glass plate is, however, necessary here in order to obtain high qual-

ity copies. Applying an index-matching liquid for each copy makes this method more time-consuming.

Normally, an HeNe laser is used for this type of work. The holographic film from Ilford (Hotec) works very well in these types of recordings. The BIPS factor gives a nice orange-yellow color after the processing with a fixation-free rehalogenating bleach process for holograms exposed to the HeNe laser wavelength. The Ilford processing chemistry is particularly suitable here as it can be used in processing machines, which is important in any mass production system for holograms.

Finally, it is worth mentioning that *Vanin* [8.73] has published a review of all types of copying techniques, including both optical and mechanical methods, and supplying information on details of theoretical aspects of the copying process as well as practical and experimental considerations.

8.5 Archival Properties of Holograms

One important problem in holography is the permanency of the holographic recordings stored in silver-halide emulsions. The question has especially high relevance for many expensive artistic holograms produced nowadays. Since the holograms discussed here are recorded in fine-grained emulsions of the black-and-white type, the experience gained in conventional photography can be applied as guidance for the archival performance of holograms. However, most holograms differ from photographic recordings in one respect, namely that the holographic emulsion is usually bleached in order to produce phase holograms.

In conventional photography, the color stability is of great importance. In holograms the color permanence is different from that in conventional photography since the holographic recording technique is very unlike the conventional color photographic technique based on dyes in the emulsion. The only similarity between color photography and color holography can be found in the old Lippmann color photography method. Although very limited in its use, Lippmann photographs that have survived during all these years from the beginning of the century normally show no color distortions or changes. Therefore, color permanence is not the main problem as regards permanence of a holographic recording. For example, colors in transmission holograms of the rainbow type are created by white-light diffraction from superimposed gratings. This is why one does not have to be concerned about color preservation here. As regards holograms of the Lippmann reflection type, the occurring color changes are due to the variations in the emulsion thickness throughout time. The most important question concerning holographic permanency is therefore the permanence of the interference fringes stored as silver particles or silver-halide particles embedded in gelatin and the stability of the gelatin itself.

8.5.1 Archival Performance of Conventional Photographs

Conventional black-and-white photographs have been around for quite some time and a lot of experience has been gathered even as regards archival performance. American standards regarding photographic materials are found in ANSI PH 1.45-1981 "Practice for Storage of Processed Photographic Plates" and ANSI PH 1.43-1983 "Practice for Storage of Processed Safety Photographic Film". Kodak has published a book on the preservation of photographs [8.74]. The stability of the photographic image, in general, has been recently discussed by *Anderson* and *Goetting* [8.75]. Actually, the most relevant comparison between photography and holography is for black-and-white emulsions. Concerning these photographic materials, a review paper has been published by *Drago* and *Lee* [8.76]. The main factors to consider are the following:

- Processing influence.
- Emulsion surface contamination.
- Biological contamination.
- Atmospheric pollution.
- Thermal influence.
- Humidity influence.
- Irradiation influence.

All these factors have to be considered when discussing the stability of conventional photographs. As regards the processing of black-and-white photographs, the main thing to remember is that thiosulfate ions from the emulsion must be removed after fixing. Thiosulfate ions remaining in the emulsion from incomplete washing, or residual silver-thiosulfate complexes from the use of exhausted fixing baths can result in fading or yellowing of the image. If the film is properly fixed and carefully washed - which can often be improved upon by using a hypo-clearing agent - the danger of emulsion degradation is minimized. For fine-grained black-and-white microfilms used for archival purposes the American ANSI PH 1.41-1984 recommends the residual thiosulfate concentration in the emulsion to be less than 0.007 g/m^2. However, small quantities of residual thiosulfate in an emulsion can actually minimize the risks of changes caused by heat and temperature. The use of a Kodak Hypo Eliminator HE-1 (not the clearing agent) is not recommended for emulsions on glass or film, since it can form gas bubbles in the emulsion which can cause blisters [8.77].

Emulsion surface contamination can also be caused by lacquers applied to it. This problem can be reduced by selecting lacquers containing chemical components which cannot penetrate moist gelatin or produce peroxides. Suitable solvents are such as hydrocarbons and halogenated hydrocarbons. Peroxides can be formed from oil-based paints, which is why such lacquers should not be considered for emulsion protection.

The danger of biological contamination must also be studied since gelatin is attractive to living organisms, such as fungus or mold. These organisms are more likely to form if the photographic material is stored in a warm, highly humid environment. The biodegradation of photographic gelatin has been treated by *Stickley* [8.77]. This degradation occurs as a result of the hydrolysis of peptide bonds on the polypeptide chains of the gelatin molecule. There are several micro-organisms that can attack these bonds. Some fungi (e.g., Aureobasidium sp) are capable of liquefying gelatin. Bacteria (e.g., Bacillius sp and Pseudomonas sp) can also give rise to biodegradation of the gelatin. The manufacturers are adding preservatives or biocides to the emulsion in order to limit this type of problem. If conventional film is to be stored under conditions of high humidity and temperature, it can be rinsed in a solution of 5 g/ℓ of vioform (powdered iodochlorhydroxyquin) before drying. This treatment gives an effective protection against most growths. Kodak [8.74] recommened the following treatment for black-and-white negatives. After the final wash, soak the negative in a 1%-water solution of zink fluosilicate. (Zink fluosilicate is extremely poisonous). Black-and-white prints can be treated in a 1% solution of diisobutyl-cresoxy-ethoxy-ethyl-dimethyl benzylammonium chloride (Hyamine 1622) [8.78] according to Kodak. The treatment time is 3 to 5 minutes in a solution of a temperature not lower than 21° C.

Polluted air is another factor that can deteriorate the quality of the material. In particular, oxides of sulfur and nitrogen present in the atmosphere are apt to affect silver images, but also peroxy radicals and acidic atmospheric gases can have a negative influence. Local oxidation of silver can generate microblemishes in a photographic emulsion. These microspots are generated by initial local oxidation of silver forming mobile silver ions which migrate to the surface where they are reduced to metallic silver by either hydrogen peroxide or light. The silver ions can then be converted to silver sulfide by hydrogen sulfide often present in the atmosphere. For conventional photographic products, toning (e.g., by using the Kodak Rapid Selenium Toner) is often recommended to reduce microspot formation and improve the long-term stability.

The influence of high temperature combined with high humidity can cause serious damages to photographic emulsions. Therefore, archival storage of photographic products must be done in a temperature of less than 24° C and a humidity of less than 50% RH (but above 25%).

If the finished photographic emulsion is exposed to excessive radiation (visible, UV or IR) the effect on the images can be severe. Color photographs are particularly affected by radiation which causes color changes and general image fading. Photographs can be UV protected by applying a UV absorbing coating, e.g. by lamination.

Finally, the influence of the substrate is also important in photography. Here, the best choice from the point of view of stability is, of course, the glass plate. However, most photographs are produced on flexible film materials, therefore the long-term stability of these materials is important. The archival quality of modern film bases has been treated by *Brems* [8.79]. The best material from this aspect is the polyester base.

8.5.2 Archival Performance of Holograms

Holograms have only been around for about 30 years which is why the information on their archival performance must be based on the experience gathered from conventional photography, combined with artificial ageing experiments on various holograms. There are two very different aspects concerning the archival problems in holography:

(i) The first case, when the holographic recording is made mainly in order to be stored for future use, storing should present no real problem, provided that the holographic material is kept in total darkness and in an environment specifically designed for archival photography. In particular, fixed amplitude holograms on glass plates should be able to last at least 100 years or more, if one is to go by the experience gained from the storage of photographic black-and-white negative plates. A slight oxidation of the silver image that may occur at times should not be regarded as any serious damage. This will simply turn the amplitude hologram into a partly phase hologram. The signal-to-noise ratio can, however, be affected, which is caused by grain growth during oxidation. In addition, the image resolution can eventually be affected, too, which can cause a problem in, e.g., high-resolution hologrammetry. For holograms in which the image is not focussed in the plane of the emulsion, microspots and local defects, such as scratches for example, will not be as damaging as in photography. Amplitude holograms (fixed and well washed) formed in silver-halide emulsions can therefore be regarded as excellent media for the long-term storage of information.

(ii) The second aspect of the hologram's long-term storage and performance is much more difficult to define. In this category we find mainly bleached holograms which are often permanently displayed. A great deal of these holograms are produced on film materials. To make things worse, sometimes the emulsion is coated by a black paint, sometimes it is not protected at all. Another problem unique in holography, is the printout effect occurring in bleached silver-halide emulsions. The combined effect of all these factors must be taken into consideration when predicting the lifetime of a hologram. This is, however, almost impossible to do. Under the worst conditions a hologram can degrade within less than a year, while under benevolent conditions it can last for many decades. In the following, some guidelines are given concerning the archival storage and performance problems connected with bleached display holograms. The printout problems (including printout protection) have already been treated in Sect. 5.2.5, and will not be further discussed here. Most of the bleached holograms produced nowadays are not fixed and therefore the problems connected with the use of thiosulfate discussed previously do not apply to these holograms. The problems concerning the appearance of microspots in rainbow, image-plane or Denisyuk display holograms should be considered with some seriousness since these will affect the overall appearance of the holographic image.

Brown and *Jacobson* [8.80-82] have discussed the archival performance of holograms recorded in silver-halide emulsions. The main thing to keep in mind here is the protection of the emulsion and the effect of continuous illumination of holograms. The environmental conditions in holographic museums and galleries must be such that at all times they satisfy the humidity, temperature and air-pollution requirements to guarantee a long-term stability of the holograms. This is very often not at all complied with in places exhibiting holograms. For example, the air conditioning is often turned off after the museum's closing time. It is more ridiculous if we realize that air conditioning is actually more important for holograms which are on permanent display, than for the visitors staying in the exhibition rooms for a short time only. The present author's own experience of early reflection holograms (unprotected emulsion) exhibited and stored a long time in air-polluted areas of Germany shows that emulsion degradation has occurred in them. Other holograms produced at the same time but exhibited in other, less polluted areas are still intact. Also sealed transmission holograms exhibited in the unfavorable environment, for example in high temperature/humidity conditions in Latin America, have degraded rapidly, as reported by *Berkhout* [8.22]. These holograms have completely been destroyed by fungal spores and microorganisms attacking the emulsion because of the high temperature and humidity surrounding them during a long period of time. However, many problems associated with these environmental aspects can be planned for and remedied when exhibiting or storing holograms over extended periods of time, even though they are often connected with a substantial expense.

A problem more difficult to control is a hologram's exposure to radiation. It goes without saying that all exposure to UV and other unnecessary electromagnetic radiation should be avoided. However, exposure to visible light is part of the reconstructing process of the holographic image but that very thing constitutes a serious problem for holograms on permanent display. Another problem one has to contend with is connected with the hologram production process and has to do with emulsion protection. It is believed that recording holograms on glass plates where the emulsion is sealed with another glass plate by optical cement represents the safest way of protection. This protects the emulsion from air pollution, the UV radiation and humidity variations. However, sealing does not protect the emulsion completely from micro-organism attacks or printout problems. Also, there is a risk that the optical cement will affect the long-term stability. In addition, one has no access to the emulsion of the sealed hologram for corrections should a problem occur later. Another problem associated with reflection holograms is heat absorption of the black coating applied to the emulsion by the light. The type of lacquer applied to a holographic emulsion is important here as well as the choice of the chemical components in contact with the emulsion. One must carefully consider also the amount of the IR radiation that will be absorbed by the black coating during the illumination of the hologram. Heating an emulsion in the presence of moisture or oxidizing gases may encourage deterioration of the emulsion. If possible,

316

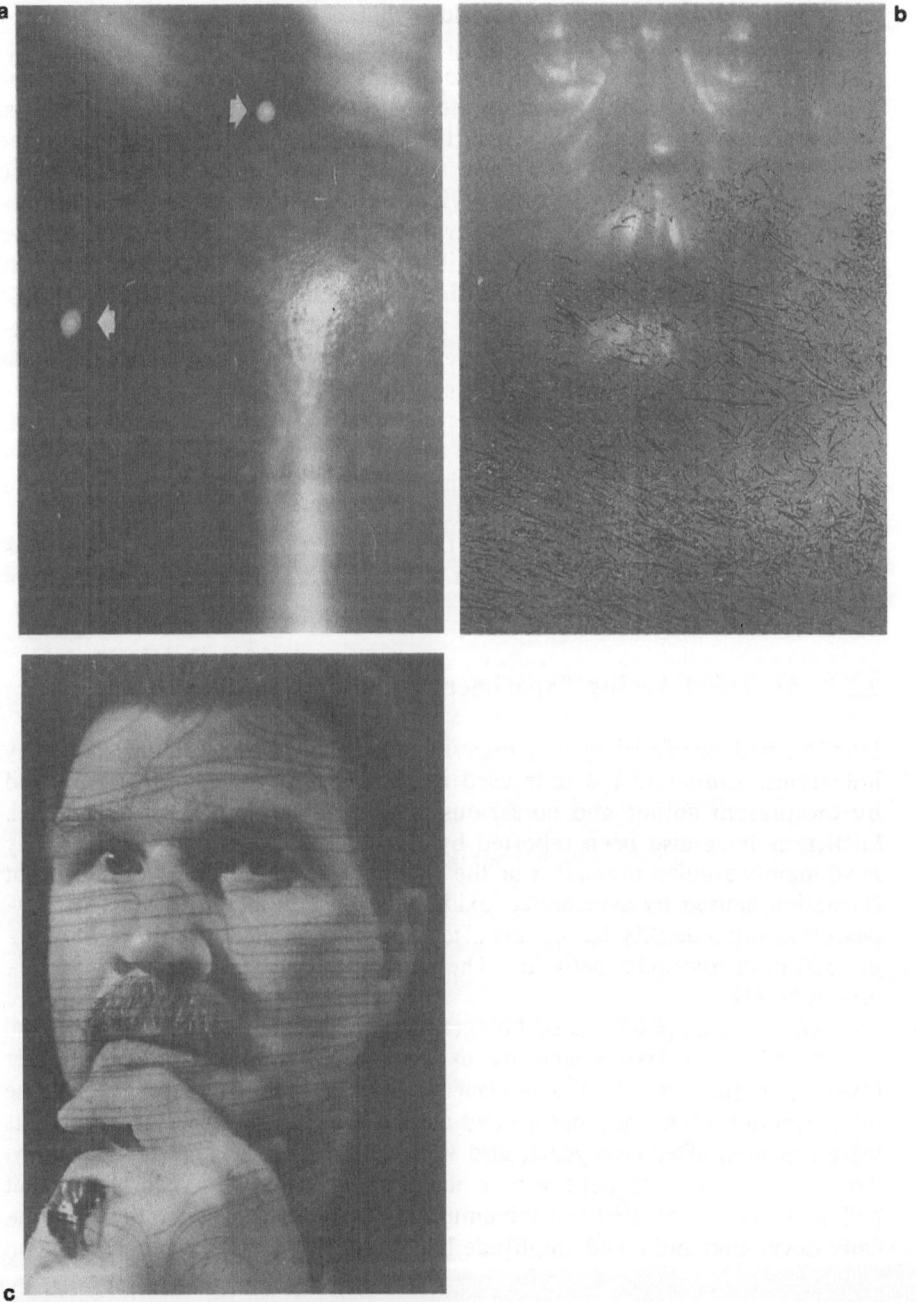

Fig.8.4. Holographic emulsion defects. Emulsion spots (a), cracked emulsion on glass plate (b), emulsion pattern in laminated film (c)

cold light (containing no IR radiation) should be used to illuminate holograms. Even in this area, the present author has encountered many examples of damages to reflection holograms (unsealed, unpainted, unprotected emulsion) that were continuously illuminated for over five years, which caused cracks and blisters in the emulsion. Again, a hologram of the same object, produced at the same time, but kept in darkness showed no emulsion defects. This and the previous example demonstrate how important emulsion protection is. The answer to the question of whether to leave the light on continuously or switch it off at the end of the day is not quite obvious. Even though limiting the time of illumination may seem desirable as it will reduce printout (as well as save the electricity), the cycle of heating and cooling the gelatin layer every day may also cause emulsion stress which can eventually become a serious problem and result in the destruction of gelatin. Rapid temperature cycles are definitely not recommended. It is difficult to give general advice as to how to best deal with this problem. One thing is certain and that is the fact that the preservation of the hologram image in a hologram (sealed or unsealed) which is never illuminated is very good, assuming that the other important storing conditions have been observed. In Fig.8.4 some of the defects just described are illustrated.

8.5.3 Artificial Ageing Experiments

Unpublished, artificial ageing experiments on protected reflection display holograms, similar to the tests used in photography, have been performed by the present author and no serious image degradation has been noticed. Such tests have also been reported by *Brown* and *Jacobson* [8.80-82]. They have mainly studied the effect of the residual thiosulfate ions and microspot formation caused by accelerated oxidative emulsion attack. Microspots appeared in high-density holograms after four days in an atmosphere containing 500 ppm hydrogen peroxide. The temperature was 50°C and the humidity 80% RH.

Aliaga et al. [8.83] tested holograms processed with a solution-physical developer and a fixing step by exposing them to the Chilean summer (average temperature 30°C and clear skies) for a month without any change of diffraction efficiency before and after the treatment. The measurements were repeated after five years, also with no observed change in efficiency. There is no reason to believe that holograms processed in such a way that colloidal silver is obtained in the emulsion should be any worse than chemically developed and fixed amplitude holograms if the former are also fixed.

8.5.4 Recovering Affected Holograms

Holograms suffering from printout can often be rebleached, provided the emulsion was not permanently sealed. Such treatment is suitable for transmission holograms mainly. However, sometimes such a rescue effort can prove to be even more destructive than the printout effect itself. For example, when Fe-EDTA bleached holograms that had been exhibited for a long time where rebleached, the emulsion separated completely from the substrate. A disastrous effect like this is caused by a combination of factors affecting the gelatin. The best could well be to avoid any attempts to re-bleach the hologram. Fe-EDTA bleached hologram suffering from printout darkening have been treated simply by continuing the illumination using strong light, which causes solarization to take place, thus making them transparent again.

Another disastrous occurrence took place when the emulsion of re-versely bleached holograms which had been exhibited previously collapsed completely when soaked in water, and the image turned from orange into a deep blue color after drying. This was definitely due to the effect of radiation since the color along the edges of the emulsion that had been hidden from light by a metallic frame was not affected.

As regards defects other than printout, often very little can be done to reverse the destruction of the hologram. Sometimes microspots can be re-duced by bleaching and after that the emulsion is sealed to prevent the ap-pearance of more spots. If the emulsion has been painted the paint can sometimes be lifted off from the emulsion by the application of a strong adhesive tape to the paint. Color changes in reflection holograms (unsealed or not permanently sealed) are often caused by storing the hologram under the wrong humidity conditions (too high or too low humidity). The holo-gram will often slowly recover if it is placed in an environment containing the right humidity. However, if the surrounding humidity was very low (often combined with high temperature) during an extended period of time the color will have become more or less permanent, which is caused by the hardening of the gelatin layer. It is then very difficult to recover the color even if the hologram is kept in an environment of high humidity for a long time.

Biodegradation is regarded as an irreversible damage to the hologram. In these cases when often a very large area or even the whole hologram has been affected, the image was either completely disappeared or showed a very high level of noise.

When discussing the possibilities of recovering badly affected holo-grams one must mention that also a new (light-sensitive) holographic emul-sion can be saved, treated and reused even when it has been accidentally exposed to white light or laser light. *Wesly* [8.84] mentioned that the latent image formed during the unwanted exposure can be erased by simply bleaching the material in a rehalogenating bleach. The latent image consists of silver specks formed on the surface of the silver-halide grains. If the sil-ver atoms are brought in contact with an oxidizer and if bromine ions are provided at the same time, the latent image will be converted to silver bro-mide again. *Wesly* investigated the Agfa materials and two different bleach baths, and found that the ferric sodium EDTA bleach could restore the emulsion to the same sensitivity as that of the original emulsion. Using the PBQ-2 bleach, an increased sensitivity over the original 8E75 HD emulsion sensitivity was experienced. The latent image caused by the exposure, which was erased with this method represented an optical density >4 had the emulsion been developed. This means, that a box of large, expensive plates that had been accidentally opened in daylight, for example, should not be regarded as destroyed and the plates can be erased and reused.

8.5.5 Conclusions

In general, it is not possible to guarantee that a holographic image will last forever. However, if necessary precautions as regards storing and exhibiting holograms are taken and the requested procedures of sealing and protecting the emulsion are performed during manufacturing, it is believed that holograms can have a rather long life-time. In particular, color holograms should perform much better that color photographs. There is definitely no reason to expect that the colors of transmission holograms should change over time. Reflection holograms, if properly protected, should also show no color degradation. Actually, it has been suggested that important color photographs should be stored holographically in order to preserve their colors. *Ih* [8.85] has discussed the possibility of storing 2D color transparencies in Fourier color holograms. The use of the rainbow holographic technique for the same purpose was described by *Yu* et al. [8.86]. In both cases color information has been geometrically encoded in a black-and-white emulsion. The reproduction of colors is done in the first case by using the same wavelengths at the reconstruction as for the recording, and in the second case, by observing the image from the rainbow hologram at a well defined point in space in front of the holographic plate.

9. Color, Infrared and Ultraviolet Holography

This chapter describes the various existing methods of recording holograms in true colors and the ways of creating pseudo-color holograms, so popular among artists. The possibility of recording holograms of objects in which the images show colors identical to those of the original object are very important if display holography is to become more acceptable. However, there are many problems associated with the recording of high-quality color holograms, which will be discussed in the first part of the chapter. The second part takes up infrared and ultraviolet holography. This is another important field of holography, where the main interest is in applications of Holographic Optical Elements (HOEs). The possibility of recording holograms in these parts of the spectrum using silver-halide emulsions as well as other materials is discussed.

9.1 Color Holography

9.1.1 General Considerations

Even after 30 years since the appearance of the first monochromatic holograms the possibilities of recording high-quality holograms in true colors are still very limited. Although various special techniques of today allow for the production of holograms exhibiting several different colors, in most cases the colors displayed in these holograms are not the true, original colors of the holographed object. The following sections outline the history of the development of color holography [9.1-81] and discuss some of the special techniques for the recording of pseudo- or multicolor holograms [9.94-120].

The problem of finding a suitable name for the various techniques used for recording holograms of more than one color has been a hotly debated issue in the holographic world. Suggestions such as *full color*, *natural color*, or *true color* holograms have been forwarded for lifelike holographic images but have not received a full acclaim. Perhaps the most logical name that comes to mind in analogy with color photography, color movies, and color television would be *color holography*. However, this term is sometimes objected to on the grounds that some colors of the objects we normally see around are sometimes impossible to record holographically since holograms can only *reproduce colors* of objects *created by the laser light scattered* (light consisting of various laser wavelengths). Colors we see are often the result of fluorescence which cannot be recorded in a hologram. For example, some dyed and plastic objects achieve their bright, saturated colors by fluorescence. This limitation in color holography does not seem to be, however, very dramatic. If we compare, for example, color images on various TV screens in a store showing the same picture or color prints produced from the same negative but on different occasions, we shall usually think that each TV image or a photographic print is alone a true color recording.

Jeong and *Wesly* [9.66, 67] gave a heuristic definition of a color hologram, which relates directly to the colors obtained through the laser recording process: "A hologram is said to have true color, if it recreates an image which has the same combination of wavelengths and their relative intensities as those laser wavelengths detected from the object during recording".

Hubel [9.65, 73], who was rather unsatisfied with this definition, gave a more quantitative and exact definition: "A holographic technique is said to reproduce 'true' colors if the average vector length of a standard set of colored surfaces is less than 0.015 chromaticity coordinate units, and the gamut area obtained by these surfaces is within 40% of the reference gamut. Average vector length and gamut area should both be computed using a suitable, white-light standard reference illuminant".

If a color hologram shows colors at the reconstruction which are in good agreement with the Hubel definition, the observer will most likely claim that the color rendition is quite accurate. The problems of recording color holograms will be further discussed in Sect. 9.1.3.

The problem of finding a suitable name for all the other types of color techniques is more difficult. Here, there exist methods for creating colors that give an impression of a true color in the finished image (e.g., multiple recorded stereograms or rainbow holograms) although the recordings could have well been made from objects with completely different colors. However, they can also be made from three sets of color-separated photographs. By using the rainbow technique it has actually been possible to mass-produce embossed holograms with either "natural" or artificial colors. It is also very common to make multiple-exposed color reflection holograms using a single wavelength laser where the emulsion thickness has been changed in between the recordings of special objects. These techniques will be described in Sect. 9.2.

9.1.2 History

The first methods for recording color holograms were established a long time ago. *Leith* and *Upatnieks* proposed multicolor wavefront reconstruction in one of their early papers [9.1]. *Mandel* [9.2] pointed out that it might be possible to record color holograms directly using a polychromatic laser and an off-axis setup. *Lohmann* [9.3] included polarization as an extension of the suggested technique. These first methods concerned mainly transmission holograms recorded with three different wavelengths from a laser or lasers, combined with different reference directions to avoid cross-talk. The color hologram was then reconstructed using the same three laser wavelengths from the corresponding reference directions. Color holograms of a quite high quality could be made this way, but the complicated and expensive reconstruction setup prevented this technique from becoming popular. The first transmission color hologram was made by *Pennington* and *Lin* [9.4] as mentioned in Chap. 1. They used the 15 μm-thick Kodak 649-F emulsion with a spectral bandwidth of about 10 nm. This narrow bandwidth essentially eliminated cross-talk between the two colors (633 and 488nm) at the reconstruction. In general, the cross-talk problem was solved by *Collier* and *Pennington* [9.11] who used spatial multiplexing and coded reference beams which made it possible to record color holograms in thin media.

The Lippmann color technique can be rather obviously used in holographic color recordings as well. *Lin* et al. [9.5] made the first color reflection hologram that could be reconstructed in white light. They recorded a reflection hologram of a color transparency illuminated with two wavelengths (633 and 488nm). The material here was the Kodak 649-F plate which was processed without fixing in order to avoid shrinkage. This technique seems to be the most promising one for the actual recording of color holograms and will be further discussed later on. However, the three-beam transmission technique can eventually become more applicable if inexpensive multicolor semiconductor lasers appear in the future. Very few improvements in color holography were made during the 1960's and the 1970's, although several papers on color holography were published [9.1-24]. During the following decade, several new and improved techniques appeared [9.25-81]. A review of various transmission and reflection techniques for color holography can be found in [9.33]. Concerning reflection color holography, an extensive contribution was made by *Hubel* and *Solymar* [9.65].

9.1.3 Color Recording in Holography

a) Color Transmission Holography

Methods for recording transmission holograms with artificial or natural colors will not be treated here at all. The principles governing the recording of transmission holograms are based mainly on the geometry of the recording setup and are very little affected by the material used for the recording. Most materials, including silver-halide materials, work very well for transmission holograms. *Ruzek* and *Muzik* [9.21] compared Kodak, Agfa, and a Czechoslovakian material VRE-52 with regard to spectral sensitivity in recording color transmission holograms. For more details the reader is directed to the references, in particular, [9.23, 27, 30, 32-35, 38, 42, 45, 47, 53, 58, 64, 70, 76, 79].

b) Color Reflection Holography

Color reflection holography presents no problems as regards the geometry of the recording setup but the final result is highly dependent on the recording material used and the processing techniques applied. The single-beam Denisyuk recording scheme has produced the best results so far. Color holograms have been recorded in single-layer silver-halide emulsions [9.7, 8, 12, 17, 66, 67, 71, 73, 81], or in two separate silver-halide emulsions in a sandwich [9.22, 25, 26, 29, 37, 39, 40, 52, 59]. Pure dichromated gelatin emulsions [9.24, 61, 78], or a DCG in combination with a silver-halide emulsion in a sandwich [9.43] have also been utilized. Even three-layer emulsions for color holography have been proposed and manufactured [9.69]. Photopolymer recording materials have also been experimented with [9.15, 57, 66-68, 77, 80].

There are at least four fundamental problems associated with the recording color reflection holograms in silver-halide emulsions which are the most convenient materials for large-format color holograms because of their high sensitivity:

• Scattering occurring in the blue part of the spectrum found in *commercial* silver-halide emulsions makes them rather unsuitable for recording color holograms. Therefore, in the most successful color holograms the blue part has so far been always recorded in dichromated gelatin [9.24, 43].

• Another problem is that multiple exposures of a single emulsion will reduce the diffraction efficiency of each individual recording [9.36, 82]. The diffraction efficiency of a multiple-exposed single-layer emulsion varies inversely as the square of the total number of recordings.

• The third problem is the shrinkage of the emulsion that often takes place during processing, causing a wavelength shift. White-light-illuminated reflection holograms normally show an increased bandwidth upon reconstruction, thus affecting the color rendition.

• The fourth problem, which is to some extent also related to the material, is to choose the laser wavelengths to be used and to decide upon the best combination of these in order to obtain the best possible color rendition of the object [9.28, 48, 62, 63, 65, 73, 74].

The problems just mentioned have already been discussed by *Lin* and *LoBianco* [9.12]. They recorded reflection holograms on Kodak 649-F plates using 632.8, 514.5, and 488 or 476 nm wavelengths. They also had a specially prepared version of the 649-F emulsion with a lower silver-halide concentration than in the normal emulsion. This new emulsion proved to be successful and holograms recorded in it had rather nice colors. It was suggested that the shrinkage problem could be resolved with a TEA treatment of the plate after the processing. For the reconstruction of the color holograms a zirconium arc lamp was recommended, rather than thermal lamps which have a rather low luminosity in the blue end of the spectrum, where also the sensitivity of the human eye is low. In addition, the diffraction efficiency of the hologram is poor in the blue part and the scattering noise is also highest there.

Noguchi [9.17] tried to make a quantitative colorimetric comparison between recorded color holograms of color-test targets. He used the standard 649-F emulsion and four primary wavelengths for the recording. The reproduction of the blue colors caused him, however, some problems.

Kubota and *Ose* [9.24] demonstrated that a good color reflection hologram could be recorded in a dichromated gelatin emulsion, which gives high efficiency and low-noise blue reconstruction. *Hariharan* [9.25] introduced the sandwich-recording technique to improve image luminance compared to the earlier triple-exposed 649-F emulsions. He employed Agfa 8E75 for the red (633 nm) recording and the 8E56 emulsion for the green (515 nm) and blue (488 nm) recordings. To compensate for the shrinkage of the emulsion, *Hariharan* suggested that sorbitol $[CH_2OH(CHOH)_4CH_2OH]$ should be utilized instead of TEA. Triethanolamine, used as a permanent swelling agent introduced to a finished hologram increases the printout tendency of the bleached emulsion. The two plates were cemented together with Kodak Optical Cement HE-80.

The sandwich technique has now become the primary method of recording improved-quality color reflection holograms. *Sobolev* and *Serov* [9.26], *Smaev* et al. [9.39], and *Sainov* et al. [9.40], all used that principle. To avoid the shrinkage, water-soluble substances ($10 \div 15\%$ water solutions of sodium sulfate) added to the silver-halide emulsion at both the recording and the reconstruction stages of the color hologram were employed by *Sainov* et al. [9.40]. The most successful sandwich recording technique was demonstrated by *Kubota* [9.43] with a dichromated gelatin plate for the green (515 nm) and the blue (488 nm) components, and an Agfa 8E75 plate for the red (633 nm) component of the image. The processing of the silver-halide plate consisted of development in D-19, and fixing and bleaching in 3% bromine water. The shrinkage problem (caused by fixing the plate) was solved by treating the plate in isopropyl alcohol, the treatment originally suggested by *Cooke* and *Ward* [9.83]. This consists in that after the regular processing steps, the plate is soaked in hot water (48° C) to which 3-wt.% ammonium EDTA (EDTA·2NH$_4$) is added. The plate is then soaked in a 70% isopropyl alcohol bath of 48° C and finally in a bath of 100% isopropyl concentration and of the same temperature. In all the three baths the plates are soaked for three minutes. In this way the shrinkage of the emulsion can be accurately controlled and the diffraction efficiency of the hologram increased. Since the DCG plate is completely transparent to the red light, the silver-halide plate (containing the red image) is mounted behind the DCG plate in relation to the observer.

Hubel and *Ward* [9.52] have been using Ilford silver-halide materials for recording color reflection holograms. The sandwich technique (SP 672T for blue and green, and SP 673T for red) was applied with the recording illuminations at 458, 528 and 647 nm. Although the Ilford blue/green material works better with regard to scattering than the Agfa 8E56 HD material, the holograms produced on Ilford material are suffering from the blue recording scattering noise. The color rendition is, however, good, and *Hubel* and *Ward* have, so far, produced the best results on the Western commercial silver-halide materials.

9.1.4 Recent Progress in Color Holography

The present status of color reflection holography is that it is a relatively novel, developing area where a lot of improvements can be done. As mentioned in the previous subsections only a few good color holograms have been recorded so far. In the following sections various considerations relevant to the quality of color reflection holograms will be discussed.

9.1.5 Choice of Laser Wavelengths

Any practical color reproduction process uses only a limited number of primary colors (at least three). Other colors are obtained as a result of mixing the primaries. There are two fundamental ways of mixing colors: the *additive* and the *subtractive* method. In color holography the additive method applies for both transmission and reflection holograms

The problem of choosing the optimal primary laser wavelengths for color holography is illustrated in the 1931 CIE (Commission Internationale de l'eclairage) chromaticity diagram (Fig.9.1) and the 1976 CIE version (Fig.9.2). The diagram is a useful device for predicting the colors that can be matched by additively mixing a set of primary colors. Spectral colors are located along the horseshoe-shaped curve in the diagram. All visible colors are represented by points situated inside the diagram. Fully saturated colors are placed along the periphery of the curve. The straight line joining the extremities of the curve (extreme red and blue) is the locus of purple. In the center of the area, white is located. The colors along the curve between the white point and the red region represent colors generated by a black-body radiator, such as a hot filament at a certain temperature [K]. By mixing different spectral colors, all possible colors can be synthesized.

Some definitions used in connection with discussing color could be useful to know [9.84, 85]:

- *Hue* is that attribute of visual sensation that has given rise to color names, e.g., blue, green, yellow, red.
- *Saturation* which is also referred to as the purity of color, can be seen as that degree of purity in a given color.
- *Brightness* (or luminosity) is the attribute of a visual sensation according to which a given area appears to emit, or transmit or reflect more or less light.
- *Lightness* (or relative brightness) is the achromatic (colorless) continuum that goes from white, through gray, to black.
- *Luminance factor* is the ratio of the luminance of a reflecting or transmitting surface, viewed from a given direction, to that of a perfect diffuser receiving the same illumination.
- *Chromaticity co-ordinates* are the ratios of each of the three tristimulus values of a sample color to the sum of the tristimulus values.
- *Colorimetry* is a numerical expression of colors.
- *Primary colors* are three colors of constant chromaticity used to specify

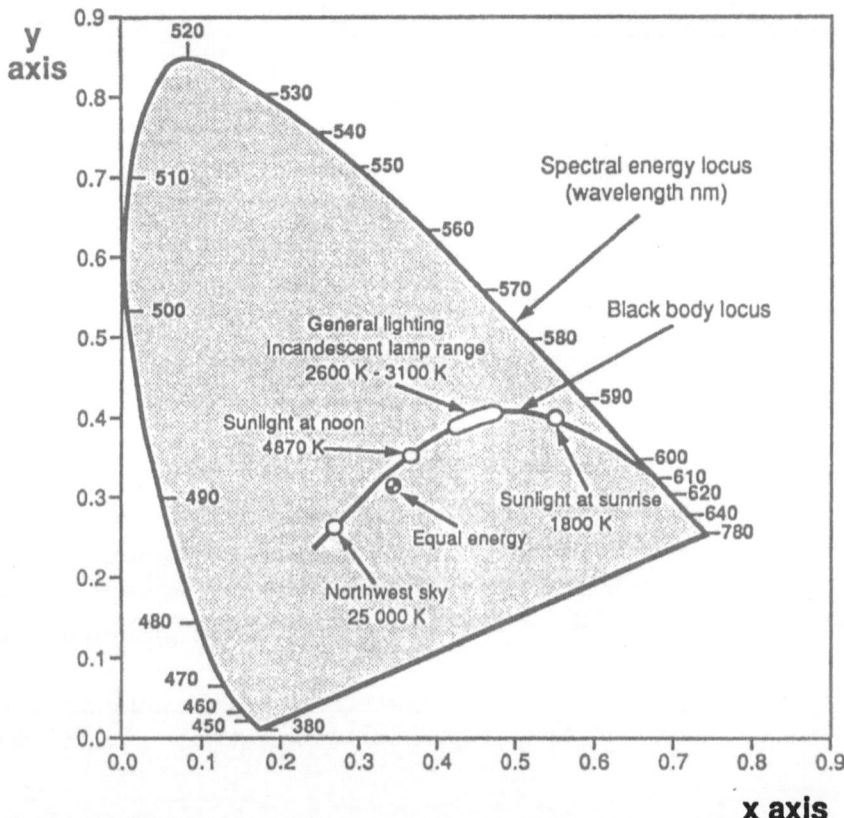

Fig.9.1. The 1931 CIE chromaticity diagram, illustrating the chromaticities of various light sources

an unknown color by such amounts that are required in an additive mixture to match the unknown color. Any three colors can serve as primary colors provided none of them can be matched by additive mixture of the other two.

• *Tristimulus values* are the amounts of the primaries required to establish a match with a sample. This is done either by the addition to the sample of all three primaries, or of only one primary to the sample to match any pair of primaries, or by the addition of any pair of primaries to the sample to match the remaining primary.

Chromaticity coordinates for any given color are computed from the tristimulus values X, Y, Z as follows:

$$x = \frac{X}{X + Y + Z} = \frac{Red}{Red + Green + Blue}, \tag{9.1}$$

$$y = \frac{Y}{X + Y + Z} = \frac{Green}{Red + Green + Blue}, \tag{9.2}$$

327

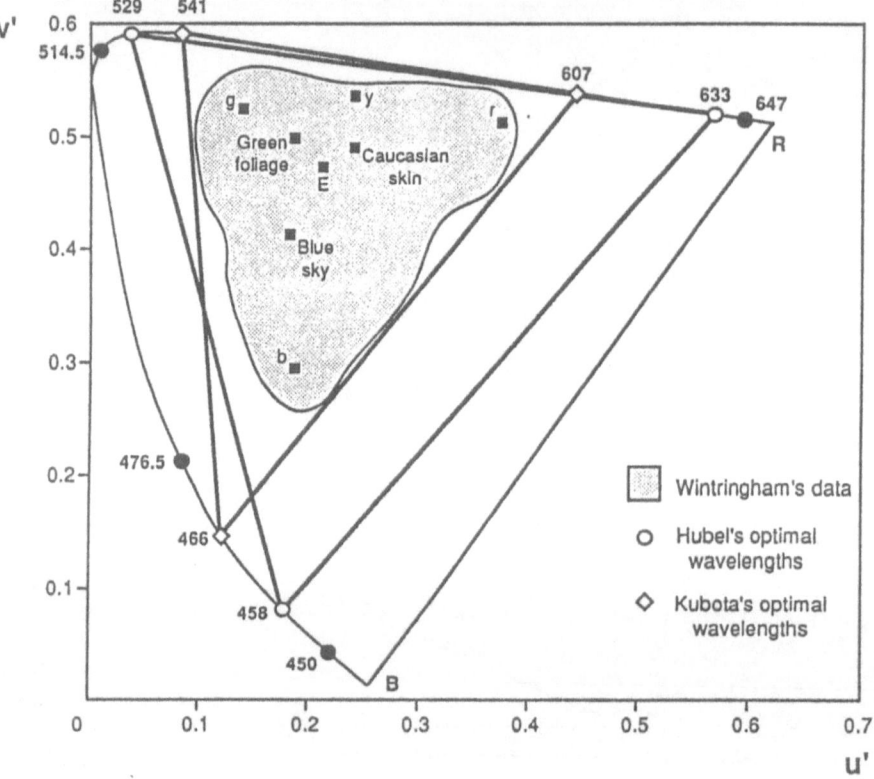

Fig.9.2. The 1976 CIE uniform scales chromaticity diagram, showing the gamut of surface colors and positions of common laser wavelengths. Hubel's and Kubota's three optimal color-recording laser wavelengths are also indicated

$$z = \frac{Z}{X + Y + Z} = \frac{\text{Blue}}{\text{Red} + \text{Green} + \text{Blue}}. \tag{9.3}$$

Chromaticity is given as a point on a rectangular co-ordinate, where chosen x and y denote ordinate and abscissa, respectively. Two of the three primaries (corresponding to X and Y) are selected in such away that their luminance factors are zero. Accordingly, the luminance factor of any color is given directly by its Y-tristimulus value. z is normally not plotted because $z = 1 - (x+y)$.

Color is approximately specified in the CIE system by x, y and Y.

After the three different primary spectral colors have been selected, a triangle is made by joining the three points corresponding to the spectral colors in the diagram. The colors within the area covered by the triangle correspond to all the colors that can be produced by an appropriate mixture of the chosen spectral colors. It may seem that the main aim in choosing the recording wavelengths for color holograms is to cover a very large area of the chromaticity diagram. However, there are many other considerations to be taken into account when choosing the wavelengths for color holograms as discussed by *Buimistryuk* and *Dmitriev* [9.28], *Bazargan* [9.48, 74], *Kubota* and *Nishimura* [9.62, 63], and *Hubel* [9.65, 73].

Bazargan's [9.48, 74] discussion is based on *Wintringham's* gamut of surface colors [9.86]. Surface colors refer to colors of natural and man-made objects. Normally, these colors are of low saturation. Such colors are also found in most of the objects considered for display color holography. *Pointer* [9.87] has extended the gamut of surface colors to include some of the highly saturated fabric dyes which were introduced after *Wintringham's* publication. So far, the recording wavelengths for color holography have often been 476.5, 514.5 and 632.8 nm. These wavelengths cover the *Wintringham* data sufficiently well. One factor to bear in mind when working with silver-halide materials is that a slightly longer blue wavelength might give higher-quality holograms because of reduced Rayleigh scattering during the recording. However, if the scattering problem is not considered, *Hubel* [9.65, 73] has made an important observation. He explained that only half of the relative efficiency is needed if a short wavelength (450 nm) is used as the blue primary compared with the 480 nm wavelength in order to balance to white with the other primary wavelengths. This observation is very important for multiple exposures of holographic single-layer silver-halide emulsions. Using a short blue wavelength the diffraction efficiency of this recording can be rather low compared to the red and green recordings and still produce a good color hologram. So what is needed is a low-scattering recording material which will make it possible to use the optimal short blue wavelength.

Another important factor to consider is the reflectivity of the object at the primary spectral wavelengths. *Thornton* [9.88] has shown that the reflectivity of an object at three wavelength bands, peaked at 450, 540 and 610 nm, has a very high bearing on color reconstruction. These wavelengths can also be considered optimal for recording color holograms and are in good agreement with investigations performed in the former USSR [9.28]. Using these wavelengths, a better rendition of the yellow part of the spectrum is obtained. The popular set of wavelengths (476.5, 514.5 and 632.8 nm), although covering the surface colors defined by *Wintringham*, often causes a distortion in the yellow colors, so that they are reconstructed as brown or pink [9.32].

The luminosity of the color image is affected by the drop in luminous efficiency with very short or very long recording wavelengths. This important factor has been pointed out by *Bazargan* [9.32] and should be remembered when choosing the wavelengths. This is also one of the reasons why trying to cover the largest possible triangle may not be the best procedure. *Hubel* [9.50] does not hold with this view and has argued that a larger triangle than is normally used is needed in order to compensate for the color desaturation (color shifting towards white) that takes place when reconstructing color reflection holograms in white light. *Hubel* and *Ward* [9.52] used the 528 nm wavelength (instead of 514.5 nm) of the argon laser, (combined with 458 and 647 nm), which was the main reason why their holograms gave good yellow rendition. *Hubel* [9.73] has continued to investigate the optimum wavelengths, both theoretically and experimentally. According to *Hubel*'s color rendering analysis these wavelengths are 464, 527 and 606 nm for the case of the sandwich silver-halide recording technique. If instead the calculations are performed to maximize the gamut area, the following combination of wavelengths are obtained: 456, 532 and 624 nm. *Hubel* suggested that 458, 529 and 633 nm is the optimal wavelength combination for practical color holography recordings. *Kubota* and *Nishimura* [9.62,63] approached the wavelength problem from a slightly different point of view. They calculated the optimal trio based on a reconstructing light source at 3400 K, a 6 μm thick emulsion with a refractive index of 1.63 and an angle of 30° between the object and the reference beam. *Kubota* and *Nishimura* obtained the following wavelengths: 466.0, 540.9 and 606.6 nm; *Bazargan* [9.74] found the ideal wavelengths to be: 450, 540 and 610 nm. Only experiments will establish which combination will be the best in practice.

Another factor that will determine the recording wavelengths is the availability of wavelengths in argon, krypton, helium neon, and helium cadmium lasers. These are the CW lasers currently in use for holographic recordings. In Table 9.1, the various CW laser wavelengths are listed.

Table 9.1. Wavelengths from common CW lasers

Wavelength [nm]	Laser	Single line power [mW]
442	Helium cadmium	<100
458[a]	Argon ion	<500
468	Krypton ion	<250
476	Krypton ion	<500
477	Argon ion	<500
488	Argon ion	$<2 \cdot 10^3$
497	Argon ion	
502	Argon ion	
514	Argon ion	$<7 \cdot 10^3$
521	Krypton ion	
529[a]	Argon ion	<600
531	Krypton ion	<350
532	Nd:YAG (CW diode pumped)	<80
540	Krypton ion	
543	Green neon	<5
633[a]	Helium neon	<75
647	Krypton ion	$<2 \cdot 10^3$

[a] recommended combination of wavelengths by *Hubel* [9.73]

As regards pulsed lasers it is hard to find lasers producing blue wavelengths with sufficient quality for holographic work. At Imperial College in London, *Bazargan*, in cooperation with M. Damzen and W. McGuigan, recorded a pulsed color hologram using a frequency-doubled Nd:YAG laser, where one third of the pulse was Raman shifted to the blue part of the spectrum [9.46, 74]. The single-beam reflection hologram showing a hand holding a credit card was, however, of rather poor quality. It seems that the quality of the green wavelength that can be obtained from a frequency-doubled Nd:YAG laser is not as good as the quality of the red light generated by commercial ruby lasers. Therefore, it is also difficult to record good monochrome reflection holograms with green Nd:YAG lasers. However, improvements on this can be expected in the future. In the former USSR pulsed xenon lasers have shown promise for color holography. A multicolor pulsed dye laser for endoscopic purposes has been developed at the present author's university by *Schacham* et al. [9.89] and *Marhic* et al. [9.90]. Further improvements concerning the temporal coherence of this laser are necessary for holographic work.

9.1.6 Color Recording Materials and Their Processing

Currently, a combination of different recording materials seems to be the most promising way of approach, for example, the DCG silver-halide sandwich, or photopolymer silver-halide sandwich. Also pure DCG or photopolymer materials can be considered if these materials can be sufficiently sensitized for the red part of the spectrum. Since this book is mainly concerned with the silver-halide recording materials, the question is whether such materials alone can be used for color holograms. At present, it seems that the commercially-produced Western materials are on the borderline, which has been demonstrated by *Hubel* and *Ward* [9.52]. On the other hand, the results obtained on Russian silver-halide materials show that good quality color holograms can be obtained, which indicates that it is possible to manufacture ultra-fine-grained emulsions suitable for color holograms. As mentioned in Chap.2, these materials are, however, difficult to produce on an industrial scale. This is, of course, the main difficulty in the commercial development of color holography. They can be produced in a holographic laboratory which is currently also the only possible place for the production of various other materials, such as dichromated gelatin plates, for example. The photopolymers from duPont are commercially available, but their red sensitivity is low and they are still in a development phase. Silver-halide emulsions produced along the lines indicated in Chap.2 and processed using colloidal development or the new bleaching techniques discussed in Chap.5 will be strong candidates for color reflection holograms. Also, the ultra-fine-grained materials from the former USSR may very well be introduced on the Western market in the future. Experiments with such emulsions for color holography have been reported by *Watson* [9.56, 81], and earlier *Sainov* [9.31] described the use of Bulgarian materials for color holograms.

Concerning color work on Western commercial materials, the work by *Hubel* and *Ward* [9.52] has shown progress with holographic silver-halide materials from Ilford. *Hubel* and *Ward* [9.59] compared different silver-halide materials produced by Western companies as potential recording materials for color holography. The results indicate that Ilford materials are highly promising for color reflection holograms. One thing to watch out for when using these materials is the shrinkage that will occur when the Built-In Pre-Swelling agent (BIPS) is removed during processing. Such shrinkage is highly undesirable in color holography, which is why hardening developers, such as a pure pyrogallol, for example, should be used to reduce or eliminate it. To eliminate the shrinkage caused by the BIPS factor the Ilford material has to be presoaked in water before using it for color recordings.

Additional improvements that can be made when employing the existing commercial materials concern the processing phase. Here the new amidol-based bleach presented in Chap.5 is interesting. The three-step processing technique introduced by *Phillips* [9.91,92] and described in Chap.7 can also offer advantages for color holography. This process was used by *Jeong* and *Wesly* [9.66,67] for recording color holograms on Agfa plates.

Hubel [9.60] has treated the problem of emulsion shrinkage and the resulting wavelength shift as well as the desaturation problem causing difficulties in holographic color reproduction. He introduced a simple model in order to simulate bandwidth variations between the recording phase and the white-light reconstruction. The white-light reconstruction of a color hologram shows a decreased signal-to-noise ratio and an increased bandwidth compared to those found in the laser beams. *Hubel* demonstrated that his theoretical model fits his experimental results and concluded that the desaturation is caused primarily by noise and partly by increased bandwidth. He points out that the magnitude of the desaturation is large and remains as the main problem with reproducing color holographic images in silver-halide materials.

The Kodak 649-F emulsion which was used in many early color experiments has the advantage of being panchromatic. The thickness of the emulsion is about 12 μm, which is another reason that it can perform rather well when the emulsion is subject to several exposures. Both Ilford and Agfa have indicated the possibility of introducing panchromatic holographic materials on the market, provided that the noise problem associated with the blue scattered light can be solved. *Hubel* [9.73] reported that he had successfully recorded a color hologram in one single-layer panchromatic emulsion specially prepared for him by Ilford.

9.1.7 Color Recording Techniques

The most promising recording technique for color reflection holograms is the single-beam Denisyuk setup, which can provide color holograms with a very large field of view and full parallax. The last mentioned are important considerations for possible commercial applications of such holograms. The

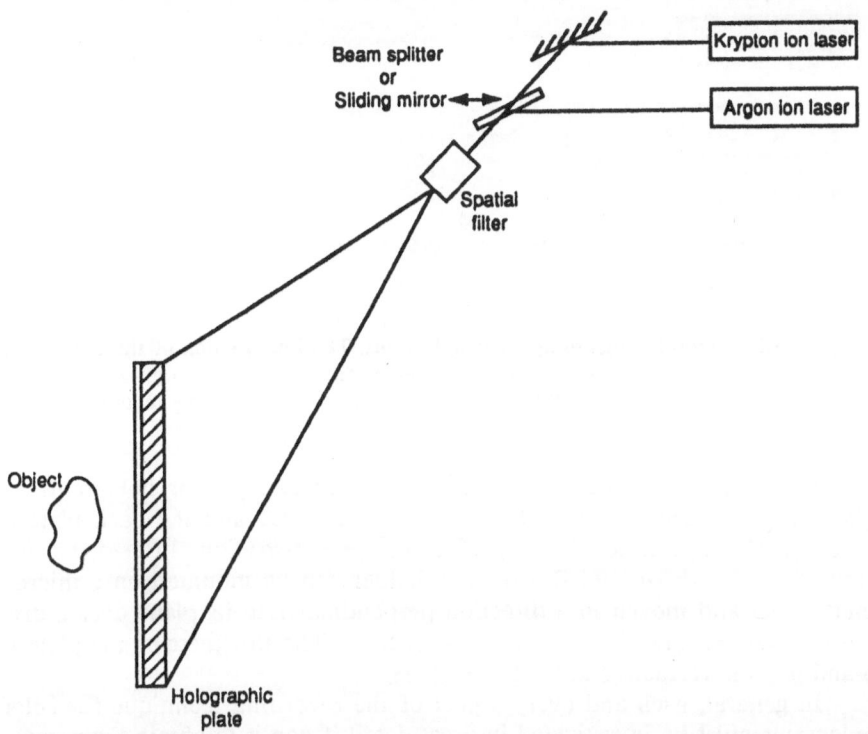

Fig.9.3. Setup for recording color holograms. The most promising recording technique for color reflection holograms is the single-beam Denisyuk setup, which can provide color holograms with a very large field of view and full parallax. The different laser wavelengths necessary for the recording can be delivered through the same beam expander and spatial filter

different laser wavelengths necessary for the recording can be delivered through the same beam expander and spatial filter (Fig.9.3).

The illumination of the object can also be made by combining a set of single-mode fibers, where the light output from each fiber is adjusted to match the spectral sensitivity of the recording material (Fig.9.4).

Single-layer or multiple-layer recording materials and sequential as well as simultaneous exposure of the material can be considered. For recording color holograms, so far, sequential exposure combined with the sandwich-recording technique has produced best results. Sequential exposures of a single-layer emulsion using different wavelengths requires careful adjustment of the individual exposures in order to obtain the correct color balance. This problem was treated by *Shevtsov* [9.36]. He investigated superimposed exposures at different wavelengths recorded in bleached reflection single-layer holograms. The total diffraction efficiency of the color hologram was expressed as a function of the ratio of the exposures, the total exposure, and the dynamic range of the recording medium. Experiments with the PE-2 emulsion were also performed.

333

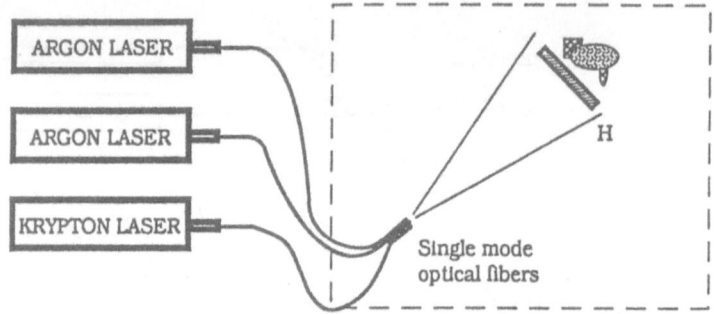

Fig.9.4. Fiber setup for recording color holograms. The illumination of the object can be also obtained by combining a set of single-mode fibers, where the light output from each fiber is adjusted to match the spectral sensitivity of the recording material [9.66]

If the sandwich method is utilized for recording color holograms on glass plates, a compensating glass plate must be used and the back plate is recorded through it to obtain perfect image registration. Instead, as described by *Hariharan* [9.25], the plate holder can be mounted on a micrometer slide and moved in a direction perpendicular to its plane over a distance equal $d[1-(1/n)]$ between the recordings. The thickness of the plate is d and n is the refractive index of the glass.

In general, each and every aspect of the recording technique for color holograms must be investigated in great detail if one is to obtain high-quality holograms. The most crucial of all the factors is, however, the recording material itself. In addition, the choice of a processing method suitable for silver-halide materials is extremely important. Finally, even though there are still many questions to be answered and the final solution of all the problems connected with color holography is not at all obvious there is no doubt that high-quality color holograms will appear sooner or later, whether they be recorded on silver halides or other materials.

9.2 Pseudo Color Holography

In the previous section the ways and possibilities of recording color holograms of real objects were discussed. This section discusses various methods of recording holographic images containing two or more colors, but where the colors of the final hologram frequently do not reflect the real colors of the objects originally used for the hologram recording. Sometimes computer-generated images are taken advantage of recording holograms in several colors. Although the colors obtained in the final product can sometimes resemble "true" colors, the techniques applied here do not represent a *color recording method* and such types of holograms are therefore best called *pseudocolor* or *multicolor* holograms. Sometimes the true or natural color holograms are also referred to as multicolor holograms, but, in accordance with the conclusions drawn from the discussion in the previous section, these should simply be called *color holograms*.

9.2.1 Pseudocolor Holograms

Pseudocolor holograms can be divided into two main groups. To the first group belong white-light pseudocolor transmission holograms based on *Benton's* original rainbow technique [9.93]. The other group contains white-light pseudocolor reflection holograms.

a) Pseudocolor Transmission Holograms

The pseudocolor transmission technique is based on multiple, white-light rainbow holograms in which different spectra are superimposed at the position of the observer. Here, natural colors can be obtained if three different wavelengths are used for the recording of the master hologram. By using a single-wavelength laser and combining different holograms recorded with different reference angles it is possible to create artificial colors in the final rainbow hologram. Often, these holograms are very impressive as regards both their colors and brightness. One problem with these holograms is that the colors vary depending on the vertical observation angle, which, particularly for the "natural" color holograms, means that they only show "true" colors at one particular observation level. So far, this type of holograms is the only one that has been produced in large quantities, since these holograms can be easily embossed. Many applications can be found for this type of hologram, e.g., in art, commerce and security. The colors of transmission holograms are created mainly by geometrical considerations of the recording setup. As transmission holograms they are often mirror-backed for reflection reconstruction. The influence of the recording material is normally of much less importance. For that, the reader is referred to the references concerning these techniques. The multicolor technique for rainbow holograms was introduced in 1977 by *Tamura* [9.94,95]. *Grover* and *Tremblay* [9.96] demonstrated the possibilities of creating natural color rainbow holograms. *Benton* et al. [9.97] and *Benton* [9.98] gave detailed descriptions of the pseudocolor/natural color recording technique. *Benton* has also developed computer programs facilitating the rather complicated calculations that are necessary to perform in order to obtain a perfect color mixture in the final hologram.

b) Pseudocolor Reflection Holograms

The other pseudocolor technique applies to reflection holograms and is therefore much more material dependent than that applied to the previous group. However, even here the recording geometry will affect the colors of holograms. It is difficult to produced pseudocolor reflection holograms on a large scale and they are therefore used mainly for artistic applications in limited editions.

The technique of creating different colors is based on the fact that the color in a reflection hologram is obtained by the reflection of light from the recorded interference layers within the emulsion. The distance between these layers will determine the color. A certain distance between the interference fringes is generated during the recording of a reflection hologram depending on the laser wavelength used. However, this distance can be manipulated in many ways by various processing methods, which means that different colors can be obtained in the finished hologram. Even by using only a single-wavelength radiation for the recording, the produced hologram can contain many different colors. Preswelling the emulsion before the recording will result in a shrinkage after processing, which will create colors of a shorter wavelength in the hologram than the color of the laser used. However, these methods are often very time-consuming and thus they are mainly of interest in holographic art applications. As regards the recording materials, the material most frequently used here is silver-halide emulsion. However, color control of dichromated gelatin has been reported [9.99], as well as pseudocolor processing [9.100]. The following sections describe some of the methods applied to silver-halide emulsions in the recording of pseudocolor reflection holograms.

The first note concerning the possibilities of obtaining different colors by double exposure and emulsion thickness manipulation in between exposures was presented in 1979 by *Blyth* [9.101]. In a paper by *Hariharan* [9.102] the pseudocolor reflection process was carefully described. The technique became popular among artists and several papers on the reflection pseudocolor method have been published [9.103-120]. *McGrew* [9.103] has given information as regards the recording geometry for reflection pseudo-color holograms. Normally, the preswelling of the emulsion is performed by immersing the plates in various concentrations of TriEthanolAmine (TEA) solutions. This treatment also acts as a hypersensitizing method, which is an advantage when using HeNe lasers for the recording. (Refer to Chap.6 in which several hypersensitizing methods are described). For permanent swelling of the bleached emulsion after its processing, TEA is not recommended because of its sensitizing effect which will increase the emulsion printout dramatically. *Hariharan* [9.102] recommended that sorbitol be used for this purpose instead of TEA. Glycerin instead of TEA for a pseudocolor work was described by *Lessing* [9.106]. Presoaking the plate in a 10% glycerin solution results in a 17% increase in thickness of the dried emulsion. Water-alcohol solutions of glycerin were employed by *Vanin* and *Vorobjev* [9.115] to produce pseudocolor holograms on the Russian PFG-03 material. They gave recommendations for suitable glycerin concentrations to obtain desired colors. In their colloidal processing technique which is based on the GP-2 developer, a fixing step is also included causing additional shrinkage after processing. An empiric formula for this process was given

$$\lambda_r = 613 - 6.65\Gamma \qquad\qquad (9.4)$$

where λ_r is the reconstructed wavelength after processing, and Γ is the glycerin concentration in percent. The formula assumes that the recording wavelength is 633 nm. *Vanin* and *Vorobjev* discussed also image distortions in pseudocolor holograms and treated, in particular, image plane holograms transferred from transmission or reflection masters.

As regards the concentration of TEA in which the material is to be treated in order to obtain different colors, this depends on the kind of material used, the recording wavelength, and the processing method applied. Normally, separate tests must be carried out in each and every case. In the following Tables 9.2 and 3 data from *Kaufman's* [9.105,113] and *Smith* and *Cvetkovich's* [9.107] experiments are presented.

Table 9.2. Kaufman's pseudocolor procedure. (Recording wavelength: 633nm)

Color	Wavelength range [nm]	TEA concentration 10 min treatment [%]
red	700–620	0
deep orange	620–580	1.2
yellow	570–580	3.5
yellow-green	510–570	5.0
green	487–510	10.0
blue	430–487	13.5
violet	400–430	17.0

Processing: D-19 + PBQ or EDTA bleach

The practical procedure for the pseudocolor TEA-treatment method has been described by *Moore* [9.104], *Kaufman* [9.105,113,116], and *Smith* and *Cvetkovich* [9.107]. These papers contain important information on all the steps performed in the recording process of pseudocolor reflection holograms. In particular, the interested reader is referred to [9.113] which considers this technique for the recording of holograms up to the size of 50 by 60 cm².

In most cases, the laser used in the pseudocolor work is the HeNe laser, sometimes a krypton laser, too. Normally, image-plane reflection glass holograms are produced from separate transmission masters for each color. *Cvetkovich* [9.109] described a method for making pseudocolor film holograms by a sequential contact-copying technique from different reflection master plates. This paper mentions also that a scanning technique can be used. More details concerning general copying of holograms are found in Chap.8. Recently, *Lieberman* [9.111] has successfully applied the copying technique to produce high-quality film copies using a gray-scale separation

Table 9.3. The Smith and Cvetkovich pseudocolor procedure. (Recording wavelength: 647nm)

Color λ peak [nm]	TEA concentration 5 min treatment [%]	Bandwidth [nm]
646.5	No soak	29
641.5	Water only	27
617.5	1.96	27
604.5	3.8	27
587	5.6	25
565	7.4	22
547	9.1	22
520	10.7	20
505	12.2	18
485	13.7	17
475	15.2	17
462	16.6	19
450	18.0	19
439	19.3	22
431	20.0	19
427	21.8	22
421	23.0	21

Processing: GP62 + PBQ bleach

technique also described by him [9.117]. For three or more colors, three master holograms must be made: one to reconstruct red, and the two others, preswelled, to produce green and blue. For example, if part of the image is to be red, the corresonding area of the model is painted white for the red recording and black for both the green and blue recordings. If a secondary color such as magenta is desired, the model must be painted white on both the red and blue exposures, and black on the green exposure. Other colors are created by mixing recordings where the object has been painted in different gray-scale values.

The transfer setup must be arranged in such a way that the necessary adjustments of the reference angles between the different recordings are carefully considered in accordance with the guidelines drawn up by *McGrew* [9.103]. He was the first to introduce the geometrical "hinge point" approach which facilitates the necessary calculations. In Fig.9.5-8, the technique is briefly illustrated. The reason the direction of the reference beam must be adjusted in relation to the copy plate is that the exposure of the emulsion in the swollen state must be compensated for. The emulsion will reconstruct the shorter-wavelength images from the final holographic plate

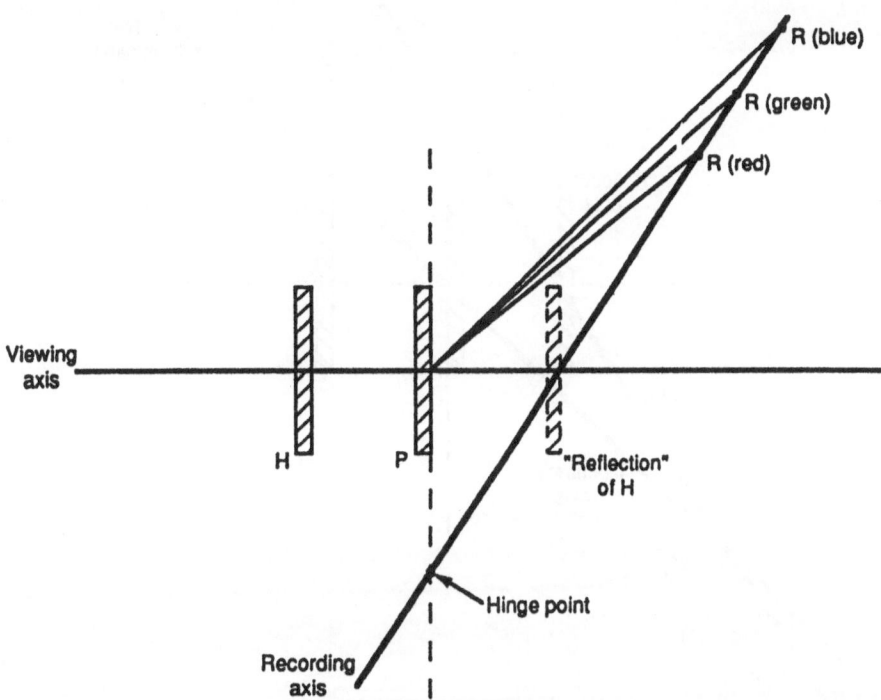

Fig.9.5. Diagram for making pseudocolor reflection holograms using different reference sources. The "reflection" of H (master holograms with the various image components) is an imaginary plate placed at the same distance from the recording plate P as the actual H, but at the opposite side. The reflection of H is in line with the reference source positions R(red), R(green), and R(blue). The reference sources and the reflection of H lie on a straight line in which the recording plate P lies. That point of intersection is the "hinge point" [9.103]

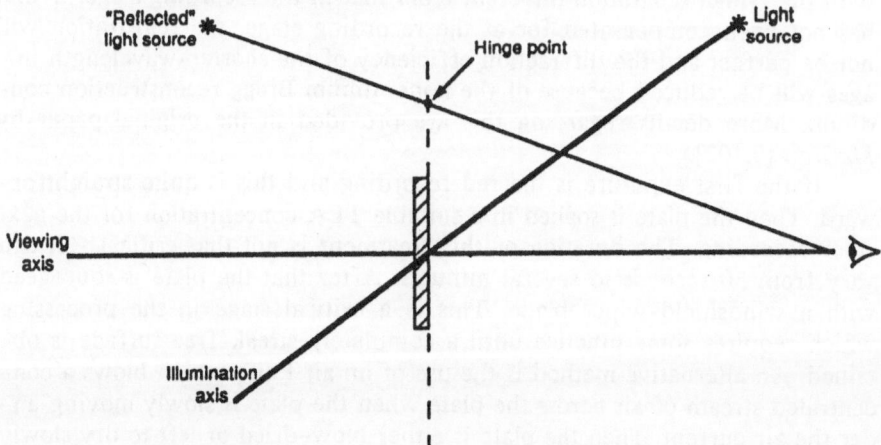

Fig.9.6. Locating the hinge point for a reflection hologram. The hinge point is found by drawing a line from the imaginary "reflected" illumination source position to the position of the observer, and finding the intersection at the hologram plane [9.103]

339

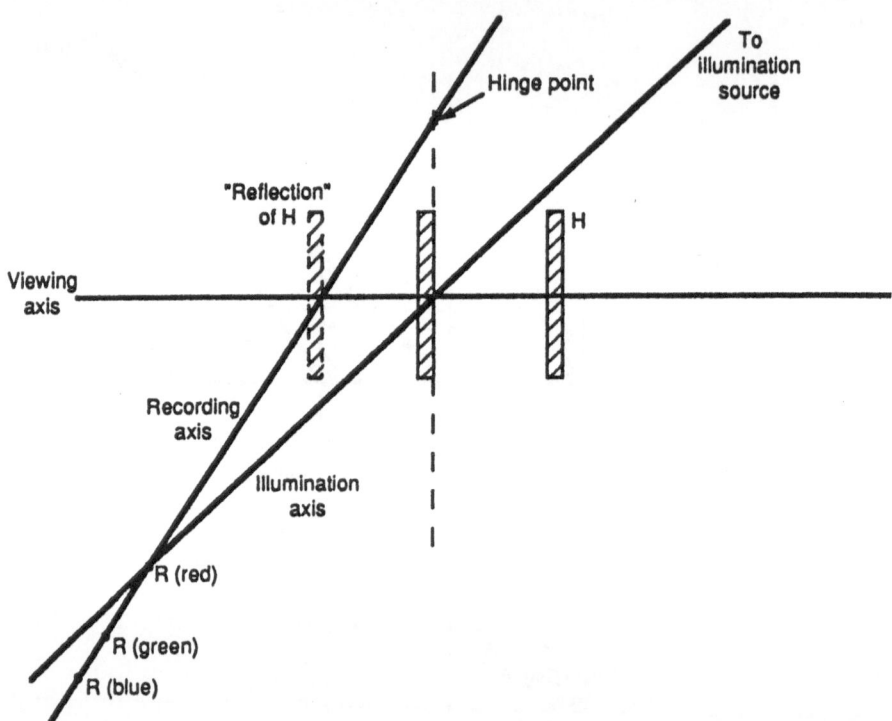

Fig.9.7. Using the hinge point found in Fig.9.5 and the illumination axis to design the recording setup for a reflection hologram. The hinge point is found first, and is then used to locate the line along which the reference source and the H-plates are to be positioned (the recording axis) [9.103]

with the fringe separation different from that in the recording stage. If that has not been compensated for at the recording stage, the registration will not be perfect and the diffraction efficiency of the shorter-wavelength images will be reduced because of the nonoptimum Bragg reconstruction condition. More details regarding this are provided in the original paper by *McGrew* [9.103].

If the first exposure is the red recording and this is quite straightforward. Then the plate is soaked in a suitable TEA concentration for the next color recording. The duration of this treatment is not that critical - it can vary from 30 seconds to several minutes. After that the plate is squeegeed with a windshield-wiper blade. This is a critical stage in the processing which requires some practice until a completely streak-free surface is obtained. An alternative method is the use of an air knife which blows a concentrated stream of air across the plate when the plate is slowly moving under the air current. Then the plate is either blow-dried or left to dry slowly by itself (in a horizontal position to avoid uneven emulsion thickness). The glass side of the plate must be carefully cleaned before the next exposure can take place. The whole process is then repeated for each additional color

340

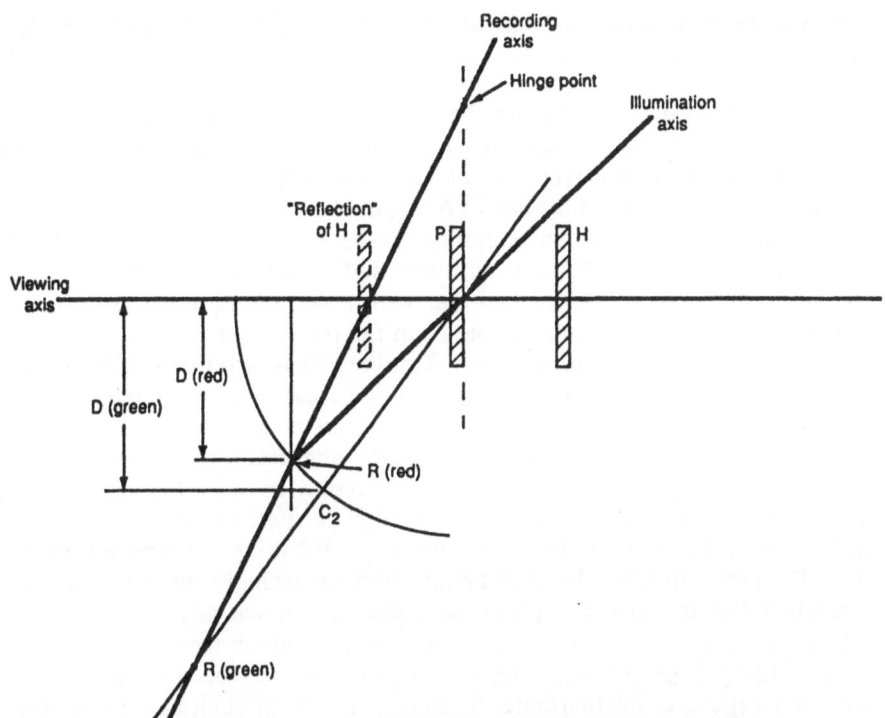

Fig.9.8. Finding reference source position at the recording axis for e.g. the green component of a reflection hologram: Find the intersection between the illuminating axis and the recording axis R(red), which is the position for the red recording, e.g. 633 nm. Draw a circle through R(red) which is centered in the middle of the recording plate P. Draw also a line from R perpendicular to the viewing axis. Measure the distance and call it D(red). Finding the reference position for the green recording D(green), at e.g. 520 nm: multiply D(red) by the ratio of the green wavelength to get D(green); D(green) = D(red)·633/520. Measure D(green) out from the viewing axis along the D(red) line. A line is drawn from D(green) parallel to the viewing axis until it intersects the circle. The intersection point is labeled C_2. A line from the center of the plate P through C_2 is extended until it intersects the recording axis. That intersecting point is the reference position for the green recording R(green) [9.103]

wanted, immersing the plate in higher concentrations of the TEA solution. It is also recommended to add a wetting agent to the TEA bath. One important factor to remember as regards the exposure times is the fact that the TEA treatment increases the sensitivity of the material. Therefore, the periods of the second and the following exposures of the TEA-treated plate must be reduced accordingly. *Moore* [9.104] has found that compared to the first exposure of the untreated plate at a given exposure time, the second exposure should be 1/2 and the third 1/4 of the first exposure time. These are only rough guidelines indicating the direction to compensate for the increased sensitivity. Only tests performed for each separate case can establish the exact color balance and wanted colors. Two factors that would counter-act the TEA treatment to some extent are first the fact that sensitizing dyes can dissolve out of the emulsion during the treatment, and second the fact that in a multiple-exposed layer the first exposure creates a higher density than the following exposures even if all are identical. However, as shown by *Moore*, these two effects are of less importance compared to the increased sensitivity caused by the TEA treatment.

To make the individual exposures more equal, it might be better to start with the highest TEA concentration and finish with the red recording for which the plate is soaked in water with a wetting agent added, only in order to remove the triethanolamine from the previous step.

As always when dealing with holograms, there are many factors contributing to the final quality of the hologram (master plate variations, copy plate emulsion hardness, humidity, TEA-processing temperature, etc.).

As regards the processing of the plates, the best way is to arrange so that the red (no preswelling) recording will replay in red. This means that a nonshrinkage processing technique should be performed. In case any shrinkage occurs during the processing (e.g., by using reversal bleaching) the TEA concentrations have to be adjusted accordingly. In any case, it is recommended to rinse the plates in water to remove any trace of TEA before processing. Therefore, the second method which terminates with the red recording is preferable, since the plate has already been washed before the final exposure. Furthermore, a tanning developer such as a pure pyro-gallol developer combined with the PBQ-rehalogenating bleach seem to be the most frequent choice so far. *Kaufman* [9.113] reported that out of safety considerations he has switched to a ferric sodium EDTA bleach. However, the new processing methods described in Chaps.5 and 7 will probably have an impact on pseudocolor processing in the future, too.

Smith [9.108] has discussed the pseudocolor technique for creating mixed colors from three primaries obtained by the TEA-swelling method. He used the three primary colors presented in Table 9.4.

Using this technique it is possible to obtain any desired colors and record achromatic as well as white images. With three masters where the three objects used for the master hologram recording had the correct grey levels

Table 9.4. Smith's primary wavelength for the pseudocolor technique

Color	TEA concentration	Wavelength [nm]	Bandwidth [nm]
red	no soak	652	23
green	12%	532	20
blue	28%	458	20

compared to the ones obtained when making black-and-white photographs of a polychrome object through three different color filters, it is actually possible to obtain a "true" color reflection hologram with a single-wavelength laser. *Lieberman* [9.111] recorded multicolor holograms based along a similar line of approach.

Walker and *Benton* [9.112] presented a rather novel approach which makes the otherwise complicated swelling procedure employed in the production of pseudocolor reflection holograms less troublesome. Their method developed at MIT is a "hands-off" in-situ swelling method in which the emulsion thickness is controlled during the exposure by immersing the material in a solvent environment. The plate can remain in the setup during the whole swelling sequence. Water, which is used as a swelling agent, is combined with different concentrations of a water-miscible solvent, such as, e.g., isopropyl alcohol. To obtain the three primary colors the material is swollen sequentially, in 8%, 18%, and 25% water solutions. For uniform swelling *Walker* and *Benton* recommended that the emulsion be pretreated in an enzyme solution, e.g., trypsin. This technique has been used in the production of reflection stereograms of computer-graphic color-separated images. It means that full color computer-generated holograms can be produced this way.

The combination of holographic stereograms and reflection volume recording has been developed by *Benton* [9.114]. There are several advantages that this combination offers, e.g., that holographic reflection stereograms are easier to exhibit than transmission holograms, that they are better than their mirrorbacked alternative, that they can be recorded as achromatic or as full-color stereograms and that one can use a dark background for the recorded image. Other important advantages are that a conventional photographic technique is applied for the master which means that all types of objects can be recorded as well as outdoor scenes. In addition, computergenerated images can easily be used for producing holograms. It is also a method for making reductions in holography without distortions. The paper mentions that the normal grating equations for transmission holograms can be combined with the "geometrical optics" of the volume recording media. This means that most of the experience gathered from the transmission stereogram technique can be directly applied here.

It should be mentioned that the emulsion thickness manipulation techniques described above can be also utilized to create achromatic and blackand-white images. *Orr* and *Tryner* [9.118-120] at Richmond Holographic Studios, UK, have produced impressive large-format, black-and-white holograms, with a technique they alone developed. In this case only two colors are necessary in order to create white. In the CIE diagram, the colors should be chosen in such a way that a straight line drawn between them should pass right through the white region in the center of the diagram. Obviously, there are many possibilities of choosing the primary colors which will satisfy this condition. *Orr* and *Trayner* used yellow and blue, which they achieved by preswelling the emulsion in different concentrations of triethanolamine.

Fig.9.9. Black-and-white hologram portrait: "Kate McGougan & Stephen Jones hat". Hologram designed and produced by Richmond Holographic Studios, Ltd. in 1985

The emulsion was first presoaked in the following bath in order to obtain the yellow image:

Triethanolamine (TEA)	17 parts
Wetting agent (Kodak 600 Photoflo)	2 parts
Distilled water	100 parts .

Soaking time > 1 minute. The film was then squeegeed and dried in warm air (30° C).

After the first exposure the film was then soaked in a higher concentration TEA bath for the blue exposure. This solution consisted of

Triethanolamine (TEA)	33 parts
Wetting agent (Kodak 600 Photoflo)	2 parts
Distilled water	100 parts .

Soaking time > 1 minute.

After being dried, the second exposure was made. In order to get perfect registration of the images the reference beam was changed between the two recordings according to the directions given by *McGrew* [9.103]. In general, the exposure times for the two recordings, the change of the ref-

344

erence beam as well as the TEA concentrations have to be fine-tuned to get a perfect result. The guidelines above were carried out for the Agfa 8E75 HD film. The material was processed in a pure pyrogallol developer and bleached in an EDTA bleach. An example of such a hologram is shown in Fig.9.9.

The pseudocolor process is rather time-consuming. *Cvetkovich* [9.109] mentioned that it takes one to four days to make masters and the first acceptable copy of a new subject. Additional copies take three hours per copy since they all have to go through the same complicated steps. Using *Lieberman*'s film copying technique [9.111], one can generate approximately 300 pseudocolor holograms, 30 by 40 cm^2 per month. The production rate can be somewhat increased by using machine processing and other improvements.

Despite the tedious procedure, several artists have been able to record several excellent-quality holograms with this technique.

9.3 Infrared Holography

The possibilities of making holograms in the near infrared and the infrared parts of the spectrum have been widely investigated and the results presented in various papers [9.121-180].

There are many IR lasers that can be used for this type of recording, e.g., Nd:YAG, CO_2 and diode lasers such as GaAlAs or GaInAsP. The main problem with infrared recording is a lack of suitable recording materials. However, several different materials, including silver-halide materials, have been suggested and tested.

The main applications of IR holography are

- In the production of Holographic Optical Elements (HOEs).
- In plasma diagnosis.
- In nondestructive testing of three-dimensional objects which are opaque in visible light, but transparent in infrared light, e.g. germanium and gallium arsenide objects;
- In holographic interferometry investigations using CO_2 lasers, where the long wavelength makes the separation between the fringes more suitable for some objects and measurements.

Two regions of the infrared spectrum have been studied with respect to the wavelengths of the existing lasers. The first is the near-infrared part, i.e., 700 to 1100 nm. The second part of the IR spectrum corresponds to the wavelength of the CO_2 laser, i.e. 10.6 μm.

The first IR holograms were made by *Lowenthal* et al. [9.121] employing a photographic gelatin layer as the recording material for the 10.6 μm wavelength. *Chivian* et al. [9.122] used an in-line setup and a CO_2 laser for the recording. A thermochromic material (Cu_2HgI_4) constituted the recording material for the 10.6 μm radiation. *Izawa* and *Kamiyama* [9.123] recorded off-axis IR holograms in a polymeric film containing a photo-chromic spiropirans painted on a glass substrate. The possibility of real-time IR holography at 10.6 μm was demonstrated by *Simpson* and *Deeds* [9.124]. They took advantage of a mixture of cholesteric liquid crystals as the recording medium.

9.3.1 IR Recordings in Silver-Halide Materials

a) Conventional Sensitized Mode of Operation

In the UV and the visible parts of the spectrum the most sensitive holo-graphic material with a sufficiently high resolving power is the silver-ha-lide film. The high sensitivity of silver-halide materials depends on the fact that during the exposure only the latent image is recorded (normally with low energy) and the subsequent development process converts the latent-image specks to a silver image. The only other holographic materials with similarly high sensitivity and resolving power that can be considered for IR recordings are thermoplastic recording plates or film.

Silver halides, as such, are not sensitive to the red part of the spectrum. To improve their sensitivity to radiation in this region, sensitizing dyes with strong absorption properties in this part of the spectrum must be employed. There is no clarity as to how this process really works, but it is believed that a photoelectron transfer from the dye to the silver-halide crystal takes place. The minimum energy of a photon to eject a photoelectron is 2.3 eV, which is possible only in the UV and blue parts of the spectrum. In the red and the infrared parts of the spectrum dyes must be used. Dyes for sensitiz-ing the emulsion up to about 1300 nm can be found. Further in the IR region, e.g. at 10.6 μm, the photon has an energy of 0.1 eV only, which is far too low, even with a dye, to eject a photoelectron. When an electron has been released it can move through the crystal lattice. At an imperfection of the crystal the electron is trapped and combined with an interstitial silver ion. This single silver atom will trap another liberated electron and the cycle repeats itself several times during the exposure, as explained in Chap.2.

In the deep red and the near-infrared parts of the spectrum ordinary holographic silver-halide materials from Agfa and Kodak can be utilized. These materials have a spectral sensitivity of up to somewhere between 750 and 850 nm. Holograms using these materials have been recorded with diode lasers (e.g., a 5-mW CSP-type GaAlAs) emitting 748 nm [9.125]. The Kodak holographic plate 120-01 had a sensitivity of two-thirds at 748 nm of the sensitivity at 633 nm.

In another investigation performed at Kodak by *Lysogorski* and *Lun-gershausen* [9.126], a 30-mW laser diode of the GaAlAs double heteros-tructure V-channeled Substrate, Inner Stripe (VSIS) type, was used. The wavelength of 780 nm was utilized and the plates were Kodak 120-02 as well as an experimental IR-sensitive film similar to the holographic film SO-173. At 780 nm the required exposure was 1 mJ/cm² for the 120-02 plates.

A slightly longer wavelength (787.3 nm) was employed by *Davis* and *Brownell* [9.127] in an investigation comparing Agfa and Kodak materials. It was found that both Kodak 120-02 and Agfa 8E75 needed about 6 mJ/cm² exposure. The faster Agfa 10E75 plate required only 0.8 mJ/cm². Notice that the exposure had to be increased six times as compared to the previous investigation where the 120-02 plates were used. The reason for this was probably that the wavelength was about 10 nm longer in the second study. It may be interesting to mention that the thermoplastic recording technique was also investigated in the same study. Here, the standard Newport HC-300 thermoplastic camera was tested for IR holography. The sensitivity was the same (0.8 mJ/cm²) as for the Agfa 10E75 silver-halide material.

The Agfa 8E75 HD material has been applied for recording reflection gratings at 750 nm by *Hart* et al. [9.128]. They found that the sensitivity of the material at this wavelength was about 20 times lower compared to the sensitivity at 633 nm. The same material was employed by *Wang* and *Kostuk* [9.129] at 820 nm for making bleached planar transmission holograms. At this wavelength they found a sensitivity of 60 mJ/cm². They observed that the 8E75 HD material has a useful IR sensitivity which extends to 840 nm. This means that this material is suitable for holographic recordings using GaAs lasers.

The standard Kodak I-N spectroscopic IR material was tested by *Sugaya* and *Iwamoto* [9.130]. They recorded holograms at 865.4 nm on this low-resolution material (about 30 line pairs/mm) with an inter-beam angle of 0.5°. The IR holograms reconstructed with a HeNe laser were of rather a poor quality, having a signal-to-noise ratio of only 10. Later, *Gilbreath* and *Clement* used the standard Kodak Spectroscopic IV-N material for recording holograms at 842.4 nm [9.131]. The IV-N material has a slightly higher resolving power of about 200 line pairs/mm and a very high sensitivity. They found that an exposure of only 56 nJ/cm² was needed at the given IR-wavelength. Both Kodak materials mentioned here require small recording angles.

An experimental Kodak material is the hypersensitized IR material 4-Z, which *Roychoudhuri* and *Thompson* [9.132] used for IR holograms. In their case holograms were made with a pulsed Nd-glass laser. The 4-Z emulsion has a resolution of only about 100 lines/mm. The spectral sensitivity reaches up to about 1200 nm.

Silver-halide materials from the former USSR have also been used for IR holography. *Ryabova* et al. [9.133] reported that the IAE-3 was studied at 840 nm and IAE-6 at 1060 nm. Both materials have high resolving power and good sensitivity. The IR holograms exposed with a GaAs laser at 840 nm were recorded on a special, sensitized holographic material from the I.V. Kurchatov Institute of Atomic Energy (IAE) [9.134]. IR-sensitized Russian material was also investigated by *Kalashnikov* et al. [9.135] for the recording of Fourier holograms at 850 nm.

A gallium-arsenide injection laser at 860 nm was used by *Zakharov* et al. [9.136] to record high-resolution microholograms on the IAE material. A

GaAlAs-GaAs double heterostructure laser has been employed by *Vorobev* et al. [9.137] to record holograms at 870-890 nm with I-880G plates. These plates have a resolution of about 1000 lines/mm and a sensitivity of 10^{-5} J/cm^2.

The IAE materials were also applied by *Gnatyuk* et al. [9.138] to record IR holograms exposed with a Q-switched Nd:YAG laser at 1054 nm. They reported a diffraction efficiency of 1% with this high-resolution material (>$1000\ell p$/mm, sensitivity $50mJ/cm^2$). In another investigation by *Dukhovnyi* et al. [9.139], domestic holographic materials have been tested, using an Nd:YAG laser at 1064 nm, free-running, Q-switched and mode locked. Two interesting results have been obtained from this investigation. The first one shows that the shortest pulses ($2 \cdot 10^{-10}$s) gave rise to a sensitivity which was approximately one order of magnitude higher than the sensitivity of the same material exposed to longer pulses. The second observation shows that a 0.5 % solution of TriEthanolAmine (TEA) highly increased the sensitivity of the material at the wavelength used. TEA is a well-known hypersensitizing agent for holography in the visible part of the spectrum.

Conventional silver-halide materials utilized for IR holography seem to work rather well in the region of 700 to 800 nm. Their sensitivity is, however, lower here than in the normal, visible part of the spectrum. The use of TEA hypersensitization can eventually increase sensitivity. If recordings are to be made in the region of 800 to 1100 nm, the material has to be sensitized using special IR dyes. The upper limit for dye-sensitized silver-halide materials lies at 1300 nm. Conventional holographic materials, for example Agfa 10E75, can be sensitized with different dyes in the following solution:

Distilled water	600 mℓ
Methanol	300 mℓ
1% Borax ($Na_2B_4O_7 \cdot 10H_2O$)	100 mℓ
Allocyanide solution in methanol 1:4000	8 mℓ .

The material has to be treated for 5 to 10 minutes, which is followed by a short wash and fast drying. The dye used here is called *allocyanide* by Agfa, whereas the same dye is called *neocyanide* by Kodak. This dye will increase the spectral sensitivity up to 900 nm. To obtain sensitivity between 900 and 1100 nm, a dye called *xenocyanide* can be used.

b) Infrared Presensitition Holography

It is possible to use silver-halide materials for IR recordings in the far end of the IR spectrum in a different way compared to the conventional principle. *InfraRed Presensitization Photography* (IRPP) is a technique which makes use of silver-halide materials for recordings in the far end of the infrared part of the spectrum [9.140-142]. As mentioned earlier, the energy of IR photons is normally far too low to eject photoelectrons. However, exposure to the IR radiation has a certain effect on the material. It is known that heat can increase the sensitivity of silver-halide materials. Therefore, the emulsion of the photographic material is thermally heated during the production process (Ostwald ripening). This step determines the potential light sensitivity of the material. During a second heat-treatment (the after-ripening), which takes place in the presence of sensitizing dyes, the material undergoes an important increase in its sensitivity. Heat treatment taking place after manufacturing can be regarded as a second after-ripening. For example, baking of the photographic material is regarded as a hypersensitization method in which the silver-ion concentration is increased to a significant level and which brings about sensitization of the material. (Compare Chap.6).

The IRPP technique is a hypersensitization procedure which works in the following way. The IR exposure causes an increase in the number or mobility of interstitial silver ions (Frenkel defects). These defects constitute sensitivity centers that will trap the arriving photoelectrons if the material is exposed to photons of sufficient energy. In this manner the IR exposure sensitizes the grains of the emulsion. Immediately after the IR exposure the material is exposed to visible light. The second exposure gives rise to formation of the latent image in the emulsion. After development, the more sensitive grains will produce darker areas than grains that have been exposed to very little or no IR light. To that extent we may say that the IR exposure is responsible for the formation of the image in the emulsion. The way in which the IR exposure affects the emulsion here is the opposite to the one which occurs when the well-known Herschel effect takes place, in which visible exposure followed by IR exposure results in desensitization.

In an investigation performed by *Juyal* et al. [9.143] this particular method was used for the recording of interferograms with a CO_2 laser and Kodak 649-F spectroscopic plates. A 4-Watt laser of 10.6 μm wavelength was used and the IR exposure of 20 seconds was immediately followed by a visible two seconds exposure provided by a tungsten lamp of 60 Watts. Diffuse light was achieved by covering the lamp with a translucent paper. The lamp was placed one meter away from the plate. Pulses from a CO_2 laser were used by *Gorlin* et al. [9.144] for IR recordings of photographic FT-SK film applying the same method. They used 70 ms IR pulses combined with 15 μs pulses of actinic light. A resolution and sensitivity investigation of this recording principle was also performed. They found that it was possible to obtain a density of 2.5 above a 0.58 fog level. The dynamic range was at least 224 and the minimum detectable energy was 5 mJ/cm^2.

The advantage of this method is that ordinary, high-resolution silver-halide materials can be used for IR recordings at very long wavelengths; the disadvantage - that the IR exposure energy must be several orders of magnitude higher than is usually required when the material is exposed to visible light.

c) Infrared Desensitization Holography

Graube [9.145] utilized the reversed effect (IR/white light in the opposite order compared to the description above) to record IR holograms on different popular holographic emulsions, such as Kodak 649-F, Kodak V-F,

Agfa 10E56, and Agfa 10E75, with a CW Nd:YAG laser (GTE Sylvania 605). First, he exposed the plates to a 100-W incandescent bulb at a distance of 40 cm for 10 seconds. After that, the plates were exposed to intersecting beams of the 1.06 μm IR radiation to give an exposure of 9000 J/cm^2! The material was developed in Kodak D-19, fixed, washed and dried. Some plates were bleached in bromine vapor and the diffraction efficiency of the gratings was measured using a HeNe laser. An efficiency of 25% was obtained from gratings exposed at 4000 J/cm^2 for 2000 seconds. The disadvantage of this low-sensitivity technique is somewhat compensated by the high energy that can be obtained from Nd:YAG lasers.

It is obvious that the IR radiation can both sensitize or desensitize a silver-halide emulsion. *Frazier* et al. [9.146] discussed this fact in a paper concerning conventional IR photography. They mentioned that the time delay between the nonactinic (IR) and actinic (white light) exposure is important. Extremely short IR pulses can photographically be recorded via this principle, as demonstrated by *Mitchel* et al. [9.147]. Concerning holographic emulsions *Mikhailov* et al. [9.148] as well as *Starobogatov* and *Nicolaev* [9.149] discussed the influence of short nonactinic pulses mixed with short actinic pulses.

9.3.2 IR Recordings in Non-Silver Materials

A number of non-silver materials and recording techniques can be considered for recording holograms in the infrared part of the electromagnetic spectrum [9.150-180]. The following summary gives a brief survey of the existing techniques making use of non-silver materials for hologram recording.

One technique for recording infrared holograms is a direct absorption of thermal energy in the material which means the induction of a spatially distributed temperature field on the surface of a material caused by the interference pattern to be recorded. This temperature field can then produce a physical or chemical change to yield a recording. At a relatively high average power from a CO_2 laser, material can be locally evaporated or sublimated from the surface of the recording material and thus record the interference pattern directly. Another possibility rests in the fact that temperature variations caused by thermal radiation are capable of selectively producing a chemical reaction or a change in some material property that can record the interference pattern. The normal holographic thermoplastic recording technique is also applicable in the IR region.

a) Two-Photon Recording Process

Another interesting method is the two-photon recording technique [9.150]. Here, two different light sources are employed. The first source, in the visible or UV, excites the recording material to a metastable state. The light from the first source does not contribute to the formation of a hologram and therefore it is not necessary for it to be monochromatic or coherent.

The second source, which emits the IR light, is used to produce the hologram. This light is absorbed by the metastable state, resulting in irreversible photochemistry. The system is self-developing since turning off the UV source stops the recording process but allows for the reconstruction of the hologram immediately afterwards, with the IR laser light.

Bräuchle et al. [9.151] used this particular technique for the recording of holograms in cyanoacrylate polymers in which a photoactive α-diketone was dissolved. Holograms were recorded at 752 nm and 1064 nm.

In an investigation by *Gerbig* et al. [9.152], a BiAcetyl (BA) dissolved in a PolyCyanoAcrylate (PCA) matrix was taken as a recording material. A Hg lamp (200W) was used for the UV illumination and a Krypton laser (752.5nm or 799.3nm) was used for the IR exposure. They also demonstrated the possibility of recording holograms (only $20\mu m$ in diameter) in the BA-PCA material, with a GaAlAs laser at 797 nm. A two-photon process was utilized by *Carré* et al. [9.153]. This process works only when both the UV and the IR are on at the same time. They prepared a mixture of a liquid and a solid monomer in which a dye (diphenyl isobenzofuran) was dissolved. IR exposure from a CW ring dye laser at 715 nm and 1.4 W/cm^2 was used. The UV radiation was provided by a mercury arc lamp at 365 nm and 2 mW/cm^2. This method [9.145] was later improved upon and a recording technique known as the two-photon two-product process was worked out and described by *Lougnot* et al. [9.154].

As already mentioned in the introduction, *Izawa* and *Kamiyama* [9.123] developed a photochromic emulsion for IR recording which consists of photochromic spiropirans (6'-methoxy,8'-nitro-1,3,3,-trimethylindolino-benzopyrylospiran) with methyl-methacrylate dissolved in tetrahydrofuran. The solution is coated on glass substrates leaving a 10 to 50 μm thick coating after drying.

b) PMMA and TAC Materials

Barkhudarov et al. [9.155] have been rather successful in recording high-efficiency IR holograms with PolyMethylMethAchrylate (PMMA) and films of TriAcetateCellulose (TAC). They used a pulsed CO_2 together with industrial TAC films of 0.2 mm thickness. Their paper specifies the sensitivities for both PMMA and TAC and compares the two, indicating that the use of TAC films ensures a higher diffraction efficiency than that of PMMA [9.156]. For the CO_2 laser radiation the TAC material has a sensitivity of about 0.8 J/cm^2 and the PMMA material 1.3 J/cm^2. In another investigation the TAC material was used, as outlined in [9.157]. Holograms were made using a 30 J pulsed CO_2 laser with a pulse length of 100 ns; the sensitivity of the material was about 0.8 J/cm^2. Three-dimensional objects were recorded and reconstructed with a HeNe laser.

c) Thin Solid and Liquid Layers

Thin layers ($10\div80$nm) of bismuth, antimony and cadmium vacuum deposited on glass or paraffin on glass substrates for IR recording have been described by *Decker* et al. [9.158]. The use of bismuth film has been reported by *Olsen* [9.159], where the film had a resolving power of 1730 lines/mm at 1060 nm and required an exposure of 50 mJ/cm². The material requires no development and is sensitive far into the infrared region. The bismuth film was compared with the PMMA and TAC materials in an investigation by *Kukharchik* et al. [9.160]. A similar effect to the one obtained in the earlier described IRPP technique was obtained in bismuth silicon oxide ($Bi_{12}SiO_{20}$) reported by *Kamshilin* and *Miteva* [9.161].

PolyVinyl Alcohol (PVA) on a glass substrate can be used for the recording of holograms at 10.6 μm, as described by *Durasov* et al. [9.162]. A PVA thickness of 30 μm and an exposure of 470 mJ/cm² were reported.

A thin oil film (about 1 μm thick) on a glass substrate has been used to record holograms at 10.6 μm [9.163, 164]. Applications of this recording technique for holographic interferometry have been reported by *Lewandowski* et al. [9.165, 166].

The thermohydrodynamic technique for recording IR holograms in thin liquid layers has been treated by *Belkin* et al. [9.167].

d) Thermoplastic Materials

Thermoplastic materials frequently used for holographic recordings in the visible part of the spectrum are also suitable for IR holography, as mentioned by *Colburn* et al. [9.168] and *Lo* et al. [9.169]. The Kalvar film, which is a thermoplastic resin primarily intended for UV recordings (developed by heat) has been recommended for IR recordings in a reversed Kalvar process [9.170].

e) Additional Materials

Another material for 10.6 μm holography is commercial wax (Takiwax), reported by *Beaulieu* et al. [9.171, 172]. The heated wax was dissolved in xylene and poured on a glass plate (thickness 10÷30 μm) heated to 65° C before the recording. The sensitivity was 0.2 J/cm².

Finally, the following materials have also been investigated: a thermochromic cuprous mercuric iodide (Cu_2HgI_4) as a recording material for 10.6 μm. [9.123, 173, 174]; cholesteric liquid crystals [9.124, 175, 176]; a thermosensitive material FTIROS (Russian acronym for "Phase transformation interference reversible reflection of light") [9.177]; two other materials with high sensitivity (about 1÷10 mJ/cm²) are CdTe and CdHgTe absorptive semiconducting thin films [9.178]. The use of vanadium oxide as a recording medium has also been reported [9.179]. *Slinger* et al. [9.180] reported on photodoped chalcogenides as a potential infrared holographic recording material.

9.4 Ultraviolet Holography

Publications on hologram recording in the ultraviolet part of the spectrum are scarce [9.181-196]. This is due to the fact that for a long time very few UV lasers had sufficient light quality for hologram recording and, in addition, suitable recording materials are difficult to find.

At the present time there are several UV lasers that can be used for hologram recording, e.g., nitrogen lasers, frequency tripled or quadrupled Nd:YAG, frequency doubled ruby, and Rare Gas Halide (RGH) lasers, such as ArF and XeF.

The recording materials considered suitable for UV holography are the photoresist, the photopolymer materials, and the dichromated gelatin emulsion. Silver-halide materials have also been used. They have sufficient sensitivity but the high scattering occurring already in the blue part of the spectrum will, of course, become even worse here. Another limitation of the silver halides is the fact that the UV light is highly absorbable in gelatin which will limit the recording depth of the emulsion. Despite these facts, silver-halide emulsions have been used for the recording of UV holograms with some success. In addition to the materials just mentioned various other materials can be considered for UV holography, about which later on.

The main applications of UV holography are:

- In the production of Holographic Optical Elements (HOEs), e.g., diffraction gratings with a grating period of 100 nm; to be used for IR applications.
- In holographic high-resolution imaging, where sub-micron resolution would be possible.
- In microcircuit lithography, offering a high resolution and a large numerical aperture imaging system.
- In holographic interferometry investigations using UV lasers, where the short wavelength makes the method more sensitive.

9.4.1 UV Recordings in Silver-Halide Materials

In the UV and the visible parts of the spectrum the most sensitive recording materials are the photographic silver-halide emulsions. For holography, however, their use is somewhat limited by their resolving power and high scattering, making them not really very suitable in these recording regions.

The first publication on UV holography was probably that of *Wuerker* et al. [9.181]. They generated the second harmonic (347nm) from a Q-switched ruby laser by sending the pulse through a KDP (potassium dihydrogen phosphate) crystal. A double UV pulse was applied to record an interferogram in which a double amount of interference fringes was obtained at 347 nm when compared to a normal 694 nm double-pulse recording. *Wuerker* et al. used the Kodak 649-F emulsion for the recording. The holograms were reconstructed with a HeNe laser. One interesting observation here was that the angular orientation selectivity of the UV hologram was lower for the UV recording than for the red recording when the reverse should apply since a shorter wavelength was used here (in a layer of a given thickness). The difference noted here can be attributed to the absorption power of the gelatin, making the effective thickness of the emulsion for the UV recording only 2.5 μm. The thickness of a fixed 649-F emulsion is about 12 μm.

Sasaki and *Hirose* [9.182] reported the use of a pulsed N_2 laser at 337 nm for the recording of in-line Fraunhofer holograms on the Japanese photographic material Fuji Minicopy film HR II. The holograms were developed in the Fuji Copinal developer. The test object was a 94 nm platinum wire.

Excimer-recorded UV holograms were reported by *Brannon* and *Asmus* [9.183]. Kodak High Resolution Plate 1A and Kodak Special Plate 125-02 were used for the recording. The high energy available from the e-beam pumped XeF laser was such that the UV sensitivity of the recording materials was not the main concern here. To limit the demand for resolution, the angle between the reference and the object beams was set to 13°. Of the two, the 1A material gave holograms of the highest quality.

Attwood et al. [9.184,185] studied laser plasma interactions using UV holographic interferometry. A frequency-tripled neodymium laser emitting 355-nm 115-ps pulses was employed in the earlier investigation and later a frequency quadrupled neodymium laser emitting 266 nm 15 ps pulses.

9.4.2 UV Recordings in Non-Silver Materials

As already mentioned, photoresist is one of the materials suitable for UV hologram recording. This and other materials that have been used in UV hologram recording will briefly be discussed in this subsection.

The well-known Shipley AZ 1350 photoresist has been used by *Shank* and *Schmidt* [9.186] for the recording of holographic gratings with a grating period of 110 nm, formed by two intersecting 325 nm HeCd beams from the 10 mW laser. The exposed area of the photoresist was about 20 mm² and the exposure time was about 7 s.

Another Shipley resist (AZ 2400) was employed by *Ross* et al. [9.187] to record resolution test masks using a commercial line-narrowed 249 nm KrF eximer laser. The photoresist layer was 1 μm thick and it was spun on a glass substrate. At this wavelength the resist has an absorption depth of about 0.3 μm. A holographic recording exposure of about 5 mJ/cm² was needed here.

Dichromated gelatin is a holographic recording material with high resolving power and which is sensitive in the blue and the UV parts of the spectrum. It has been used for the recording of UV holographic gratings by *Sosnowski* and *Kogelnik* [9.188]. A 5-mW HeCd laser was operated at 325 nm and with the intersecting beams at an angle of 20°. The emulsion thickness was 12 μm, sensitized with 1% ammonium dichromate only. Normally, a 5% ammonium dichromate concentration is used for visible-light recordings, but at this wavelength the emulsion's UV absorption is too high. A UV exposure of about 50 mJ/cm^2 was needed to obtain a diffraction efficiency of 90% (reconstructed in red light).

PolyMethyl MethAcrylate (PMMA) can be employed in both the IR and the UV parts of the spectrum. *Bjorklund* et al. [9.189] used this material in the Vacuum UltraViolet spectral region (VUV) to record gratings and Fraunhofer holograms, using the, so far, shortest UV wavelength. The ninth harmonic of the Nd:YAG laser used in the experiment meant a wavelength of only 118 nm. Multiple 12 ps pulses were used. The material was prepared by spinning a 3% solution of the PMMA in methyl ethyl ketone at 5000 rpm onto quartz flats with a diameter of 254 mm. The thickness of the layer was 140 nm. The material was then baked for half an hour at 170°C. After that the material was ready for the exposure to the UV pulses. A cumulative exposure of about 100 mJ/cm^2 was necessary, which corresponds to 5000 separate pulses. Holographic gratings (area 2mm^2) with a finest fringe spacing of 85 nm were produced. It was found that at this UV wavelength the penetration depth into the PMMA layer was less than 60 nm. Far-field Fraunhofer holograms of spherical particles ranging from 1.305 to 0.365 μm were also successfully recorded in the PMMA material, with the 118 nm wavelength.

In another investigation the PMMA material was applied for holographic gratings in deep ultraviolet. *Anderson* et al. [9.190] used an ArF eximer laser emitting a wavelength of 193 nm, with a coherence length of about 2 mm. The gratings recorded in the PMMA material had a grating period of 125 nm.

Some rather unusual recording materials have been recommended as suitable for the Extreme UltraViolet (EUV) holography. *Underwood* and *Barbee* [9.191] introduced the Layered Synthetic Microstructures (LSM) as potential recording materials for EUV holography. They presented the formalism for computing the intensity of Bragg reflection from structures layered on an atomic scale. These theoretical structures may very well push the research into X-ray and EUV holography a long step forward once they can be applied technologically.

Another potential candidate for UV holography, and one which has already met with some practical applications, is a layer of TS-diacetylene single crystals, described by *Richter* et al. [9.192]. They recorded holographic gratings with grating periods of 370 to 4400 nm, employing two interfering 257 nm laser beams and a frequency-doubled argon-ion laser, combined with an ADP (Ammonium Dihydrogen Phosphate) crystal. The above-mentioned high spatial resolution material has the sensitivity of about 220 mJ/cm^2 and requires no processing. To increase the sensitivity of the material, the hologram can be annealed for 6 hours at 60°C after the exposure. The diffraction efficiency of a recorded grating was 35%.

Finally, it should be mentioned that holograms intended for UV (or IR) reconstruction can be recorded at the visible wavelength. This fact can be important for some HOE applications. As regards gratings recorded in bleached, silver-halide materials, *Heaton* and *Solymar* [9.193] reported that a UV reconstruction was possible for gratings recorded in Agfa 8E56 HD emulsion at 514 nm. Normally, if the wavelength difference is larger the violation of the Bragg condition will make the reconstruction from a volume hologram unfeasible. Therefore, the method used here was to overexpose the material and enhance its nonlinearities during the processing. Such a procedure induces the refractive-index modulation at twice the frequency of the interference pattern. The Bragg condition is then satisfied since the grating vector is twice as large the vector of the recorded pattern. Based on such reasoning, the experiment succeeded in that a diffraction efficiency of a few percent was obtained in the region between 370 and 420 nm for gratings recorded at 514 nm. *Jannson* [9.194] employed dichromated gelatin to record HOEs which were reconstructed in the XUV region (10÷100nm). Another investigation performed by *Savant* et al. [9.195] utilized a completely new polymeric material for holography based on the graft concept. The material consists of an extremely dense photochemical crosslinking network of hydroxyalkyl acrylate-methacrylate, hard dichromated gelatin and transparent PVA. They demonstrated that a high-efficiency (up to 28% in the region of 11 to 83nm) UV/XUV volume diffraction elements could be produced, recorded in visible light (488nm).

In a recent publication *Yen* et al. [9.196] described a holographic configuration that employs two matched fused silica phase gratings and an ArF eximer laser for recording 100-nm gratings in polymethyl methacylate.

10. Recipes and Formulary

Table 10.1. Conventional photographic developers

CONSTITUENTS	AGFA 80	KODAK D-8	KODAK D-19	KODAK D-19b	KODAK D-76	KODAK D-82
Metol	2.5 g		2 g	2.2 g	2 g	14 g
Hydroquinone	10 g	45 g	8 g	8.8 g	5 g	14 g
Sodium sulfite (anhydrous)	100 g	90 g	90 g	72 g	100 g	52.5 g
Sodium carbonate (anhydrous)			52.5 g	48 g		
Sodium hydroxide		37.5 g				17.6 g
Sodium tetraborate (Borax)					2 g	
Potassium carbonate	60 g					
Potassium bromide	4 g	30 g	5 g	4 g		8.8 g
KODAK Antifog #1						0.2 g
Water (distilled)	1 l	1 l	1 l	1 l	1 l	1 l
Dilution	Use undiluted	2 parts dev + 1 part water	Use undiluted	Use undiluted	Use undiluted	Use undiluted
Developing time at 20°C	5 min	4-5 min	4-5 min	4-5 min	4-5 min	4-5 min

Table 10.2. Special photographic developers

CONSTITUENTS	KODAK SD-48 Tanning developer	LUMIERE Physical developer	POTA Wide-latitude developer	V-23 High-adjacency-effect developer
Part A:				
Ascorbic acid	1 g			
Metol				
Hydroquinone				3 g
Phenidone			1.5 g	8 g
Pyrocatechol	40 g			
Sodium sulfite (anhydrous)	8 g	180 g	30 g	
Sodium sulfate	100 g			
Potassium bromide				5 g
Potassium metabisulfite				30 g
Silver nitrate sol. (1/10)	15 g*	75 ml		
Water (distilled)	1 l	1 l	1 l	1 l
Part B:				
Metol		20 g		
Potassium carbonate				50 g
Potassium bromide				1 g
Sodium hydroxide	20 g			
Sodium sulfate	100 g			
Sodium sulfite (anhydrous)		20 g		5 g
Water (distilled)	1 l	1 l		1 l
Working solution:	Mix part A + part B	Mix 5 parts A + part B	Part A	Parts A and B should <u>not</u> be mixed
Developing time at 20°C	5 min	3 min	5 min	4 min in part A then 4 min in part B

*Normally omitted when used for holography

Table 10.3. Pyrogallol developers

CONSTITUENTS	LIPPMANN PHOTOGRAPHY					HOLOGRAPHY		
	Lippmann	Lumière	Valenta 1	Valenta 2	Lehmann	Neuhauss	Andreeva Sukhanov	van Renesse
Part A:								
Pyrogallol	1 g	1 g	1 g	4 g	1 g	1 g	1 g	1 g
Alcohol	10 ml							
Nitric acid (conc.)				6 drops	3 drops			
Water (distilled)		100 ml	100 ml	400 ml	100 ml	100 ml	100 ml	100 ml
Part B:								
Ammonia (s.w. 0.91)		5 ml	67 ml	14 ml	30 ml (25% sol.)	5 ml	30 ml	
Ammonium sulfite				12 g				
Ammonium carbonate	50 g							
Sodium carbonate		10 g	20 g	10 g	15 g	20 g	20 g	6 g
Potassium bromide	15 drops (10% sol.)							
Water (distilled)	300 ml	100 ml	200 ml	400 ml	150 ml	200 ml	240 ml	100 ml
Working solution: Mix A + B + water	A B 7.5 ml + 20 ml + 30 ml	A B 10 ml + 15 ml + 70 ml	A B 10 ml + 20 ml + 70 ml	A B 10 ml + 2/30 ml + 120/140 ml	A B 3 ml + 6 ml + 100 ml	A B 10 ml + 15 ml + 70 ml	A B 25 ml + 50 ml + 925 ml	A B 100 ml + 100 ml + 0
Developing time at 20°C:					1-3 min		9-12 min	2 min

Table 10.4. Russian holographic developers

CONSTITUENTS	GP	GP-2	GP-3	GP-8	GP-9	GP-11	FMG-1	PRG-1	UP-2
Methylphenidone	0.2 g	0.2 g		0.2 g	0.2 g	0.2 g	0.3 g		
Phenidone			0.2 g						
Metol							2 g		5 g
Hydroquinone	5 g	5 g	5 g	5 g	5 g	5 g	5 g	15 g	6 g
Sodium sulfite (anhydrous)	100 g	100 g	100 g	100 g	100 g	70 g	40 g	19 g	40 g
Sodium carbonate									31 g
Potassium carbonate							20 g	120 g	
Potassium hydroxide	5 g	5 g	25 g	10.6 g	20.3 g	7.8 g			
Ammonium thiocyanate		12 g*	45 g	24 g	48 g	13.5 g			
Potassium thiocyanate							3 g		
Sodium tetraborate (Borax)							15 g		
Potassium bromide							2 g	11 g	4 g
1-phenyl-5-mercaptotetrazole							0.2 g		
5-nitrobenzimidazole	1.2 g								
Water (distilled)	1 l	1 l	1 l	1 l	1 l	1 l	1 l	1 l	1 l
Dilution: (ml stock solution + ml water)	10 ml stock solution + 400 ml distilled water	15 ml stock solution + 400 ml distilled water	15 ml stock solution + 400 ml distilled water	60 ml stock solution + 400 ml distilled water	60 ml stock solution + 400 ml distilled water	15 ml stock solution + 400 ml distilled water	15 ml stock solution + 400 ml distilled water	40 ml stock solution + 260 ml distilled water	Undiluted
Developing time at 20°C	40 min	12-24 min (no agitation)	5-15 min	4-10 min	2-4 min	16-20 min	10-25 min	25-30 min	3-12 min

*can be substituted with 24 g potassium thiocyanate

Table 10.5. CW holographic developers

CONSTITUENTS	AGFA GP 61	ADUROL (Phillips)	MAA-3 (James & Vanselow)	MAA-6 (Skillman)	PAAP (Benton)	MAS (Saxby)
Metol	6 g		2.5 g	0.45 g		5 g
Ascorbic acid		10 g	10 g	3 g	18 g	20 g
Hydroquinone	7 g					
Chlorohydroquinone		2 g				
Phenidone	0.8 g				0.5 g	
Sodium sulfite (anhydrous)	30 g	30 g				
Sodium metaborate (Kodalk)		10 g				
Sodium tetraborate (Borax)				5 g		
Sodium phosphate (dibasic)						
Sodium carbonate (anhydrous)	60 g	60 g	55.6 g		28.4 g	20 g
Sodium hydroxide					12 g	6.5 g
Potassium bromide	2 g	5 g		1 g		1-2 g
Tetra sodium EDTA	1 g					
Water (distilled)	1 l	1 l	1 l	1 l	1 l	1 l
Dilution	Use undiluted	Use undiluted	Use undiluted	Use undiluted	Use undiluted	Use undiluted
Developing time	2 min at 20°C	3-4 min at 21-23°C	4 min at 20°C	4 min at 20°C	4 min at 20°C	12 min at 20°C
Hologram type	Transmission	Transmission	Reflection	---	Reflection	Reflection-master

Table 10.6. CW holographic developers

CONSTITUENTS	AGFA GP 62	CW-C2 (Cooke-Ward)	HOLODEV 602 (Bjelkhagen)	PYROCHROME (van Renesse)	PYROCHROME PLUS (Saxby)	ILFORD (Wood)	ILFORD (Phillips)	PYROME (Saxby)
Part A:								
Metol	15 g						5 g	5 g
Pyrogallol	7 g		50 g	10 g	20 g	12 g	10 g	15 g
Catechol		20 g				12 g		
Ascorbic acid		10 g						
Phenidone					1.2 g			
Sodium sulfite (anhydrous)	20 g	10 g	30-130 g					
Potassium metabisulfite			50 g					
Potassium bromide	4 g				30 g			
Urea		100 g					10 g	1-2 g
Tetra sodium EDTA	2 g							
Water (distilled)	1 l	1 l	1 l	1 l	1 l	1 l	1 l	1 l
Part B:								
Sodium carbonate	60 g	60 g	85 g	60 g	130 g	60 g	120 g	6.5 g
Sodium metaborate							4-20 g*	
Sodium hydroxide								
Water (distilled)	1 l	1 l	1 l	1 l	1 l	1 l	1 l	1 l
Dilution	Mix A + B + 2 parts water	Mix A + B	Mix A + B	Mix A + B	Mix A + B	Mix A + B	Mix A + B	Mix A + B
Development time	2 min at 20°C	2 min at 20°C	2.5 min at 25°C	2 min at 20°C	2-6 min at 20°C	3 min at 20°C	2 min at 20°C	6 min at 20°C
Hologram type	Reflection	Reflection	Reflection	Reflection	Reflection	Reflection	Reflection	Reflection

*Restraint of induction. Less metaborate - more stain.

Table 10.7. Solution-physical developers for holography

CONSTITUENTS	KODAK DK-20 (Smith)	IEDT-developer (Benton-James)	CPA-1 (Aliaga et al.)	Broadband AGFA (Leclère et al.)	Broadband ILFORD (Renotte et al.)
Metol	5 g				
Hydroquinone			0.65 g		
p-Phenylenediamine dihydrochloride		18.1 g			
Phenidone			0.02 g		
Pyrogallol			1.4 g	3 g	3 g
Potassium hydroxide					
Sodium hydroxide				2.3 g	2.5 g
Sodium metaborate (Kodalk)	2 g	*			
Sodium sulfite (anhydrous)	100 g	50 g	13 g		
Ammonium thiocyanate			3.1 g	5.5 g	4.4 g
Potassium thiocyanate	1 g				
Sodium thiocyanate	0.5 g	0.25 g			
Potassium bromide		1 g			
Water (distilled)	1 l	1 l	1 l	1 l	1 l
Dilution	Use undiluted	Undiluted	Use undiluted	Use undiluted	Use undiluted
Development time at 20° C	5 min	32 min	2 min	3 min	3 min
Hologram Type	Transmission	Reflection	Reflection	Reflection	Reflection

*Add Sodium metaborate to obtain a pH of 8.

Table 10.8. Developers for pulse holograms

CONSTITUENTS	D-19 Phenidone (Bjelkhagen)	RCA (Phillips)	Quephe (Unterseher)	Boone's pet (Boone)	SM-6 (Bjelkhagen)
Phenidone	1.5-2 g*	2 g	2 g		6 g
Hydroquinone	8.8 g	8 g	8 g		
Metol	2.2 g			2.5 g	
Ascorbic acid			10 g	10 g	18 g
Sodium sulfite (anhydrous)	72 g	30 g	30 g		12 g
Sodium hydroxide				45 g	
Sodium carbonate (anhydrous)	48 g	60 g	60 g		28.4 g
Sodium phosphate (dibasic)					
Disodium EDTA			2 g	2 g	
Potassium bromide				0.5 g	
Water (distilled)	1 l	1 l	1 l	1 l	1 l
Dilution	Use undiluted	Use undiluted	Use undiluted	Use undiluted	Use undiluted
Developing time	4-5 min at 20°C	2-5 min at 21-23°C	2-4 min at 19°C	4 min at 20°C	2 min at 20-25°C
Hologram type	Transmission	Transmission	Transmission/ Reflection	Reflection	Transmission/ Reflection

*1.5 g for the Agfa 10E material; 2 g for the Agfa 8E material

Table 10.9. Bleach solutions used in photographic intensifiers

CONSTITUENTS	Mercuric Intensifier Kodak IN-1	Chromium Intensifier Kodak IN-4	Mercuric Intensifier Ilford IN-2
Mercuric chloride	22.5 g		10 g
Mercuric iodide		90 g	
Potassium dichromate			
Potassium bromide	22.5 g		
Sodium sulfite (anhydrous)			100 g
Hydrochloric acid (C.P.)		64 ml	
Water (distilled)	1.5 l	1 l	1 l

Table 10.10. Conventional bleaches and reducers

CONSTITUENTS	Permanganate Kodak R-2	Farmer's Kodak R-4a	Haddon's Reducer	Dichromate Kodak R-9	Dichromate Kodak R-10	Persulfate Kodak R-15	Copper Kodak EB-2
Part A:							
Ammonium dichromate							
Potassium dichromate				18.8 g	20 g		
Potassium ferricyanide		37.5 g	10 g				
Potassium permanganate	52.5 g						
Potassium persulfate						30 g	
Cupric sulfate							120 g
Citric acid (monohydrate)							150 g
Sulfuric acid (conc.)					14 ml		
Potassium bromide							7.5 g
Water (distilled)	1 l	500 ml	1 l	1 l	1 l	1 l	1 l
Part B:							
Ammonium thiocyanate			20 g				
Sodium thiosulfate		480 g					
Sodium chloride					45 g*		
Hydrogen peroxide							30 ml
Sulfuric acid (conc.)	32 ml			24 ml		1.5 ml	
Water (distilled)	1 l	2 l	1 l	1 l	1 l	1 l	1 l
Dilution	Mix part A + B	30 ml part A + 120 ml Part B + 850 ml water	Mix part A + B	Mix part A + B	Mix part A + B + 10 parts water	Mix part A + B	Mix part A + B
Type	Reversal	Fixing	Fixing	Reversal	Rehalogenating	Reversal	Rehalogenating

*or potassium bromide 92 g, or potassium iodate 128 g ((for holography: potassium iodate 2 g))

Table 10.11a. Rehalogenating bleaching solutions for holography

CONSTITUENTS	Ferric nitrate I (Phillips)	Ferric nitrate II (Phillips)	Ferric nitrate (AGFA-GP431)	Ferric EDTA I (Phillips)	Ferric EDTA II (Phillips)	Ferric EDTA III (Phillips)
Ferric nitrate	150 g	50-100 g	150 g			
Ferric sodium EDTA				30-100 g		
Ferric sulfate					30 g	12 g
Disodium EDTA		15 g*			15-30 g	12 g
Sulfuric acid				10 ml	10 ml**	
Sodium bisulfate						50 g
Potassium bromide	30 g	30 g	30 g	30 g	30 g	30 g
Phenosafranine	300 mg		300 mg			
Glycerol	17.9 ml					
Ethyl alcohol			200 ml			
Isopropyl alcohol	0.5 l					
Water (distilled)	0.5 l	1 l	0.8 l	1 l	1 l	1 l
Dilution	Dilute 1 + 4	Undiluted	Dilute 1 + 4	Undiluted	Undiluted	Undiluted No agitation
Hologram Type	Transmission	Transmission/ Reflection	Transmission	Transmission/ Reflection	Transmission/ Reflection	Transmission/ Reflection
Processing type	Post-fixation	Post-fixation	Post-fixation	Fixation-free	Fixation-free	Fixation-free

*Without NaEDTA: Transmission-bleach; with NaEDTA: Reflection-bleach (fixation-free)
**Can be substituted with sodium hydrogen sulfate 30 g

Table 10.11b. Rehalogenating bleaching solutions for holography

CONSTITUENTS	PBQ-1 (Phillips)	PBQ-2 (Cooke & Ward)	PBQ-3 (Phillips)	PBQ AGFA GP 432	PBQ AGFA GP 433*	Copper 1 (Blyth)	Copper 2 (Blyth)
Cupric sulfate (pentahydrate)							35 g
Cupric bromide						130 g	
p-benzoquinone	2 g	2 g	2 g	2 g	2 g		
Acetic acid (glacial)						75 ml	10 ml
Boric acid	1.5 g			1.5 g	3 g		
Citric acid		15 g	15 g				
Sodium tetraborate							
Potassium bromide	30 g	50 g	30 g	50 g	30 g		110 g
Potassium iodide							
Potassium dichromate			2 g				
Water (distilled)	1 l	1 l	1 l	1 l	1 l	1 l	1 l
Dilution	Use undiluted	Use undiluted	Use undiluted	Use undiluted	Use undiluted	Use undiluted	Use undiluted
Hologram Type	Transmission/ Reflection	Transmission/ Reflection	Transmission/ Reflection	Transmission/ Reflection	Transmission/ Reflection	Reflection	Reflection

*Bleach for color shifting to a longer wavelength

Table 10.11c. Rehalogenating bleaching solutions for holography (based on oxidized developing agents)

CONSTITUENTS	PBU-quinol* Bjelkhagen-Phillips	PBU-amidol Bjelkhagen-Phillips	PBU-metol Bjelkhagen-Phillips	PBU-ascorbic acid** Bjelkhagen-Phillips
Amidol		1 g		
Ascorbic acid				5 g
Metol			1 g	
Hydroquinone (Quinol)	1 g			
Potassium persulfate	10 g	10 g	10 g	20 g
Citric acid or Sodium hydrogen sulfate	50 g	50 g	50 g	50 g
Cupric bromide	1 g	1 g	1 g	1 g
Potassium bromide	20 g	20 g	20 g	20 g
Water (distilled)	1 l	1 l	1 l	1 l
Dilution	Undiluted (does not work until 6 h after preparation)	Undiluted (does not work until 30 min after preparation)	Undiluted (does not work until 30 min after preparation)	Undiluted (does not work until 6 h after preparation)
Hologram type	Transmission/ Reflection	Transmission/ Reflection	Transmission/ Reflection	Transmission/ Reflection
Processing type	Fixation-free rehalogenation	Fixation-free rehalogenation	Fixation-free rehalogenation	Fixation-free rehalogenation

* can replace the regular PBQ type bleach
** This bleach introduces a slight emulsion shrinkage after processing.

Table 10.11d. Direct rehalogenating bleaching solutions

CONSTITUENTS	Bromine bleach (Benton)	Iodine bleach (Benton)	Iodine bleach (Berkhout)
Bromine	5 ml		
Iodine		2-5 g	5 g
Ethyl alcohol		750 ml*	
Methyl alcohol			900 ml
Water (distilled)	500 ml	250 ml	100 ml
Dilution	Use undiluted	Use undiluted	Use undiluted
Hologram type	Transmission	Transmission	Transmission
Processing type	Post-fixation	Post-fixation	Post-fixation

*or Methyl alcohol

Table 10.12. Reversal bleaching solutions for holography

CONSTITUENTS	Pyrochrome (van Renesse)	Pyrochrome (Boone)	Pyrochrome (Blyth)	Pyrochrome (Saxby)	Pyrochrome (Phillips)	Permanganate (Benton)	Permanganate (Phillips)	PSSB (Phillips)
Ammonium dichromate	4 g	3 g			20 g			
Ferric nitrate (or: sulfate)								
Potassium dicromate			5 g	15 g				30 g
Potassium persulfate								20 g
Potassium permanganate						3 g	0.5 g	
Nitric acid							0.25 ml	
Sulfuric acid	4 ml	1.5 ml			0.5 ml	10 ml		
Sodium hydrogen sulfate			80 g	15 g				30 g
Water (distilled)	1 l	1 l	1 l	1 l	1 l	1 l	1 l	1 l
Dilution	Use undiluted or Dilute: 1 + 4	Use undiluted	Use undiluted	Use undiluted	Use undiluted	Use undiluted	Use undiluted	Use undiluted
Hologram type	Reflection/ Transmission	Reflection/ Transmission	Reflection/ Transmission	Reflection/ Transmission	Reflection/ Transmission	Reflection	Reflection	Reflection/ Transmission

Table 10.13. Special holographic processing solutions

CONSTITUENTS	Stopbath Kodak SB-1a	Fixer (Nonhardening) Kodak F-24	Hardener Kodak SH-1	Hardener Kodak SH-5	Clearing bath Kodak CB-6
Sodium thiosulfate (hypo)		240 g			
Sodium sulfite (anhydrous)		10 g		50 g	15 g
Sodium bisulfite		25 g			
Sodium carbonate (monohydrate)			6 g	12 g	
Formaldehyde (37% sol.)			10 ml	5 ml	
Kodak antifog No. 2 (0.5%)				40 ml*	
Calgon					0.5 g
Acetic acid (glacial)	50 ml				
Water (distilled)	1 l	1 l	1 l	1 l	1 l
Processing time	30 sec	3 min	3 min	3 min	1 min

*optional

373

Table 10.14. Stain removers and clearing baths for holography

CONSTITUENTS	Stain remover Kodak S-13	Stain remover (Phillips)	Clearing bath (Buschmann)	Clearing bath (Phillips)
Solution A:				
Potassium permanganate	2.5 g	10 g		
Sulfuric acid (conc.)	8 ml			
Sodium sulfite			100 g	
Water (distilled)	1 l	1 l	1 l	
Solution B:				
Sodium bisulfite	10 g	10 g		
Sodium metabisulfite			2 g	
Sodium hydroxide				10 g
Water (distilled)	1 l	1 l	1 l	1 l
Processing time	1 min in Sol. A 1 min in Sol. B	1 min in Sol. A 1 min in Sol. B	1 min in mixed Sol. A + B	1 min in Sol. B

Table 10.15. Conversion tables

Avoirdupois to Metric Weight

Pounds	Ounces	Grains	Grams	Kilograms
1	16	7000	453.6	0.4536
0.0625	1	437.5	28.35	0.02835
		1	0.0648	
	0.03527	15.43	1	0.001
2.205	35.27	15430	1000	1

U.S. Liquid to Metric Measure

Gallons	Quarts	Ounces (Fluid)	Drams (Fluid)	Cubic Centimeters	Liters
1	4	128	1024	3785	3.785
0.25	1	32	256	946.3	0.9463
		1	8	29.57	0.02957
		0.125	1 (60 mins.)	3.697	0.003697
		0.03381	0.2705	1	0.001
0.2642	1.057	33.81	270.5	1000	1

Conversion Factors

Grains per 32 fluid oz multiplied by   0.06847   = grams per liter
Ounces per 32 fluid oz multiplied by 29.96   = grams per liter
Pounds per 32 fluid oz multiplied by 479.3   = grams per liter

Grams per liter multiplied by 14.60   = grains   per 32 fluid oz
Grams per liter multiplied by 0.03338 = ounces per 32 fluid oz
Grams per liter multiplied by 0.002086 = pounds per 32 fluid oz

Grams per liter approximately equals ounces per 30 quarts
Grams per liter approximately equals pounds per 120 gallons
Ounces (fluid) per 32 oz multiplied by 31.25 = cubic centimeters per liter
Cubic centimeters per liter multiplied by 0.032 = ounces (fluid) per 32 oz
cm x 0.3937 = inches   inches x 2.5400 = cm

Temperature Conversion

From °Celcius (Centigrade) → °Fahrenheit

$$°F = \frac{9 \cdot °C}{5} + 32$$

From °Fahrenheit → °Celcius (Centigrade)

$$°C = \frac{5}{9} (°F - 32)$$

CHEMICAL	FORMULA	MW	CAS #
Acetic acid	CH_3COOH	60.05	64-19-7
Acetone	$(CH_3)_2CO$	58.08	67-64-1
Acridine orange (see (Dimethyl amino)-acridine hydrocloride hydrate)			
Adurol (see Chlorohydroquinone)			
Aluminum chloride	$AlCl_3$	133.34	7446-70-0
Aluminum sulfate	$Al_2(SO_4)_3$	342.15	10043-01-3
Aluminum sulfate, octadecahydrate	$Al_2(SO_4)_3 \cdot 18H_2O$	666.42	7784-31-8
Amidol (see Diaminophenol dihydrochloride)			
o-Aminophenol	$NH_2C_6H_4OH$	109.13	95-55-6
p-Aminophenol	$NH_2C_6H_4OH$	109.13	123-30-8
p-Aminophenol hydrochloride	$NH_2C_6H_4OH \cdot HCl$	145.59	51-78-5
Ammonia (anhydrous)	NH_3	17.03	7664-41-7
Ammonia (liquid) (see Ammonium hydroxide)			
Ammonium alum	$NH_4Al(SO_4)_2 \cdot 12H_2O$	453.33	7784-26-1
Ammonium bromide	NH_4Br	97.94	12124-97-9
Ammonium carbonate	$(NH_4)_2CO_3$	96.09	506-87-6
Ammonium chloride	NH_4Cl	53.49	12125-02-9
Ammonium chromate	$(NH_4)_2CrO_4$	152.09	7788-89-9
Ammonium citrate, dibasic	$(NH_4)_2HC_6H_5O_7$	226.19	3012-65-5
Ammonium dichromate	$(NH_4)_2Cr_2O_7$	252.10	7789-09-5
Ammonium hydroxide	NH_4OH	35.05	1336-21-6
Ammonium hyposulfite (see Ammonium thiosulfate)			

CHEMICAL	FORMULA	MW	CAS #
Ammonium persulfate	$(NH_4)_2S_2O_8$	228.19	7727-54-0
Ammonium rhodanide (see Ammonium thiocyanate)			
Ammonium sulfate	$(NH_4)_2SO_4$	132.12	7783-20-2
Ammonium sulfite	$(NH_4)_2SO_3$	134.20	10196-04-0
Ammonium thiocyanate	NH_4SCN	76.12	1762-95-4
Ammonium thiosulfate	$(NH_4)_2S_2O_3$	148.21	7783-18-8
Auramine O	---	303.84	2465-27-2
L-Ascorbic acid	$CH_2OHCHOH(CHCOH:COHCOO)$	176.12	50-81-7
L-Ascorbic acid, sodium salt	$CH_2OHCHOH(CHCOH:COHCOO)Na$	198.11	134-03-2
Benzene	C_6H_6	78.11	71-43-2
Benzodiazole (see Benzimidazole)			
Benzimidazole	$C_7H_6N_2$	118.14	51-17-2
Benzoic acid	C_6H_5COOH	122.12	65-85-0
Benzole (see Benzene)			
p-Benzoquinone	$C_6H_4O_2$	108.10	106-51-4
Benzotriazole	$C_6H_4NHN_2$	119.13	95-14-7
Benzyl alcohol	$C_6H_5CH_2OH$	108.14	100-51-6
Borax (see Sodium tetraborate, decahydrate)			
Boric acid	H_3BO_3	61.83	10043-35-3
Bromine	Br_2	159.81	7726-95-6
Calcium carbonate	$CaCO_3$	100.09	471-34-1
Calcium hydroxide	$Ca(OH)_2$	74.09	1305-62-0
Calcium hypochlorite	$Ca(OCl)_2$	142.99	7778-54-3
Calgon (see Sodium hexametaphosphate)			

CHEMICAL	FORMULA	MW	CAS #
Calomel (see Mercurous chloride)			
Carbamide (see Urea)			
Carbon tetrachloride	CCl_4	153.82	56-23-5
Carbonyl diamide (see Urea)			
Catechol	C_6H_4-1,2-$(OH)_2$	110.11	120-80-9
Caustic soda (see Sodium hydroxide)			
Cellulose triacetate	$[C_6H_7O_2(OOCCH_3)_3]_{n\ 250<n<300}$	$(288.12)_n$	9012-09-3
Ceric (Cerium IV) sulfate	$Ce(SO_4)_2$	332.24	13590-82-4
Chlorohydroquinone	ClC_6H_3-1,4-$(OH)_2$	144.56	615-67-8
Chrome alum	$KCr(SO_4)_2 \cdot 12H_2O$	499.43	7789-99-0
Chromic acid	CrO_3	100.01	1333-82-0
Citric acid	$HOC(COOH)(CH_2COOH)_2$	192.12	77-92-9
Citric acid, monohydrate	$HOC(COOH)(CH_2COOH)_2 \cdot H_2O$	210.14	5949-29-1
Corrosive sublimate (see Mercuric chloride)			
Cupric bromide	$CuBr_2$	223.37	7789-45-9
Cupric chloride	$CuCl_2$	134.45	7447-39-4
Cupric chloride, dihydrate	$CuCl_2 \cdot 2H_2O$	170.48	10125-13-0
Cupric sulfate	$CuSO_4$	159.60	7758-98-7
Cupric sulfate, pentahydrate	$CuSO_4 \cdot 5H_2O$	249.68	7758-99-8
Decahydronaphthalene	$C_{10}H_{18}$	138.25	91-17-8
Decalin (see Decahydronaphthalene)			
2,4-Diaminophenol dihydrochloride	$(NH_2)_2C_6H_3OH \cdot 2HCl$	197.06	137-09-7
3,7-Diamino-5-phenyl phenazinium chloride	$C_{18}H_{15}ClN_4$	322.81	81-93-6
Diethanolamine	$(CH_2OHCH_2)_2NH$	105.14	111-42-2

CHEMICAL	FORMULA	MW	CAS #
Diethylamine	$(C_2H_5)_2NH$	73.14	109-89-7
1,1'-Diethyl-2,2'-carbo cyanine bromide	$C_{23}H_{25}N_2Br$	433.40	2670-67-9
1,1'-Diethyl-2,2'-carbo cyanine chloride	$C_{23}H_{25}N_2Cl$	388.94	2768-90-3
1,1'-Diethyl-2,2'-carbo cyanine iodide	$C_{25}H_{25}N_2I$	480.39	605-91-4
1,1'-Diethyl-2,2'-cyanine iodide	$C_{23}H_{23}N_2I$	454.36	977-96-8
1,1'-Diethyl-2,4'-cyanine iodide	$C_{23}H_{23}N_2I$	454.36	634-21-9
Diethylene glycol	$(HOCH_2CH_2)_2O$	106.12	111-46-6
Diethylenetriamine	$(NH_2CH_2CH_2)_2NH$	103.17	111-40-0
Diethylenetriaminepentaacetic acid (DPTA)	$(CH_2COOH)_5N_3(CH_2)_4$	393.35	67-43-6
DTPA pentasodium salt	$(CH_2COOH)_5N_3(CH_2)_4Na_5$	508.30	140-01-2
3,3'-Diethylthiadicarbocyanine iodide	$C_{23}H_{23}IN_2S_2$	518.47	905-97-5
2,4-Dihydroxybenzophenone	$(HO)_2C_6H_3COC_6H_5$	214.22	131-56-6
Elon (see p-Methylaminophenol sulfate)			
Erythrosin B (see Tetraiodfluoresine disodium salt)			
Eosin Y (see Tetrabromofluorescein disodium salt)			
Ethanol (see Ethyl alcohol)			
Ethanolamine	$NH_2CH_2CH_2OH$	61.08	141-43-5
Ethyl acetate	$CH_3COOCH_2CH_3$	88.11	141-78-6
Ethyl alcohol	C_2H_5OH	46.07	64-17-5
Ethylamine	$C_2H_5NH_2$	45.08	75-04-7
Ethylenediamine	$NH_2CH_2CH_2NH_2$	60.10	107-15-3
Ethylenediamine sulfate	$NH_2CH_2CH_2NH_2H_2SO_4$	158.17	25723-52-8

CHEMICAL	FORMULA	MW	CAS #
Ethylenediaminetetraacetic acid (EDTA)	$(HOOCCH_2)_2NCH_2CH_2N$ $(CH_2COOH)_2$	292.24	60-00-4
EDTA sodium salt	$(EDTA)Na$	314.24	17421-79-3
EDTA disodium salt	$(EDTA)Na_2$	372.23	139-33-3
EDTA trisodium salt	$(EDTA)Na_3$	376.20	150-38-9
EDTA tetra sodium salt	$(EDTA)Na_4$	380.17	64-02-8
EDTA tetra sodium salt, tetrahydrate	$(EDTA)Na_4 \cdot 4H_2O$	452.24	464-02-8
EDTA ferric sodium salt	$(EDTA)FeNa$	367.07	15708-41-5
Ethylene glycol	$HOCH_2CH_2OH$	62.07	107-21-1
Ferric ammonium sulfate	$FeNH_4(SO_4)_2 \cdot 12H_2O$	482.19	7783-83-7
Ferric chloride	$FeCl_3$	160.20	7705-08-0
Ferric chloride, hexahydrate	$FeCl_3 \cdot 6H_2O$	270.32	10025-77-1
Ferric nitrate, nonahydrate	$Fe(NO_3)_3 \cdot 9H_2O$	404.00	7782-61-8
Ferric sulfate	$Fe_2(SO_4)_3$	399.90	10028-22-5
Ferrous sulfate, heptahydrate	$FeSO_4 \cdot 7H_2O$	278.02	7782-63-0
Formaldehyde	$HCHO$	30.03	50-00-0
Formalin (see Formaldehyde)			
Formic acid	$HCOOH$	46.03	64-18-6
Gallic acid	$3,4,5 - (HO)_3C_6H_2COOH$	170.12	149-91-7
Gelatin	---	15000 - 250000	9000-70-8
Glycerin (see Glycerol)			
Glycerol	$CH_2OHCHOHCH_2OH$	92.09	56-81-5
Glycin (photographic)	$HOC_6H_4NH(CH_2COOH)$	167.16	122-87-2
Gold chloride, trihydrate	$HAuCl_4 \cdot 3H_2O$	398.83	16961-25-4
Hydrazine	$NH_2 \cdot NH_2$	32.05	302-01-2
Hydrobromic acid	HBr	80.91	10035-10-6

CHEMICAL	FORMULA	MW	CAS #
Hydrochloric acid	HCl	36.46	7647-01-0
Hydrofluoric acid	HF	20.01	7664-39-3
Hydrogen peroxide	H_2O_2	34.01	7722-84-1
Hydroquinone	C_6H_4 -1,4 -$(OH)_2$	110.11	123-31-9
Hydroxylamine	NH_2OH	33.03	7803-49-8
Hydroxylamine hydrochloride	$NH_2OH \cdot HCl$	69.49	5470-11-1
Hydroxylamine sulfate	$(NH_2OH)_2 \cdot H_2SO_4$	164.14	10039-54-0
Hypo (see Sodium thiosulfate, penta-hydrate)			
Iodine	I_2	253.81	7553-56-2
β-Ionone	$C_{13}H_{20}O$	192.30	79-77-6
Isocyanine iodide (see 1,1'-Diethyl-2,4'-cyanine iodide)			
Isopropanol (see Isopropyl alcohol)			
Isopropyl alcohol	$(CH_3)_2CHOH$	60.10	67-63-0
Kodak Anti-fog #1 (see Benzotriazole)			
Kodak Anti-fog #2 (see 6-Nitro-benzimidazole)			
Kodak Anti-cal #3 (see Dihydroxybenzophenone)			
Kodalk (see Sodium metaborate)			
Kryptocyanine (see 1,1'-Diethyl-4,4'-carbocyanine iodide)			
Lactic acid	$CH_3CH(OH)COOH$	90.08	598-82-3
Mercaptoacetic acid	$HSCH_2COOH$	92.11	68-11-1
Mercuric chloride	$HgCl_2$	271.50	7487-94-7
Mercuric iodide	HgI_2	454.40	7774-29-0
Mercurous chloride	Hg_2Cl_2	472.09	7546-30-7

CHEMICAL	FORMULA	MW	CAS #
Methanol (see Methyl alcohol)			
3-Methoxycatechol	$CH_3OC_6H_3$-1,2-$(OH)_2$	140.14	934-00-9
Methyl alcohol	CH_3OH	32.04	67-56-1
Methylamine	CH_3NH_2	31.06	74-89-5
Methylamine hydrochloride	$CH_3NH_2 \cdot HCl$	67.52	593-51-1
p-Methylaminophenol sulfate	$(HOC_6H_4NHCH_3)_2 \cdot H_2SO_4$	344.32	55-55-0
1-Methyl-2-p-dimethylamino styryl pyridine	$C_{17}H_{21}N_2I$	380.28	3785-01-1
Methyl phenidone	$C_{10}H_{12}N_2O$	176.19	2654-57-1
Metol (see p-Methylaminophenol sulfate)			
Naphtha	---	---	MX8030-31-7
Nigrosin	---	---	8005-03-6
Nitric acid	HNO_3	63.01	7697-37-2
5-Nitrobenzimidazole	$5-NO_2(C_7H_5N_2)$	163.14	94-52-0
6-Nitrobenzimidazole nitrate	$5-NO_2(C_7H_5N_2) \cdot H$	226.10	27896-84-0
2,2-Oxydiethanol (see Diethylene glycol)			
Orange II	$C_{16}H_{11}N_2O_4SNa$	350.30	633-96-5
Oxalic acid	$(-COOH)_2$	90.03	144-62-7
Paraformaldehyde	$(CH_2O)_x$	$(30.03)_x$	30525-89-4
Phenidone (A) (see 1-Phenyl-3-pyrazolidone)			
Phenidone (B) (see Methyl phenidone)			
Phenosafranine (see 3,7-Diamino-5-phenylphenazinium chloride)			
o-Phenylendiamine	$C_6H_4(NH_2)_2$	108.14	95-54-5
p-Phenylendiamine	$C_6H_4(NH_2)_2$	108.14	106-50-3
p-Phenylendiamine dihydrochloride	$C_6H_4(NH_2)_2 \cdot 2HCl$	181.07	624-18-0

CHEMICAL	FORMULA	MW	CAS #
1-Phenyl-3-pyrazolidone	$C_6H_5\text{-}C_3H_5N_2O$	162.19	92-43-3
Pinacyanol bromide (see 1,1'-Diethyl-2,2'-carbocyanine bromide)			
Pinacyanol chloride (see 1,1'-Diethyl-2,2'-carbocyanine chloride)			
Pinacyanol iodide (see 1,1'-Diethyl-2,2'-carbocyanine iodide)			
Phosphoric acid	H_3PO_4	97.99	7664-38-2
Pinaflavole (see 1-Methyl-2-p-dimethylaminoststyryl pyridine)			
Polyester	$CH_3O[OC\text{-}C_6H_4\text{-}COOCH_2CH_2O]_n \bullet H$	---	---
Potassium alum	$KAl(SO_4)_2 \bullet 12H_2O$	474.40	7784-24-9
Potassium biborate (see Potassium tetraborate)			
Potassium bisulfate (see Potassium hydrogen sulfate)			
Potassium borohydride	KBH_4	53.94	13762-51-1
Potassium bromide	KBr	119.01	7758-02-3
Potassium carbonate	K_2CO_3	138.21	584-08-7
Potassium chloride	KCl	74.56	7447-40-7
Potassium chromate	K_2CrO_4	194.21	7789-00-6
Potassium citrate	$K_3C_6H_5O_7 \bullet H_2O$	324.40	6100-05-6
Potassium cyanide	KCN	65.11	151-50-8
Potassium dichromate	$K_2Cr_2O_7$	294.19	7778-50-9
Potassium ferricyanide	$K_3Fe(CN)_6$	329.26	13746-66-2
Potassium ferrocyanide, trihydrate	$K_4Fe(CN)_6 \bullet 3H_2O$	422.39	14459-95-1
Potassium hydrogen sulfate	$KHSO_4$	136.20	7646-93-7
Potassium hydroxide	KOH	56.11	1310-58-3
Potassium iodate	KIO_3	214.00	7758-05-6
Potassium iodide	KI	166.01	7681-11-0
Potassium metabisulfite	$K_2S_2O_5$	222.33	16731-55-8

CHEMICAL	FORMULA	MW	CAS #
Potassium nitrate	KNO_3	101.11	7757-79-1
Potassium permanganate	$KMnO_4$	158.03	7722-64-7
Potassium persulfate	$K_2S_2O_8$	270.33	7727-21-1
Potassium phosphate, dibasic	K_2HPO_4	174.17	7758-11-4
Potassium phosphate, dibasic, trihydrate	$K_2HPO_4 \cdot 3H_2O$	228.24	16788-57-1
Potassium pyrosulfate	$K_2S_2O_7$	254.30	7790-62-7
Potassium pyrosulfite (see Potassium metabisulfite)			
Potassium rhodanide (see Potassium thiocyanate)			
Potassium sulfate	K_2SO_4	174.27	7778-80-5
Potassium sulfite	K_2SO_3	158.27	10117-38-1
Potassium tetraborate	$K_2B_4O_7 \cdot 4H_2O$	305.50	12045-78-2
Potassium thiocyanate	KSCN	97.18	333-20-0
Propanol (see Propyl alcohol)			
Propyl alcohol	$CH_3CH_2CH_2OH$	60.10	71-23-8
Pseudocyanine iodide (see 1,1'-Diethyl-2,2'-cyanine iodide)			
Pyrocatechol (see Catechol)			
Pyrogallol	$C_6H_3-1,2,3-(OH)_3$	126.11	87-66-1
Quinaldine blue (see 1,1'-Diethyl-2,2'-carbocyanine chloride)			
Quinol (see Hydroquinone)			
Quinone (see p-Benzoquinone)			
Rubinol (see Methoxycatechol)			
Sensitol red (see 1,1'-Diethyl-2,2'-carbocyanine iodide)			

CHEMICAL	FORMULA	MW	CAS #
Silver bromide	AgBr	187.80	7785-23-1
Silver chloride	AgCl	143.32	7783-90-6
Silver iodide	AgI	234.77	7783-96-2
Silver nitrate	$AgNO_3$	169.87	7761-88-8
Silver sulfide	Ag_2S	247.80	21548-73-2
Silver thiocyanate	AgSCN	165.95	1701-93-5
Sodium acetate	CH_3COONa	82.03	127-09-3
Sodium ascorbate	$C_6H_7O_6Na$	198.11	134-03-2
Sodium bicarbonate	$NaHCO_3$	84.01	144-55-8
Sodium bisulfate	$NaHSO_4$	120.06	7681-38-1
Sodium bisulfate, monohydrate	$NaHSO_4 \cdot H_2O$	138.07	10034-88-5
Sodium bisulfite	$NaHSO_3$	104.07	7631-90-5
Sodium borate	$Na_2B_4O_7$	201.26	1344-90-7
Sodium borate, decahydrate	$Na_2B_4O_7 \cdot 10H_2O$	381.37	1303-96-4
Sodium borohydride	$NaBH_4$	37.83	16940-66-2
Sodium bromide	NaBr	102.89	7647-15-6
Sodium carbonate, anhydrous	Na_2CO_3	105.99	497-19-8
Sodium carbonate, monohydrate	$Na_2CO_3 \cdot H_2O$	124.00	5968-11-6
Sodium carbonate, decahydrate	$Na_2CO_3 \cdot 10H_2O$	286.14	6132-02-1
Sodium chloride	NaCl	58.44	7647-14-5
Sodium chromate	Na_2CrO_4	161.97	7775-11-3
Sodium chromate, tetrahydrate	$Na_2CrO_4 \cdot 4H_2O$	234.06	10034-82-9
Sodium citrate, dihydrate	$C_6H_5Na_3O_7 \cdot 2H_2O$	294.10	6132-04-3
Sodium dichromate	$Na_2Cr_2O_7$	262.01	10588-01-9
Sodium dichromate, dihydrate	$Na_2Cr_2O_7 \cdot 2H_2O$	298.00	7789-12-0
Sodium hexametaphosphate	$(NaPO_3)_6$	611.17	10124-56-8

CHEMICAL	FORMULA	MW	CAS #
Sodium hydrogen sulfate (see Sodium bisulfate)			
Sodium hydrogen sulfite (see Sodium bisulfite)			
Sodium hydrosulfite	$Na_2S_2O_4$	174.11	7775-14-6
Sodium hydroxide	NaOH	40.00	1310-73-2
Sodium hypochlorite, pentahydrate	$NaClO \cdot 5H_2O$	164.53	7681-52-9
Sodium hypophosphite, monohydrate	$NaH_2PO_2 \cdot H_2O$	105.99	10039-56-2
Sodium hyposulfite (see Sodium thiosulfate)			
Sodium iodide	NaI	149.89	7681-82-5
Sodium metabisulfite	$Na_2S_2O_5$	190.10	7681-57-4
Sodium metaborate	$NaBO_2$	65.80	7775-19-1
Sodium metaborate tetrahydrate	$NaBO_2 \cdot 4H_2O$	137.88	10555-76-7
Sodium nitrate	$NaNO_3$	84.99	7631-99-4
Sodium orthophosphate, dodecahydrate	$Na_3HPO_4 \cdot 12H_2O$	381.13	10101-89-0
Sodium permanganate, monohydrate	$NaMnO_4 \cdot H_2O$	159.94	79048-36-5
Sodium persulfate	$Na_2S_2O_8$	238.10	7775-27-1
Sodium phosphate, dibasic	Na_2HPO_4	141.96	7558-79-4
Sodium phosphate, dibasic, monohydrate	$Na_2HPO_4 \cdot H_2O$	159.97	10049-21-5
Sodium phosphate, dibasic, heptahydrate	$Na_2HPO_4 \cdot 7H_2O$	268.07	7782-85-6
Sodium pyrophosphate, decahydrate	$Na_4P_2O_7 \cdot 10H_2O$	446.06	13472-36-1
Sodium rhodanide (see Sodium thiocyanate)			

386

CHEMICAL	FORMULA	MW	CAS #
Sodium sulfate	Na_2SO_4	142.04	7757-82-6
Sodium sulfide	Na_2S	78.05	1313-82-2
Sodium sulfide, nonahydrate	$Na_2S \cdot 9H_2O$	240.20	1313-84-4
Sodium sulfite, anhydrous	Na_2SO_3	126.04	7757-83-7
Sodium sulfite, crystal, heptahydrate	$Na_2SO_3 \cdot 7H_2O$	252.06	7557-83-9
Sodium tetraborate, anhydrous	$Na_2B_4O_7$	201.27	1330-43-4
Sodium tetraborate, pentahydrate	$Na_2B_4O_7 \cdot 5H_2O$	291.29	12179-04-3
Sodium tetraborate, decahydrate	$Na_2B_4O_7 \cdot 10H_2O$	381.37	1303-96-4
Sodium thiocyanate	$NaSCN$	81.07	540-72-7
Sodium thiosulfate, anhydrous	$Na_2S_2O_3$	158.11	7772-98-7
Sodium thiosulfate, crystal, pentahydrate	$Na_2S_2O_3 \cdot 5H_2O$	248.18	10102-17-7
D-Sorbitol	$[CH_2OH(CHOH)_4CH_2OH]$	182.17	50-70-4
Stoddard solvent (See Naphtha)			
Sulfuric acid	H_2SO_4	98.07	7664-93-9
Tartaric acid	$C_4H_6O_6$	150.09	87-69-4
Tartaric acid, monohydrate	$C_4H_6O_6 \cdot H_2O$	168.10	147-73-9
Tetrabromofluorescein disodium salt	$C_{20}H_8Br_4O_5Na_2$	691.89	17372-87-1
Tetraiodfluoresine disodium salt	$C_{20}H_6I_4O_5Na_2$	879.87	568-63-8
Thiocarbamide (see Thiourea)			
Thioglycolic acid (see Mercaptoacetic acid)			
Thiourea	H_2NCSNH_2	76.12	62-56-6
Toluene	$C_6H_5CH_3$	92.14	108-88-3
Triethanolamine	$(HOCH_2CH_2)_3N$	149.19	102-71-6
Urea	NH_2CONH_2	60.06	57-13-6

CHEMICAL	FORMULA	MW	CAS #
Vitamin C (see Ascorbic acid)			
Washing soda (see Sodium carbonate, decahydrate)			
Water	H_2O	18.016	7732-18-5
m-Xylene	$C_6H_4(CH_3)_2$	106.17	108-38-3
o-Xylene	$C_6H_4(CH_3)_2$	106.17	95-47-6
p-Xylene	$C_6H_4(CH_3)_2$	106.17	106-42-3
Zinc bromide	$ZnBr_2$	225.18	7699-45-8
Zinc chloride	$ZnCl_2$	136.29	7646-85-7
Zinc sulfate, monohydrate	$ZnSO_4 \cdot 1H_2O$	175.45	7733-02-0
Zinc sulfate, heptahydrate	$ZnSO_4 \cdot 7H_2O$	287.54	7446-20-0

References

1.1 S.S. Shushurin: On the history of holography. Sov. Phys. USPEKHI 14, 655-657 (1972)

1.2 M. Wolfke: Über die Möglichkeit der optischen Abildung von Molekulargittern [Transl.: On the possibility of the optical recording of a molecular lattice]. Physik. Z. 21, 495-497 (1920)

1.3 M.A. Cotton: Réseaux obtenus par la photographie des ondes stationnaires [Transl.: Patterns obtained from the photography of stationary waves]. Soc. Franc. Phys., Séance du 5 Juillet, 70*-73* (1901)

1.4 D. Gabor: A new microscopic principle. Nature 161, 777-778 (1948)

1.5 D. Gabor: Microscopy by reconstructed wave-fronts. Proc. Roy. Soc. (London) A 197, 454-487 (1949)

1.6 D. Gabor: Microscopy by reconstructed wave-fronts: II. Proc. Phys. Soc. (London) 64, Pt.6, 449-469 (1951)

1.7 D. Gabor: The outlook for holography. Optik 28, 437-441 (1968)

1.8 E.N. Leith, J. Upatnieks: Reconstructed wavefronts and communication theory. J. Opt. Soc. Am. 52, 1123-1130 (1962)

1.9 E.N. Leith, J. Upatnieks: Wavefront reconstruction with continuous-tone transparencies. J. Opt. Soc. Am. 53, 522A (1963)

1.10 E.N. Leith, J. Upatnieks: Wavefront reconstruction with continuous-tone objects. J. Opt. Soc. Am. 53, 1377-1381 (1963)

1.11 E.N. Leith, J. Upatnieks: Wavefront reconstruction with diffused illumination and three-dimensional objects. J. Opt. Soc. Am. 54, 1295-1301 (1964)

1.12 Yu.N. Denisyuk: Dokl. Akad. Nauk. (Academy of Sciences Reports, USSR) 144, 1275-1278 (1962) [Transl.: Photographic reconstruction of the optical properties of an object in its own scattered radiation field. Sov. Phys. Doklady 7, 543-545 (1962)]

1.13 Yu.N. Denisyuk: On the reproduction of the optical properties of an object by the wave field of its scattered radiation. Opt. Spectrosc. (USSR) 14, 279-284 (1963)

1.14 Yu.N. Denisyuk, I.R. Protas: Improved Lippmann photographic plates for recording stationary light waves. Opt. Spectrosc. (USSR) 14, 381-383 (1963)

1.15 G.W. Stroke: *An Introduction to Coherent Optics and Holography* (Academic, New York 1966)

1.16 H.M. Smith (ed.): *Principles of Holography* (Wiley Interscience, New York 1969)

1.17 R.J. Collier, C.B. Burckhardt, L.H. Lin: *Optical Holography* (Academic, New York 1971)

1.18 R.K. Erf (ed.): *Holographic Nondestructive Testing* (Academic, New York 1974)

1.19 H.J. Caulfield (ed.): *Handbook of Optical Holography* (Academic, New York 1979)

1.20 C.M. Vest: *Holographic Interferometry* (Wiley, New York 1979)

1.21 W. Schumann, M. Dubas: *Holographic Interferometry*, Springer Ser. Opt. Sci., Vol.16 (Springer, Berlin, Heidelberg 1979)

1.22 G. von Bally (ed.): *Holography in Medicine and Biology*, Springer Ser. Opt. Sci., Vol.18 (Springer, Berlin, Heidelberg 1979)

1.23 Yu.I. Ostrovsky, M.M. Butusov, G.V. Ostrovskaya: *Interferometry by Holography*, Springer Ser. Opt. Sci., Vol.20 (Springer, Berlin, Heidelberg 1980)
1.24 W. Schumann, J.-P. Zürcher, D. Cuche: *Holography and Deformation Analysis*, Springer Ser. Opt. Sci., Vol.46 (Springer, Berlin, Heidelberg 1985)
1.25 Yu.I. Ostrovsky, V.P. Shchepinov, V.V. Yakovlev: *Holographic Interferometry in Experimental Mechanics*, Springer Ser. Opt. Sci., Vol.60 (Springer, Berlin, Heidelberg 1991)
1.26 N.H. Abramson: *The Making and Evaluation of Holograms* (Academic, New York 1981)
1.27 P. Hariharan: *Optical Holography, Principles, Techniques and Applications* (Cambridge Univ. Press, New York 1986)
1.28 G. Saxby: *Practical Holography* (Prentice Hall, New York 1988)
1.29 G. Saxby: *The Manual of Practical Holography* (Focal, Butterworth-Heinemann, Oxford 1991)
1.30 H.M. Smith (ed.): *Holographic Recording Materials*, Topics Appl. Phys., Vol.20 (Springer, Berlin, Heidelberg 1977)
1.31 N.I. Kirillov: [Transl.: High Resolution Photographic Materials for Holography and Their Processing Methods (in Russian)] (Nauka, Moscow 1979)
1.32 V.G. Komar, O.B. Serov: [Transl.: Display Holography and the Holographic Motion Picture (in Russian)] (Isskustvo, Moscow 1987)
1.33 G.A. Sobolev (ed.): [Transl.: High-efficiency materials for hologram recording (in Russian)] (Nauka, Leningrad 1988)
1.34 S.A. Benton: Intra-emulsion diffusion-transfer processing of volume dielectric holograms. J. Opt. Soc. Am. 64, 1393A (1974)
1.35 Yu.N. Denisyuk: Holography. Sov. J. Opt. Technol. 45, 745-748 (1978)
1.36 M.G. Lippmann: La photographie des couleurs. Comptes Rendus Hebdomadaires des Séances de l'Academie des Sciences 112, 274-275 (1891)
1.37 M.G. Lippmann: Photographies colorées du spectre, sur albumine et sur gélatine bichromatées. Comptes Rendus Hebdomadaires des Séances de l'Academie des Sciences 115, 575 (1892)
1.38 M.G. Lippmann: Sur la théorie de la photographie des couleurs simples et composées par la méthode interférentielle. J. Physique 3 (No.3), 97–107 (1894)
1.39 H.E. Ives: An experimental study of the Lippmann color photograph. Astrophysical J. 27, 325-352 (1908)
1.40 P. Connes: Silver salts and standing waves: The history of interference colour photography. J. Optics (Paris) 18, 147-166 (1987)
1.41 H. Nareid: A review of the Lippmann colour process. J. Photogr. Sci. 36, 140-147 (1988)
1.42 H. Nareid, H.M. Pedersen: Modelling of spectral response and tone reproduction in Lippmann photography and reflection holography, in *Practical Holography IV*, ed. by S.A. Benton. Proc. SPIE 1212, 63-72 (1990)
1.43 N.J. Phillips, H. Heyworth, T. Hare: On Lippmann's photography. J. Photogr. Sci. 32, 158-169 (1984)
1.44 N.J. Phillips: Links between photography and holography: The legacy of Gabriel Lippmann, in *Applications of Holography*, ed. by L. Huff. Proc. SPIE 523, 313-318 (1985)
1.45 N.J. Phillips, R.A.J. van der Werf: The creation of efficient reflective Lippmann layers in ultra-fine grain silver halide materials using non-laser sources. J. Photogr. Sci. 33, 22-28 (1985)
1.46 N.J. Phillips, D. Martens: Aspects of the copying of holograms using incoherent light, in *Progress in Holographic Applications*, ed. by J. Ebbeni. Proc. SPIE 600, 123-126 (1985)
1.47 J. Oliva, A. Fimia, J.A. Quintana: Copia de hologramas con luz parcialmente coherente. Opt. Pura y Aplicada 13, 129-134 (1980)

1.48 J. Oliva, A. Fimia, J.A. Quintana: Diffuse-object holograms in dichromated gelatin. Appl. Opt. **21**, 2891-2893 (1982)

1.49 P. Miller: Sodium vapour copies. Holographics Int'l (No.3), 21 (1988)

1.50 R.E. Brooks: Low-angle holographic interferometry using Tri-X Pan film. Appl. Opt. **6**, 1418-1419 (1967)

1.51 L.H. Tanner: Some applications of holography in fluid mechanics. J. Sci. Instrum. **43**, 81-83 (1966)

1.52 J.R. Fienup: Kodachrome as a holographic material. J. Opt. Soc. Am. **65**, 1220A (1975)

1.53 G.W. Stroke, D. Brumm, A. Funkhouser: Three-dimensional holography with "lensless" Fourier-transform holograms and coarse P/N Polaroid film. J. Opt. Soc. Am. **55**, 1327-1328 (1965)

1.54 G.W. Stroke, A. Funkhouser, C. Leonard, G. Indebetouw, R.G. Zech: Hand-held holography. J. Opt. Soc. Am. **57**, 110 (1967)

1.55 J.C. Albergotti: Instant holograms. Am. J. Phys. **35**, 1092 (1967)

1.56 R. Bamler, H. Glünder: Holographic recording on Polaroid's Polaplan CT. Opt. Commun. **56**, 321-324 (1986)

1.57 K. Biedermann, L. Ek: A recording and display system for hologram interferometry with low resolution imaging devices. J. Phys. E **8**, 571-576 (1975)

1.58 L. Ek, K. Biedermann: Fringe contrast in a system for hologram interferometry with low resolution imaging devices. J. Phys. E **8**, 691-696 (1975)

1.59 W.T. Cathey: Three-dimensional wavefront reconstruction using a phase hologram. J. Opt. Soc. Am. **55**, 457 (1965)

1.60 G.L. Rogers: Experiments in diffraction microscopy. Proc. Roy. Soc. Edinburgh A **63**, 193-221 (1951)

1.61 G.L. Rogers: Phase-contrast holograms. J. Opt. Soc. Am. **55**, 1181 (1965)

1.62 G.W. Stroke: Lensless Fourier-transform method for optical holography. Appl. Phys. Lett. **6**, 201-203 (1965)

1.63 G.W. Stroke, A.E. Labeyrie: White-light reconstruction of holographic images using the Lippmann-Bragg diffraction effect. Phys. Lett. **20**, 368-370 (1966)

1.64 A.S. Hoffman, J.G. Doidge, D.G. Mooney: Inverted reference-beam hologram. J. Opt. Soc. Am. **55**, 1559 (1965)

1.65 E.N. Leith: White-light holograms. Scientific Am. **235**, 80-95 (October 1976)

1.66 K.S. Pennington, L.H. Lin: Multicolor wavefront reconstruction. Appl. Phys. Lett. **7**, 56-57 (1965)

1.67 L.H. Lin, K.S. Pennington, G.W. Stroke, A.E. Labeyrie: Multicolor holographic image reconstruction with white-light illumination. Bell Syst. Tech. J. **45**, 659-661 (1966)

1.68 P. Kirkpatrick: History of holography, in *Holography*, ed. by B.J. Thompson. Proc. SPIE **15**, 9-12 (1968)

1.69 S.A. Benton: Hologram reconstructions with extended incoherent sources. J. Opt. Soc. Am. **59**, 1545-1546 (1969)

1.70 J.C. Urbach: Advances in hologram recording materials, in *Holography*, ed. by B.J. Thompson. Proc. SPIE **15**, 17-41 (1968)

1.71 R.L. Kurtz, R.B. Owen: Holographic recording materials - A review. Opt. Eng. **14**, 393-401 (1975)

1.72 J.W. Eastes: Materials research for holographic recording, Report No.1, ETL-0088, U.S. Army Engineer Topographic Laboratories, Fort Belvoir, VA 22060, USA (1976)

1.73 R.A. Bartolini, H.A. Weakliem, B.F. Williams: Review and analysis of optical recording media. Opt. Eng. **15**, 99-108 (1976)

1.74 R.A. Bartolini: Optical recording media review, in *Optical Storage Materials and Methods*, ed. by L. Beiser, D. Chen. Proc. SPIE **123**, 2-9 (1977)

1.75 J.W. Gladden: Review of photosensitive materials for holographic recordings, ETL-0128, U.S. Army Engineer Topographic Laboratories, Fort Belvoir, VA 22060, USA (1978)

1.76 P. Hariharan: Holographic recording materials: Recent developments. Opt. Eng. 19, 636-641 (1980)

1.77 L. Solymar, D.J. Cooke: Holographic recording materials, in *Volume Holography and Volume Gratings* (Academic, London 1981) pp.254-304

1.78 N.J. Phillips: Classical and new materials for holography, Tutorial T.15, E.P.S. - Europtica - SPIE (1989)

1.79 R.D. Rallison, S.E. Bialkowski: Survey of properties of volume holographic materials, in *Practical Holography III*, ed. by S.A. Benton. Proc. SPIE 1051, 68-75 (1989)

Chapter 2

2.1 C.E.K. Mees: *The Theory of The Photographic Process*, 2nd edn. (Macmillan, New York 1959)

2.2 C.B. Neblette: *Photography, its Materials and Processes*, 6th edn. (Van Nostrand, New York 1962)

2.3 J.M. Sturge (ed.): *Neblette's Handbook of Photography and Reprography*, 7th edn. (Van Nostrand Reinhold, New York 1977)

2.4 T.H. James (ed.): *The Theory of The Photographic Process*, 4th edn. (Macmillan, New York 1977)

2.5 S.A. Frecska: Characteristics of the Agfa-Gevaert type 10E70 holographic film. Appl. Opt. 7, 2312-2314 (1968)

2.6 J.A. Armstrong: Fresnel holograms: Their imaging properties and aberrations. IBM J. Res. & Dev. 9, 171-178 (1965)

2.7 A.A. Pistolkors: Resolving power of a hologram. Sov. Phys. Dokl. 12, 79-81 (1967)

2.8 B.P. Konstantinov, A.N. Zaidel, V.B. Konstantinov, Yu. I. Ostrovskii: Coherent light photography: Experimental technique and resolving power. Sov. Phys. - Tech. Phys. 11, 1279-1281 (1967)

2.9 A. Kozma, J.S. Zelenka: Effect of film resolution and size in holography. J. Opt. Soc. Am. 60, 34-43 (1970)

2.10 R.J. Bieringer, J.A. Ringlien: Diffraction-limited holography. Appl. Opt. 10, 1632-1635 (1971)

2.11 O.B. Gusev, V.B. Konstantinov: Method for determining the spatial resolution and contrast of holograph systems. Sov. Phys. - Tech. Phys. 16, 170-173 (1971)

2.12 V.M. Ginzburg, B.I. Fedorovskii: Resolving power of holograms for available photographic materials. Sov. Phys. - Tech. Phys. 15, 1733-1736 (1971)

2.13 E.B. Champagne, N.G. Massey: Resolution in holography. Appl. Opt. 8, 1879-1885 (1969)

2.14 D. Gabor, W.P. Goss: Interference microscope with total wavefront reconstruction. J. Opt. Soc. Am. 56, 849-858 (1966)

2.15 B.R. Russell: Resolution limitations in holographic images. Appl. Opt. 8, 971-973 (1969)

2.16 T. Kubota: The bending of interference fringes inside a hologram. Optica Acta 26, 731-743 (1979)

2.17 A.D. Gara, F.T.S. Yu: Effect of emulsion thickness variations on wavefront reconstruction. Appl. Opt. 10, 1324-1328 (1971)

2.18 R.F. Majkowski, A.D. Gara: Accuracy of holographic images. Appl. Opt. 11, 1867-1869 (1972)

2.19 J.N. Butters, D. Denby, J.A. Leendertz: A method for reducing movement in holographic emulsions. J. Phys. E 2, 116-117 (1969)

2.20 A.A. Friesem, J.L. Walker: Experimental investigation of some anomalies in photographic plates. Appl. Opt. 8, 1504-1506 (1969)

2.21 D.E. Duffy: Reducing photographic emulsion shrinkage for real-time holographic interferometry. J. Phys. E 3, 561-562 (1970)

2.22 M.I. Dzyubenko, A.P. Pyatikop, V.V. Shevchenko: Increasing the diffraction efficiency of reflecting three-dimensional holograms by preventing emulsion shrinkage. Sov. Phys. - Tech. Phys. 20, 965-966 (1975)

2.23 A.L. Ingalls: The effect of film thickness variations on coherent light. Photogr. Sci. Eng. 4, 135-140 (1960)

2.24 M. Matsumura: Analysis of wave-front aberrations caused by deformation of hologram media. J. Opt. Soc. Am. 64, 677-681 (1974)

2.25 M. Matsumura: Evaluation of deformation tolerance of the hologram medium. J. Opt. Soc. Am. 64, 928-933 (1974)

2.26 O.D.D. Soares, A.M.P.P. Leite: Effect of plate substrate quality on high resolution holographic imaging, in *Optical Information Storage*, ed. by K.G. Leib. Proc. SPIE 177, 44-49 (1979)

2.27 O.D.D. Soares, A.M.P.P. Leite: Holographic image degradation due to wavefront defects, in *Optics and Photonics Applied to Three-Dimensional Imagery*, ed. by M. Grosmann, P. Meyrueis. Proc. SPIE 212, 17-21 (1980)

2.28 A.A. Verbovetskii, V.B. Fedorov: Effect of shrinkage of the recording medium on the aberrations of binary-information images reconstructed from a hologram stack. Sov. Phys. - Tech. Phys. 22, 1387-1391 (1977)

2.29 P.C. Gupta, A.K. Aggarwal: Evaluation of photographic emulsions using speckle patterns. Opt. Commun. 25, 19-22 (1978)

2.30 Z. Jaroszewicz: Evaluation of holographic emulsion movement using phase difference amplification. Opt. Laser Technol. 20, 251-254 (1988)

2.31 M. Born, E. Wolf: *Principles of Optics*, 4th edn. (Pergamon, New York 1970)

2.32 J.C. Dainty (ed.): *Laser Speckles and Related Phenomena*, 2nd edn. Topics Appl. Phys., Vol.9 (Springer, Berlin, Heidelberg 1984)

2.33 Y. Kawagoe, N. Takai, N., T. Asakura: Speckle reduction by a rotating aperture at the Fourier transform plane. Optics and Lasers in Eng. 3, 197-218 (1982)

2.34 T.T. Baker: *Photographic Emulsion Technique* (American Photographic Publ., Boston 1941)

2.35 V.L. Zelikman, S.M. Levi: *Making and Coating Photographic Emulsion* (Focal, London 1964)

2.36 G.F. Duffin: *Photographic Emulsion Chemistry* (Focal, London, New York 1966)

2.37 T.T. Hill: Laboratory-scale photographic emulsion technique. J. Chem. Education 43, 492-498 (1966)

2.38 N.J. Phillips: Display holography: A frontier of high resolution photography. J. Photogr. Sci. 31, 134-142 (1983)

2.39 Yu.N. Denisyuk, I.R. Protas: Improved Lippmann photographic plates for recording stationary light waves. Opt. Spectrosc. (USSR) 14, 381-383 (1963)

2.40 H.E. Ives: An experimental study of the Lippmann color photograph. Astrophysical J. 27, 325-352 (1908)

2.41 Sigma Chemical Co., P.O. Box 14508, St. Louis, Montana 63178, USA.

2.42 Kind & Knox Gelatin, Inc., P.O. Box 927, Sioux City, Iowa 51102, USA.

2.43 R.J. Croome: Some aspects of the use of high molecular weight syntetic and natural polymers as additives to or replacements for gelatin in photographic products. J. Photogr. Sci. 30, 181-186 (1982)

2.44 E. Valenta: *Die Photographie in Natürlichen Farben*, Encyklopädie der Photographie (Knapp Verlag, Halle a.S. 1912) Heft 2

2.45 B.E. Jones: *Encyclopedia of Photography* (Cassell, London 1911) Reprinted edition (Arno Press, New York 1974) pp.340–342

2.46 R. Neuhauss: *Die Farbenphotographie nach Lippmann's Verfahren*, Encyklopädie der Photographie (Knapp Verlag, Halle a.S. 1898) Heft 33

2.47 R. Neuhauss: Laboratory journals and work notes 1894–1908. (in German) (Preus Fotomuseum, Horten, Norway 1990)

2.48 B.H. Crawford: The preparation of ultra-fine grain photographic emulsions. J. of Sci. Instrum. **31**, 333–335 (1954)

2.49 B.H. Crawford: *Small-Scale Preparation of Fine-Grain (Colloidal) Photographic Emulsions*. NPL, Notes on Appl. Sci. No.20 (Her Majesty's Stationery Office, London 1960)

2.50 I.H. Leubner, R. Jagannathan, J.S. Wey: Formation of silver bromide crystals in double-jet precipitation. Photogr. Sci. Eng. **24**, 268–272 (1980)

2.51 P. Demers: Improvements in the making of special photographic emulsions for nuclear physics. Science **110**, 380 (1949)

2.52 R.W. Berriman: Crystal growth during the formation of a silver-bromide dispersion in gelatin. J. Photogr. Sci. **12**, 121–133 (1964)

2.53 G.D. Dew, L.A. Sayce: On the production of diffraction gratings I. The copying of plane gratings. Proc. Roy. Soc. A **207**, 278–286 (1951)

2.54 Yu.N. Denisyuk, Z.A. Zagorskaya, A.M. Nizhin, S.B. Shevchenko: Method for obtaining light-sensitive media for use in holography. Sov. J. Opt. Technol. **50**, 597–598 (1983)

2.55 H. Thiry: Preparation and properties of ultra-fine grain AgBr emulsions. J. Photogr. Sci. **35**, 150–154 (1987)

2.56 H.M. Liu: Silver halide holographic plates sensitized to blue and green light. BME-C-99-report, Northwestern University, Evanston, IL 60208, USA (1987)

2.57 J.E. Shave, H.M. Liu: Silver halide holographic plates sensitized to blue light. BME-C-99-report, Northwestern University, Evanston, IL 60208, USA (1988)

2.58 R.D. Specialties, P.O.Box 206, Webster, NY 14580, USA.

2.59 A. Bonmati, J.J. Crespo, M. Pardo: Preparation of photographic emulsions for recording transmission holograms. Optica Pura y Aplicada **19**, 19–22 (1986)

2.60 Yu.N. Denisyuk: Holographic art with recording in three-dimensional media on the basis of Lippmann photographic plates, in *1st Europ. Cong. Optics Applied to Metrology*, ed. by M. Grosmann, P. Meyrueis. Proc. SPIE **136**, 365–368 (1978)

2.61 Yu.N. Denisyuk: Holographic art with recording in three-dimensional media on the basis of Lippmann photographic plates. Optica Applicata **8**, 49–53 (1978)

2.62 Yu.N. Denisyuk: Holography for artistic purposes with three-dimensional recording and Lippmann plates. Sov. Phys. - Tech. Phys. **23**, 954–957 (1978)

2.63 Z.A. Zagorskaya: High-resolution photographic emulsion for recording three-dimensional holograms. Sov. J. Opt. Tech. **40**, 134 (1973)

2.64 N.I. Kirillov, N.V. Vasilieva, V.L. Zielikman: Preparation of concentrated photographic emulsions by means of their successive freezing and thawing (in Russian). Zh. Nauchn. Prikl. Fotogr. Kinematogr. **15**, 441–443 (1970)

2.65 N.I. Kirillov, N.V. Vasilieva, V.L. Zielikman: A method for the concentration of the hard phase of the photographic emulsion by consecutive freezing and thawing (in Russian). Uspkhi Nauchno i Fotografii **16**, 204–211 (1972)

2.66 R.V. Ryabova, E.S. Barinova, O.V. Myatezh, A.N. Zaborov, E.V. Romash, D.I. Chernyi: High-resolution photographic material of Institute of Atomic Energy for holography in intersecting beams, in *Three-Dimensional Holography: Science, Culture, Education*, ed. by T.H. Jeong, V.B. Markov. Proc. SPIE **1238**, 166–170 (1991)

2.67 J.J. Crespo, A. Bonmati, M. Pardo: Influence of Ag^+ concentration of the photographic emulsion on the efficiency of reflection holograms. Optica Pura y Aplicada **19**, 85–91 (1986)

2.68 M. Pantcheva, T. Petrova, N. Pangelova, A. Katsev: Chemical sensitization of fine-grain silver halide emulsions for holographic recording, in *Holography'89*, ed. by Y.N. Denisyuk, T.H. Jeong. Proc. SPIE 1183, 128-130 (1990)

2.69 N. Pangelova, T. Petrova, A. Katsev, M. Pantcheva: Silver halide materials for pulsed holographic recording, in *Holography'89*, ed. by Y.N. Denisyuk, T.H. Jeong. Proc. SPIE 1183, 131-133 (1990)

2.70 N.S. Gafurova, L.G. Logak, Kh.Kh. Fassakhova, R.K. Khakimova, R.K. Teplova, I.N. Zelinsky, V.T. Chernikh: Silver halide photographic material having a flexible base for the use in pulse holography. Proc. 14th Int'l Congr. on High Speed Photography and Photonics, ed. by B.M. Stepanov (Moscow, October 19-24, 1980) pp.437-439

2.71 M.T. Sprackling: The deformation of a photographic film. J. Photogr. Sci. 35, 92-96 (1987)

2.72 V.B. Konstantinov, V.I. Kochenov: Parasitic interference structures in the interference-diffraction method of studying high-resolution photographic materials. Sov. Phys. - Tech. Phys. 16, 2086-2088 (1972)

2.73 M.P. Owen, A.A. Ward, L. Solymar: Internal reflections in bleached reflection holograms. Appl. Opt. 22, 159-163 (1983)

2.74 A.K. Richter, F.P. Carlson: Holographically generated lens. Appl. Opt. 13, 2924-2930 (1974)

2.75 R. Foley, F. Wendt: Making a high efficient antihalation backing for photographic plates which eliminates interference bands produced by coherent light. Appl. Opt. 6, 977-978 (1967)

2.76 E. Wesly: Antihalation backings. Holosphere 15 (No.3), 12-13 (1987)

2.77 Universal Photonics Inc., 495 West John Street, Hicksville, NY 11801, USA.

2.78 K. Biedermann: Silver halide photographic materials, in *Holographic Recording Materials*, ed. by H.M. Smith, Topics Appl. Phys., Vol.20 (Springer, Berlin, Heidelberg 1977) p.26

2.79 N.J. Phillips: Holography as a visual medium for the bubble chamber physicist. Proc. RL 81-042 Application of Holographic Techniques to Bubble Chamber Physics, ed. by R.L. Sekulin (Rutherford and Appleton Lab., Didcot, Oxon, UK 1981) pp.278-285

2.80 O.D.D. Soares: Elimination of rear reflections in holographic plates. Am. J. Phys. 48, 409-410 (1980)

2.81 MacTac, Morgan Adhesives Co., 4560 Darrow Rd., Stow, OH 44224, USA

2.82 R.W. Gurney, N.F. Mott: The theory of the photolysis of silver bromide and the photographic latent image. Proc. Roy. Soc. (London), Ser A, 164, 151-167 (1938)

2.83 N.F. Mott: The photographic latent image. Photogr. J. 81, 62-69 (1941)

2.84 J.W. Mitchell: Photographic sensitivity. Repts. Progr. Phys. 20, 433-515 (1957)

2.85 J.W. Mitchell: The concentration theory of latent image formation. Photogr. Sci. Eng. 22, 249-255 (1978)

2.86 J.W. Mitchell: The formation of the latent image in photographic emulsion grains. Photogr. Sci. Eng. 25, 170-188 (1981)

2.87 T. Tani: Physics of the photographic latent image. Physics Today 42, 36-41 (September 1989)

2.88 E.N. Leith: Photographic film as an element of a coherent optical system. Photogr. Sci. Eng. 6, 75-80 (1962)

2.89 J.C. Wyant, M.P. Givens: Effect of the photographic gamma on the luminance of hologram reconstructions. J. Opt. Soc. Am. 58, 357-361 (1968)

2.90 A.A. Friesem, A. Kozma, G.F. Adams: Recording parameters of spatially modulated coherent wavefronts. Appl. Opt. 6, 851-856 (1967)

2.91 M. King: Figures of merit for holographic storage media. J. Opt. Soc. Am. 60, 513-517 (1970)

2.92 M. King: Measurement and application of dynamic range for holographic storage media. Appl. Opt. 11, 791-797 (1972)

2.93 K. Biedermann: A function characterizing photographic film that directly relates to brightness of holographic images. Optik 28, 160-176 (1968/69)

2.94 L.H. Lin: Method of characterizing hologram-recording materials. J. Opt. Soc. Am. 61, 203-208 (1971)

2.95 J.P. Prenel: Study of the linearity of response of a holographic photosensitive emulsion. Opt. Laser Technol. 3, 157-161 (1971)

2.96 M. Lehmann: Holography technique and practice, in *The Engineering Uses of Holography*, ed. by E.R. Robertson, J.M. Harvey (Cambridge Univ. Press, Cambridge 1970) pp.1-24

2.97 H. Nassenstein, H.T. Buschmann, H. Dedden, E. Klein, E. Moisar, H. Rieck: An investigation of properties of photographic materials for holography, in *The Engineering Uses of Holography*, ed. by E.R. Robertson, J.M. Harvey (Cambridge Univ. Press, Cambridge 1970) pp.25-38

2.98 F.T.S. Yu, J.P. Morrison: A theoretical and experimental technique of the optimum holographic process, in *Holography*, ed. by B.J. Thompson. Proc. SPIE 25, 87-92 (1971)

2.99 D.G. Falconer: Enhancement of photographically recorded imagery, in *Holography*, ed. by B.J. Thompson. Proc. SPIE 25, 191-195 (1971)

2.100 K. Biedermann: Information storage materials for holography and optical data processing. Optica Acta 22, 103-124 (1975)

2.101 N.J. Phillips, P.G. Gwynn, A.A. Ward: Modulation mechanisms in the holographic display, in *Optics and Photonics Applied to Three-Dimensional Imagery*, ed. by M. Grosmann, P. Meyrueis. Proc. SPIE 212, 10-16 (1980)

2.102 Yu.N. Denisyuk: Static and dynamic three-dimensional holograms, Sov. Phys. - Tech. Phys. 26, 946-950 (1981)

2.103 N.J. Phillips: The making of successful holograms. Proc. Int'l Symp. on Display Holography, ed. by T.H. Jeong, Lake Forest College, IL 60045, USA, Vol.I (1983) pp.27-43

2.104 D. Gabor: Microscopy by reconstructed wave-fronts. Proc. Roy. Soc. (London) A 197, 454-487 (1949)

2.105 N. Nishida: Reconstruction of negative image in holography. Appl. Opt. 7, 1862-1863 (1968)

2.106 H. Frieser: Spread function and contrast transfer function of photographic layers. Photogr. Sci. Eng. 4, 324-329 (1960)

2.107 H.T. Buschmann, H.J. Metz: Die Wellenlängenabhängigkeit der Übertragungseigenschaften photographischer Materialen für die Holographie. Opt. Commun. 2, 373-376 (1971)

2.108 H.T. Buschmann: The wavelength dependence of the transfer properties of photographic materials for holography, in *Optical and Acoustical Holography*, ed. by E. Camatini (Plenum, New York 1972) pp.151-172

2.109 H.T. Buschmann: Determination of the MTF by diffraction and intrinsic phase effects in developed silver halide emulsions for holography. Photogr. Sci Eng. 16, 425-429 (1972)

2.110 K. Biedermann, S. Johansson: Evaluation of the modulation transfer function of photographic emulsions by means of a multiple-sine-slit microdensitometer. Optik 35, 391-403 (1972)

2.111 S. Johansson, K. Biedermann: Multiple-sine-slit microdensitometer and MTF evaluation for high resolution emulsions. I: Theory and mode of operation. Appl. Opt. 13, 2280-2287 (1974)

2.112 S. Johansson, K. Biedermann: Multiple-sine-slit microdensitometer and MTF evaluation for high resolution emulsions. II: MTF data and other recording parameters of high resolution emulsions for holography. Appl. Opt. 13, 2288-2291 (1974)

2.113 A. Vander Lugt, R.H. Mitchel: Technique for measuring modulation transfer functions of recording media. J. Opt. Soc. Am. 57, 372-379 (1967)

2.114 J.C. Dainty: Methods of measuring the modulation transfer function of photographic emulsions. Optica Acta 18, 795-813 (1971)

2.115 D. Gabor, G.W. Stroke: The theory of deep holograms. Proc. Roy. Soc. A 304, 275-289 (1968)

2.116 H. Kogelnik: Coupled wave theory for thick hologram gratings. Bell Sys. Tech. J. 48, 2909-2947 (1969)

2.117 P.J. van Heerden: Theory of optical information storage in solids. Appl. Opt. 2, 393-400 (1963)

2.118 A.A. Friesem: Holograms on thick emulsions. Appl. Phys. Lett. 7, 402-403 (1965)

2.119 N. George, J.W. Matthews: Holographic diffraction gratings. Appl. Phys. Lett. 9, 212-215 (1966)

2.120 E.N. Leith, A. Kozma, J. Upatnieks, J. Marks, N. Massey: Holographic data storage in three-dimensional media. Appl. Opt. 5, 1303-1311 (1966)

2.121 W.F. Berg: The photographic emulsion layer as a three-dimensional recording medium. Appl. Opt. 8, 2407-2416 (1969)

2.122 J. Upatnieks, C. Leonard: Efficiency and image contrast of dielectric holograms. J. Opt. Soc. Am. 60, 297-305 (1970)

2.123 J. Upatnieks, C.D. Leonard: Characteristics of dielectric holograms. J. IBM. Research 14, 527-532 (1970)

2.124 A.A. Friesem, J.L. Walker: Thick absorption recording media in holography. Appl. Opt. 9, 201-214 (1970)

2.125 M. Chang, N. George: Holographic dielectric grating: Theory and practice. Appl. Opt. 9, 713-719 (1970)

2.126 W.F. Berg: The photographic emulsion layer as a three-dimensional recording medium, in *Applications of Holography*, ed. by E.S. Barrekette, W.E. Kock, T. Ose, J. Tsujiuchi, G.W. Stroke (Plenum, New York 1971) pp.9-17

2.127 V.V. Aristov, V.Sh. Shekhtman: Properties of three-dimensional holograms. Sov. Phys. USPEKHI 14, 263-277 (1971)

2.128 K. Biedermann, S.I. Ragnarsson, P. Komlos: Volume holograms in photographic emulsions of extended thickness. Opt. Commun. 6, 205-209 (1972)

2.129 Y. Ninomiya: Recording characteristics of volume holograms. J. Opt. Soc. Am. 63, 1124-1130 (1973)

2.130 M.R.B. Forshaw: Thick holograms: A survey. Opt. Laser Technol. 6, 28-35 (1974)

2.131 S.F. Su, T.K. Gaylord: Calculation of arbitrary-order diffraction efficiencies of thick gratings with arbitrary grating shape. J. Opt. Soc. Am. 65, 59-64 (1975)

2.132 S.I. Ragnarsson: Holograms recorded in extremely thick photographic emulsions. Opt. Commun. 14, 39-41 (1975)

2.133 R. Alferness: Analysis of optical propagation in thick holographic gratings. Appl. Phys. 7, 29-33 (1975)

2.134 V.G. Sidorovich: Diffraction efficiency of three-dimensional phase holograms. Sov. Phys. - Tech. Phys. 21, 742-745 (1976)

2.135 N.D. Vorzobova, A.A. Leshchev, P.M. Semenov, V.G. Sidorovich, D.I. Staselko: Method of optimizing 3-D hologram recording conditions. Opt. Spectrosc. (USSR) 45, 690-695 (1978)

2.136 L. Solymar: A two-dimensional volume hologram theory including the effect of varying average dielectric constant. Opt. Commun. 26, 158-160 (1978)

2.137 T. Kubota: Characteristics of a thick hologram grating recorded in absorptive medium. Optica Acta 25, 1035-1053 (1978)

2.138 U. Langbein, F. Lederer: Modal theory for thick holographic gratings with sharp boundaries. Optica Acta 27, 171-182 (1980)

2.139 F. Lederer, U. Langbein: Modal theory for thick holographic gratings with sharp boundaries II. Unslanted transmission and reflection gratings. Optica Acta 27, 183-200 (1980)

2.140 M.G. Moharam, T.K. Gaylord: Coupled-wave analysis of reflection gratings. Appl. Opt. 20, 240-244 (1981)

2.141 B.Ya. Zeldovich, V.V. Shkunov, T.V. Yakovleva: Theory of reconstruction of thick-layer speckle-field holograms. Sov. J. Quantum Electron. 13, 1040-1043 (1983)

2.142 Yu.L. Korzinin, V.I. Sukhanov: Light diffraction by 3-D holograms with a continuous spectrum of spatial frequencies. System of equations for coupled waves. Opt. Spectrosc. (USSR) 56, 467-469 (1984)

2.143 Yu.L. Korzinin, V.I. Sukhanov: Diffraction of light by 3-D holograms with a continuous spectrum of spatial frequencies. Opt. Spectrosc. (USSR) 56, 572-574 (1984)

2.144 L. Solymar, R.R.A. Syms: Coupled wave theory for volume holograms recorded by two strong and N weak plane waves. Opt. Quantum Electron. 16, 35-39 (1984)

2.145 E. Guibelalde: Coupled wave analysis for out-of-phase mixed thick hologram gratings. Opt. Quantum Electron. 16, 173-178 (1984)

2.146 I.A. Mikhailov: A geometrical analysis of thick holograms. Opt. Spectrosc. (USSR) 58, 374-377 (1985)

2.147 Yu.L. Korzinin, V.I. Sukhanov: Diffraction efficiency of a 3-D phase hologram of a diffuse object. Opt. Spectrosc. (USSR) 58, 86-88 (1985)

2.148 V.I. Lokshin, G.B. Semenov, A.F. Kavtrev: Investigation of amplitude and phase holograms of diffuse objects. Opt. Spectrosc. (USSR) 36, 590-593 (1974)

2.149 H.M. Smith: Effect of emulsion thickness on the diffraction efficiency of amplitude holograms. J. Opt. Soc. Am. 62, 802-806 (1972)

2.150 A. Fimia, L. Carretero, R. Fuentes: Volume influence on intermodulation noise of dielectric diffuse-object holograms. Appl. Opt. 31, 2408-2409 (1992)

2.151 H. Kiemle: Experiments on technology and performance of Lippmann-Bragg phase holograms. Opt. Technol. 1, 203-207 (1969)

2.152 P. Hariharan: Volume-phase reflection holograms. The effect of hologram thickness on image luminance. Optica Acta 26, 1443-1447 (1979)

2.153 H. Dammann: Phase holograms of diffuse objects. J. Opt. Soc. Am. 60, 1635-1639 (1970)

2.154 C. Clausen, H. Dammann: Effects of intrinsic non-linearities on efficiency and image contrast of bleached holograms. Opt. Commun. 2, 263-269 (1970)

2.155 M. Kovachev, V. Sainov, Ts. Mateeva: Diffraction efficiency of discrete-carrier holograms. Sov. J. Quantum Electron. 6, 1307-1309 (1976)

2.156 D.I. Staselko, A.L. Churaev: Method for calculating the contrast of images of diffusely scattering objects produced by thick-layer transmission phase holograms. Sov. Phys. - Tech. Phys. 31, 197-202 (1986)

2.157 S.I. Ragnarsson: Scattering phenomena in volume holograms with strong coupling. Appl. Opt. 17, 116-127 (1978)

2.158 M.R.B. Forshaw: Explanation of the "venetian blind" effect in holography, using the Ewald sphere concept. Opt. Commun. 8, 201- 206 (1973)

2.159 A.P. Yakimovich: Secondary scattering effects in volume holograms. Sov. J. Quantum Electron. 13, 1043-1046 (1983)

2.160 F. Dubois, F. De Schryver, B. Biran: Theoretical study of size effects in volume holograms. J. Opt. Soc. Am. A 8, 270-273 (1991)

2.161 J. Harthong: Alternative theory of diffraction by modulated media. J. Opt. Soc. Am. A 8, 3-10 (1991)

2.162 R.R.A. Syms: *Practical Volume Holography* (Clarendon, Oxford 1990)

2.163 C.D. Leonard, A.L. Smirl: Holographic recording with limited laser light. Appl. Opt. 10, 625-631 (1971)

2.164 J.W. Goodman: Film-grain noise in wavefront-reconstruction imaging. J. Opt. Soc. Am. **57**, 493-502 (1967)

2.165 C.W. Helstrom: Image luminance and ray tracing in holography. J. Opt. Soc. Am. **56**, 433-441 (1966)

2.166 C.B. Burckhardt: Storage capacity of an optically formed spatial filter for character recognition. Appl. Opt. **6**, 1359-1366 (1967)

2.167 S.A. Benton: A signal and noise analysis of grainy emulsions, Ph.D. Thesis (Harvard University, Cambridge, MA 1968)

2.168 A. Kozma: Effects of film-grain noise in holography. J. Opt. Soc. Am. **58**, 436-438 (1968)

2.169 J.W. Goodman, R.B. Miles, R.B. Kimball: Comparative noise performance of photographic emulsions in holographic and conventional imagery. J. Opt. Soc. Am. **58**, 609-614 (1968)

2.170 J.C. Urbach, R.W. Meier: Holographic recording materials, in *Holography*, ed. by B.J. Thompson. Proc. SPIE **15**, 55-73 (1968)

2.171 J.C. Urbach, R.W. Meier: Properties and limitations of hologram recording materials. Appl. Opt. **8**, 2269-2281 (1969)

2.172 G.N. Buinov, A.V. Lukin, K.S. Mustafin: Scattering function and image quality in holography. Opt. Spectrosc. (USSR) **28**, 410-411 (1970)

2.173 K. Biedermann: The scattered flux spectrum of photographic materials for holography. Optik **31**, 367-389 (1970)

2.174 S. Lowenthal, J. Serres, H. Arsenault: Resolution and film-grain noise in Fourier transform holograms recorded with coherent or spatially incoherent light. Opt. Commun. **1**, 438-442 (1970)

2.175 D.G. Falconer: Noise and distortion in photographic data storage. IBM J. Res. Develop. **14**, 521-526 (1970)

2.176 D.H.R. Vilkomerson: Measurements of the noise spectral power density of photosensitive materials at high spatial frequencies. Appl. Opt. **9**, 2080-2087 (1970)

2.177 W.-H. Lee, M.O. Greer: Noise characteristics of photographic emulsions used for holography. J. Opt. Soc. Am. **61**, 402-409 (1971)

2.178 K.O. Hill, G.W. Jull: Holographic noise levels in two silver halide recording media. Optica Acta **18**, 729-742 (1971)

2.179 K.A. Stetson, K. Singh: Measurement of signal-to-noise ratio in hologram reconstructions, by vibration interferograms. Opt. Laser Technol. **3**, 104-108 (May 1971)

2.180 F.T.S. Yu: Film-grain noise and signal-to-noise ratio. Optik **36**, 434-442 (1972)

2.181 H.M. Smith: Light scattering in photographic materials for holography. Appl. Opt. **11**, 26-32 (1972)

2.182 L. Celaya, S. Mallick: Film-grain noise in holography. J. Opt. Soc. Am. **69**, 278-279 (1979)

2.183 K.A. Stozharova: Study of the noise diffraction spectra of photographic films. Sov. J. Opt. Technol. **46**, 192-194 (1979)

2.184 R.L. van Renesse: Scattering properties of fine-grained bleached emulsions. Photogr. Sci. Eng. **24**, 114-119 (1980)

2.185 R.R.A. Syms, L. Solymar: Noise gratings in photographic emulsion. Opt. Commun. **43**, 107-110 (1982)

2.186 A.L. Churaev, D.I. Staselko: Light scattering by silver halide holographic photomaterials. Light-scattering functions of microcrystals and emulsion-surface relief. Opt. Spectrosc. (USSR) **61**, 371-374 (1986)

2.187 D.I. Staselko, A.L. Churaev: Light scattering by silver halide materials for holography. Effect of noise holograms and specklegrams. Opt. Spectrosc. (USSR) **61**, 518-523 (1986)

2.188 R.K. Kostuk, G.T. Sincerbox: Polarization sensitivity of noise gratings recorded in silver halide volume holograms. Appl. Opt. **27**, 2993-2998 (1988)

2.189 H. Ghandeharian, W.M. Boerner: Degradation of holographic images due to depolarization of reflected light. J. Opt. Soc. Am. **68**, 931-934 (1978)

2.190 E.L. O'Neill: *Introduction to Statistical Optics* (Addision-Wesley, Reading, MA 1963) p.113

2.191 N. Phillips, H. Heyworth, T. Hare: On Lippmann's photography. J. Photogr. Sci. **32**, 158-169 (1984)

2.192 N.J. Phillips: The role of silver halide materials in the formation of holographic images, in *Applications of Holography*, ed. by L. Huff. Proc. SPIE **532**, 29-38 (1985)

2.193 N.J. Phillips: The silver halides - The workhorse of the holography business. Proc. Int'l Symp. on Display Holography, ed. by T.H. Jeong, Lake Forest College, IL 60045, USA, Vol.III (1989) pp.35-73

2.194 L. Silberstein: Quantum theory of photographic exposure. Phil. Mag. **44**, 257-273 (1922)

2.195 L. Silberstein: On the number of quanta required for the developability of a silver halide grain. J. Opt. Soc. Am. **31**, 343-348 (1941)

2.196 G.R. Bird, R.C. Jones, A.E. Ames: The efficiency of radiation detection by photographic films: State-of-the-art and methods of improvement. Appl. Opt. **8**, 2389-2405 (1969)

2.197 R.C. Jones: Information capacity of photographic films. J. Opt. Soc. Am. **51**, 1159-1171 (1961)

2.198 C.T. Chang, J.L. Bjorkstam: Effect of nonuniform irradiance, and irradiance fluctuations, upon the response of photographic film. J. Opt. Soc. Am. **65**, 1495-1501 (1975)

2.199 C.T. Chang, J.L. Bjorkstam: Effect of the threshold detection property of photographic films on hologram linearity and efficiency. J. Opt. Soc. Am. **66**, 558-564 (1976)

2.200 C.E. Thomas: Film characteristics pertinent to coherent optical data processing systems. Appl. Opt. **11**, 1756-1765 (1972)

2.201 C.T. Chang, J.L. Bjorkstam: Amplitude hologram efficiences with arbitrary modulation depth, based upon a realistic photographic film model. J. Opt. Soc. Am. **67**, 1160-1164 (1977)

2.202 N.J. Phillips: Some meanderings through holography. J. Photogr. Sci. **33**, 162-166 (1985)

2.203 J.J. Cowan: The holographic honeycomb microlens, in *Applications of Holography*, ed. by L. Huff. Proc. SPIE **523**, 251-259 (1985)

2.204 J.J. Cowan, W.D. Slafer: The recording and replication of holographic micropatterns for the ordering of photographic emulsion grains in film systems. J. Imaging Sci. **31**, 100-107 (1987)

2.205 J.J. Cowan: Embossed volume hologram: The aztec structure, in *Holographic Systems, Components and Applications*, ed. by J.C. Dainty. Proc. IEE, London, UK, No.311, 38-44 (1989)

2.206 A. Kozma: Photographic recording of spatially modulated coherent light. J. Opt. Soc. Am. **56**, 428-432 (1966)

2.207 A.A. Friesem, J.S. Zelenka: Effects of film nonlinearities in holography. Appl. Opt. **6**, 1755-1759 (1967)

2.208 F.G. Kaspar, R.L. Lamberts: Effects of some photographic characteristics on the light flux in a holographic image. J. Opt. Soc. Am. **58**, 970-976 (1968)

2.209 G.R. Knight: An extension of effects of film nonlinearities in holography. Appl. Opt. **7**, 205-206 (1968)

2.210 J.W. Goodman, G.R. Knight: Effects on film nonlinearities on wavefront-reconstruction images of diffuse objects. J. Opt. Soc. Am. **58**, 1276-1283 (1968)

2.211 A. Kozma: Analysis of the film non-linearities in hologram recording. Optica Acta **15**, 527-551 (1968)

2.212 O. Bryngdahl, A. Lohmann: Nonlinear effects in holography. J. Opt. Soc. Am. **58**, 1325-1334 (1968)

2.213 M. De Belder: Quality criteria of photographic materials for use in holography. Photogr. Sci. Eng. **13**, 351-360 (1969)

2.214 A. Kozma, G.W. Jull, K.O. Hill: An analytical and experimental study of nonlinearities in hologram recording. Appl. Opt. **9**, 721-731 (1970)

2.215 F.J. Tischer: Analysis of ghost images in holography by the use of Chebyshev polynomials. Appl. Opt. **9**, 1369-1374 (1970)

2.216 C.H.F. Velzel: Influence of non-linear recording on image formation in holography. Opt. Commun. **3**, 133-136 (1971)

2.217 C.R. Bendall, B.D. Guenther, R.L. Hartman: Thick amplitude holograms: Effect of nonlinear recording. Appl. Opt. **11**, 2992- 2993 (1972)

2.218 C.H.F. Velzel: Image contrast and efficiency of non-linearly recorded holograms of diffusely reflecting objects. Optica Acta **20**, 585-606 (1973)

2.219 R. Güther, S. Kusch: Ein Beitrag zum Intermodulationsrauschen in der Volumenholograpie. Exp. Technik in der Physik **22**, 119-141 (1974)

2.220 G. Goldmann: Effects of film nonlinearities on the signal-to-ghost ratio of storage holograms. Opt. Quantum Electron. **8**, 355-365 (1976)

2.221 K. Chalasinska-Macukow, J. Slaby, T. Szoplik: Comparision of nonlinear effects in amplitude and phase holograms recorded in photographic materials. Opt. Commun. **35**, 332-336 (1980)

2.222 N. Sultanova, H. Kasprzak: Variations of film nonlinearities in respect to the reconstruction wavelength. Optik **67**, 155-164 (1984)

2.223 N. Sultanova, H. Kasprzak: Influence of film nonlinearity on the Rayleigh criterion of resolution and energy concentration. Optica Applicata **14**, 443-450 (1984)

2.224 C.W. Slinger, R.R.A. Syms, L. Solymar: Non linear recording in silver halide planar volume holograms. Appl. Phys. B **36**, 217-224 (1985)

2.225 N.J. Phillips, H. Heyworth: Some problems of computer controlled holography, in *Progress in Holographic Applications*, ed. by J. Ebbeni, Proc. SPIE **600**, 116-122 (1985)

2.226 H. Kasprzak, N. Sultanova, H. Podbielska: Nonlinear effects of the recording material on the image quality of a Fourier hologram. J. Opt. Soc. Am. A **4**, 843-846 (1987)

2.227 A.A. Benken, A.P. Mikhalchenko, A.L. Churaev: Nonlinear recording of thin phase holograms on silver halide recording media. Opt. Spectrosc. (USSR) **66**, 652-654 (1989)

2.228 N. Sultanova, T. Staneva: Variation of the optical transfer function with the nonlinearity of absorption recording media, in *Holography'89*, ed. by Y.N. Denisyuk, T.H. Jeong. Proc. SPIE **1183**, 134-142 (1990)

Chapter 3

3.1 Agfa NDT Technical Information 21.7271(587)LI (1987)

3.2 Agfa NDT Technical Information 21.7280(587)LI (1987)

3.3 Agfa NDT Technical Information 21.7281(787)LI (1987)

3.4 Agfa NDT Technical Information NDT/1286/2012/GB/LI, (1987)

3.5 R. Phelan: Technical Information, NDT Systems Holography, H-155 (1985)

3.6 G.L. Rogers: Safelight for Agfa 10E75 and 8E75 plates. Appl. Opt. **17**, 1846 (1978)

3.7 EncapSulite International, P.O. Box 539, Stafford, TX 77477, USA

3.8 J.R. Feroe: EncapSulite safelights for holography: Test results, Lab notes from Holoshpere **17** (No.2-3), 1-2 (1990)

3.9 Ilford Technical Information, Holographic Film, Publication No.15717 (1988)

3.10   Ilford Technical Information, Holographic Film, Publication No.15718 (1986)

3.11   G.P. Wood: New silver halide materials for the mass production of holograms, in *Practical Holography*, ed. by T.H. Jeong. Proc. SPIE 615, 74-80 (1986)

3.12   G.P. Wood: Ilford holography - 1986 progress report, in *Practical Holography II*, ed. by T.H. Jeong. Proc. SPIE 747, 62-66 (1987)

3.13   G.P. Wood: Ilford holography - A progress report, in *Holography Yearbook 89/90* (Wittig, Hückelhoven, Germany 1989) pp.1.119-1.124

3.14   G.P. Wood: Progress in holographic materials: Silver halide film and chemistry, in *Practical Holography III*, ed. by S.A. Benton. Proc SPIE 1051, 44-50 (1989)

3.15   G.P. Wood: Ilford holographic consumables - World strategy. Proc. Int'l Symp. on Display Holography, ed. by T.H. Jeong, Lake Forest College, IL 60045, USA, Vol.III (1989) pp.105-114

3.16   G.P. Wood: A new rehalogenating bleach for the production of Lippmann holograms, in *Practical Holography IV*, ed. by S.A. Benton. Proc SPIE 1212, 46-54 (1990)

3.17   Eastman KODAK Company, Data Release-KODAK Holographic Plate, Type 120-01, 120-02, and KODAK Holographic Film SO-173, P-311 (1984)

3.18   Eastman KODAK Company, Characteristics of KODAK Plates for Scientific and Technical Applications, P-140 (1985)

3.19   Eastman KODAK Company, Scientific Imaging with KODAK Films and Plates, P-315 (1987)

3.20   Eastman KODAK Company, KODAK Scientific Imaging Products, Publication L-10 (1989)

3.21   L.G. Logak, H.H. Fassakhova, N.E. Antonova, L.A. Minina, R.K. Gainutdinov: Ultra-fine grain silver halide photographic materials for holography on the flexible film base, in *Three-Dimensional Holography: Science, Culture, Education*, ed. by T.H. Jeong, V.B. Markov. Proc. SPIE 1238, 171-176 (1991)

3.22   HP-490 Holographic Plates, Bulgarian Academy of Sciences, Sofia (1983)

3.23   HP-650 Holographic Plates, Bulgarian Academy of Sciences, Sofia (1983)

3.24   V.C. Sainov: Basic characteristics and applications of reflection holograms. Proc. Int'l Symp. on Display Holography, ed. by T.H. Jeong, Lake Forest College, IL 60045, USA, Vol.I (1983) pp.55-69

3.25   A. Katsev, M. Mazakova, M. Pancheva, N. Pangelova, Ts. Petrova, V. Razsolkov: Silver halide light-sensitive plates HP-490 and developer FHP-3 for holographic applications. J. Signal AM 9, 107-111 (1981)

Chapter 4

4.1   J. Eggert: The present status of photographic development in theory and practice. Photogr. Sci. Eng. 1, 94-103 (1958)

4.2   T.H. James: The mechanism of photographic development. Photogr. Sci. Eng. 1, 141-145 (1958)

4.3   W. Jaenicke: The mechanism of photographic development. Photogr. Sci. Eng. 6, 185-196 (1962)

4.4   L.F.A. Mason: *Photographic Processing Chemistry* (Focal, London 1966)

4.5   L.R. Solman: The role of physical development in determining the sensitometric response of maximum-resolution emulsion. J. Photogr. Sci. 14, 171-178 (1966)

4.6   C.R. Berry: A growth mechanism for filamentary silver. Photogr. Sci. Eng. 13, 65-68 (1969)

4.7   D.C. Shuman, T.H. James: Kinetics of physical development. Photogr. Sci. Eng. 15, 42-47 (1971)

4.8   D.C. Skillman: Silver particle growth analysis from development of extremely fine grains. Photogr. Sci. Eng. 19, 28-37 (1975)

4.9   L.K.H. van Beek: Special properties of physical development processes. Photogr. Sci. Eng. 20, 88-91 (1976)
4.10 Yu.N. Denisyuk, I.R. Protas: Improved Lippmann photographic plates for recording stationary light waves. Opt. Spectrosc. (USSR) 14, 381-383 (1963)
4.11 D.G Falconer: Role of the photographic process in holography. Photogr. Sci. Eng. 10, 133-139 (1966)
4.12 B.J. Pernick, D. Yustein, C. Bartolotta: Film transmittance-exposure characteristics for 649-F at 6328 Å. Appl. Opt. 7, 714-715 (1968)
4.13 K. Biedermann, K.A. Stetson: Adjusting development time to influence the characteristics of holograms. Photogr. Sci. Eng. 13, 361-370 (1969)
4.14 M.E. Cox, R.G. Buckles: Influence of selected processing variables on holographic film parameters: Kodak SO-243. Appl. Opt. 10, 916-921 (1971)
4.15 R.L. Lamberts, C.N. Kurtz: Reversal bleaching for low flare light in holograms. Appl. Opt. 10, 1342-1347 (1971)
4.16 Yu.E. Usanov, M.M. Yermolayev: The development of photographic films for holograms. Sov. J. Opt. Technol. 39, 758-760 (1972)
4.17 K. Biedermann, S. Johansson: Development effects and the MTF of high- resolution photographic materials for holography. J. Opt. Soc. Am. 64, 862-870 (1974)
4.18 H.M. Smith, C.A. Callari: Some holographic effects caused by the presence of silver halide solvent in the developer. Photogr. Sci. Eng. 19, 130-135 (1975)
4.19 N.J. Phillips, D. Porter: An advance in the processing of holograms. J. Phys. E 9, 631-634 (1976)
4.20 Yu.E. Usanov, N.L. Kosobokova, G.P. Tikhomirov: Investigation of the dependence of the diffraction efficiency of holograms on the sizes of the developed silver particles. Sov. J. Opt. Technol. 44, 528-530 (1977)
4.21 L. Joly, R. Vanhorebeek: Development effects in white-light reflection holography. Photogr. Sci. Eng. 24, 108-113 (1980)
4.22 R.L. van Renesse: Scattering properties of fine-grained bleached emulsions. Phot. Sci. Eng. 24, 114-119 (1980)
4.23 A. Fimia, M. Pardo, J.A. Quintana: Improvement of image quality in bleached holograms. Appl. Opt. 21, 3412-3413 (1982)
4.24 N.J. Phillips: The making of successful holograms. Proc. Int'l Symp. on Display Holography, ed. by T. H. Jeong, Lake Forest College, IL 60045, USA, Vol.I (1983) pp.27-43
4.25 M.M. Ermolaev, L.V. Selyavko, V.P. Smaev: Effect of developing solutions on the photographic image quality. Sov. J. Opt. Technol. 51, 542-545 (1984)
4.26 L. Joly, R. Phelan, M. Redzikowski: Processing for reflection holography, in Practical Holography, ed. by T.H. Jeong, L.E. Ludman. Proc. SPIE 615, 66-73 (1986)
4.27 R. De Winne, R. Phelan: Processing of Agfa-Gevaert Holotest 8E56 HD for reflection holography. Proc. Int'l Symp. on Display Holography, ed. by T. H. Jeong, Lake Forest College, IL 60045, USA, Vol.II (1986) pp.263-277
4.28 J. Blyth: Notes on processing holograms with solvent bleach. Proc. Int'l Symp. on Display Holography, ed. by T.H. Jeong, Lake Forest College, IL 60045, USA, Vol.II (1986) pp.325-331
4.29 J. Crespo, A. Fimia, J.A. Quintana: Fixation-free methods in bleached reflection holography. Appl. Opt. 25, 1642-1645 (1986)
4.30 P. Hariharan, C.M. Chidley: Photographic phase holograms: The influence of developer composition on scattering and diffraction efficiency. Appl. Opt. 26, 1230-1234 (1987)
4.31 M. Austin: A comparison between hydroquinone, metol and phenidone as developing agents for transmission holography. J. Photogr. Sci. 36, 148- 152 (1988)

4.32 N. Phillips: The silver halides - The workhorse of the holography business. Proc. Int'l Symp. on Display Holography, ed. by T. H. Jeong, Lake Forest College, IL 60045, USA, Vol.III (1989) pp.35-73

4.33 E. Valenta: *Die Photographie in natürlichen Farben*, Encyklopädie der Photographie (Knapp Verlag, Halle a.S. 1912) Heft.2

4.34 G.C. Alletag: Degradation of phenidone in developer solutions during storage. Photogr. Sci. Eng. 2, 213-218 (1958)

4.35 T.H. James, W. Vanselow: The rate of solution of silver halide grains in a developer. PSA Tech. Quart. 135-143 (November 1955)

4.36 M.R.V. Sahyun: Mechanisms of development restraint-chemistry of 6-nitrobenzimidazole and benzotriazole. Photogr. Sci. Eng. 18, 383-387 (1974)

4.37 H.M. Smith: Holography, in *SPSE Handbook of Photographic Science and Engineering*, ed. by T. Woodlief (Wiley-Interscience, New York 1973) Sect.22, pp.1310-1311

4.38 Tetenal Photowerk, Hamburg-Berlin, Postfach 2029, D-W-2000 Norderstedt, Germany.

4.39 P. Demers: Cosmic ray phenomena at minimum ionization in a new nuclear emulsion having a fine grain, made in the laboratory. Cdn. J. Phys. 32, 538-554 (1954)

4.40 J.F. Garfield: Apparatus and a laboratory for processing thick nuclear track emulsions. Photogr. Sci. Eng. 2, 85-90 (1958)

4.41 R. Diotallevi, E. Lamanna, A. Lucci, F. Meddi, G. Rosa: The laboratory for nuclear emulsion processing at Rome. Nota Interna No.775, Physics Institute, University of Rome (1981)

4.42 F.G. Kaspar, R.L. Lamberts, C.D. Edgett: Comparision of experimental and theoretical holographic image radiance. J. Opt. Soc. Am. 58, 1289-1295 (1968)

4.43 M. Austin: Some effects of sulphite ions in holographic development. J. Photogr. Sci. 33, 191-196 (1985)

4.44 G.F. van Veelen, J.F. Willems: The superadditivity with hydroquinone of photographic developing agents forming positively charged semiquinones: III. The influence of sodium sulfite and general considerations. Photogr. Sci. Eng. 7, 113-122 (1963)

4.45 H. Thiry: Qualitative comparision of two types of development on two holographic materials. J. Photogr. Sci. 33, 159-160 (1985)

4.46 W. Spierings: "Pyrochrome" processing yields color-controlled results with silver-halide materials. Holosphere 10 (No.7-8), 1-7 (1981)

4.47 G. Saxby: Jottings from the UK. Holosphere 12 (No.5), 9 (1983)

4.48 D.J. Cooke, A.A. Ward: Reflection-hologram processing for high efficiency in silver-halide emulsions. Appl. Opt. 23, 934-941 (1984)

4.49 R. Aliaga, H. Chuaqui: Diffraction efficiency and signal to noise measurements on 8E75 HD using catechol developer. Proc. Int'l Symp. on Display Holography, ed. by T. H. Jeong, Lake Forest College, IL 60045, USA, Vol.II (1986) pp.315-324

4.50 S.A. Benton: Photographic materials and their handling, in *Handbook of Optical Holography*, ed. by H.J. Caulfield (Academic, New York 1979) Chap.9, pp.349-366

4.51 G.M. Haist: *Monobath Manual* (Morgan & Morgan, Hastings-on Hudson, New York 1966)

4.52 S.I. Ragnarsson: A new holographic method of generating a high efficiency, extended range spatial filter with application to restoration of defocussed images. Physica Scripta 2, 145-153 (1970)

4.53 P. Hariharan, C.S. Ramanathan, G.S. Kaushik: Monobath processing for holography. Appl. Opt. 12, 611-612 (1973)

4.54 P. Hariharan: Simplified processing techniques for holography. Photogr. Sci. Eng. 24, 105-107 (1980)

4.55 P. Hariharan, B.S. Ramprasad: Rapid *in situ* processing for real-time holographic interferometry. J. Phys. E **6**, 699–701 (1973)

4.56 H.F. Dietrich, R.J. Raine, R.N. O'Brien: A 5-minute monobath for Kodak 649-F plates used in holography and holographic interferometry. J. Photogr. Sci. **24**, 120–123 (1976)

4.57 G.M. Haist, J.R. King, L.H. Bassage: Organic silver-complexing agents for photographic monobaths. Photogr. Sci. Eng. **5**, 198–203 (1961)

4.58 LTI, Laser Technology, Inc., 1055 W. Germantown Pike, Norristown, PA 19404, USA.

4.59 Keystone Scientific Co., P.O.Box 22, Thorndale, PA 19372, USA.

4.60 H.M. Smith, M.H. Sewell, J.R. King: Real-time holographic interferometry: A system. Appl. Opt. **15**, 729–733 (1976)

4.61 M.M. Ermolaev, L.V. Selyavko, V.P. Smaev: Contact-diffusion processing of holographic photographic plates. Sov. J. Opt. Technol. **51**, 33–34 (1984)

4.62 O.V. Andreeva, V.I. Sukahanov: Production of unbleached volume holograms with high diffraction efficiency. Opt. Spectrosc. **30**, 424–425 (1971)

4.63 H. Lehmann: *Beiträge zur Theorie und Praxis der direkten Farbenphotographie mittels Stehender Lichtwellen nach Lippmanns Methode* (Trömer, Freiburg i.Br. 1906) p.52

4.64 S.A. Benton: Intra-emulsion diffusion-transfer processing of volume dielectric holograms. J. Opt. Soc. Am. **64**, 1393–1394A (1974)

4.65 T.H. James: Kinetics of development by *p*-phenylenediamine in neutral and moderately alkaline solutions. J. Photogr. Sci. **6**, 49–56 (1958)

4.66 S.A. Benton: Development effects in holographic imaging, in *Photo-and Electro-Imaging*, ed. by M. Sukigara. Proc. SPSE-Tokyo Symp.'77 (1978) pp.23–28

4.67 J. Ruzek, P. Fiala: Reflection holographic portraits. Opt. Acta **26**, 1257–1264 (1979)

4.68 R. Aliaga, H. Chuaqui, P. Pedraza: Solution physical development of Agfa-Gevaert emulsions for holography. Opt. Acta **30**, 1743–1748 (1983)

4.69 R. Aliaga, H. Chuaqui: Microscopic and macroscopic measurement of holographic emulsions, in *Holography Applications*, ed. by J. Ke, R. Pryputniewicz. Proc. SPIE **673**, 491–494 (1986)

4.70 R. Aliaga, H. Chuaqui, P. Pedraza: Archival and wide exposure latitude process for holography. Appl. Opt. **29**, 2861–2863 (1990)

4.71 W. Spierings: Practical considerations on solution physical development. Proc. Int'l Symp. on Display Holography, ed. by T. H. Jeong, Lake Forest College, IL 60045, USA, Vol.II (1986) pp.299–302

4.72 A. Bonmati, J. Crespo, M. Pardo: Factors influencing the colloidal silver formation in reflection holograms, in *Image Detection and Quality*, ed. by L.F. Guyot. Proc. SPIE **702**, 101–104 (1986)

4.73 P. Fiala, G. Loncar, J. Ruzek, T. Jerie: Study of holographic gratings formed by ultra-fine silver particles, in *Industrial Applications of Laser Technology*, ed. by W.F. Fagan. Proc. SPIE **398**, 185–192 (1983)

4.74 P. Fiala, G. Loncar, J. Ruzek, T. Jerie: High efficiency display hologram making. Proc. Int'l Symp. on Display Holography, ed. by T. H. Jeong, Lake Forest College, IL 60045, USA, Vol.II (1986) pp.287–297

4.75 P. Leclère, Y. Renotte, Y. Lion: Improvement of recording capability of silver halide emulsions by physical development, in *Holographic Systems, Components and Applications*, ed. by J.C. Dainty. Proc. IEE **311**, 233–235 (1989)

4.76 P. Leclère, Y. Renotte, Y. Lion: Improvement of holographic performances of silver halide emulsions by a solution physical development. J. Photogr. Sci. **39**, 33–37 (1991)

4.77 Y. Renotte, Y. Lion, P. Leclère, P. Wojtaszczyk: Amelioration des performances des emulsions aux halogenures d'argent utilisées en holographie, in *Proc. OPTO'91* (ESI publications, Paris 1991) pp.325–331

4.78 W.B. Yuan: A new unbleached hologram (in Chinese). Acta Electronica Sinica 9 (No.4), 36-38 (1981)

4.79 W.B. Yuan: A practical holographic ultrafiche file storage (HUFS) system, in *Practical Holography II*, ed. by T.H. Jeong. Proc. SPIE 747, 135-138 (1987)

4.80 L. Zhao, S.Q. Yan, D.L. Wang, J.X. Xiao, H.J. Zhang, J.H. Dai: A method of increasing the diffraction efficiency of hologram (in Chinese). Acta Physica Sinica 30, 143-146 (1981)

4.81 Z.S. Zho, Z.Z. Tang: Diluting developer is a method to produce phase holograms (in Chinese). J. Applied Laser (China) 2, 18-19 (1982)

4.82 C.P. Zhang, G.Y. Zhang, Z.F. Li, Z.S. Liu, X.Y. Li, M.R. Shang: Investigation of spectrum characteristics of dilute developed hologram plate of silver halide. Kexue Tongbao 29, 1017-1022 (1984)

4.83 X.H. Zhang, W.B. Yuan: A new technique for improving diffraction efficiency of the unbleached hologram (in Chinese). J. Applied Laser (China) 5, 31-33 (1985)

4.84 X.W. Ni: The developer for higher diffraction efficiency of holograms (in Chinese). J. Applied Laser (China) 5, 231-232 (1985)

4.85 C.P. Zhang, G.Y. Zhang, H. Cong, L. Zhao, S.Q. Yan, X.L. Bi: Calculation and measurement of the diffraction efficiency of a dilute developed silver halide holographic plate. Opt. Acta 32, 679-687 (1985)

4.86 G. Windischbauer, M. Küster, G. Keck, H. Bachinger: Investigations of photographic layers for holography. Proc. Applications of Holography, ed. by J.Ch. Viénot, J. Bulabois, J. Pasteur (Besançon, France 1970) pp.11-12

4.87 G. Windischbauer, G. Keck, M. Küster, G. Ranninger: Untersuchungen zur holographischen Mikroskopie. Optik 34, 382-386 (1972)

4.88 A.A. Verbovetskii, L.P. Vakhtanova, E.A. Gruz, K.S. Bogomolov: Parameters of bleached holograms with binary data under physical development. Opt. Spectrosc. (USSR) 43, 74-75 (1977)

4.89 D.H.R. Vilkomerson, D. Bostwick: Some effects of emulsion shrinkage on a hologram's image space. Appl. Opt. 6, 1270-1272 (1967)

4.90 H.I. Bjelkhagen: Investigation on the resolution that can be obtained with the Baltay holographic arrangement for the 15-foot bubble chamber. Proc. Photonics Applied to Nuclear Physics: 2. CERN 85-10 (1985) pp.7-49

4.91 R. Naon, H. Bjelkhagen, R. Burnstein, L. Voyvodic: A system for viewing holograms. Nucl. Instr. and Meth. A 283, 24-36 (1989)

4.92 D.J. Young: A method of permanently controlling the thickness and profile of a processed photographic emulsion. J. Photogr. Sci. 23, 190-192 (1975)

4.93 D.K. Angell: Controlling emulsion thickness variations in silver halide (sensitized) gelatin, in *Holographic Optics: Design and Applications*, ed. by I. Cindrich. Proc. SPIE 883, 106-113 (1988)

4.94 A guide to Dow Corning silane coupling agents. Dow Corning, No. 23-012B-85 (1985)

4.95 C. Hernandez, C. Bainer, D. Courjon: Role of double-exposure holography on the shrinkage effect in silver halide gelatin. Optica Acta 32, 469-477 (1985)

Chapter 5

5.1 G.L. Rogers: Experiments in diffraction microscopy. Proc. Roy. Soc. Edinburgh A 63, 193-221 (1951)

5.2 Yu.N. Denisyuk: On the reproduction of the optical properties of an object by the wave field of its scattered radiation. Opt. Spectrosc. (USSR) 14, 279-284 (1963)

5.3 W.T. Cathey: Three-dimensional wavefront reconstruction using a phase hologram. J. Opt. Soc. Am. 55, 457 (1965)

5.4   N. George, J.W. Matthews: Holographic diffraction gratings. Appl. Phys. Lett. 9, 212-215 (1966)
5.5   A.P. Komar, M.V. Stabnikov, B.G. Turukhano: Reconstruction of the images of transparent and refracting objects using phase holograms. Sov. Phys. Dokl. 11, 712-713 (1967)
5.6   H. Hannes: Interferometrische Messungen an Phasenstrukturen für die Holographie. Optik 26, 363-380 (1967/68)
5.7   V. Russo, S. Sottini: Bleached holograms. Appl. Opt. 7, 202 (1968)
5.8   L.F. Collins: Diffraction theory description of bleached holograms. Appl. Opt. 7, 1236-1237 (1968)
5.9   J.N. Latta: The bleaching of holographic diffraction gratings for maximum efficiency. Appl. Opt. 7, 2409-2416 (1968)
5.10   H. Kiemle, W. Kreiner: Lippmann-Bragg-Phasenhologramme mit hohem Wirkungsgrad. Phys. Lett. A 28, 425-426 (1968)
5.11   J. Upatnieks, C. Leonard: Diffraction efficiency of bleached, photographically recorded interference patterns. Appl. Opt. 8, 85-89 (1969)
5.12   H. Kiemle: Phase holograms in photographic emulsions for digital data storage. Opt. Technol. 1, 146-149 (1969)
5.13   D.H. McMahon, A.R. Franklin: Efficient, high-quality, R-10 bleached holographic diffraction gratings. Appl. Opt. 8, 1927-1929 (1969)
5.14   F.T.S. Yu: A note on phase holograms. Appl. Opt. 8, 2350-2351 (1969)
5.15   M. Young, F.H. Kittredge: Amplitude and phase holograms exposed on Agfa-Gevaert 10E75 plates. Appl. Opt. 8, 2353-2354 (1969)
5.16   C.B. Burckhardt, E.T. Doherty: A bleach process for high-efficiency low-noise holograms. Appl. Opt. 8, 2479-2482 (1969)
5.17   W.T. Cathey: Phase holograms, phase-only holograms, and kinoforms. Appl. Opt. 9, 1478-1479 (1970)
5.18   H.W. Lorber: A theory of granularity and bleaching for holographic information recording. IBM J. Res. Develop. 14, 515-520 (1970)
5.19   D.H. Kelly: System analysis of the photographic process I: A three-stage model. J. Opt. Soc. Am. 50, 269-276 (1960)
5.20   M. Chang, N. George: Holographic dielectric grating: Theory and practice. Appl. Opt. 9, 713-719 (1970)
5.21   J. Upatnieks, C. Leonard: Efficiency and image contrast of dielectric holograms. J. Opt. Soc. Am. 60, 297-305 (1970)
5.22   J. Upatnieks, C.D. Leonard: Characteristics of dielectric holograms. J. IBM. Res. Develop. 14, 527-532 (1970)
5.23   K.S. Pennington, J.S. Harper: Techniques for producing low-noise, improved efficiency holograms. Appl. Opt. 9, 1643-1650 (1970)
5.24   K.S. Pennington, J.S. Harper: Techniques for producing low-noise, improved efficiency holograms (Erratum). Appl. Opt. 9, 2590 (1970)
5.25   M. Lehmann, J.P. Lauer, J.W. Goodman: High efficiencies, low noise, and suppression of photochromic effects in bleached silver halide holography. Appl. Opt. 9, 1948-1949 (1970)
5.26   M. Lehmann: *Holography* (Focal, New York 1970) pp.141-144
5.27   H. Dammann: Phase holograms of diffuse objects. Proc Applications of Holography, ed. by J.-Ch. Viénot, J. Bulabois, J. Pasteur (Besançon, France 1970) pp.11-14
5.28   A. Schmackpfeffer, W. Järisch, W.W. Kulcke: High-efficiency phase-hologram gratings. IBM J. Res. Develop. 14, 533-538 (1970)
5.29   R.L. Lamberts, C.N. Kurtz: Bleached holograms with reduced flare light. J. Opt. Soc. Am. 60, 724A (1970)
5.30   Reversal bleach process for producing phase holograms on Kodak Spectroscopic plates, Type 649-F, KODAK Pamphlet No. P-230, Rochester, NY 14650, USA (1970)

5.31 S.A. Benton: Granularity effects in phase holograms. J. Opt. Soc. Am. 61, 649A (1971)

5.32 R.L. van Renesse, N.J. van der Zwaal: Refractive index and thickness variations of the photographic emulsion. Opt. Laser Technol. 3, 41- 44 (1971)

5.33 A.A. Verbovetskii, V.B. Fedorov: Diffraction efficiency of bleached holograms. Opt. Spectr. 31, 339-340 (1971)

5.34 H.T. Buschmann: Bleichprozesse zur Erzeugung rauscharmer, lichtstarker Phasenhologramme. Optik 34, 240-253 (1971)

5.35 P. Hariharan: Reversal processing technique for phase holograms. Opt. Commun. 3, 119-121 (1971)

5.36 P. Hariharan, C.S. Ramanathan, G.S. Kaushik: Simplified processing technique for photographic phase holograms. Opt. Commun. 3, 246-247 (1971)

5.37 R.L. Lamberts, C.N. Kurtz: Reversal bleaching for low flare light in holograms. Appl. Opt. 10, 1342-1347 (1971)

5.38 P. Barlai: Phasenhologramme in photographischen Emulsionen mit hohem Wirkungsgrad im blaugrünen Spektralbereich. Z. Naturforsch. 27a, 544 (1972)

5.39 V.I. Bobrinev, V.K. Kozlova, M.A. Maiorchuk: Holograms of high diffraction efficiency. Sov. J. Quantum Electron. 1, 553-554 (1972)

5.40 R.L. Lamberts: Characterization of a bleached photographic material. Appl. Opt. 11, 33-41 (1972)

5.41 G.C. Righini, V. Russo, S. Sottini: Low noise and good efficiency volume holograms. Appl. Opt. 11, 951-953 (1972)

5.42 H. Thiry: New technique of bleaching photographic emulsions and its application to holography. Appl. Opt. 11, 1652-1653 (1972)

5.43 A.A. Verbovetskii, V.B. Fedorov: Diffraction efficiency of a bleached hologram at 0.44 μm. Sov. Phys. - Tech. Phys. 17, 176-177 (1972)

5.44 P. Hariharan, G.S. Kaushik, C.S. Ramanathan: Reduction of scattering in photographic phase holograms. Opt. Commun. 5, 59-61 (1972)

5.45 P. Hariharan, G.S. Kaushik, C.S. Ramanathan: Simplified, low-noise processing technique for photographic phase holograms. Opt. Commun. 6, 75-76 (1972)

5.46 H.T. Buschmann: Kontrasterhöhung von Phasenhologrammen durch Beschichtung der Platten. Opt. Commun. 6, 290-294 (1972)

5.47 P. Hariharan: Bleached reflection holograms. Opt. Commun. 6, 377-379 (1972)

5.48 R.L. van Renesse, F.A.J. Bouts: Efficiency of bleaching agents for holography. Optik 38, 156-168 (1973)

5.49 A. Graube: Advances in bleaching methods for photographically recorded holograms. Appl. Opt. 13, 2942-2946 (1974)

5.50 R.L. van Renesse: Advances in bleaching methods for photographically recorded holograms (comments). Appl. Opt. 14, 1763-1764 (1975)

5.51 N.J. Phillips, D. Porter: An advance in the processing of holograms. J. Phys. E 9, 631-634 (1976)

5.52 N.J. Phillips, D. Porter: Organically accelerated bleaches: Their role in holographic image formation. J. Phys. E 10, 96-98 (1977)

5.53 S.Y. Feng, H.M. Lai, L.K. Su: Phase holograms and the action of bleaching. Appl. Opt. 16, 1800-1801 (1977)

5.54 F.T.S. Yu: Phase holograms and the action of bleaching (comments). Appl. Opt. 16, 1802 (1977)

5.55 M.J. Landry, G.S. Phipps, C.E. Robertson: Measurement of diffraction efficiency, SNR, and resolution of single- and multiple-exposure amplitude and bleached holograms. Appl. Opt. 17, 1764-1770 (1978)

5.56 J.W. Eastes: Materials research for holographic recording: Report No. 2, Bleaching methods for photographically recorded holograms, U.S. Army Corps of Engineers, Engineer Topographic Laboratories, Fort Belvoir, VA 22060, USA, Report ETL-0156 (1978)

5.57 P. Hariharan: Volume-phase reflection holograms. The effect of hologram thickness on image luminance. Opt. Acta 26, 1443-1447 (1979)

5.58 S.A. Benton: Photographic materials and their handling, in *Handbook of Optical Holography*, ed. by H.J. Caulfield (Academic, New York 1979) Chap.9, pp. 349-366

5.59 N.J. Phillips, A.A. Ward, R. Cullen, D. Porter: Advances in holographic bleaches. Phot. Sci. Eng. 24, 120-124 (1980)

5.60 J.H. Sterner, F.L. Oglesby, B. Anderson: Quinone vapors and their harmful effects. J. Ind. Hygiene and Toxicology 29 (No.2), 60-73 (1947)

5.61 L. Joly and R. Vanhorebeek: Development effects in white-light reflection holography. Phot. Sci. Eng. 24, 108-113 (1980)

5.62 R.L. van Renesse: Scattering properties of fine-grained bleached emulsions. Phot. Sci. Eng. 24, 114-119 (1980)

5.63 W. Spierings: "Pyrochrome" processing yields color-controlled results with silver-halide materials. Holosphere 10 (No.7-8), 1-7 (1981)

5.64 M. Pardo, F. Anton, C. Pastor: Rendimiento y senal - Ruido de hologramas obtenidos con diferentes procesos de blanqueo. Optica Pura y Aplicada 14, 61-65 (1981)

5.65 G. Liebmann: Die Messung der Bleicheffektivität an Mikratplatten ORWO LP2. J. Signal. AM 9, 121-127 (1981)

5.66 J. Slaby, L. Sirko: Essential features of the bleached silver halide holographic materials. Opt. Commun. 37, 165-168 (1981)

5.67 L. Sirko, J. Slaby: Wavefront phase modulation in bleached holographic emulsion. Opt. Commun. 41, 407-410 (1982)

5.68 J. Oliva, M. Pardo, J.A. Quintana: High SNR in bleached silver-halide holography. Appl. Opt. 21, 171 (1982)

5.69 A. Fimia, M. Pardo, J.A. Quintana: Improvement of image quality in bleached holograms. Appl. Opt. 21, 3412-3413 (1982)

5.70 A.V. Alekseev-Popov, S.A. Gevelyuk: Contributions of amplitude and phase modulation to diffraction efficiency in three-dimensional reflective holograms. Sov. Phys. - Tech. Phys. 27, 1289-1291 (1982)

5.71 R.R.A. Syms, L. Solymar: Planar volume phase holograms formed in bleached photographic emulsions. Appl. Opt. 22, 1479-1496 (1983)

5.72 N.J. Phillips: Colour reflection holography, in *Three-Dimensional Imaging*, ed. by J. Ebbeni, A. Monfils. Proc. SPIE 402, 19-24 (1983)

5.73 L. Joly: Grain growth during rehalogenating bleaching. J. Photogr. Sci. 31, 143-147 (1983)

5.74 D.I. Staselko, A.L. Churaev: Investigation of the phase characteristics of holographic recording media. Opt. Spectrosc. (USSR) 57, 411-415 (1984)

5.75 R.R.A. Syms, L. Solymar: The effects of swelling on volume holograms formed in bleached photographic emulsion. Opt. Acta 31, 149-157 (1984)

5.76 M. Quintanilla, A.M. de Frutos, I. Arias: Characterization of volume and phase holographic gratings. Appl. Opt. 23, 214-217 (1984)

5.77 D.J. Cooke, A.A. Ward: Reflection-hologram processing for high efficiency in silver-halide emulsions. Appl. Opt. 23, 934-941 (1984)

5.78 R.L. van Renesse, J.W. Burgmeijer: Application of pulsed reflection holography to material testing. Opt. Eng. 24, 1086-1092 (1985)

5.79 D.I. Staselko, A.L. Churaev: Method for calculating the contrast of images of diffusely scattering objects produced by thick-layer transmission phase holograms. Sov. Phys. - Tech. Phys. 31, 197-202 (1986)

5.80 P. Hariharan: Bleached photographic phase holograms: The influence of drying procedures on diffraction efficiency. Opt. Commun. 56, 318-320 (1986)

5.81 R. De Winne, R. Phelan: Processing of Agfa-Gevaert Holotest 8E56 HD for reflection holography. Proc. Int'l Symp. on Display Holography, ed. by T. H. Jeong, Lake Forest College, IL 60045, USA, Vol.II (1986) pp.263-277

5.82 J. Blyth: Notes on processing holograms with solvent bleach. Proc. Int'l Symp. on Display Holography, ed. by T.H. Jeong, Lake Forest College, IL 60045, USA, Vol.II (1986) pp.325-331

5.83 J. Crespo, A. Fimia, J.A. Quintana: Fixation-free methods in bleached reflection holography. Appl. Opt. 25, 1642-1645 (1986)

5.84 N. Phillips: Benign bleaching for health holography. Holosphere 14, (No.4), 21-22 (1986)

5.85 J. Blyth: A novel approach to colour processing. Wavefront 2 (No.2), 23 (1987)

5.86 R.C. Sehlin: Persulfate bleach and motion-picture film processes. J. Soc. Motion Picture and Television Eng. 91, 158-163 (1982)

5.87 J.E. Crisante, W.A. Szafranski: Kodak persulfate bleach for process ECN-2. J. Soc. Motion Picture and Television Eng. 91, 1058-1065 (1982)

5.88 J.A. Keiler, G. Pollakowski: Persulfate/quinone bleach - Environmental and economic aspects. J. Soc. Motion Picture and Television Eng. 95, 220-223 (1986)

5.89 P. Hariharan, C.M. Chidley: Photographic phase holograms: The influence of developer composition on scattering and diffraction efficiency. Appl. Opt. 26, 1230-1234 (1987)

5.90 P. Hariharan, C.M. Chidley: Rehalogenating bleaches for photographic phase holograms: The influence of halide type and concentration on diffraction efficiency and scattering. Appl. Opt. 26, 3895-3898 (1987)

5.91 P. Hariharan, C.M. Chidley: Photographic phase holograms: Spatial frequency effects with conventional and reversal bleaches. Appl. Opt. 27, 3065-3067 (1988)

5.92 P. Hariharan, C.M. Chidley: Rehalogenating bleaches for photographic phase holograms 2: Spatial frequency effects. Appl. Opt. 27, 3852-3854 (1988)

5.93 P. Hariharan, C.M. Chidley: Bleached reflection holograms: A study of color shifts due to processing. Appl. Opt. 28, 422-424 (1989)

5.94 A.A. Ward, L. Solymar: Diffraction efficiency limitations of holograms recorded in silver-halide emulsions. Appl. Opt. 28, 1850- 1855 (1989)

5.95 J. Thomas: Reversal techniques for high quality reflection holograms. C99 report, Biomedical Engineering, Northwestern University, Evanston, IL 60208, USA (1989)

5.96 D.Y. Qiu, C.C. Jiang: Mathematical method to predict the diffraction efficiency of holograms. Opt. Laser Technol. 21, 47-50 (1989)

5.97 G. Ackermann, J. Eichler, C. Schneeweiss-Wolter: Exposure, developing and bleaching of holographic layers. Laser und Optoelektronik 21 (No.4), 56-59 (1989)

5.98 N. Phillips: The silver halides - The workhorse of the holography business. Proc. Int'l Symp. on Display Holography, ed. by T. H. Jeong, Lake Forest College, IL 60045, USA, Vol.III (1989) pp.35-73

5.99 L. Joly, P. Jacobs: Spectral response of reflection gratings on Holotest 8E75 HD plates. Proc. Int'l Symp. on Display Holography, ed. by T.H. Jeong, Lake Forest College, IL 60045, USA, Vol.III (1989) pp.115-126

5.100 Z.S. Hegedus, P. Hariharan: Some new developments in display holography, in *Three-Dimensional Holography: Science, Culture, Education*, ed. by T.H. Jeong, V.B. Markov. Proc. SPIE 1238, 480-488 (1991)

5.101 P. Hariharan: Rehalogenating bleaches for photographic phase holograms 3: Mechanism of material transfer. Appl. Opt. 29, 2983-2985 (1990)

5.102 P. Hariharan: Basic processes involved in the production of bleached holograms. J. Photogr. Sci. 38, 76-81 (1990)

5.103 M.H. Jeong, M.J. Lee, I.W. Lee: Rehalogenating properties of photographic plate for phase holograms, in *Optics in Complex Systems*, ed. by F. Lanzl, H.-J. Preuss, G. Weigelt. Proc. SPIE 1319, 307 (1990)

410

5.104 R.K. Kostuk: Effects of bleach constituents on the performance of silver-halide holograms, in *Practical Holography IV*, ed. by S.A. Benton. Proc. SPIE 1212, 55-62 (1990)

5.105 R.K. Kostuk: Factorial optimization of bleach constituents for silver halide holograms. Appl. Opt. 30, 1611-1616 (1991)

5.106 A.L. Churaev, V.V. Artyomova: Characteristic curves and phase-exposition characteristics of photographic photomaterials, in *Three-Dimensional Holography: Science, Culture, Education*, ed. by T.H. Jeong, V.B. Markov. Proc. SPIE 1238, 158-165 (1991)

5.107 H.I. Bjelkhagen, N. Phillips, W. Ce: Chemical symmetry - Developers that look like bleach agents for holography, in *Practical Holography IV*, ed. by S.A. Benton. Proc. SPIE 1461, 321-328 (1991)

5.108 R.K. Kostuk, J.W. Goodman: Refractive index modulation mechanism in bleached silver halide holograms. Appl. Opt. 30, 369-371 (1991)

5.109 S. Kumar, K. Singh: Bleached phase holograms using Agfa-Gevaert 10E75 NAH plates: Influence of different developers and developer composition on the diffraction efficiency, scattering and stability. Optik 86, 99-103 (1990)

5.110 S. Kumar, K. Singh: Bleached phase holograms using Agfa-Gevaert 10E75 NAH plates. Opt. & Laser Technol. 23, 37-41 (1991)

5.111 S. Kumar, K. Singh: Study of parameters of amplitude and bleached holograms recorded and reconstructed at 442 nm using photographic emulsion. Optik 88, 45-49 (1991)

5.112 S. Kumar, K. Singh: Comparative study of maximum diffraction efficiency of bleached holograms at different read-beam angles using 633 nm and 442 nm wavelengths. Optik 90, 75-79 (1992)

5.113 A. Beléndez, I. Pascual, A. Fimia: Influences of recording geometry parameters on diffraction efficiency in bleached silver halide transmission holograms. J. Mod. Opt. 39, 1855-1861 (1992)

5.114 T.A. Shankoff: Phase holograms in dichromated gelatin. Appl. Opt. 7, 2101-2105 (1968)

5.115 D.H. McMahon, W.T. Maloney: Measurements of the stability of bleached photographic phase holograms. Appl. Opt. 9, 1363-1368 (1970)

5.116 A.J. Chenoweth: Humidity testing of bleached holograms. Appl. Opt. 10, 913-915 (1971)

5.117 F.P. Laming, S.L. Levine, G. Sincerbox: Lifetime extension of bleached holograms. Appl. Opt. 10, 1181-1182 (1971)

5.118 P. Hariharan, C.S. Ramanathan: Suppression of printout effect in photographic phase holograms. Appl. Opt. 10, 2197-2199 (1971)

5.119 S.L. Norman: Dye-induced stabilization of bleached holograms. Appl. Opt. 11, 1234-1239 (1972)

5.120 D.J. Locker: Analysis of the effects of environment of the photolytic darkening of silver halide microcrystals. Photogr. Sci. Eng. 18, 242-247 (1974)

5.121 N. Nishida: Bleached phase hologram containing nonsilver metal compound. Appl. Opt. 13, 2769-2770 (1974)

5.122 T. Inagaki, J. Nakajima, Y. Nishimura: Characteristics of bleached hologram according to exposure stability to light. Fujitsu Sci. & Tech. J. (Jpn.) 10 (No.1), 135-155 (1974)

5.123 G. Hüttmann: Holography with a frequency-doubled Nd:YAG laser, in *Holography Techniques and Applications*, ed. by W.P.O. Jüptner. Proc. SPIE 1026, 14-21 (1988)

5.124 D. Vila, E. Wesly: Controlling the effects of ultra-violet light on holographic emulsions. Proc. Int'l Symp. on Display Holography, ed. by T.H. Jeong, Lake Forest College, IL 60045, USA, Vol.III (1989) pp.141-148

5.125 N.D. Vorzobova, R.V. Rjabova, A.I. Schvarzvald: Holographic characteristics of IAE and PFG-01 photoplates for colored pulsed holography, in *Three-Dimensional Holography: Science, Culture, Education*, ed. by T.H. Jeong, V.B. Markov. Proc. SPIE **1238**, 476-477 (1991)

5.126 L.P. Vakhtangova, B.I. Shapiro, E.A. Gruz, K.S. Bogomolov: The method of phase hologram stabilization. Proc. 3rd. USSR All-Union Conf. on Holography (Uljanovsk, USSR 1978) p.329

5.127 S. Kumar, K. Singh: Stability improvement in bleached phase holograms. Opt. & Laser Technol. **23**, 225-227 (1991)

5.128 R.E. Jacobson, P. Baxter: Factors influencing print-out in bleached holograms, in *Practical Holography VI*, ed. by S. Benton. Proc. SPIE **1667**, 243-255 (1992)

5.129 V. Weiss, E. Millul, A.A. Friesem: Photolytically stable bleached silver halide holograms for archival storage. Proc. From Gallileos Ochialino to Optoelectronics (Padova, Italy 1992)

5.130 J.M.C. Jonathan, R. Kinany: Generation of uniaxial medium from bleached photographic plates. Opt. Commun. **27**, 61-64 (1978)

Chapter 6

6.1 R.E. Brooks, L.O. Heflinger, R.F. Wuerker, R.A. Briones: Holographic photography of high-speed phenomena with conventional and Q-switched ruby lasers. Appl. Phys. Lett. **7**, 92-94 (1965)

6.2 A.D. Jacobson, F.J. McClung: Holograms produced with pulsed laser illumination. Appl. Opt. **4**, 1509-1510 (1965)

6.3 L.D. Siebert: Front-lighted pulse laser holography. Appl. Phys. Lett. **11**, 326-328 (1967)

6.4 L.D. Siebert: Large-scene front-lighted hologram of a human subject. Proc. IEEE **56**, 1242-1243 (1968)

6.5 R.G. Zech, L.D. Siebert: Pulsed laser reflection holograms. Appl. Phys. Lett. **13**, 417-418 (1968)

6.6 D.A. Ansley: Techniques for pulsed laser holography of people. Appl. Opt. **9**, 815-821 (1970)

6.7 G. Harigel, C. Baltay, M. Bregman, M. Hibbs, A. Schaffer, H. Bjelkhagen, J. Hawkins, W. Williams, P. Nailor, R. Michaels, H. Akbari: Pulse stretching in a Q-switched ruby laser for bubble chamber holography. Appl. Opt. **25**, 4102-4110 (1986)

6.8 R. Bunsen, H. Roscoe: Photochemische Untersuchungen. Ann. Physik Chemie **117**, 529-562 (1862)

6.9 M. Hercher, B. Ruff: High-intensity reciprocity failure in Kodak 649-F plates at 6943 Å. J. Opt. Soc. Am. **57**, 103-105 (1967)

6.10 H. Nassenstein, H. Dedden, H.J. Metz, H.E. Rieck, D. Schultze: Physical properties of holographic materials. Photogr. Sci. Eng. **13**, 194-199 (1969)

6.11 R.O. Rice, J.D. Macomber: Reciprocity failure in Kodak HIE film exposed to ruby-laser pulses. J. Opt. Soc. Am. **65**, 1489-1494 (1975)

6.12 G. Hüttmann: Holography with a frequency-doubled Nd:YAG laser, in *Holography Techniques and Applications*, ed. by W.P.O. Jüptner. Proc. SPIE **1026**, 14-21 (1988)

6.13 N.D. Vorzobova, D.I. Staselko: Exposure characteristics of high-resolution silver-halide photographic materials for recording three-dimensional holograms by means of a pulsed laser. Sov. J. Opt. Technol. **44**, 249-250 (1977)

6.14 N.D. Vorzobova, D.I. Staselko: Diffraction efficiency of 3-D holograms recorded with short exposures. Opt. Spectrosc. (USSR) **45**, 90-93 (1978)

6.15 A.A. Benken, D.I. Staselko: Light scattering in the formation of a latent image by pulsed laser radiation. Sov. Phys. - Tech. Phys. **27**, 896- 898 (1982)

6.16 M. Pantcheva, T. Petrova, N. Pangelova, A. Katsev: Chemical sensitization of fine-grain silver halide emulsions for holographic recording, in *Holography'89*, ed. by Y.N. Denisyuk, T.H. Jeong. Proc. SPIE 1183, 128-130 (1990)

6.17 A. Hautot: Reciprocity characteristics of silver bromide and silver chloride emulsions. Photogr. Sci. Eng. 4, 254-256 (1960)

6.18 H.E. Spencer, L.E. Brady, J.F. Hamilton: Study of the mechanism of sulfur sensitization by a development-center technique. J. Opt. Soc. Am. 54, 492-497 (1964)

6.19 H.E. Spencer, R.E. Atwell: Sulfur sensitization and high-intensity reciprocity failure of silver bromide grains. J. Opt. Soc. Am. 54, 498-505 (1964)

6.20 H.E. Spencer, R.E. Atwell: Development centers and high-intensity reciprocity failure. J. Opt. Soc. Am. 56, 1095-1101 (1966)

6.21 P. Binfield, R. Galloway, J. Watson: Reciprocity failure in continous wave holography. Appl. Opt. 32 (1993) in print

6.22 K.M. Johnson, L. Hesselink, J.W. Goodman: Holographic reciprocity law failure. Appl. Opt. 23, 218-227 (1984)

6.23 C.D. Leonard, A.L. Smirl: Holographic recording with limited laser light. Appl. Opt. 10, 625-631 (1971)

6.24 H.J. Caulfield, S. Lu, J.L. Harris: Biasing for single-exposure and multiple-exposure holography. J. Opt. Soc. Am. 58, 1003-1004 (1968)

6.25 N. Nishida, M. Sakaguchi: Improvement of nonuniformity of the reconstructed beam intensity from a multiple-exposure hologram. Appl. Opt. 10, 439-440 (1971)

6.26 H. Akahori, K. Sakurai: Information search using superimposed holograms. Appl. Opt. 10, 665-666 (1971)

6.27 H. Akahori, K. Sakurai: Information search using holography. Appl. Opt. 11, 413-415 (1972)

6.28 M.J. Beesley, J.G. Castledine: The use of photoresist as a holographic recording medium. Appl. Opt. 9, 2720-2724 (1970)

6.29 C.S. Vikram, R.S. Sirohi: Performance of pre- or postexposed holograms. Appl. Opt. 10, 2195-2196 (1971)

6.30 M.J. Landry, G.S. Phipps: Holographic characteristics of 10E75 plates for single- and multiple-exposure holograms. Appl. Opt. 14, 2260-2266 (1975)

6.31 J.J.A. Couture, R.A. Lessard: Intermittent characteristic curves for Kodak 649F plates at 514.5 nm. Appl. Opt. 18, 3644-3651 (1979)

6.32 J.J.A. Couture, R.A. Lessard: Diffraction efficiency of specular multiplexed holograms recorded on Kodak 649F plates. Appl. Opt. 18, 3652-3660 (1979)

6.33 G.S. Phipps, C.E. Robertson, F.M. Tamashiro: Reprocessing of nonoptimally exposed holograms. Appl. Opt. 19, 802-811 (1980)

6.34 A.A. Hoag, W.C. Miller: Application of photographic materials in astronomy. Appl. Opt. 8, 2417-2430 (1969)

6.35 K. Biedermann: Attempts to increase the holographic exposure index of photographic materials. Appl. Opt. 10, 584-595 (1971)

6.36 H.I. Bjelkhagen: Holographic recording materials and the possibility to increase their sensitivity, CERN, Geneva, Switzerland, EF-report 84-7 (1984)

6.37 N.I. Kirillov: [Transl.: *High Resolution Photographic Materials for Holography and Their Processing Methods* (in Russian)] (Nauka, Moscow 1979)

6.38 Yu.N. Denisyuk, I.R. Protas: Improved Lippmann photographic plates for recording stationary light waves. Opt. Spectrosc. (USSR) 14, 381-383 (1963)

6.39 J. Crespo, A. Fimia, J.A. Quintana: Fixation-free methods in bleached reflection holography. Appl. Opt. 25, 1642-1645 (1986)

6.40 K. Biedermann, N.-E. Molin: Combining hypersensitization and rapid *in situ* processing for time-average observation in real-time hologram interferometry. J. Phys. E 3, 669-680 (1970)

6.41 W. Spierings: Common sense and nonsense in holographic research. Proc. Int'l Symp. on Display Holography, ed. by T.H. Jeong, Lake Forest College, IL 60045, USA, Vol.II (1986) pp.245-253

6.42 T. Cvetkovich, A. Eijnde, W. Spierings: Hypersensitization, L.A.S.E.R. News 3 (No.1), 5-6 (1986)

6.43 M. Todorova, A. Stankova, I. Markov, M. Pancheva: Photographic action of Pb(II)-EDTA. J. Photogr. Sci. 35, 196-199 (1987)

6.44 M. Todorova, A. Stankova: Hypersensitization of gelatin silver chlorobromide photographic layers with thallium(I)-EDTA. Photogr. Sci. Eng. 28, 207-210 (1984)

6.45 M. Chang, N. George: Holographic dielectric grating: Theory and practice. Appl. Opt. 9, 713-719 (1970)

6.46 N.N. Yaroslavskaya, O.V. Andreeva, V.I. Sukhanov: Thermal method of increasing the sensitivity of photographic materials for recording three-dimensional holograms. Sov. J. Opt. Technol. 42, 542-543 (1975)

6.47 N.L. Kosobokova, Yu.E. Usanov: The effect of heating on the properties of LOI-2 holographic plates. Sov. J. Opt. Technol. 42, 745-747 (1975)

6.48 T.A. Babcock, P.M. Ferguson, W.C. Lewis, T.H. James: A novel form of chemical sensitization using hydrogen gas. Photogr. Sci. Eng. 19, 49-55 (1975)

6.49 T.A. Babcock, P.M. Ferguson, W.C. Lewis, T.H. James: Chemical sensitization using hydrogen gas, Pt.2. With other types of chemical sensitization. Photogr. Sci. Eng. 19, 211-214 (1975)

6.50 G.A. Janusonis: Sensitization of photographic film in elevated-temperature hydrogen gas and its development, Pt.1. Photogr. Sci. Eng. 22, 297-306 (1978)

6.51 R. Sliva: Photography in astronomy. Hypersensitizing, Pt.1. Astronomy No.4, 39-42 (1981)

6.52 R. Sliva: Photography in astronomy. Hypersensitizing, Pt.2. Astronomy No.5, 48-50 (1981)

6.53 R.L. Scott, A.G. Smith, R.J. Leacock: The use of forming gas in hypersensitizing Kodak spectroscopic plates. Am. Astro. Soc. Phot. Bull. (No.15) 2 (1977) p.12

6.54 J. Rothstein: Enhancement of photographic speed and sensitivity by electric fields. Photogr. Sci. Eng. 4, 5-11 (1960)

6.55 A. Shepp, L. Corben: Double latensification. Photogr. Sci. Eng. 8, 69-76 (1964)

6.56 T.H. James, W. Vanselow, R.F. Quirk: Gold and mercury latensification and hypersensitization for direct and physical development. PSA J. 14, 349-353 (1948)

6.57 T.H. James: Electron injection by developing agents - Latensification of internal image. Photogr. Sci. Eng. 10, 344-349 (1966)

6.58 B.S. Askins: Photographic image intensification by autoradiography. Appl. Opt. 15, 2860-2865 (1976)

6.59 A. Beiser: Latent image fading in nuclear emulsions. Phys. Rev. 81, 153 (1951)

6.60 W.L. McLaughlin, M. Ehrlich: Film badge dosimetry: How much fading occurs? Nucleonics 12 (No.10), 34-36 (1954)

6.61 R.A. Armistead, F.B. Galimba: Latent-image fading of three commercially available fine grained emulsions. Photogr. Sci. Eng. 17, 42-46 (1973)

6.62 H. Akbari, H.I. Bjelkhagen: Holographic latent-image fading, Fermilab, Batavia, IL 60510, USA, E-632 report (March 24, 1987)

6.63 S.A. Benton: Photographic materials and their handling, in *Handbook of Optical Holography*, ed. by H.J. Caulfield (Academic, New York 1979) Chap.9, pp. 349-366

6.64 N.J. Phillips, D. Porter: An advance in the processing of holograms. J. Phys. E 9, 631-634 (1976)

6.65 N. Phillips: The silver halides - The workhorse of the holography business. Proc. Int'l Symp. on Display Holography, ed. by T. H. Jeong, Lake Forest College, IL 60045, USA, Vol.III (1989) pp.35-73

6.66 F. Unterseher: Integrating pulse holography with varied holographic techniques. Proc. Int'l Symp. on Display Holography, ed. by T.H. Jeong, Lake Forest College, Lake Forest, IL 60045, USA, Vol.III (1989) pp.403-419

6.67 P.M. Boone: Some problems associated with processing Agfa-Gevaert 8E75 HD sheet film for reflection holography, in *Progress in Holographic Applications*, ed. by J. Ebbeni. Proc. SPIE 600, 172-177 (1985)

6.68 P.M. Boone: Practical problems associated with processing Agfa-Gevaert 8E75 HD sheet film for reflection holography. Proc. Int'l Symp. on Display Holography, ed. by T.H. Jeong, Lake Forest College, Lake Forest, IL 60045, USA, Vol.II (1986) pp.333-340

6.69 P.M. Boone: Some recent Belgian holographic achievments. Proc. Int'l Symp. on Display Holography, ed. by T.H. Jeong, Lake Forest College, Lake Forest, IL 60045, USA, Vol.III (1989) pp.267-278

6.70 S.D. Nikolaev, I.O. Starobogatov: Formation of highly coherent radiation in a free-running ruby laser for pulsed holography. Opt. Spectrosc. (USSR) 61, 682-685 (1986)

6.71 A.V. Aristov, N.D. Vorzobova, D.A. Kozlovskii, M.B. Levin, D.I. Staselko, V.L. Strigun, A.S. Cherkasov: Image hologram recording using a pulsed dye laser. Opt. Spectrosc. (USSR) 61, 90-91 (1986)

6.72 I.M. Kliot-Dashinskaya, D.I. Staselko, A.L. Churaev: Brightness and contrast of holographic images of small particles. Opt. Spectrosc. (USSR) 48, 180-183 (1980)

6.73 I.M. Kliot-Dashinskaya, D.I. Staselko, V.L. Strigun: Investigation of the recording of graphic reflection holograms using a pulsed ruby laser. Opt. Spectrosc. (USSR) 58, 377-380 (1985)

6.74 I.M. Kliot-Dashinskaya, V.I. Michailova, G.P. Paltsev, V.L. Strigun: Display holography with pulsed ruby laser. Recording and copying, in *Three-Dimensional Holography: Science, Culture, Education*, ed. by T.H. Jeong, V.B. Markov. Proc. SPIE 1238, 465-469 (1991)

6.75 D.A. Delwiche, J.D. Clifford, W.R. Weller: Printing motion-picture films immersed in a liquid, Pt.III: Evaluation of liquids. J. Soc. Motion Picture and Television Eng. 67, 678-686 (1958)

6.76 E. Sklar: Local inhomogeneities in the refractive index of gelatin containing a silver image treated with a tanning bleach. Photogr. Sci. Eng. 13, 29-31 (1969)

6.77 R.K. Kostuk, J.W. Goodman: Refractive index modulation mechanism in bleached silver halide holograms. Appl. Opt. 30, 369-371 (1991)

6.78 G.P. Paltsev, K.A. Stozharova: Immersion liquids for holograms recorded on photographic film. Sov. J. Opt. Technol. 38, 43-45 (1971)

6.79 H. Nassenstein: Interference, diffraction and holography with surface waves ("subwaves") II. Optik 30, 44-55 (1969)

6.80 A.L. Ingalls: The effect of film thickness variations on coherent light. Photogr. Sci. Eng. 4, 135-140 (1960)

6.81 O. Bryngdahl: Can detrimental effects in photographic volume holography be compensated for? Appl. Opt. 11, 195 (1972)

6.82 R.R.A. Syms, and L. Solymar: Planar volume phase holograms formed in bleached photographic emulsions. Appl. Opt. 22, 1479-1496 (1983)

6.83 M. Richardson: Index-matching fluid. Holosphere 13 (No.3), 23 (1985)

6.84 M. Quintanilla, A.M. de Frutos, I. Arias: Characterization of volume and phase holographic gratings. Appl. Opt. 23, 214-217 (1984)

6.85 D.R. Wuest, R.S. Lakes: Color control in reflection holograms by humidity. Appl. Opt. 30, 2363-2367 (1991)

6.86 A.G. Tull: Tanning development and its application to dye transfer images. J. Photogr. Sci. 11, 1-26 (1963)

6.87 H.M. Smith: Photographic relief images. J. Opt. Soc. Am. 58, 533-539 (1968)

6.88 H.M. Smith: Production of photographic relief images with arbitrary profile. J. Opt. Soc. Am. 59, 1492-1494 (1969)

6.89 J.H. Altman: Microdensitometry of high resolution plates by measurement of the relief image. Photogr. Sci. Eng. 10, 156-159 (1966)

6.90 J.H. Altman: Pure relief images on type 649-F plates. Appl. Opt. 5, 1689-1690 (1966)

6.91 A.K. Rigler: Wavefront reconstruction by reflection. J. Opt. Soc. Am. 55, 1693 (1965)

6.92 W.T. Cathey: Spatial phase modulation of wavefronts in spatial filtering and holography. J. Soc. Opt. Am. 56, 1167-1171 (1966)

6.93 H. Hannes: Interferometric measurements of phase structures in photographs. J. Opt. Soc. Am. 58, 140-141 (1968)

6.94 V. Russo, S. Sottini: Bleached holograms. Appl. Opt. 7, 202 (1968)

6.95 G.B. Brandt, A.K. Rigler: Reflection holograms of focused images. Phys. Lett. A 25, 68-69 (1967)

6.96 M.M. Butusov, A.I. Ioffe: Investigation of parameters of holographic periodic multiple-imaging structures. Sov. J. Quantum Electron. 6, 519-521 (1976)

6.97 L.N. Beinarovich, N.P. Larionov, A.V. Lukin, K.S. Mustafin: Production of high-quality copies of holograms. Opt. Spectrosc. (USSR) 30, 186-187 (1971)

6.98 A.D. Galpern, M.M. Ermolaev, I.V. Kalinina, L.V. Selyavko, V.P. Smaev: Investigation of the possibility of producing phase-relief holograms on silver-halide photographic materials. Sov. J. Opt. Technol. 52, 256-257 (1985)

6.99 A.D. Galpern, I.V. Kalinina, L.V. Selyavko, V.P. Smaev: Obtaining relief-phase holograms on PE-2 photographic plates and their copying. Opt. Spectrosc. (USSR) 60, 644-645 (1986)

6.100 A.D. Galpern, A.A. Paramonov, L.V. Selyavko, V.P. Smaev, Yu.V. Solomatin, N.S. Shelekhov: Recording and copying of iridescent holograms of false-color images. Sov. J. Opt. Technol. 53, 222-224 (1986)

6.101 A.D. Galpern, V.P. Smaev, L.V. Selyavko, N.S. Szelehov: Relief-phase holograms on silver halide materials and their copying using the thermo- polymerization method (in Russian), in *High Efficiency Materials for Hologram Recording*, ed. by G.A. Sobolev (Academy of Sciences USSR, Leningrad 1988) pp.100-107

6.102 E.B. Brui, S.N. Koreshev: Characteristics of the use of thin layers of PE-2 photoemulsion for obtaining low-frequency relief hologram structures. Opt. Spectrosc. (USSR) 67, 403-405 (1989)

6.103 S.N. Koreshev, S.V. Gil: Profile of low-frequency relief hologram structures obtained on thin layers of PE-2 photoemulsion. Opt. Spectrosc. (USSR) 68, 247-249 (1990)

6.104 T. Ahlhorn, H. Kreye: Verfahren zur Herstellung von holografischen Prägematrizen, in *Jahrbuch Oberflächentechnik*, Bd.47 (Metall-Verlag, Berlin 1991) pp.376-381

6.105 M.T. Gale, K. Knop: *Surface-Relief Images for Color Reproduction* (Focal, London 1980)

6.106 A.M. Farberov, V.P. Smaev, A.D. Galpern, L.N. Vasileva, T.M. Sinitsyna: Fabrication of a metallic matrix for high-volume copying of phase-relief holograms. Sov. J. Opt. Technol. 55, 168-169 (1988)

6.107 A.D. Galpern, V.P. Smaev: Methods of recording and printing decorative relief-phase holograms. Sov. J. Opt. Technol. 55, 694-701 (1988)

6.108 K.S. Pennington, J.S. Harper, F.P. Laming: New phototechnology suitable for recording phase holograms and similar information in hardened gelatin. Appl. Phys. Lett. 18, 80-84 (1971)

6.109 J.W. Gladden, J.W. Eastes: Materials research for holographic recording. (Report No.3, Hardened gelatin holographic recording materials), ETL-0197, U.S. Army Engineer Topographic Laboratories, Fort Belvoir, VA 22060, USA (1979)

6.110 W.R. Graver, J.W. Gladden, J.W. Eastes: Phase holograms formed by silver halide (sensitized) gelatin processing. Appl. Opt. 19, 1529-1536 (1980)

6.111 B.J. Chang, K. Winick: Silver-halide gelatin holograms, in *Recent Advances in Holography*, ed. by T.C. Lee, P.N. Tamura. Proc. SPIE 215, 172-177 (1980)

6.112 B.J. Chang, K. Winick: Silver-halide gelatin holograms. Holosphere 10 (No.6), 1-5 (1981)

6.113 A. Fimia, M. Pardo, J.A. Quintana: Noise reduction in holographic images reconstructed with blue light. Appl. Opt. 22, 3318 (1983)

6.114 P.G. Boj, A. Fimia, J.A. Quintana: Silver-halide gelatin for the fabrication of holographic optical elements, in *Image Detection and Quality*, ed. by L.F. Guyot. Proc. SPIE 702, 105-108 (1986)

6.115 A. Fimia, I. Pascual, A. Beléndez: Silver halide sensitized gelatin as a holographic storage medium, in *Laser Technologies in Industry*, ed. by O.D.D. Soares, S.P. Almeida. Proc. SPIE 952, 288-291 (1988)

6.116 A. Fimia, I. Pascual, C. Vázquez, A. Beléndez: Silver-halide sensitized holograms and their applications, in *Holographic Optics II: Principles and Applications*, ed. by C.M. Morris. Proc. SPIE 1136, 53-57 (1989)

6.117 S.A. Benton: Photographic materials and their handling, in *Handbook of Optical Holography*, ed. by H.J. Caulfield (Academic, New York 1979) Chap.9, pp. 349-36

6.118 R.A. Ferrante: Silver halide gelatin spatial frequency response. Appl. Opt. 23, 4180-4181 (1984)

6.119 P. Hariharan: Silver halide sensitized gelatin holograms: Mechanism of hologram formation. Appl. Opt. 25, 2040-2042 (1986)

6.120 D.K. Angell: Improved diffraction efficiency of silver halide (sensitized) gelatin. Appl. Opt. 26, 4692-4702 (1987)

6.121 D.K. Angell: Controlling emulsion thickness variations in silver halide (sensitized) gelatin, in *Holographic Optics: Design and Applications*, ed. by I. Cindrich. Proc. SPIE 883, 106-113 (1988)

6.122 A guide to Dow Corning silane coupling agents, Dow Corning, No. 23-012B-85 (1985)

6.123 V. Weiss, E. Millul: Bleached silver halide holographic recording materials, in *Holographic Techniques and Applications*, ed. by W.P.O. Jüptner. Proc. SPIE 1026, 55-61 (1988)

6.124 V. Weiss, Y. Amitai, A.A. Friesem, E. Millul: Silver halide sensitized gelatin holographic recording materials, in *Sixth Meeting in Israel on Optical Engineering*, ed. by R. Finkler, J. Shamir. Proc. SPIE 1038, 110-114 (1988)

6.125 Yu.Ye. Usanov, Ye.A. Vavilova, N.L. Kosobokova, M.K. Shevtsov: Reflection silver-halide gelatin holograms, in *Three-Dimensional Holography: Science, Culture, Education*, ed. by T.H. Jeong, V.B. Markov. Proc. SPIE 1238, 178-182 (1991)

6.126 N.L. Kosobokova, Yu.Ye. Usanov, M.K. Shevtsov: Properties of volume reflection silver halide gelatin holograms, in *Three-Dimensional Holography: Science, Culture, Education*, ed. by T.H. Jeong, V.B. Markov. Proc. SPIE 1238, 183-188 (1991)

6.127 Yu.E. Usanov, M.K. Shevtsov, N.L. Kosobokova, E.A. Kirienko: Mechanism for forming a microvoid structure and methods for obtaining silver-halide gelatin holograms. Opt. Spectrosc. (USSR) 71, 375-379 (1991)

6.128 I. Pascual, A. Beléndez, A. Fimia: Reflection holographic optical elements in silver halide sensitized gelatin. Opt. Appl. 21, 239-244 (1991)

6.129 A. Fimia, A. Beléndez, I. Pascual: Silver halide (sensitized) gelatin in Agfa-Gevaert plates: The optimized process. J. Mod. Opt. 38, 2043-2045 (1991)

6.130 A. Fimia, A. Beléndez, I. Pascual: Influence of R-10 bleaching on latent image formation in silver halide sensitized gelatin. Appl. Opt. **31**, 3203-3205 (1992)

6.131 A. Fimia, I. Pascual, A. Beléndez: Optimized spatial frequency response in silver halide sensitized gelatin. Appl. Opt. **31**, 4625-4627 (1992)

6.132 N.J. Phillips, R.D. Rallison, C.A. Barnett, S.R. Schicker, Z.A. Coleman: Dichromated gelatin - some heretical comments, in *Practical Holography VII: Imaging and Materials*, ed. by S.A. Benton. Proc. SPIE **1914**, 101-114 (1993)

6.133 M. Mazakova, M. Pancheva, P. Kandilarov, P. Sharlandjiev: Dichromated gelatin for volume holographic recording with high sensitivity. Pt.I. Opt. Quantum Electr. **14**, 311-315 (1982)

6.134 M. Mazakova, M. Pancheva, P. Kandilarov, P. Sharlandjiev: Dichromated gelatin for volume holographic recording with high sensitivity. Pt.II. Opt. Quantum Electr. **14**, 317-320 (1982)

6.135 N. Nishida: Bleached phase hologram containing nonsilver metal compound. Appl. Opt. **13**, 2769-2770 (1974)

6.136 H. Nassenstein, J. Eggers: Über einen neuen Hologrammtyp mit wellenlängenselektiver Rekonstruktion. Phys. Lett. A **28**, 141-142 (1968)

6.137 G.I. Lashkov, V.I. Sukhanov: Use of dispersion photorefraction due to processes in which triplet states participate to record 3-D phase holograms. Opt. Spectrosc. (USSR) **44**, 590-594 (1978)

6.138 R. Röhler, K. Krusehe, J. Marangos: Holography with chromogen developed photographic emulsions. Opt. Commun. **25**, 169-172 (1978)

6.139 V.I. Sukhanov, O.V. Andreeva, M.V. Khazova: Use of dispersive refraction to record phase holograms in photographic film through dye substitution for silver. Sov. Tech. Phys. Lett. **9**, 355-356 (1983)

6.140 V.I. Sukhanov, M.V. Khazova, A.M. Kursakova, O.V. Andreeva: Bulk capillary recording media with latent image. Opt. Spectrosc. (USSR) **65**, 282-284 (1988)

6.141 V.I. Sukhanov, M.V. Khazova, A.M. Kursakova, O.V. Andreeva, T.S. Tsekhomskaya, G.P. Roskova: Writing volume phase holograms in light-sensitive systems with a capillary structure. Sov. Tech. Phys. Lett. **14**, 465-466 (1988)

6.142 V.I. Sukhanov: Heterogeneous recording media, in *Three-Dimensional Holography: Science, Culture, Education*, ed. by T.H. Jeong, V.B. Markov. Proc. SPIE **1238**, 226-230 (1991)

6.143 O.V. Andreeva: Analysis of the Focar-type silver halide heterogeneous media, in *Three-Dimensional Holography: Science, Culture, Education*, ed. by T.H. Jeong, V.B. Markov. Proc. SPIE **1238**, 231-234 (1991)

6.144 R. Neuhauss: *Die Farbenphotographie nach Lippmann's Verfahren*, Encyklopädie der Photographie (Knapp Verlag, Halle a.S. 1898) Heft 33

6.145 H.E. Ives: An experimental study of the Lippmann color photograph. Astrophysical J. **27**, 325-352 (1908)

6.146 M. Akagi, T. Kaneko, T. Ishiba: Electron micrographs of hologram cross sections. Appl. Phys. Lett. **21**, 93-95 (1972)

6.147 H.T. Buschmann: Bleichprozesse zur Erzeugung rauscharmer, lichtstarker Phasenhologramme. Optik **34**, 240-253 (1971)

6.148 L. Joly, R. Vanhorebeek: Development effects in white-light reflection holography. Photogr. Sci. Eng. **24**, 108-113 (1980)

6.149 R. Aliaga, H. Chuaqui: Microscopic and macroscopic measurement of holographic emulsions, in Proc. *Int'l Conf. on Holography Applications*, ed. by J. Ke, R. Pryputniewicz. Proc. SPIE **673**, 491-494 (1986)

6.150 T. Kubota: Cross-sectional view of Lippmann hologram gratings. Appl. Opt. **27**, 4358-4360 (1988)

6.151 J.P.F. Eichler, W. Wolff, H.E. Wolf, L.F.S. Coelho, S. de Barros, A.M. Borges, G. Ackermann: Processing of holographic AgBr films studied by X-ray fluorescence analysis. Appl. Opt. **30**, 1201-1205 (1991)

Chapter 7

7.1　N.J. Phillips: Patterns of interference: Information and fog in holographic iamges. Opt. Eng. 30, 1299-1305 (1991)

7.2　A.A. Ward, J.C.W. Newell, L. Solymar: Image blurring in display holograms and in holographic optical elements, in *Progress in Holographic Applications*, ed. by J. Ebbeni, Proc. SPIE 600, 57-65 (1985)

7.3　A.A. Ward, L. Solymar: Image distortions in display holograms. J. Photogr. Sci. 34, 62-76 (1986)

7.4　T. Kubota: Image sharpening of Lippmann hologram by compensation of wavelength dispersion, in *Practical Holography III*, ed. by S.A. Benton. Proc. SPIE 1051, 12-17 (1989)

7.5　C.S. Vikram, M.L. Billet: Optimizing image-to-background irradiance ratio in far-field in-line holography. Appl. Opt. 23, 1995-1998 (1984)

7.6　R. Bexon, M.G. Dalzell, M.C. Stainer: In-line holography and the assessment of aerosols. Opt. Laser Technol. 8, 161-165 (1976)

7.7　P. Dunn, J.M. Walls: Improved microimages from in-line absorption holograms. Appl. Opt. 18, 263-264 (1979)

7.8　P. Dunn, J.M. Walls: Absorption and phase in-line holograms: A comparison. Appl. Opt. 18, 2171-2174 (1979)

7.9　H.I. Bjelkhagen: Investigation on the resolution that can be obtained with the Baltay holographic arrangement for the 15-foot bubble chamber. Proc. Photonics Applied to Nuclear Physics II, CERN 85-10 (1985) pp.7-49

7.10　J. Watson, P.W. Britton: Applications of optical holography to underwater visual inspection, in *Optics in Engineering Measurement*, ed. by W.F. Fagan. SPIE 599, 26-31 (1986)

7.11　R. Naon, H. Bjelkhagen, R. Burnstein, L. Voyvodic: A system for viewing holograms. Nucl. Instr. and Meth. A 283, 24-36 (1989)

7.12　H.I. Bjelkhagen, J. Chang, K. Moneke: High-resolution contact Denisyuk holography. Appl. Opt. 31, 1041-1047 (1992)

7.13　R. Fusek, K. Harding, J. Harris, J. Murphy: Holographic documentation camera for component study evaluation, in *High Power Lasers and Applications*, ed. by C.C. Tang. Proc. SPIE 270, 186-195 (1981)

7.14　B.A. Tozer, R. Glanville, A.L. Gordon, M.J. Little, J.M. Webster, D.G. Wright: Holography applied to inspection and measurement in an industrial environment. Opt. Eng. 24, 746-753 (1985)

7.15　S. Kumar, K. Singh: Bleached phase holograms exposed on Agfa-Gevaert 10E75 NAH plates. Opt. & Laser Technol. 23, 37-41 (1991)

7.16　N. Phillips: The silver halides - The workhorse of the holography business. Proc. Int'l Symp. on Display Holography, ed. by T. H. Jeong, Proc. Lake Forest College, IL 60045, Vol.III (1989) pp.35-73

7.17　N.J. Phillips: New recommendations for the processing of Ilford plates, Physics Department, Loughborough University Note (July 1989)

7.18　R. Berkhout: Working with Kodak plates 120-01, making white light transmission holograms. Proc. Int'l Symp. Display Holography, ed. by T.H. Jeong, Lake Forest College, IL 60045, USA, Vol.III (1989) pp.127-129

7.19　S.A. Benton: Photographic materials and their handling, in *Handbook of Optical Holography*, ed. by H.J. Caulfield (Academic, New York 1979) Chap.9, pp.349-366

7.20　J.O. Bolstad: Holograms and spatial filters processed and copied in position. Appl. Opt. 6, 170 (1967)

7.21　D.H. Casler, H.D. Pruett: Multaneous exposure-development of holograms on 649-F film. Appl. Phys. Lett. 10, 341-342 (1967)

7.22　K. Biedermann, N.-E. Molin: Combining hypersensitization and rapid *in situ* processing for time-average observation in real-time hologram interferometry. J. Phys. E 3, 669-680 (1970)

7.23 LTI, Laser Technology, Inc., 1055 W. Germantown Pike, Norristown, PA 19404, USA.

7.24 Keystone Scientific Co., P.O.Box 22, Thorndale, PA 19372, USA.

7.25 D.B. Neumann, R.C. Penn: Objection motion compensation using reflection holography. J. Opt. Soc. Am. 62, 1373A (1972)

7.26 R.L. van Renesse, J.W. Burgmeijer: Application of Denisyuk pulsed holography to material testing, in *Industrial Applications of Laser Technology*, ed. by W.F. Fagan. Proc. SPIE 398, 138-148 (1983)

7.27 N. Phillips: Bridging the gap between Soviet and Western holography. Speaking notes, Holography Workshop, Lake Forest College, IL 60045, USA (1990)

7.28 N. Phillips: Bridging the gap between Soviet and Western holography, in *Holography, Commemorating the 90th Aniversary of the Birth of Dennis Gabor*, ed. by P. Greguss, T.H. Jeong. SPIE Institute Volume IS 8, 206-214 (1991)

7.29 J. Blyth: A novel approach to colour processing. Wavefront 2 (No.2), 23 (1987)

7.30 G. Hüttmann: Holography with a frequency-doubled Nd:YAG laser, in *Holography Techniques and Applications*, ed. by W.P.O. Jüptner. Proc. SPIE 1026, 14-21 (1988)

7.31 G. Saxby: Bypass holograms: A family of stable optical configurations for holography in unpromising environments, in *Holographics Int'l'92*, ed. by Yu. Denisyuk, F. Wyrowski. Proc. SPIE 1732, 411-422 (1993)

7.32 J.M. Heaton, L. Solymar: Wavelength and angular selectivity of high diffraction efficiency reflection holograms in silver halide photographic emulsion. Appl. Opt. 24, 2931-2936 (1985)

7.33 D.J. Cooke, A.A. Ward: Reflection-hologram processing for high efficiency in silver-halide emulsions. Appl. Opt. 23, 934-941 (1984)

7.34 P.M. Boone: Secondary effects in processing holograms, in *Practical Holography III*, ed. by S.A. Benton. Proc. SPIE 1051, 52-59 (1989)

7.35 P.M. Boone: Secondary effects in processing silver halide holograms. Part II, in *Holography'89*, ed. by Y.N. Denisyuk, T.H. Jeong. Proc. SPIE 1183, 193-200 (1990)

Chapter 8

8.1 N. Phillips: Benign bleaching for health holography. Holosphere 14 (No.4), 21-22 (1986)

8.2 D.B. Coblitz, J.A. Carney: Dye removal from holographic films. Appl. Opt. 13, 1994-1995 (1974)

8.3 A. Green, G.I.P. Levenson: Emulsion swelling during washing, etc. J. Photogr. Sci. 22, 194-197 (1974)

8.4 A. Green, G.I.P. Levenson: The swelling of thin, gelatin layers. J. Photogr. Sci. 30, 79-83 (1982)

8.5 K. Biedermann, N.-E. Molin: Combining hypersensitization and rapid *in situ* processing for time-average observation in real-time hologram interferometry. J. Phys. E 3, 669-680 (1970)

8.6 P.E. Perkins: A review of the effects of squeegees in continuous processing machines. J. Soc. Motion Picture and Television Eng. 79, 121-123 (1970)

8.7 L.I. Edgcomb, J.S. Zankowski: Molded squeegee blades for photographic processing. J. Soc. Motion Picture and Television Eng. 79, 123-126 (1970)

8.8 P. Hariharan: Bleached photographic phase holograms: The influence of drying procedures on diffraction efficiency. Opt. Commun. 56, 318-320 (1986)

8.9 Yu.N. Denisyuk, I.R. Protas: Improved Lippmann photographic plates for recording stationary light waves. Opt. Spectrosc. (USSR) 14, 381-383 (1963)

8.10 L.H. Lin, C.V. LoBianco: Experimental techniques in making multicolor white light reconstructed holograms. Appl. Opt. 6, 1255-1258 (1967)

8.11 N. Nishida: Correction of the shrinkage of a photographic emulsion with trieth-anolamine. Appl. Opt. 9, 238-240 (1970)

8.12 P. Hariharan: Pseudocolour images with volume reflection holograms. Opt. Commun. 35, 42-44 (1980)

8.13 D.J. Young: A method of permanently controlling the thickness and profile of a processed photographic emulsion. J. Photogr. Sci. 23, 190-192 (1975)

8.14 M.I. Dzyubenko, V.A. Krishtal, A.P. Pyatikop, V.V. Shevchenko: Parameters of three-dimensional holograms on high-resolution emulsions. Opt. Spectrosc. (USSR) 38, 429-431 (1975)

8.15 M.I. Dzyubenko, A.P. Pyatikop, V.V. Shevchenko: Increasing the diffraction efficiency of reflecting three-dimensional holograms by preventing emulsion shrinkage. Sov. Phys. - Tech. Phys. 20, 965-966 (1975)

8.16 A. Kusakabe, H. Yokota, H. Katsuma: Improvement of diffraction efficiency by baking for Lippmann hologram. Topical Meeting on Holography Technical Digest, 86:5 (Optical Society of America, Washington, DC 1986) pp.104-106

8.17 C.S. Guo, L.Z. Cai: Postheat treatment of silver halide holograms. Opt. Lett. 16, 1777-1779 (1991)

8.18 D.R. Wuest, R.S. Lakes: Color control in reflection holograms by humidity. Appl. Opt. 30, 2363-2367 (1991)

8.19 O.B. Serov, A.M. Smolovich, G.A. Sobolev: Properties of three- dimensional holograms subject to emulsion swelling. Sov. Phys. - Tech. Phys. 22, 1392-1394 (1977)

8.20 H.T. Buschmann: Kontrasterhöhung von Phasenhologrammen durch Beschichtung der Platten. Opt. Commun. 6, 290-294 (1972)

8.21 MACtac, 4560 Darrow Road, Stow, OH 44224, USA

8.22 R. Berkhout: Working with Kodak plates 120-01, making white light transmission holograms. Proc. Int'l Symp. Display Holography, ed. by T.H. Jeong, Lake Forest College, IL 60045, USA, Vol.III (1989) pp.127-129

8.23 Epoxy Technology, Inc., 14 Fortune Drive, Billerica, MA 01821, USA

8.24 N.J. Phillips, D. Porter: Highly efficient holograms by simplified index matching. J. Phys. E 9, 1022 (1976)

8.25 B. Janowska, J. Szydlowska: Improved efficiency reflection holograms of diffusely reflecting objects. Opt. Appl. 9, 3-6 (1979)

8.26 P. Hariharan: Bleached reflection holograms. Opt. Commun. 6, 377-379 (1972)

8.27 Excello Color & Chemical MFG Div., 400 N. Nobel St., Chicago, IL 60622, USA

8.28 J. Blyth: Notes on processing holograms with solvent bleaches. Proc. Int'l Symp. Display Holography, ed. by T.H. Jeong, Lake Forest College, IL 60045, USA, Vol.II (1985) pp.325-331

8.29 Imagys International, Inc., Astra House, 1-3 Reading Road, Eversley, Hampshire RG27 0RP, UK

8.30 B. Cantos: Bleaching and noise affect brightness. Holosphere 13 (No.1), 26 (1985)

8.31 A. Kiraly: The Kiraly method of embedding Cibachrome display prints for archival protection, in *The Stability and Conservation of Photographic Images: Chemical, Electronic and Mechanical*, ed. by L.E. Ravich, S. Siripant. Proc. SPSE (1986) pp.141-144

8.32 G. Weiss (ed.): *Hazardous Chemical Data Book*, 2nd edn. (Noyes Data Corp., Park Ridge, NJ 07565 1986)

8.33 N. Cheung: Chemical effects in holography pose real hazards for workers. Holosphere 18, (No.12), 6-7 (1979)

8.34 M. Crenshaw: Hazards of holographic processing chemicals. Proc. Int'l Symp. Display Holography, ed. by T.H. Jeong, Lake Forest College, IL 60045, USA, Vol.II (1986) pp.257-261

8.35 G.L. Rogers: Experiments in diffraction microscopy. Proc. Roy. Soc. Edinburgh A 63, 193-221 (1951)

8.36 M.H. Horman: An application of wavefront reconstruction to interferometry. Appl. Opt. 4, 333-336 (1965)

8.37 F.B. Rotz, A.A. Friesem: Holograms with nonpseudoscopic real images. Appl. Phys. Lett. 8, 146-148 (1966)

8.38 F.S. Harris, G.C. Sherman, B.H. Billings: Copying holograms. Appl. Opt. 5, 665-666 (1966)

8.39 M.J. Landry: Copying holograms. Appl. Phys. Lett. 9, 303-304 (1966)

8.40 H.W. Rose: Resolution of images reconstructed from copied holograms. J. Opt. Soc. Am. 56, 542A (1966)

8.41 D.B. Brumm: Copying holograms. Appl. Opt. 5, 1946-1947 (1966)

8.42 Y. Belvaux: Duplication des hologrammes. Ann. Radioélectr. 22, 105-108 (1967)

8.43 C.N. Kurtz: Copying reflection holograms. J. Opt. Soc. Am. 58, 856-857 (1968)

8.44 M.J. Landry: The effect of two hologram-copying parameters on the quality of copies. Appl. Opt. 6, 1947-1956 (1967)

8.45 G.C. Sherman: Diffraction theory of hologram copying. J. Opt. Soc. Am. 57, 563A (1967)

8.46 G.C. Sherman: Hologram copying by Gabor holography of transparencies. Appl. Opt. 6, 1749-1753 (1967)

8.47 D.B. Brumm: Double images in copy holograms. Appl. Opt. 6, 588-589 (1967)

8.48 H. Rieck: Hologrammkopien mittels Rubinlaser. Optik 27, 255-258 (1968)

8.49 H. Nassenstein: Kopieren von Bildebenen-Hologrammen mit weissem Licht. Phys. Lett. A 26, 225-226 (1968)

8.50 H. Nassenstein: Über das Kopieren von Hologrammen. Optik 27, 327-334 (1968)

8.51 J.C. Palais, J.A. Wise: Improving the efficiency of very low efficiency holograms by copying. Appl. Opt. 10, 667-668 (1971)

8.52 T. Suhara, H. Nishihara, J. Koyama: The modulation transfer function in the hologram copying process. Opt. Commun. 14, 35-38 (1975)

8.53 E.G. Zemtsova, L.V. Lyakhovskaya: A study of a method of copying three-dimensional holograms. Sov. J. Opt. Technol. 43, 744-746 (1976)

8.54 V.A. Vanin: Preparation of reflection holograms by interference copying of transmission holograms. Sov. J. Quantum Electron. 8, 855- 859 (1978)

8.55 N.G. Vlasov, N.A. Lapshina, S.P. Semenov, E.G. Semenov, S.G. Egorova: Interference copying of Denisyuk holograms. Opt. Spectrosc. (USSR) 50, 326 (1981)

8.56 H.I. Bjelkhagen: Denisyuk-reflection holography: Recording and copying technique. Proc. Int'l Symp. Display Holography, ed. by T.H. Jeong, Lake Forest College, IL 60045, USA, Vol.I (1983) pp.45-48

8.57 J. Oliva, A. Fimia, J.A. Quintana: Hologram copying in dichromated gelatin with sunlight, in Three-Dimensional Imaging, ed. by J. Ebbeni, A. Monfils. Proc. SPIE 402, 57-58 (1983)

8.58 N.J. Phillips, R.A.J. van der Werf: The creation of efficient reflective Lippmann layers in ultra-fine grain silver halide materials using non-laser sources. J. Photogr. Sci. 33, 22-28 (1985)

8.59 N.J. Phillips, D. Martens: Aspects of the copying of holograms using incoherent light, in Progress in Holographic Applications, ed. by J. Ebbeni. Proc. SPIE 600, 123-126 (1985)

8.60 G.R. Chamberlin, D.J. McCartney: Contact copying of holographic transmission gratings, in Holography Techniques and Applications, ed. by W.P.O. Jüptner. Proc. SPIE 1026, 64-68 (1988)

8.61 S. Piazzolla, B.K. Jenkins, A.R. Tanguay: Single-step copying process for multiplexed volume holograms. Opt. Lett. 17, 676-678 (1992)

8.62 I. Pascual, A. Beléndez, A. Fimia: Holographic system for copying holograms by using partially coherent light. Appl. Opt. 31, 3312-3319 (1992)

8.63 S.J.S. Brown: Continuous wave - pulse transfer for high security holograms, in *Progress in Holographic Applications*, ed. by J. Ebbeni. Proc. SPIE 600, 145-150 (1985)

8.64 S.J.S. Brown: Automated holographic mass production, in *Practical Holography*, ed. by T.H. Jeong, J.E. Ludman. Proc. SPIE 615, 46-49 (1986)

8.65 D.P. Towers, P.J. Bryanston-Cross, T.R. Judge: Development of a pulse laser system for the production and copying of holographic stereograms, in *Holographic Systems, Components and Applications*, ed. by J.C. Dainty. Proc. IEE 311, 51-55 (1989)

8.66 H.I. Bjelkhagen, M. Epstein, M.E. Marhic: Holicon Corporation: Products and services. Proc. Int'l Symp. on Display Holography, ed. by T.H. Jeong, Lake Forest College, IL 60045, USA, Vol.III (1989) pp.587-590

8.67 A.D. Galpern, V.P. Smaev, B.K. Rozhkov: Recording and copying of multicolor holograms. Opt. Spectrosc. (USSR) 67, 536-538 (1989)

8.68 J.C. Palais: Scanned beam holography. Appl. Opt. 9, 709-711 (1970)

8.69 T.Sh. Imedadze, S.D. Kakichashvili: Scanning method of receiving high-effective reflective holograms on bichromate gelatine, in *Three-Dimensional Holography: Science, Culture, Education*, ed. by T.H. Jeong, V.B. Markov. Proc. SPIE 1238, 439-441 (1991)

8.70 Z. Qu, Q. Feng, E. Wesly, T.H. Jeong: Scanning holography and its applications, in *Int'l Symp. on Display Holography*, ed. by T.H. Jeong. Proc. SPIE 1600, 187-198 (1992)

8.71 T.J. Cvetkovich: Techniques for the replication of multicolor reflection holograms, in *Applications of Holography*, ed. by L. Huff. Proc. SPIE 523, 47-51 (1985)

8.72 T.J. Cvetkovich: Particulars on contact copying reflection holograms. Proc. Int'l Symp. on Display Holography, ed. by T.H. Jeong, Lake Forest College, IL 60045, USA, Vol.II (1986) pp.237-243

8.73 V.A. Vanin: Hologram copying (review). Sov. J. Quantum Electron. 8, 809-818 (1978)

8.74 Preservation of Photographs. Kodak Publication No.F-30 (Eastman Kodak Co., Rochester, NY 1979)

8.75 S. Anderson, R. Goetting: Environmental effects on the image stability of photographic products. J. Imaging Technol. 14, 111-115 (1988)

8.76 F.J. Drago, W.E. Lee: Review of the effects of processing on the image stability of black-and-white silver materials. J. Imaging Technol. 12, 57-65 (1986)

8.77 F.L. Stickley: The biodegradation of gelatin and its problems in the photographic industry. J. Photogr. Sci. 34, 111-112 (1986)

8.78 Rohm and Haas Co., Independence Matt West, Philadelphia, PA 19106, USA

8.79 K.A.H. Brems: The archival quality of film bases. J. Soc. Motion Picture and Television Eng. 97, 991-993 (1988)

8.80 K.C. Brown, R.E. Jacobson: Archival permanence of holograms? J. Photogr. Sci. 33, 177-182 (1985)

8.81 K.C. Brown, R.E. Jacobson: The archival permanence of holograms in silver halide materials, in *The Stability and Conservation of Photographic Images: Chemical, Electronic and Mechanical*, ed. by L.E. Ravich, S. Siripant. Proc. SPSE (1986) pp.95-107

8.82 R.E. Jacobson, K.C. Brown: Archival properties of holograms, in *Practical Holography III*, ed. by S.A. Benton. Proc. SPIE 1051, 60-67 (1989)

8.83 R. Aliaga, H. Chuaqui, P. Pedraza: Archival and wide exposure latitude process for holography. Appl. Opt. 29, 2861-2863 (1990)

8.84 E. Wesly: Recycling of holographic plates. Proc. Int'l Symp. on Display Holography, ed. by T.H. Jeong, Lake Forest College, IL 60045, USA, Vol.III (1989) pp.131-140

8.85 C.S. Ih: Archival characteristics of Fourier color holograms. Appl. Opt. 17, 1059- 1065 (1978)
8.86 F.T.S. Yu, A. Tai, H. Chen: Archival storage of color films by rainbow holographic technique. Opt. Commun. 27, 307-310 (1978)

Chapter 9

9.1   E.N. Leith, J. Upatnieks: Wavefront reconstruction with diffused illumination and three-dimensional objects. J. Opt. Soc. Am. 54, 1295-1301 (1964)
9.2   L. Mandel: Color imagery by wavefront reconstruction. J. Opt. Soc. Am. 55, 1697-1698 (1965)
9.3   A.W. Lohmann: Reconstruction of vectorial wavefronts. Appl. Opt. 4, 1667-1668 (1965)
9.4   K.S. Pennington, L.H. Lin: Multicolor wavefront reconstruction. Appl. Phys. Lett. 7, 56-57 (1965)
9.5   L.H. Lin, K.S. Pennington, G.W. Stroke. A.E. Labeyrie: Multicolor holographic image reconstruction with white-light illumination. Bell Syst. Tech. J. 45, 659-661 (1966)
9.6   A.A. Friesem, R.J. Fedorowicz: Recent advances in multicolor wavefront reconstruction. Appl. Opt. 5, 1085-1086 (1966)
9.7   J. Upatnieks, J. Marks, R. Fedorowicz: Color holograms for white light reconstruction. Appl. Phys. Lett. 8, 286-287 (1966)
9.8   G.W. Stroke, R.G. Zech: White-light reconstruction of color images from black-and-white volume holograms recorded on sheet film. Appl. Phys. Lett. 9, 215-217 (1966)
9.9   E. Marom: Color imagery by wavefront reconstruction. J. Opt. Soc. Am. 57, 101-102 (1967)
9.10 A.A. Friesem, R.J. Fedorowicz: Multicolor wavefront reconstruction. Appl. Opt. 6, 529-536 (1967)
9.11 R.J. Collier, K.S. Pennington: Multicolor imaging from holograms formed on two-dimensional media. Appl. Opt. 6, 1091-1095 (1967)
9.12 L.H. Lin, C.V. LoBianco: Experimental techniques in making multicolor white light reconstructed holograms. Appl. Opt. 6, 1255-1258 (1967)
9.13 A.A. Friesem, R.J. Fedorowicz: Multicolor holography, in Holography, ed. by B.J. Thompson. Proc. SPIE 15, 41-48 (1968)
9.14 A.A. Friesem. J.L. Walker: Thick absorption recording media in holography. Appl. Opt. 9, 201-214 (1970)
9.15 E.T. Kurtzner, K.A. Haines: Multicolor images with volume photopolymer holograms. Appl. Opt. 10, 2194-2195 (1971)
9.16 S. Tatuoka: Color image reconstruction by image plane holography. Jpn. J. Appl. Phys. 10, 1742-1743 (1971)
9.17 M. Noguchi: Color reproduction by multicolor holograms with white-light reconstruction. Appl. Opt. 12, 496-499 (1973)
9.18 R.A. Lessard, S.C. Som. A. Boivin: New technique of color holography. Appl. Opt. 12, 2009-2011 (1973)
9.19 C.P. Grover, M. May: Multicolor wave-front reconstruction of partially diffusing plane objects. J. Opt. Soc. Am. 63, 533-537 (1973)
9.20 R.A. Lessard, P. Langlois. A. Boivin: Orthoscopic color holography of 3-D objects. Appl. Opt. 14, 565-566 (1975)
9.21 J. Ruzek. J. Muzik: Some problems of colour holography. Tesla Electronics 9, 60-61 (1976)
9.22 P. Hariharan, W.H. Steel, Z.S. Hegedus: Multicolor holographic imaging with a white-light source. Opt. Lett. 1, 8-9 (1977)

9.23 H. Chen. A. Tai, F.T.S. Yu: Generation of color images with one-step rainbow holograms. Appl. Opt. 17, 1490-1491 (1978)

9.24 T. Kubota, T. Ose: Lippmann color holograms recorded in methylene-blue-sensitized dichromated gelatin. Opt. Lett. 4, 289-291 (1979)

9.25 P. Hariharan: Improved techniques for multicolor reflection holograms. J. Optics (Paris) 11, 53-55 (1980)

9.26 G.A. Sobolev, O.B. Serov: Recording color reflection holograms. Sov. Tech. Phys. Lett. 6, 314-315 (1980)

9.27 L. Huff, R.L. Fusek: Color holographic stereograms. Opt. Eng. 19, 691-695 (1980)

9.28 G. Ya. Buimistryuk. A. Ya. Dmitriev: Selection of laser emission wavelengths to obtain color holographic images (in Russian). Izv. VUZ Priborostr. (USSR) 25, 79-82 (1982)

9.29 Yu. N. Denisyuk, S.V. Artemev, Z.A. Zagorskaya. A.M. Kursakova, M.K. Shevtsov, T.V. Shedrunova: Color reflection holograms from bleached PE-2 photographic plates. Sov. Tech Phys. Lett. 8, 259-260 (1982)

9.30 F.T.S. Yu. J.A. Tome, F.K. Hsu: Dual-beam encoding for color holographic construction. Opt. Commun. 46, 274-277 (1983)

9.31 V.C. Sainov: Basic characteristics and applications of reflection holograms. Proc. Int'l Symp. on Display Holography, ed. by T.H. Jeong, Lake Forest College, IL 60045, USA, Vol.I (1983) pp.55-69

9.32 K. Bazargan: Review of colour holography, in Optics in Entertainment, ed. by C. Outwater. Proc. SPIE 391, 11-18 (1983)

9.33 P. Hariharan: Colour holography. Progress in Optics 20, 263-324 (North-Holland, Amsterdam 1983)

9.34 W.J. Molteni: Natural color holographic stereograms by superimposing three rainbow holograms, in Optics in Entertainment II, ed. by C. Outwater. Proc. SPIE 462, 14-19 (1984)

9.35 F.T.S. Yu, F.K. Hsu: White-light Fourier holography. Opt. Commun. 52, 384-389 (1985)

9.36 M.K. Shevtsov: Diffraction efficiency of phase holograms for exposure superposition. Sov. J. Opt. Technol. 52, 1-3 (1985)

9.37 V.Z. Bryskin, E.M. Znamenskaya. A.M. Kursakova, V.P. Smaev, T.V. Schedrunova: Photographic material for recording a holographic image in the green region of the spectrum. Sov. J. Opt. Technol. 52, 436-437 (1985)

9.38 F.T.S. Yu, G. Gerhart: White light transmission color holography: A review. Opt. Eng. 24, 812-819 (1985)

9.39 V.P. Smaev, V.Z. Bryskin, E.M. Znamenskaya. A.M. Kursakova, I.B. Shakhova: Features of the recording of holograms on a two-layer photographic material. Sov. J. Opt. Technol. 53, 287-290 (1986)

9.40 V. Sainov, S. Sainov, H. Bjelkhagen: Color reflection holography, in Practical Holography, ed. by T.H. Jeong. J.E. Ludman. Proc. SPIE 615, 88-91 (1986)

9.41 K. Bazargan: A new method of colour holography. Int'l Conf. on Holography Applications, ed. by J. Ke and R. Pryputniewicz. Proc. SPIE 673, 68-70 (1986)

9.42 F.T.S. Yu: White-light color holography and its applications, in Int'l Conf. on holography Applications, ed. by J. Ke and R. Pryputniewicz. Proc. SPIE 673, 222-234 (1986)

9.43 T. Kubota: Recording of high quality color holograms. Appl. Opt. 25, 4141-4145 (1986)

9.44 A.D. Galpern, B.K. Rozhkov, V.P. Smaev, Yu. A. Vavilova: Diffraction characteristics of color transmission holograms. Opt. Spectrosc. (USSR) 62, 810-812 (1987)

9.45 A.D. Galpern, B.K. Rozhkov, V.P. Smaev, Recording of rainbow holograms. Opt. Spectrosc. (USSR) 63, 226-229 (1987)

9.46 S. Bains: Natural colour pulse hologram made. Holographics Int'l No.1, 7 (1987)

9.47 K. Ohnuma, F. Iwata: Color rainbow hologram and color reproduction. Appl. Opt. 27, 3859-3863 (1988)

9.48 K. Bazargan: Choice of laser wavelengths for recording true-colour holograms, in *Holographphic Systems, Components and Applications*, ed. by J.S. Dainty. Proc. IEE 311, 49-50 (1989)

9.49 K. Bazargan: Colour controversy, Pt.1. Holographics Int'l No.5, 4 (1989)

9.50 P. Hubel: Colour controversy continues. Holographics Int'l No.5, 5 (1989)

9.51 K. Bazargan: Design of a one-step full-colour holographic recording system, in *Practical Holography III*, ed. by S. Benton. Proc. SPIE 1051, 6-11 (1989)

9.52 P.M. Hubel. A.A. Ward: Color reflection holography, in *Practical Holography III*, ed. by S. Benton. Proc. SPIE 1051, 18-24 (1989)

9.53 K. Ohnuma, T. Nishihara, F. Iwata: Full color rainbow hologram using a photoresist plate, in *Practical Holography III*, ed. by S. Benton. Proc. SPIE 1051, 25-30 (1989)

9.54 J. Hecht: 3-color hologram recorded in solvent. Lasers & Optronics 8 (No.5), 18 (1989)

9.55 R.R. Erickson: "Full color" holography. Holosphere 16 (No.4), 12-15 (1989)

9.56 J. Watson: Color holography with Soviet emulsions. Holosphere 16 (No.4), 20-21 (1989)

9.57 T.H. Jeong, E. Wesly: True color holography on Du Pont photopolymer material. Holosphere 16 (No.4), 22-23 (1989)

9.58 N.G. Vlasov. A.N. Zaborov. A.V. Yanovskii: Production of color specimens by rainbow holography. Opt. Spectrosc. (USSR) 67, 243-245 (1989)

9.59 P.M. Hubel. A.A. Ward: A comparison of silver halide emulsions for color reflection display holography. Proc. Int'l Symp. on Display Holography, ed. by T.H. Jeong, Lake Forest College, Il 60045, USA, Vol.III (1989) pp.149-166

9.60 P.M. Hubel: Effects of bandwidth and peak replay wavelength shifts on color holograms, in *Holography'89*, ed. by Yu. Denisyuk, T.H. Jeong. Proc. SPIE 1183, 183-190 (1990)

9.61 T. Mizuno, T. Goto, M. Goto, K. Matsui, T. Kubota: High efficient multicolor holograms recorded in methylene blue sensitized dichromated gelatin, in *Practical Holography IV*, ed. by S. Benton. Proc. SPIE 1212, 40-45 (1990)

9.62 T. Kubota, M. Nishimura: Recording and demonstration of cultural assets by color holography (I) - Analysis for the optimum color reproduction. J. Soc. Photogr. Sci. Tech. Jpn. 53, 291-296 (1990)

9.63 T. Kubota, M. Nishimura: Recording and demonstration of cultural assets by color holography (II) - Recording method of hologram for optimizing the color reproduction. J. Soc. Photogr. Sci. Tech. Jpn. 53, 297-302 (1990)

9.64 C. Fan, C.C. Jiang, L. Guo: Lensless true rainbow hologram with He-Ne laser, in *Optics in Complex Systems*, ed. by F. Lanzl, H.-J. Preuss, G. Weigelt. Proc. SPIE 1319, 314 (1990)

9.65 P.M. Hubel, L. Solymar: Color reflection holography: Theory and experiment. Appl. Opt. 30, 4190-4203 (1991)

9.66 T.H. Jeong, E. Wesly: Progress in true color holography, in *Practical Holography IV*, ed. by S. Benton. Proc. SPIE 1212, 183-189 (1990)

9.67 T.H. Jeong, E. Wesly: Progress in true color holography, in *Three- Dimensional Holography: Science, Culture, Education*, ed. by T.H. Jeong, V.B. Markov. Proc. SPIE 1238, 298-305 (1991)

9.68 J.F. Zhang, C.R. Ma, H.Y. Lang: Colour reflection holograms with photopolymer plates, in *Three-Dimensional Holography: Science, Culture, Education*, ed. by T.H. Jeong, V.B. Markov. Proc. SPIE 1238, 306-310 (1991)

9.69 V.P. Smaev. A.D. Galpern, Yu. A. Vavilova: Three-layer material for the registration of coloured holograms, in *Three-Dimensional Holography: Science, Culture, Education*, ed. by T.H. Jeong, V.B. Markov. Proc. SPIE 1238, 311-315 (1991)

9.70 J.F. Zhang, M.W. Yu, S.Q. Tang, Z.F. Zhu: Chromaticity and colour fidelity of images with multicolour rainbow holograms, in *Three- Dimensional Holography: Science, Culture, Education*, ed. by T.H. Jeong, V.B. Markov. Proc. SPIE 1238, 401-405 (1991)

9.71 V.G. Bespalov, V.N. Krylov, V.N. Sizov: Pulsed laser system for recording large-scale colour hologram, in *Three-Dimensional Holography: Science, Culture, Education*, ed. by T.H. Jeong, V.B. Markov. Proc. SPIE 1238, 457-461 (1991)

9.72 N.D. Vorzobova, V.N. Sizov, R.V. Rjabova: Monochromatic and two-color recording of holographic portraits with the use of pulsed lasers, in *Three-Dimensional Holography: Science, Culture, Education*, ed. by T.H. Jeong, V.B. Markov. Proc. SPIE 1238, 462-464 (1991)

9.73 P.M. Hubel: Recent advances in color reflection holography, in *Practical Holography V*, ed. by S.A. Benton. Proc. SPIE 1461, 167-174 (1991)

9.74 K. Bazargan: Factors affecting the choice of optimum recording wavelengths in true-color holography, in *Int'l Symp. on Display Holography*, ed by T.H. Jeong. Proc. SPIE 1600, 178-181 (1992)

9.75 P. St.-Hilaire, S.A. Benton, M. Lucente, P.M. Hubel: Color images with MIT holographic video display, in *Practical Holography VI*, ed. by S.A. Benton. Proc. SPIE 1667, 73-84 (1992)

9.76 M.A. Klug, M.W. Halle, P.M. Hubel: Full color ultragrams, in *Practical Holography VI*, ed. by S.A. Benton. Proc. SPIE 1667, 110-119 (1992)

9.77 P.M. Hubel, M.A. Klug: Color holography using multiple layers of Du Pont photopolymer, in *Practical Holography VI*, ed. by S.A. Benton. Proc. SPIE 1667, 215-224 (1992)

9.78 S. Namba, K. Kurokawa, T. Fujita, T. Mizuno, T. Kubota: Improvement of the transmittance of methylene blue sensitized dichromated gelatin, in *Practical Holography VI*, ed. by S.A. Benton. Proc. SPIE 1667, 233-238 (1992)

9.79 C. Jiang, C. Fan, L. Guo: New true-color rainbow holography of 3-D object, in *Practical Holography VI*, ed. by S.A. Benton. Proc. SPIE 1667, 239-242 (1992)

9.80 T. Nishihara, A. Sato, F. Iwata: Full color Lippmann hologram, in *Holographics Int'l'92*, ed. by Yu. Denisyuk, F. Wyrowski. Proc. SPIE 1732, 405-410 (1993)

9.81 G. Ross, J. Watson: Single stage white-light colour holograms, in *Holographics Int'l'92*, ed. by Yu. Denisyuk, F. Wyrowski. Proc. SPIE 1732, 424-429 (1993)

9.82 M. Chomát: Diffraction efficiency of multiple-exposure thick absorption holograms. Opt. Commun. 2, 109-110 (1970)

9.83 D.J. Cooke. A.A. Ward: Reflection-hologram processing for high efficiency in silver-halide emulsions. Appl. Opt. 23, 934-941 (1984)

9.84 D.L. MacAdam: *Color Measurement, Theme and Variations*, 2nd edn., Springer Ser. Opt. Sci., Vol.27 (Springer, Berlin, Heidelberg 1985)

9.85 G.A. Agoston: *Color Theory and its Application in Art and Design*, 2nd edn., Springer Ser. Opt. Sci., Vol.19 (Springer, Berlin, Heidelberg 1987)

9.86 W.T. Wintringham: Color television and colorimetry. Proc. IRE 39, 1135-1172 (1951)

9.87 M.R. Pointer: The gamut of real surface colours. Color Res. Appl. 5 (No.3), 145-155 (1980)

9.88 W.A. Thornton: Luminosity and color-rendering capability of white light. J. Opt. Soc. Am. 61, 1155-1163 (1971)

9.89 S.E. Schacham, M.E. Marhic, M. Epstein: Efficient white laser illuminators for plastic optical fibers. Appl. Opt. 16, 1041-1044 (1977)

9.90 M.E. Marhic, M. Epstein, R. Haidle: White-light flashlamp-pumped dye laser for photography through endoscopes. Opt. Commun. 45, 21-25 (1983)

9.91 N. Phillips: Bridging the gap between Soviet and Western holography, in *Holography, Commemorating the 90th Anniversary of the Birth of Dennis Gabor*, ed. by P. Greguss, T.H. Jeong. SPIE Institute Volume IS 8, 206-214 (1991)

9.92 N. Phillips: Bridging the gap between Soviet and Western holography. Speaking notes, Holography Workshop, Lake Forest College, IL 60045, USA (1990)

9.93 S.A. Benton: Hologram reconstructions with extended incoherent sources. J. Opt. Soc. Am. 59, 1545-1546 (1969)

9.94 P.N. Tamura: Multicolor image from superposition of rainbow holograms, in *Clever Optics*, ed. by N. Balasubramanian. J.C. Wyant. Proc. SPIE 126, 59-66 (1977)

9.95 P.N. Tamura: Pseudocolor encoding of holographic images using a single wavelength. Appl. Opt. 17, 2532-2536 (1978)

9.96 C.P. Grover, R. Tremblay: Multicolor wave front reconstruction in white light. Appl. Opt. 19, 3044-3046 (1980)

9.97 S.A. Benton, H.S. Mingace, W.R. Walter: One-step white-light transmission holography, in *Recent Advances in Holography*, ed. by T.C. Lee, P.N. Tamura. Proc. SPIE 215, 156-161 (1980)

9.98 S.A. Benton: Topics in advanced display holography. Workshop Notes, Lake Forest Holography Workshops, Lake Forest College, IL (July 1989)

9.99 S.P. McGrew: Color control in dichromated gelatin reflection holograms, in *Recent Advances in Holography*, ed. by T.C. Lee, P.N. Tamura. Proc. SPIE 215, 24-31 (1980)

9.100 H. Owen. A.E. Hurst: Multicolour image holograms in dichromated gelatin. Proc. Int'l Symp. on Display Holography, ed. by T.H. Jeong, Lake Forest College, IL 60045, USA, Vol.III (1989) pp.95-103

9.101 J. Blyth: "Pseudoscopic" moldmaking handy trick for Denisyuk holographers. Holosphere 8 (No.11), 5, (1979)

9.102 P. Hariharan: Pseudocolour images with volume reflection holograms. Opt. Commun. 35, 42-44 (1980)

9.103 S. McGrew: A graphical method for calculating pseudocolor hologram recording geometries. Proc. Int'l Symp. on Display Holography, ed. by T.H. Jeong, Lake Forest College, IL 60045, USA, Vol.I (1983) pp.171-183

9.104 L. Moore: Pseudo-color reflection holography. Proc. Int'l Symp. on Display Holography, ed. by T.H. Jeong, Lake Forest College, IL 60045, USA, Vol.I (1983) pp.163-169

9.105 J.A. Kaufman: Previsualization and pseudo-color image plane reflection holograms. Proc. Int'l Symp. on Display Holography, ed. by T.H. Jeong, Lake Forest College, IL 60045, USA, Vol.I (1983) pp.195-207

9.106 R. Lessing: Farbige Weisslichtholographie, in *Optronics in Engineering*, ed. by W. Waidelich, Proc. Laser 83 Optoelektronik (Springer, Berlin, Heidelberg 1984) pp.155-158

9.107 S.L. Smith, T. Cvetkovich: Multi-color holography with a single frequency laser utilizing triethanolamine as a pre-exposure agent, in *Optics in Entertainment II*, ed. by C. Outwater. Proc. SPIE 462, 8-13 (1984)

9.108 S.L. Smith: Application of the tri-color theory of additive color mixing to the full color reflection hologram, in *Applications of Holography*, ed. by L. Huff. Proc. SPIE 523, 42-46 (1985)

9.109 T.J. Cvetkovich: Techniques for the replication of multicolor reflection holograms, in *Applications of Holography*, ed. by L. Huff. Proc. SPIE 523, 47-51 (1985)

9.110 M.M. Crenshaw: Pseudo-color reflection holography - An artists perspective, in *Practical Holography II*, ed. by T.H. Jeong. Proc. SPIE 747, 104-107 (1987)

9.111 L. Lieberman: Full color from holographic images, Holosphere 16 (No. 4), 24-25 (1989)

9.112 J.L. Walker, S.A. Benton: In-situ swelling for holographic color control, in *Practical Holography III*, ed. by S.A. Benton. Proc. SPIE 1051, 192-199 (1989)

9.113 J.A. Kaufman: Update of pseudo-color reflection techniques. Proc. Int'l Symp. on Display Holography, ed. by T.H. Jeong, Lake Forest College, IL 60045, USA, Vol.III (1989) pp.367-378

9.114 S.A. Benton: The principles of reflection holographic stereograms. Proc. Int'l Symp. on Display Holography, ed. by T.H. Jeong, Lake Forest College, IL 60045, USA, Vol.III (1989) pp.593-608

9.115 V.A. Vanin, S.P. Vorobjev: Pseudocolor reflection hologram properties recorded using monochrome photographic materials, in *Three-Dimensional Holography: Science, Culture, Education*, ed. by T.H. Jeong, V.B. Markov. Proc. SPIE 1238, 324-331 (1991)

9.116 J.A. Kaufman: Large format pseudo-color reflection holograms on film, in *Int'l Symp. on Display Holography*, ed. by T.H. Jeong. Proc. SPIE 1600, 38-43 (1992)

9.117 L. Lieberman: Paint with light - Artistic manipulation of color in multicolor reflection holograms, in *Int'l Symp. on Display Holography*, ed. by T.H. Jeong. Proc. SPIE 1600, 224-228 (1992)

9.118 E. Orr, D. Trayner: Deep-image reflection holograms in black and white and additional colors, Holosphere, 15 (No.4), 14-17 (1987)

9.119 E. Orr, D. Trayner: Getting it down in black & white. Holographics Int'l No.3, 13-15 (1988)

9.120 E. Orr, D. Trayner: Deep image reflection holograms in black and white and additional colours. Proc. Int'l Symp. on Display Holography, ed. by T.H. Jeong, Lake Forest College, IL 60045, USA, Vol.III (1989) pp.379-388

9.121 S. Lowenthal, E. Leiba, M. Lucas. A. Werts: Holographie en lumière infrarouge à 10 μm. Comp. Rend. Acad. Sc. Paris, B 266, 1363-1366 (1968)

9.122 J.S. Chivian, R.N. Claytor, D.D. Eden: Infrared holography at 10.6 μm. Appl. Phys. Lett. 15, 123-125 (1969)

9.123 T. Izawa, M. Kamiyama: Infrared holography with organic photochromic films. Appl. Phys. Lett. 15, 201-203 (1969)

9.124 W.A. Simpson, W.E. Deeds: Real-time visual reconstruction of infrared holograms. Appl. Opt. 9, 499-501 (1970)

9.125 K. Tatsuno. A. Arimoto: Hologram recording by visible diode lasers. Appl. Opt. 19, 2096-2097 (1980)

9.126 C.D. Lysogorski. A.W. Lungershausen: Holography in the 780 nm range, in *Practical Holography II*, ed. by T.H. Jeong. Proc. SPIE 747, 139-142 (1987)

9.127 J.A. Davis, M.F. Brownell: Laser-diode holographic recording. Opt. Lett. 11, 196-197 (1986)

9.128 S. Hart, G. Mendes, K. Bazargan, S. Xu: Deep-red holography using a junction laser and silver-halide holographic emulsion. Opt. Lett. 13, 955-957 (1988)

9.129 L. Wang, R.K. Kostuk: Direct formation of planar holograms and noise gratings at 820 nm in bleached silver-halide emulsions. Opt. Lett. 14, 919-921 (1989)

9.130 T. Sugaya. A. Iwamoto: Holograms produced with double-heterojunction laser illumination. Opt. Commun. 10, 37-38 (1974)

9.131 G.C. Gilbreath. A.E. Clement: Absorption holograms made with mode-stabilized laser diodes for use in a satellite communications link. Opt. Lett. 12, 648-650 (1987)

9.132 C. Roychoudhuri, B.J. Thompson: Infrared holography with 4-Z emulsion. Opt. Commun. 10, 23-25 (1974)

9.133 R.V. Ryabova, Yu.A. Bykovsky, V.A. Elkhov. A.I. Larkin, V.L. Potapov, L.N. Gnatiuk, M.L. Gurari, S.N. Marchenko: IAE-3 and IAE-6 high resolution photographic materials for recording laser radiation in the IR-region. Opt. Commun. 18, 406-409 (1976)

9.134 Yu.A. Bykovskii, V.L. Velichanskii, V.A. Elkhov, Yu.P. Zakharov. A.I. Larkin, V.A. Maslov, R.A. Ryabova, D.M. Samoilovich, V.L. Smirnov: Coherence of the radiation of a pulsed single-mode injection semiconductor laser. Sov. Phys. Dokl. 17, 359-361 (1972)

9.135 S.P. Kalashnikov, I.I. Klimov, V.V. Nikitin, G.I. Semenov: Recording of Fourier holograms using radiation emitted from pulse semiconductor lasers. Sov. J. Quantum. Electron. 7, 946-949 (1977)

9.136 Yu.P. Zakharov, V.V. Kostryukov, M.A. Maiorchuk, V.V. Nikitin, R.V. Ryabova, V.D. Samoilov, D.M. Samoilovich, K.A. Seiranova: Injection laser recording of high-resolution microholograms. Sov. J. Quantum. Electron. 4, 103 (1974)

9.137 A.V. Vorobev, V.A. Elkhov, I.I. Klimov, V.N. Morozov, G.T. Pak, Yu.M. Popov, R.P. Shidlovskii, I.V. Yashumov: Recording of holograms using semiconductor laser radiation and a holographic selector. Sov. J. Quantum Electron. 10, 1557-1558 (1980)

9.138 L.N. Gnatyuk, M.L. Gurari, S.N. Marchenko, R.V. Ryabova: Recording of holograms at the 10.6 μm wavelength. Sov. J. Quantum. Electron. 5, 261-262 (1975)

9.139 A.M. Dukhovnyi. A.E. Korolev, R.V. Ryabova, D.I. Staselko: Diffraction efficiency of holograms recorded in the IR region by pulses of $2 \cdot 10^{-10}$ - 15 sec duration. Opt. Spectrosc. (USSR) 49, 510-513 (1980)

9.140 J.M. Geary: Preliminary model for infrared presensitization photography. Opt. Eng. 26, 337-341 (1987)

9.141 J.M. Geary: Infrared presensitization photography and sensitizing dyes. Opt. Eng. 26, 675-678 (1987)

9.142 J.M. Geary: Infrared presensitization photography and ionic conductivity. Opt. Eng. 27, 321-324 (1988)

9.143 D.P. Juyal, S.P. Gupta, R. Hradaynath: Recording of interferograms on normal high resolution plates using a CO_2 laser at 10.6 μm. Appl. Opt. 22, 2152-2154 (1983)

9.144 G.B. Gorlin, L.G. Paritskii, T.V. Tisnek: Photographic system for recording 10.6 μm radiation. Sov. Phys. - Tech. Phys. 32, 93-94 (1987)

9.145 A. Graube: Infrared holograms recorded in high-resolution photographic plates with the Herschel reversal. Appl. Phys. Lett. 27, 136-137 (1975)

9.146 G.F. Frazier, T.D. Wilkerson. J.M. Lindsay: Infrared photography at 5 μm and 10 μm. Appl. Opt. 15, 1350-1352 (1976)

9.147 G.R. Mitchel, B. Grek, T.W. Johnston, F. Martin, H. Pépin: Nanosecond photography at 10.6 μm using silver halide film. Appl. Opt. 18, 2422- 2426 (1979)

9.148 V.N. Mikhailov, O.V. Grinevitskaya, Z.A. Zagorskaya, V.I. Mikhailova: Study of electronic stage of the Herschel effect in holographic emulsions with different types of chemical sensitization, in *Three- Dimensional Holography: Science, Culture, Education*, ed. by T.H. Jeong, V.B. Markov. Proc. SPIE 1238, 144-152 (1991)

9.149 I.O. Starobogatov, S.D. Nicolaev: Research of fast stages of latent image formation in holographic photoemulsions influenced by ultrashort radiation pulses, in *Three-Dimensional Holography: Science, Culture, Education*, ed. by T.H. Jeong, V.B. Markov. Proc. SPIE 1238, 153-157 (1991)

9.150 G.C. Bjorklund, C. Bräuchle, D.M. Burland, D.C. Alvarez: Two-photon holography with continuous-wave lasers. Opt. Lett. 6, 159-161 (1981)

9.151 C. Bräuchle, U.P. Wild, D.M. Burland, G.C. Bjorklund, D.C. Alvarez: A new class of materials for holography in the infrared, IBM J. Res. Develop. 26, 217-227 (1982)

9.152 V. Gerbig, R.K. Grygier, D.M. Burland, G. Sincerbox: Near-infrared holography by two-photon photochemistry. Opt. Lett. 8, 404-406 (1983)

9.153 C. Carré, D. Ritzenthaler, D.J. Lougnot. J.P. Fouassier: Biphotonic process for recording holograms with continuous-wave lasers in the near infrared. Opt. Lett. 12, 646-647 (1987)

9.154 D.J. Lougnot. J.-P. Fouassier, C. Carré, P. van de Walle: Holographic recording in the near infra-red: The two photon two product process, in *Holography Techniques and Applications*, ed. by W.P.O. Jüptner. Proc. SPIE 1026, 22-28 (1988)

9.155 E.M. Barkhudarov, V.R. Berezovskii, M.I. Brodzeli. A.M. Gilels, I.A. Eligulashvili, T.N. Makharadze, M.I. Taktakishvili, T.Ya. Chelidze: Recording on triacetatecellulose of infrared holograms in the 10.6 μm band. Opt. Spectrosc. (USSR) 48, 453–454 (1980)

9.156 E.M. Barkhudarov, V.R. Berezovskii, M.O. Mdivnishvili, M.I. Taktakishvili, T.Ya. Chelidze: The time and exposure charateristics of IR hologram recording in the 10.6-μm region in polymer recording media. Opt. Spectrosc. (USSR) 58, 530–532 (1985)

9.157 E.M. Barkhudarov, V.R. Berezovskii, M.O. Mdivnishvili, M.I. Taktakishvili, N.L. Tsintsadze, T.Ya. Chelidze: Three-dimensional holography at 10.6 μm. Sov. Tech. Phys. Lett. 9, 427–428 (1983)

9.158 G. Decker, H. Herold, H. Röhr: Holography and holographic interferometry with pulsed high-power infrared lasers. Appl. Phys. Lett. 20, 490–492 (1972)

9.159 J.N. Olsen: Picosecond infrared holography on bismuth film. Appl. Phys. Lett. 24, 220–221 (1974)

9.160 P.D. Kukharchik, V.G. Belkin. A.S. Skripko, V.M. Greben: Thermooptic method for recording infrared holograms. Sov. Tech. Phys. Lett. 7, 317–318 (1981)

9.161 A.A. Kamshilin, M.G. Miteva: Effect of infra-red irradiation on holographic recording in bismuth silicon oxide. Opt. Commun. 36, 429–433 (1981)

9.162 V.M. Durasov. A.S. Rubanov, I.V. Stashkevich. A.V. Chalei: Infrared hologram recording in polyvinyl alcohol films. Sov. Tech. Phys. Lett. 9, 506–507 (1983)

9.163 M. Cormier, M. Blanchard, M. Rioux, R. Beaulieu: Holographie en infrarouge sur de minces couches d'huile. Appl. Opt. 17, 3622–3626 (1978)

9.164 M. Rioux, M. Blanchard, M. Cormier, R. Beaulieu: Use of the TEM_{10} laser mode for IR holography at 10.6 μm. Appl. Opt. 17, 3864–3865 (1978)

9.165 J. Lewandowski, B. Mongeau, M. Cormier: Real-time interferometry using IR holography on oil films. Appl. Opt. 23, 242–246 (1984)

9.166 J. Lewandowski, B. Mongeau, M. Cormier. J. Lapierre: Infrared holographic interferometry. Appl. Opt. 25, 3291–3296 (1986)

9.167 V.G. Belkin, P.D. Kukharchik. A.S. Skripko: Thermohydrodynamic recording of IR holograms. Sov. Phys. - Tech. Phys. 31, 809–810 (1986)

9.168 W.S. Colburn, L.M. Ralston. J.C. Dwyer: Holographic recording in thermoplastic at 1.15 μm. Appl. Phys. Lett. 23, 145–146 (1973)

9.169 D.S. Lo, L.H. Johnson, R.W. Honebrink: Infrared laser heating for thermoplastic recording, in *Optical Storage Materials and Methods*, ed. by L. Beiser, D. Chen. Proc. SPIE 123, 32–36 (1977)

9.170 J.F. Forkner, D.D. Lowenthal: A photographic recording medium for 10.6-μm laser radiation. Appl. Opt. 6, 1419–1420 (1967)

9.171 R. Beaulieu, R.A. Lessard, M. Cormier, M. Blanchard, M. Rioux: Infrared holography on commercial wax at 10.6 μm. Appl. Phys. Lett. 31, 602–603 (1977)

9.172 R. Beaulieu, R.A. Lessard, M. Cormier, M. Blanchard, M. Rioux: Pulsed IR holography on Takiwax films. Appl. Opt. 17, 3619–3621 (1978)

9.173 R.R. Roberts, T.D. Black: Infrared holograms recorded at 10.6 μm and reconstructed at 0.6328 μm. Appl. Opt. 15, 2018–2019 (1976)

9.174 J.M. Yang, D.W. Sweeney: Infrared holography using the thermochromic material Cu_2HgI_4. Appl. Opt. 18, 2398–2406 (1979)

9.175 J.L. Fergason: Liquid crystals in nondestructive testing. Appl. Opt. 7, 1729–1737 (1968)

9.176 T. Sakusabe, S. Kobayashi: Infrared holography with liquid crystals. Jpn. J. Appl. Phys. 10, 758–761 (1971)

9.177 B.P. Zakharchenya, F.A. Chudnovskii, Z.I. Shteingol'ts: Infrared holography in FTIROS with a CO_2 laser. Sov. Tech. Phys. Lett. 9, 32–33 (1983)

9.178 E.N. Salkova: New nonsilver media for optical recording, in OPTIKA'84, ed. by G. Lupkovics. A. Podmaniaczky. Proc. SPIE 473, 292–293 (1984)

431

9.179 A.A. Bugaev, B.P. Zakharchenya, F.A. Chudnovskii: Use of vanadium oxide films as a holographic recording medium. Sov. J. Quantum Electron. 9, 855-857 (1979)

9.180 C.W. Slinger, A. Zakery, P.J.S. Ewen, A.E. Owen: Photodoped chalcogenides as potential infrared holographic media. Appl. Opt. 31, 2490-2498 (1992)

9.181 R.F. Wuerker, L.O. Heflinger, R.A. Briones: Holographic interferometry with ultraviolet light. Appl. Phys. Lett. 12, 302-303 (1968)

9.182 A. Sasaki, T. Hirose, S. Kishi, K. Fukuda: In-line Fraunhofer holography using N_2 laser as a light source. Jpn. J. Appl. Phys. 18, 1197-1198 (1979)

9.183 J.H. Brannon. J.F. Asmus: UV hologram recording with an excimer laser. Appl. Phys. Lett. 38, 299-300 (1981)

9.184 D.T. Attwood, L.W. Coleman, D.W. Sweeney: Holographic microinterferometry of laser-produced plasmas with frequency-tripled probe pulses. Appl. Phys. Lett. 26, 616-618 (1975)

9.185 D.T. Attwood, D.W. Sweeney. J.M. Auerbach, P.H.Y. Lee: Interferometric confirmation of radiation-pressure effects in laser-plasma interactions. Phys. Rev. Lett. 40, 184-187 (1978)

9.186 C.V. Shank, R.V. Schmidt: Optical technique for producing 0.1-μm periodic surface structures. Appl. Phys. Lett. 23, 154-155 (1973)

9.187 I.N. Ross, G.M. Davis, D. Klemitz: High-resolution holographic image projection at visible and ultraviolet wavelengths. Appl. Opt. 27, 967-972 (1988)

9.188 T.P. Sosnowski, H. Kogelnik: Ultraviolet hologram recording in dichromated gelatin. Appl. Opt. 9, 2186-2187 (1970)

9.189 G.C. Bjorklund, S.E. Harris. J.F. Young: Vacuum ultraviolet holography. Appl. Phys. Lett. 25, 451-452 (1974)

9.190 E.H. Anderson, K. Komatsu, H.I. Smith: Achromatic holographic lithography in the deep ultraviolet. J. Vac. Sci. Technol. B 6, 216-218 (1988)

9.191 J.H. Underwood, T.W. Barbee: Layered synthetic microstructures as Bragg diffractors for X rays and extreme ultraviolet: Theory and predicted performance. Appl. Opt. 20, 3027-3034 (1981)

9.192 K.H. Richter, W. Güttler, M. Schwoerer: UV-holographic gratings in TS- diacetylene single crystals. Appl. Phys. A 32, 1-11 (1983)

9.193 J.M. Heaton, L. Solymar: Reflection holograms replayed at infrared and ultraviolet. Opt. Commun. 62, 151-154 (1987)

9.194 T.P. Jannson: Holographic optical elements and grazing incidence mirrors in XUV region, in Grazing Incidence Optics for Astronomical and Laboratory Applications, ed by S. Bowyer, J.C. Green. Proc. SPIE 830, 112-119 (1987)

9.195 G. Savant, T. Jannson, Y. Qiao: Super-high resolution holographic materials for UV and XUV applications, in Practical Holography III, ed. by S. Benton. Proc. SPIE 1051, 148-155 (1989)

9.196 A. Yen, E.H. Anderson, R.A. Ghanbari, M.L. Schattenburg, H.I. Smith: Achromatic holographic configuration for 100-nm period lithography. Appl. Opt. 31, 4540-4544 (1992)

Subject Index

Springer Series in Optical Sciences

Editorial Board: A. L. Schawlow A. E. Siegman T. Tamir

Managing Editor: H. K. V. Lotsch